Readings in the
Philosophy of Technology

Readings in the Philosophy of Technology

Edited by
David M. Kaplan

ROWMAN & LITTLEFIELD PUBLISHERS, INC.
Lanham • Boulder • New York • Toronto • Oxford

ROWMAN & LITTLEFIELD PUBLISHERS, INC.

Published in the United States of America
by Rowman & Littlefield Publishers, Inc.
A wholly owned subsidiary of The Rowman & Littlefield Publishing Group, Inc.
4501 Forbes Boulevard, Suite 200, Lanham, Maryland 20706
www.rowmanlittlefield.com

PO Box 317
Oxford
OX2 9RU, UK

British Library Cataloguing in Publication Information Available

Library of Congress Cataloging-in-Publication Data

Kaplan, David M.
  Readings in the philosophy of technology / edited by David M. Kaplan.
    p. cm.
  Includes bibliographical references and index.
  ISBN 0-7425-1488-9 (alk. paper)—ISBN 0-7425-1489-7 (pbk.: alk. paper)
  1. Technology—Philosophy. I. Title.
T14 .R39 2004
601—dc22   2003023573

Printed in the United States of America

♾ ™ The paper used in this publication meets the minimum requirements of American
National Standard for Information Sciences—Permanence of Paper for Printed Library
Materials, ANSI/NISO Z39.48-1992.

For Asher, Ellie, and Claire

Contents

Acknowledgments

I'd like to express my gratitude to the people who helped me edit this anthology: Don Ihde, Morton Winston, and Andrew Light for their insight and advice at the start of this project; Jon Bain for helping me decide what to put in the *Technology and Science* section; Michael McGandy for his invaluable copyediting; Becca Rosen for her unsolicited proofreading; Charles Buckley for his cover art ideas; and everyone at Rowman & Littlefield who helped make this book, especially Jonathan Sisk, Stephen Driver, Cheryl Adam, John Wehmueller, Tessa Fallon, and, above all, Eve DeVaro Fowler. I think we all did a great job on this book.

Introduction

Our lives are filled with technologies. They are everywhere. We live in them. We prepare food with them. We wear them as clothes. We read and write with them. We work and play with them. We manufacture and purchase them. And we constantly cope with them in one way or another whether we realize it or not.

Human life is thoroughly mediated by technology. It is hard even to imagine a life that didn't involve at least some tools, devices, or implements. Today, it is even harder to imagine a life without complex technological systems of energy, transportation, waste management, and production. Our world is largely a constructed environment; our technologies and technological systems form the background, context, and medium for lives. We rely on what we make in order to survive, to thrive, and to live together in societies. Sometimes the things we make improve our lives; sometimes they make our lives worse. Technological devices and systems shape our culture and the environment, alter patterns of human activity, and influence who we are and how we live. In short, we make and use a lot of stuff—and stuff matters.

Philosophy of technology is a critical, reflective examination of the nature of technology as well as the effects and transformation of technologies in human knowledge, activities, societies, and environments. The aim of philosophy of technology is to understand, evaluate, and criticize the ways in which technologies reflect as well as change human life individually, socially, and politically. It also examines the transformations effected by technologies on the natural world of nonhuman life and the broader ecospheres. The assumption underlying the philosophy of technology is that the devices and substances we make and use transform our experience in ways that are philosophically relevant. That is, technology not only enlarges and extends our capacities and effects changes in the natural and social worlds but does so in ways that are interesting with respect to fundamental areas of philosophical inquiry. Technology poses unique practical and conceptual problems of epistemology, metaphysics, moral philosophy, and political philosophy. The task for a philosophy of technology is to analyze the phenomenon of technology, its significance, and the ways that it mediates and transforms our experience.

In the area of epistemology, technology raises questions about the nature of knowledge, such as how to determine what counts as technical knowledge and technical explanation, or what the relationship is between technological experimentation and scientific discovery. Technology raises metaphysical questions about what is considered to be real, what is natural and what is artificial, and what is human and what is nonhuman. Technology raises moral questions about appropriate uses and consequences of devices, the desirability and permissibility of technological means, and whether things are value laden or value neutral. Finally, technology raises political questions about how we should live together in societies, who decides what technologies are developed and how they are to be administered, and how technologies alter our social relations as citizens, consumers, patients, and workers. These are just some of the important and philosophically relevant questions concerning technology.

Defining what precisely counts as technology is not easy. There are so many different kinds of technologies, each designed for a different purpose, made from different materials, requiring different skills, and used in different contexts, that it is unlikely that a common set of defining properties could possibly apply to all of them. The range of objects included in the class of technologies is enormous. If every humanly made object is a technology (with the arguable exception of art) and if technologies include everything from low-tech handheld tools to high-tech satellite-communications systems, it is hard to see what such different things have in common. Other than the fact that each is a humanly made artifact, there doesn't seem to be much in common among the diversity of stuff we make. Imagine trying to teach someone the meaning of the word *technology*. Would you point out different manmade objects? What would be the distinctly technological character of each item you point to? Would you be limited to referring to artificiality as a manufactured, nonnaturally occurring object?

There is a good deal of intuitive appeal to the idea that a technology is an artificial, nonnaturally occurring object. Technology is a manufactured artifact that would not otherwise exist on Earth. The problem is that although a natural/artificial distinction can sometimes be helpful—for example, to distinguish between real and fake flowers or natural and artificial light—that distinction quickly breaks down when pressed into service to describe more complex cases. Anything modified in even the slightest way could no longer be considered natural if by *natural* we mean existing in an unchanged state if it were left alone, free from human intervention. That would rule out almost every human intervention in the world, including cooking food, farming with tools, wearing clothing, building shelter, and writing with implements, as well as countless other activities that seem to be natural for humans to do. Everything we make and do to modify our environment—including any product of culture—would be considered unnatural or artificial. That definition is unhelpfully broad and vague. Surely there is more to be said about technology than simply that it is manufactured and not naturally occurring. And if we say that it is natural for humans to make and use technology, then we have truly made a natural/artificial distinction meaningless. Every human action and creation would be considered natural. Therefore every manufactured object, no matter how high-tech and synthetic, would be a product of nature, like any other naturally occurring object. That just is false.

Another tempting way to define technology is in terms of its technical properties. Technology, it is often said, is applied science. A technology is a practical, useful device, the design of which is based on scientific (or at least rational) principles. As such it is seen to embody a kind of pure, abstract, universal rationality—in other words, a rationality governed only by natural laws and technical considerations independent of social forces. What matters most in a technology is that it works, so this line of argument goes, and what works can be determined objectively according to universally valid, scientifically established principles. Because technology is technical matter, it is also value free and neutral. It can be used for a variety of human ends and good or bad purposes. The technology itself is neutral with respect to ends;

it is simply a tool. There's no such thing as good or bad technology, only good or bad users. The technology obeys a different kind of reason, one that is value free and context free. It is precisely this indifference to ends that makes technology so practical, or so proponents of the position claim. It works everywhere—and when it breaks down, it can be fixed the same way, by anyone with the right technical know-how. Like a scientific explanation, a technical explanation applies everywhere. The same standards, the same rules, the same techniques, and the same concept of efficiency govern the creation and use of technologies. In its essence, a technology is a neutral tool that functions best when it functions effectively. Or so goes the common sense, instrumental understanding of technology.

The problem with this instrumentalist view is that it overlooks the obvious fact that if a technology is made and used by human beings, then it cannot help but reflect human ends, values, and desires. Technology can't be value neutral because people aren't value neutral. All of our goals and purposes and actions are subject to social interpretation and moral judgment. Making and using technology is no exception. Human ends and values direct technological processes, making them an entirely human affair. These human concerns are designed into things; our technologies embody our humanity. It is more helpful to think of technology as a *socially constructed reality* rather than as the application of universally valid scientific principles. That way we can begin to appreciate—and criticize—how the things we make are human creations that, in turn, greatly affect human affairs. Technological devices are a complex of the material substances and artifacts out of which things are made, the skills and techniques required to make and use things, the ends and functions that things serve, and the social contexts of development, production, distribution, and administration in which things are made and used.

Technological systems are even more complicated networks, linking small devices to massive machinery to legal institutions and social practices. Technologies and society are like pieces of puzzle; the parts are designed to fit in with each other. They fit in terms of and in relation to technological systems. For example, it is hard to makes sense of technologies like airplanes, air conditioners, and MRI machines without considering the technological systems of transportation of energy, and health care in which they function. Explaining these technologies only in terms of their technical properties does not even begin to tell the whole story about them. To do that, one would have to take a broad view of things that would show how technologies are inextricably bound to human interests, social practices, natural laws, and a very long list of other constitutive factors. Humanity and technology are situated in a circular relationship, each shaping and affecting the other. By weaving together the technical and social, we get a more complete picture of human societies and technologies as well as the ways we are both independent of and dependent upon our machines. Creating new interpretations of these relationships helps reveal the relativity and necessity behind our technological choices and thus opens up prospects for better, more informed decisions about our current and future technologies.

The contributors to this anthology examine a range of philosophical questions concerning technology. In addition to attempting to determine its essence, they also consider technology in its moral, political, epistemological, and metaphysical dimensions. Part I examines the early foundations for a philosophy of technology. It consists of readings representing early philosophy of technology, from the 1940s to the 1970s. The philosophy of technology at that time tended to offer theories explaining the historical and transcendental conditions of modern technology, which was seen as qualitatively different from earlier technology. The readings here by the forerunners of philosophy of technology tend to view technology as an independent force. Part II consists of readings by contemporary philosophers who, by contrast, tend to view technology as a contingent process that interacts with other social forces. The representatives of the current, empirical turn in philosophy of technology are accordingly a bit less ambitious than their predecessors. They seek to determine the actual conditions in which humans and technologies are related. The readings in Parts I and II seek to establish the framework for a philosophy of technology by considering the various ways that humans and machines, means and ends, as well as social values and technical reasoning relate to one another.

Part III considers some of the moral questions raised by technology. The philosophers here are less concerned with the effects, risks, and consequences of particular technologies than they are with the ways that technology transforms how we think about moral issues. They explore how technology and morality are bound together so as to transform our notions of responsibility, human rights, constitutional interpretation, and the good life. Part IV considers some of the political questions raised by technology. The philosophers in this section examine the relationship among technological development, production, and administration as the facets of technology that relate to political values and institutions. Like its moral dimensions, the political dimensions of technology are not extrinsic but intrinsic to technological practice. The readings examine the relationship of technology to our political rights, democratic practices, and social and economic justice.

Part V considers the relationship between technology and human nature. The readings in this section examine the ways that artificial aids to human activity call into question what is natural for a person to do or to be. This class of technologies includes things that radically transform us when used (e.g., medical technologies) and things that are so lifelike they blur the lines between what is natural and what is artificial (e.g., computer technologies). The readings examine the relationship of technology to personal and social identity, medical practice and social values, artificial intelligence, and our associations with artificial life.

Finally, Part VI examines the relationship between technology and science (also known as *technoscience*). The central insight shared by this group of philosophers is that technological instrumentation is essential to scientific practice. Without technology, there would be no science as we know it. The readings here examine the relationship of technology to scientific experimentation, laboratory life, scientific realism (determining what really exists), and democratic ideals encoded within technoscientific practice.

The contributors to *Readings in the Philosophy of Technology* explore the multiple ways that humanity shapes and affects technologies and, in turn, is shaped and affected by them. Studies of technology from a philosophic perspective enrich the way we experience it and deepen the way we understand technological development. The aim of this collection is to help us think critically about the ways in which technologies reflect as well as change human life on an individual, social, and cultural level. We make and use a lot of stuff—and stuff matters.

Part I

EARLY PHILOSOPHY OF TECHNOLOGY

The founders of philosophy of technology tend to lay out *transcendental* perspectives on technology, or theories that account for the very conditions of making and using instruments. They treat technology as a singular phenomenon with a logic that is radically different from that found in human culture. Corresponding to this strong distinction between humans and things are two distinct forms of knowledge. Each of these forms of knowledge is restricted to mutually exclusive domains of inquiry: explanation to the causal world of facts and laws, and understanding to the human world of intentions and desires. Technology and science belong to the realm of causal explanation, not social interpretation. Early philosophers of technology contrast the detached objectivity of technological rationality with a more primordial, humane form of experience that is connected to, not severed from, the natural and social worlds. The problem with objective, neutral, efficient "techniques" of knowledge, according to their account, is that they are at the heart of technological forces that restructure the entire social world as an object of control. Although that kind of thinking may be useful for dealing with technical matters, these philosophers worry about the fate of human beings when treated as mere things. Early philosophies of technology criticize technology so as to describe the limits of technological thinking, call attention to the dangers of it, and suggest humane alternatives to dehumanizing technological societies.

In the first reading, "Do Machines Make History?" Robert Heilbroner examines the theory of *technological determinism*—the idea that technology is what determines the course of history. Technological determinism is often attributed to Marx, who declared in *The Poverty of Philosophy,* "The hand mill gives you society with the feudal lord; the steam mill society with the industrial capitalist." Marx, of course, was no determinist. His claim is simply that a reciprocal relationship exists between historical development and technological development. The central thesis of historical materialism is that humanity produces itself through our labor; different stages in history correspond to different modes and relations of production. Technology in this reading is dialectically related to society: a circular relationship exists between the two, each influencing—but not strictly determining—the development of the other. More precisely, Marx claims that technology only influences the socioeconomic; it does not influence our lives in total.

Heilbroner makes a somewhat stronger claim regarding the way technology shapes society. He believes there is a "soft determinism" by which technology shapes the course of his-

1

tory. He tests this claim in two stages. First, he examines the "laws of motion" of technology to see if we can explain why it evolves in the sequence it does. Next, he examines how exactly the technological apparatus of a society affects social relationships. Heilbroner endeavors to show how the history of technological development is always related to the knowledge available at any given time (e.g., no electronic technologies without the accumulation of knowledge with respect to electricity), as well as how such development is relevant to the capability to solve technical problems (e.g., no steam engine without prior knowledge of how to make pistons and gears). He also finds evidence of how specific technologies structure social relations. For example, the assembly line influences the composition of the labor force and its organizational structures. This is the *soft determinism* that governs the influence of technology over social relations. Technology imposes a determinate pattern on history, *but* this imposition is not so strong to be able to predict the course of technological progress and social evolution. Technology mediates and steers a society, but it does not quite drive it.

In "Toward a Philosophy of Technology," Hans Jonas considers technology by two approaches—one formal and one material. The first examines the *formal dynamics* of technology as an "abstract whole of movement" with its own internal dynamics of change. The second examines the *substantive content* of technology in terms of the actual things made and used, as well as how they affect our world and our lives. The *formal approach* attempts to determine the processes by which modern technology drives toward innovation and novelty. The *material approach* examines the novel products themselves. A third approach employed by Jonas examines the moral dimensions of technology—in other words, the way it determines our responsibilities for the long-term effects on future generations and the environment. Jonas considers the moral dimension of technology at greater length in his essay "Technology and Responsibility" (see Part III).

The formal difference between traditional and modern technology hinges on the nature of change and development regarding technology. Traditional technology is characterized by a relatively harmonious balance between the various tools, devices, and procedures and the social purposes they served. Technological change and development were slow and gradual, tending toward an equilibrium of means in relation to established needs and goals. Modern technology, however, has distinct and opposite characteristics. It is a dynamic process, tending toward disequilibrium, driving forward for the sake of novelty and innovation almost regardless of established purposes. Four traits of modern technology are that (1) innovations produce endless more developments, creating new ends, needs, and purposes; (2) innovations spread knowledge and practical adoption of technologies everywhere; (3) rather than satisfying preexisting ends, new technological means create new ends, and technology creates new necessities of life that in turn demand new technologies to satisfy them; and 4) progress drives technological development and not the satisfaction of purposes.

Jonas describes technology as "restless." It is driven not only by contingent social forces, like competition, war, and scarcity of resources, but also by our conviction that there can be progress because we believe there has to be something better. The belief in the endless perfectibility of every human accomplishment is the underlying ontological-epistemological premise. The source of the restlessness of technology is science. Science continually unlocks new dimensions and makes new discoveries that confirm our belief in the infinity of things to be discovered and known and the promise that unending inquiry will deliver us closer to the truth. The drive for increased knowledge is fueled by and leads to technological innovations—which, in turn, propel science forward. Science and technology feed off each other; each requires and spurs the other. The result is an inversion of means and ends that creates a world of autonomous technological systems spiraling out of control. In the situation Jonas describes, it is no longer clear if technology serves humanity or humanity serves technology. Not only has technology derived new ends from the invention of new means, but it has also elevated itself to become the ultimate end: technology for technology's sake. The remainder of "Toward a Philosophy of Technology" details the material works of technology. The combination of formal dynamics (tending toward novelty and disequilibrium) and material content (that alter the makeup of the natural and social worlds) raises unique moral questions and summons us to take responsibility for our increasing powers. Jonas calls on philosophers to question and evaluate technologies that threaten the planet, limit our freedoms, and radically transform human nature.

Next is Martin Heidegger's "Question Concerning Technology," perhaps the single most important work in philosophy of technology. Heidegger endeavors to uncover the essence of technology so that we may have a "free relationship" with it—once we understand the real essence of technology, we will learn how to experience it within its own bounds. According to Heidegger, what we have until now failed to understand is that the essence of technology isn't technological: it is not something like a neutral tool. We commonly assume that technology is instrumental—in other words, it is a means to an end, a human activity geared toward manipulating and controlling things. Means produce ends just as causes produce effects. Although accurate, this definition represents only the causal, instrumental meaning of a means. Heidegger reminds us that the ancient Greeks had a broader conception of causality. A cause is that which brings something about or that which is responsible for something. The four causes articulated by Aristotle are like explanations—of what something is made out of, what it is to be a something, what produces it, and what it is for. Together the four causes are responsible for bringing something into appearance (*poiesis*). They make it present; they bring forth "out of concealment into unconcealment." The essence of technology is not a means but a way of revealing (*aletheia*). It is a kind of truth.

Yet the way that technology (*techne*) reveals is problematic. It is a "challenging revealing," one that orders and arranges nature into a "standing reserve" of energy and resources. It places an unreasonable demand on nature: that it supply us endlessly and efficiently. Even humans, the supposed masters of technology, are challenged and ordered into standing reserve as "human resources." Heidegger calls this way of revealing the world "enframing" (*Ge-stell*). It is a way of ordering people to see the world—and each other—as a mere stockpile of resources to be manipulated. *Enframing* happens both in us and in the world; it is the revelation of being (human beings and nature) as standing reserve. Particular technologies in the ordinary sense of tools and machines only respond to the enframing—they are the consequence, not the cause, of it; they merely help reveal things as standing reserve. The same is true of modern science. It, too, is derivative of enframing.

The danger of technology is twofold. First, Heidegger finds that we ourselves become mere standing reserve. Second, in our role as human standing reserve, we tend to think we are the masters of everything. But in truth we cannot see ourselves or understand the world clearly. *Ge-stell* keeps the essence of things concealed; it obscures other ways of seeing things—above all, revealing as *poiesis*. The danger is the partial, incomplete, enframing revealing. *Poeisis* is a broader form of revealing than *techne*. The essence of technology is ultimately *poiesis*. Heidegger argues we need to realize the essence of technology, stop construing technology as mere instrumentality, and overcome the illusion of our pretension to complete mastery and control over things. The "saving power" of technology is that the essence of technology is ambiguous. The very instrumentality (*techne*) that threatens us also saves us (as *poiesis*).

Heidegger exhorts us to question technology; to see through the limits of technological, enframing revealing; and to be open to possibilities for a broader, more creative way of experiencing the world.

In "Heidegger on Gaining a Free Relation to Technology," Hubert Dreyfus further clarifies Heidegger's notion of the essence of technology, what is wrong with our conventional understanding of technology, and what we can do about it. The essence of technology for Heidegger is a condition we have inherited whereby we have come to accept a technological understanding of being, a limiting and restricting understanding of reality. The technological understanding of being is to seek more and more efficiency, and this efficiency is sought for its own sake. We interpret everything, including ourselves, as resources to be dealt with as effectively and efficiently as possible. However, once we realize that we have a technological understanding of things, we have already stepped out of that framework; we are then free to appreciate that there is more to life than efficiency. This new attitude toward technology is what Heidegger calls "releasement."

We are released from the burden of seeking efficiency for its own sake, Dreyfus claims, once we understand that this kind of calculative thinking is only a historical product and that things could be different. The next step after seeing through the limits of calculative thinking is to develop meditative thinking. Dreyfus reads Heidegger as advocating a new cultural paradigm that celebrates marginal practices—in other words, practices not based on efficiency—that affirm what is ultimately more important in our lives: living closer to nature, shared values, friendship, love, music, art, and community. Dreyfus proposes that this new understanding of reality is the new god that Heidegger says (at the end of "Question Concerning Technology") is the only thing that can save us now. We need to become more open to understanding things differently and to embrace nonefficient living. The key is neither to reject nor to accept technological thinking without criticism but to take a new attitude toward it, understand its limits, and affirm a broader, more inclusive understanding of reality.

In "Some Social Implications of Modern Technology," Herbert Marcuse echoes Heidegger's belief that the essence of technology is technological rationality of efficiency, but Marcuse's orientation is more clearly social and political. Like Heidegger, Marcuse claims that technology is more than just instruments and machines; it is also a mode of organization geared toward control and domination of people and nature. But unlike Heidegger, Marcuse claims technological rationality is a political rationality that swallows up all opposition by homogenizing nature and people into neutral objects of manipulation; it dissolves traditional rationality into a technological rationality that employs efficiency as the single standard of judgment. We have come to accept mass production, mechanization, industrialization, standardization, and bureaucracy as the (seeming) embodiments of rationality and efficiency. But we accept this only because technological rationality fosters the very beliefs and attitudes that make us ready to assent to the dictates of technological efficiency. We are caught up in a systematic deception; our needs have become the needs of the technopolitical apparatus.

In the twentieth century, political power—including state capitalism, fascism, and state socialism—developed new mechanics of social cohesion that integrate individuals into society with pleasurable means of social control. Consequently, Marcuse argues, appeals to enlightened self-interest, freedom, and autonomy have come to appear quaint and irrational. We are stripped of our individuality by a technological rationality that makes conformity seem reasonable and protest seem unreasonable. The result is a "one-dimensional" universe of thought and behavior in which our capacity for critical thinking and practical resistance is disappearing (but not entirely in eclipse). Marcuse, however, holds out the possibility that technological rationality may be used as an instrument to foster democracy, autonomy, and an authentic individuality free to satisfy its true needs. The same objective, impersonal rationality that makes individualism unnecessary can be harnessed by a society to fully realize (rather than repress) human capacities. Marcuse is pessimistic about the prospects for such a transformation because the technological apparatus has incorporated and subsumed all critical and oppositional thought. Yet despite Marcuse's pessimism regarding practically achieving such a transformation, he maintains that it is, in principle, possible.

In "Technical Progress and the Social Life-World," Jürgen Habermas agrees with Marcuse that technology is a form of instrumental, calculative reason but disagrees that it is destined to subsume all forms of thought and action. Habermas attempts to find a middle ground between the realms of rational, efficient technical control and the free, communicative social world. He follows T. H. Huxley's division between the domain of literature, which reflects and examines the real world of our cares and concerns, and the domain of science, which deals with quantified regularities that lie outside of the realm of human concerns. Huxley juxtaposes the life-world of social interaction with the "worldless" universe of facts: the former is characterized by interpretive, normative, and qualitative activities; the latter is defined as objective, value-free, and quantitative. Progress would occur, Huxley estimates, when literature assimilates the results of science and science takes on the "flesh and blood" of literature. Habermas disagrees. He claims that empirical scientific knowledge is entirely different from action-oriented understanding that individuals and groups use to make sense of the world. Science can never enter literature unless it first enters our lives through the technologies with which we live. Technology mediates the objective world of science and social life-world of human concerns. We need to "translate technical progress into practical consciousness" by rethinking the relationship between technological progress and the social world.

Habermas is skeptical of traditional, nineteenth-century attempts to integrate scientific knowledge into our lives through traditional education in the sciences and humanities; rather, the transformation of knowledge into practical contexts is a political matter. The sciences and humanities have become so specialized and detached from one another than any attempt to assimilate one into the other is bound to fail. The reason is that the capacity for control made possible by the empirical sciences is not the same as the capacity for enlightened action. As a result, we need more than just scientific expertise to bring about the assimilation of technology into the practical life-world. We need to learn how to bring the realm of technical control under the power of rational discussion and democratic institutions. Habermas is thus skeptical of the tendency of Enlightenment (and Marxist) thinkers to automatically link technological progress with social emancipation, as well as of the Marcusean school that treats technology as an independent force. Technology has two sides: one technical and scientific, the other social and potentially democratic. The key in Habermas's estimation is to make the technical side serve democratic ends and not the other way around.

1

Do Machines Make History?

Robert Heilbroner

The hand mill gives you society with the feudal lord; the steam mill, society with
the industrial capitalist.

—Karl Marx, *The Poverty of Philosophy*

That machines make history in some sense—that the level of technology has a direct bearing
on the human drama—is of course obvious. That they do not make all of history, however that
word is defined, is equally clear. The challenge, then, is to see if one can say something systematic about the matter, to see whether one can order the problem so that it becomes intellectually manageable.

To do so calls at the very beginning for a careful specification of our task. There are a
number of important ways in which machines make history that will not concern us here. For
example, one can study the impact of technology on the *political* course of history, evidenced
most strikingly by the central role played by the technology of war. Or one can study the effect
of machines on the *social* attitudes that underlie historical evolution: one thinks of the effect
of radio or television on political behavior. Or one can study technology as one of the factors
shaping the changeful content of life from one epoch to another: when we speak of "life" in
the Middle Ages or today we define an existence much of whose texture and substance is
intimately connected with the prevailing technological order.

None of these problems will form the focus of this [chapter]. Instead, I propose to examine the impact of technology on history in another area—an area defined by the famous quota-
tion from Marx that stands beneath our title. The question we are interested in, then, concerns
the effect of technology in determining the nature of the *socioeconomic order*. In its simplest
terms the question is: did medieval technology bring about feudalism? Is industrial technology
the necessary and sufficient condition for capitalism? Or, by extension, will the technology of
the computer and the atom constitute the ineluctable cause of a new social order?

Even in this restricted sense, our inquiry promises to be broad and sprawling. Hence, I
shall not try to attack it head-on, but to examine it in two stages:

1. If we make the assumption that the hand mill does "give" us feudalism and the steam
   mill capitalism, this places technological change in the position of a prime mover of

From *Technology and Culture* 8 (1967): 335–45. Copyright © 1967 by the Society for the History of Technology. Reprinted by permission of Johns Hopkins University Press.

social history. Can we then explain the "laws of motion" of technology itself? Or to put the question less grandly, can we explain why technology evolves in the sequence it does?

2. Again, taking the Marxian paradigm at face value, exactly what do we mean when we assert that the hand mill "gives us" society with the feudal lord? Precisely how does the mode of production affect the superstructure of social relationships?

These questions will enable us to test the empirical content—or at least to see if there *is* an empirical content—in the idea of technological determinism. I do not think it will come as a surprise if I announce now that we will find *some* content, and a great deal of missing evidence, in our investigation. What will remain then will be to see if we can place the salvageable elements of the theory in historical perspective—to see, in a word, if we can explain technological determinism historically as well as explain history by technological determinism.

I

We begin with a very difficult question hardly rendered easier by the fact that there exist, to the best of my knowledge, no empirical studies on which to base our speculations. It is the question of whether there is a fixed sequence to technological development and therefore a necessitous path over which technologically developing societies must travel.

I believe there is such a sequence—that the steam mill follows the hand mill not by chance but because it is the next "stage" in a technical conquest of nature that follows one and only one grand avenue of advance. To put it differently, I believe that it is impossible to proceed to the age of the steam mill until one has passed through the age of the hand mill, and that in turn one cannot move to the age of the hydroelectric plant before one has mastered the steam mill, nor to the nuclear power age until one has lived through that of electricity.

Before I attempt to justify so sweeping an assertion, let me make a few reservations. To begin with, I am fully conscious that not all societies are interested in developing a technology of production or in channeling to it the same quota of social energy. I am very much aware of the different pressures that different societies exert on the direction in which technology unfolds. Lastly, I am not unmindful of the difference between the discovery of a given machine and its application as a technology—for example, the invention of a steam engine (the aeolipile) by Hero of Alexandria long before its incorporation into a steam mill. All these problems, to which we will return in our last section, refer however to the way in which technology makes its peace with the social, political, and economic institutions of the society in which it appears. They do not directly affect the contention that there exists a determinate sequence of productive technology for those societies that are interested in originating and applying such a technology.

What evidence do we have for such a view? I would put forward three suggestive pieces of evidence:

1. The Simultaneity of Invention

The phenomenon of simultaneous discovery is well known.[1] From our view, it argues that the process of discovery takes place along a well-defined frontier of knowledge, rather than in grab-bag fashion. Admittedly, the concept of "simultaneity" is impressionistic,[2] but the related phenomenon of technological "clustering" again suggests that technical evolution follows a sequential and determinate rather than random course.[3]

2. The Absence of Technological Leaps

All inventions and innovations, by definition, represent an advance of the art beyond existing base lines. Yet, most advances, particularly in retrospect, appear essentially incremental, evolutionary. If nature makes no sudden leaps, neither, it would appear, does technology. To make my point by exaggeration, we do not find experiments in electricity in the year *1500,* or attempts to extract power from the atom in the year *1700.* On the whole, the development of the technology of production presents a fairly smooth and continuous profile rather than one of jagged peaks and discontinuities.

3. The Predictability of Technology

There is a long history of technological prediction, some of it ludicrous and some not.[4] What is interesting is that the development of technical progress has always seemed *intrinsically* predictable. This does not mean that we can lay down future timetables of technical discovery, nor does it rule out the possibility of surprises. Yet I venture to state that many scientists would be willing to make *general* predictions as to the nature of technological capability twenty-five or even fifty years ahead. This too suggests that technology follows a developmental sequence rather than arriving in a more chancy fashion.

I am aware, needless to say, that these bits of evidence do not constitute anything like a "proof" of my hypothesis. At best they establish the grounds on which a prima facie case of plausibility may be rested. But I should like now to strengthen these grounds by suggesting two deeper-seated reasons why technology *should* display a "structured" history.

The first of these is that a major constraint always operates on the technological capacity of an age, the constraint of its accumulated stock of available knowledge. The application of this knowledge may lag behind its reach; the technology of the hand mill, for example, was by no means at the frontier of medieval technical knowledge, but technical realization can hardly precede what men generally know (although experiment may incrementally advance both technology and knowledge concurrently). Particularly from the mid-nineteenth century to the present do we sense the loosening constraints on technology stemming from successively yielding barriers of scientific knowledge—loosening constraints that result in the successive arrival of the electrical, chemical, aeronautical, electronic, nuclear, and space stages of technology.[5]

The gradual expansion of knowledge is not, however, the only order-bestowing constraint on the development of technology. A second controlling factor is the material competence of the age, its level of technical expertise. To make a steam engine, for example, requires not only some knowledge of the elastic properties of steam but the ability to cast iron cylinders of considerable dimensions with tolerable accuracy. It is one thing to produce a single steam machine as an expensive toy, such as the machine depicted by Hero, and another to produce a machine that will produce power economically and effectively. The difficulties experienced by Watt and Boulton in achieving a fit of piston to cylinder illustrate the problems of creating a technology, in contrast with a single machine.

Yet until a metal-working technology was established—indeed, until an embryonic machine-tool industry had taken root—an industrial technology was impossible to create. Furthermore, the competence required to create such a technology does not reside alone in the ability or inability to make a particular machine (one thinks of Babbage's ill-fated calculator as an example of a machine born too soon), but in the ability of many industries to change their products or processes to "fit" a change in one key product or process.

The necessary requirement of technological congruence[6] gives us an additional cause of sequencing. For the ability of many industries to co-operate in producing the equipment needed for a "higher" stage of technology depends not alone on knowledge or sheer skill but on the division of labor and the specialization of industry. And this in turn hinges to a considerable degree on the sheer size of the stock of capital itself. Thus the slow and painful accumulation of capital, from which springs the gradual diversification of industrial function, becomes an independent regulator of the reach of technical capability.

In making this general case for a determinate pattern of technological evolution—at least insofar as that technology is concerned with production—I do not want to claim too much. I am well aware that reasoning about technical sequences is easily faulted as *post hoc ergo propter hoc*. Hence, let me leave this phase of my inquiry by suggesting no more than that the idea of a roughly ordered progression of productive technology seems logical enough to warrant further empirical investigation. To put it as concretely as possible, I do not think it is just by happenstance that the steam mill follows, and does not precede, the hand mill, nor is it mere fantasy in our own day when we speak of the coming of the automatic factory. In the future as in the past, the development of the technology of production seems bounded by the constraints of knowledge and capability and thus, in principle at least, open to prediction as a determinable force of the historic process.

II

The second proposition to be investigated is no less difficult than the first. It relates, we will recall, to the explicit statement that a given technology imposes certain social and political characteristics upon the society in which it is found. It is true that, as Marx wrote in *The German Ideology,* "A certain mode of production, or industrial stage, is always combined with a certain mode of cooperation, or social stage,"[7] or as he put it in the sentence immediately preceding our hand mill, steam mill paradigm, "In acquiring new productive forces men change their mode of production, and in changing their mode of production they change their way of living—they change all their social relations."

As before, we must set aside for the moment certain "cultural" aspects of the question. But if we restrict ourselves to the functional relationships directly connected with the process of production itself, I think we can indeed state that the technology of a society imposes a determinate pattern of social relations on that society.

We can, as a matter of fact, distinguish at least two such modes of influence:

1. The Composition of the Labor Force

In order to function, a given technology must be attended by a labor force of a particular kind. Thus, the hand mill (if we take this as referring to late medieval technology in general) required a work force composed of skilled or semiskilled craftsmen, who were free to practice their occupations at home or in a small atelier, at times and seasons that varied considerably. By way of contrast, the steam mill—that is, the technology of the nineteenth century—required a work force composed of semiskilled or unskilled operatives who could work only at the factory site and only at the strict time schedule enforced by turning the machinery on or off. Again, the technology of the electronic age has steadily required a higher proportion of skilled attendants; and the coming technology of automation will still further change the needed mix of skills and the locale of work, and may as well drastically lessen the requirements of labor time itself.

2. The Hierarchical Organization of Work

Different technological apparatuses not only require different labor forces but different orders of supervision and coordination. The internal organization of the eighteenth-century handicraft unit, with its typical man-master relationship, presents a social configuration of a wholly different kind from that of the nineteenth-century factory with its men-manager confrontation, and this in turn differs from the internal social structure of the continuous-flow, semi-automated plant of the present. As the intricacy of the production process increases, a much more complex system of internal controls is required to maintain the system in working order.

Does this add up to the proposition that the steam mill gives us society with the industrial capitalist? Certainly the class characteristics of a particular society are strongly implied in its functional organization. Yet it would seem wise to be very cautious before relating political effects exclusively to functional economic causes. The Soviet Union, for example, proclaims itself to be a socialist society although its technical base resembles that of old-fashioned capitalism. Had Marx written that the steam mill gives you society with the industrial *manager,* he would have been closer to the truth.

What is less easy to decide is the degree to which the technological infrastructure is responsible for some of the sociological features of society. Is anomie, for instance, a disease of capitalism or of all industrial societies? Is the organization man a creature of monopoly capital or of all bureaucratic industry wherever found? The questions tempt us to look into the problem of the impact of technology on the existential quality of life, an area we have ruled out of bounds for this paper. Suffice it to say that the similar technologies of Russia and America are indeed giving rise to similar social phenomena of this sort.

As with the first portion of our inquiry, it seems advisable to end this section on a note of caution. There is a danger, in discussing the structure of the labor force or the nature of intrafirm organization, of assigning the sole causal efficacy to the visible presence of machinery and of overlooking the invisible influence of other factors at work. Gilfillan, for instance, writes, "engineers have committed such blunders as saying the typewriter brought women to work in offices, and with the typesetting machine made possible the great modern newspapers, forgetting that in Japan there are women office workers and great modern newspapers getting practically no help from typewriters and typesetting machines."[8] In addition, even where technology seems unquestionably to play the critical role, an independent "social" element unavoidably enters the scene in the *design* of technology, which must take into account such facts as the level of education of the work force or its relative price. In this way the machine will reflect, as much as mold, the social relationship of work.

These caveats urge us to practice what William James called a "soft determinism" with regard to the influence of the machine on social relations. Nevertheless, I would say that our cautions qualify rather than invalidate the thesis that the prevailing level of technology imposes itself powerfully on the structural organization of the productive side of society. A foreknowledge of the shape of the technical core of society fifty years hence may not allow us to describe the political attributes of that society, and may perhaps only hint at its sociological character, but assuredly it presents us with a profile of requirements, both in labor skills and in supervisory needs, that differ considerably from those of today. We cannot say whether the society of the computer will give us the latter-day capitalist or the commissar, but it seems beyond question that it will give us the technician and the bureaucrat.

III

Frequently, during our efforts thus far to demonstrate what is valid and useful in the concept of technological determinism, we have been forced to defer certain aspects of the problem until later. It is time now to turn up the rug and to examine what has been swept under it. Let us try to systematize our qualifications and objections to the basic Marxian paradigm:

1. Technological Progress Is Itself a Social Activity

A theory of technological determinism must contend with the fact that the very activity of invention and innovation is an attribute of some societies and not of others. The Kalahari bushmen or the tribesmen of New Guinea, for instance, have persisted in a neolithic technology to the present day; the Arabs reached a high degree of technical proficiency in the past and have since suffered a decline; the classical Chinese developed technical expertise in some fields while unaccountably neglecting it in the area of production. What factors serve to encourage or discourage this technical thrust is a problem about which we know extremely little at the present moment.[9]

2. The Course of Technological Advance Is Responsive to Social Reform

Whether technology advances in the area of war, the arts, agriculture, or industry depends in part on the rewards, inducements, and incentives offered by society. In this way the direction of technological advance is partially the result of social policy. For example, the system of interchangeable parts, first introduced into France and then independently into England, failed to take root in either country for lack of government interest or market stimulus. Its success in America is attributable mainly to government support and to its appeal in a society without guild traditions and with high labor costs.[10] The general *level* of technology may follow an independently determined sequential path, but its areas of application certainly reflect social influences.

3. Technological Change Must Be Compatible with Existing Social Conditions

An advance in technology not only must be congruent with the surrounding technology but must also be compatible with the existing economic and other institutions of society. For example, labor-saving machinery will not find ready acceptance in a society where labor is abundant and cheap as a factor of production. Nor would a mass production technique recommend itself to a society that did not have mass market. Indeed, the presence of slave labor seems generally to inhibit the use of machinery and the presence of expensive labor to accelerate it.[11]

These reflections on the social forces bearing on technical progress tempt us to throw aside the whole notion of technological determinism as false or misleading.[12] Yet, to relegate technology from an undeserved position of *primum mobile* in history to that of a mediating factor, both acted upon by and acting on the body of society, is not to write off its influence but only to specify its mode of operation with greater precision. Similarly, to admit we understand very little of the cultural factors that give rise to technology does not depreciate its role but focuses our attention on that period of history when technology is clearly a major historic force, namely Western society since 1700.

IV

What is the mediating role played by technology within modern Western society? When we ask this much more modest question, the interaction of society and technology begins to clarify itself for us:

1. The Rise of Capitalism Provided a Major Stimulus for the Development of a Technology of Production

Not until the emergence of a market system organized around the principle of private property did there also emerge an institution capable of systematically guiding the inventive and innovative abilities of society to the problem of facilitating production. Hence the environment of the eighteenth and nineteenth centuries provided both a novel and an extremely effective encouragement for the development of an *industrial* technology. In addition, the slowly opening political and social framework of late mercantilist society gave rise to social aspirations for which the new technology offered the best chance of realization. It was not only the steam mill that gave us the industrial capitalist but the rising inventor-manufacturer who gave us the steam mill.

2. The Expansion of Technology within the Market System Took On a New "Automatic" Aspect

Under the burgeoning market system not alone the initiation of technical improvement but its subsequent adoption and repercussion through the economy was largely governed by market considerations. As a result, both the rise and the proliferation of technology assumed the attributes of an impersonal diffuse "force" bearing on social and economic life. This was all the more pronounced because the political control needed to buffer its disruptive consequences was seriously inhibited by the prevailing laissez-faire ideology.

3. The Rise of Science Gave a New Impetus to Technology

The period of early capitalism roughly coincided with and provided a congenial setting for the development of an independent source of technological encouragement—the rise of the self-conscious activity of science. The steady expansion of scientific research, dedicated to the exploration of nature's secrets and to their harnessing for social use, provided an increasingly important stimulus for technological advance from the middle of the nineteenth century. Indeed, as the twentieth century has progressed, science has become a major historical force in its own right and is now the indispensable precondition for an effective technology.

It is for these reasons that technology takes on a special significance in the context of capitalism—or, for that matter, of a socialism based on maximizing production or minimizing costs. For in these societies, both the continuous appearance of technical advance and its diffusion throughout the society assume the attributes of autonomous process, "mysteriously" generated by society and thrust upon its members in a manner as indifferent as it is imperious. This is why, I think, the problem of technological determinism—of how machines make history—comes to us with such insistence despite the ease with which we can disprove its more extreme contentions.

*Technological determinism is thus peculiarly a problem of a certain historic epoch—*specifically that of high capitalism and low socialism—*in which the forces of technical change have been unleashed, but when the agencies for the control or guidance of technology are still rudimentary.*

The point has relevance for the future. The surrender of society to the free play of market forces is now on the wane, but its subservience to the impetus of the scientific ethos is on the rise. The prospect before us is assuredly that of an undiminished and very likely accelerated pace of technical change. From what we can foretell about the direction of this technological advance and the structural alterations it implies, the pressures in the future will be toward a society marked by a much greater degree of organization and deliberate control. What other political, social, and existential changes the age of the computer will also bring we do not know. What seems certain, however, is that the problem of technological determinism—that is, of the impact of machines on history—will remain germane until there is forged a degree of public control over technology far greater than anything that now exists.

NOTES

1. See Robert K. Merton, "Singletons and Multiples in Scientific Discovery: A Chapter in the Sociology of Science," *Proceedings* of the American Philosophical Society 105 (October 1961): 470–86.

2. See John Jewkes, David Sawers, and Richard Stillerman, *The Sources of Invention* (New York: Norton, 1960 [paperback edition]), p. 227, for a skeptical view.

3. "One can count 21 basically different means of flying, at least eight basic methods of geophysical prospecting, four ways to make uranium explosive; . . . 20 or 30 ways to control birth. . . . If each of these separate inventions were autonomous, i.e., without cause, how could one account for their arriving in these functional groups?" S. C. Gilfillan, "Social Implications of Technological Advance," *Current Sociology* 1 (1952): 197. See also Jacob Schmookler, "Economic Sources of Inventive Activity," *Journal of Economic History* (March 1962): 1–20; and Richard Nelson, "The Economics of Invention: A Survey of the Literature," *Journal of Business* 32 (April 1959): 101–19.

4. Jewkes et al. (see note 2) present a catalogue of chastening mistakes (p. 230 f.). On the other hand, for a sober predictive effort, see Francis Bello, "The 1960s: A Forecast of Technology," *Fortune* 59 (January 1959): 74–78; and Daniel Bell, "The Study of the Future," *Public Interest* 1 (Fall 1965): 119–30. Modern attempts at prediction project likely avenues of scientific advance or technological function rather than the feasibility of specific machines.

5. To be sure, the inquiry now regresses one step and forces us to ask whether there are inherent stages for the expansion of knowledge, at least insofar as it applies to nature. This is a very uncertain question. But having already risked so much, I will hazard the suggestion that the roughly parallel sequential development of scientific understanding in those few cultures that have cultivated it (mainly, classical Greece, China, the high Arabian culture, and the West since the Renaissance) makes such a hypothesis impossible, provided that one looks to broad outlines and not to inner detail.

6. The phrase is Richard LaPiere's in *Social Change* (New York: McGraw-Hill, 1965), p. 263 f.

7. Karl Marx and Friedrich Engels, *The German Ideology* (London: Lawrence and Wisehart, 1942), p. 18.

8. Gilfillan (see note 3), p. 202.

9. An interesting attempt to find a line of social causation is found in E. Hagen, *The Theory of Social Change* (Homewood, Ill.: Dorsey Press, 1962).

10. See K. R. Gilbert, "Machine Tools," in Charles Singer, E. J. Holmwood, A. R. Hall, and Trevor I. Williams (eds.), *A History of Technology* (Oxford University Press, 1958), IV, chap. 14.

11. See LaPiere (note 6), p. 284; also H. J. Habbakuk, *British and American Technology in the 19th Century* (Cambridge University Press, 1962), *passim.*

12. As, for example, in A. Hansen, "The Technological Determination of History," *Quarterly Journal of Economics* (1921): 76–83.

2

Toward a Philosophy of Technology

Hans Jonas

Are there philosophical aspects to technology? Of course there are, as there are to all things of importance in human endeavor and destiny. Modern technology touches on almost everything vital to man's existence—material, mental, and spiritual. Indeed, what of man is *not* involved? The way he lives his life and looks at objects, his intercourse with the world and with his peers, his powers and modes of action, kinds of goals, states and changes of society, objectives and forms of politics (including warfare no less than welfare), the sense and quality of life, even man's fate and that of his environment: all these are involved in the technological enterprise as it extends in magnitude and depth. The mere enumeration suggests a staggering host of potentially philosophic themes.

To put it bluntly: if there is a philosophy of science, language, history, and art; if there is social, political, and moral philosophy; philosophy of thought and of action, of reason and passion, of decision and value—all facets of the inclusive philosophy of man—how then could there not be a philosophy of technology, the focal fact of modern life? And at that a philosophy so spacious that it can house portions from all the other branches of philosophy? It is almost a truism, but at the same time so immense a proposition that its challenge staggers the mind. Economy and modesty require that we select, for a beginning, the most obvious from the multitude of aspects that invite philosophical attention.

The old but useful distinction of "form" and "matter" allows us to distinguish between these two major themes: (1) the *formal dynamics* of technology as a continuing collective enterprise, which advances by its own "laws of motion"; and (2) the *substantive content* of technology in terms of the things it puts into human use, the powers it confers, the novel objectives it opens up or dictates, and the altered manner of human action by which these objectives are realized.

The first theme considers technology as an abstract whole of movement; the second considers its concrete uses and their impact on our world and our lives. The formal approach will try to grasp the pervasive "process properties" by which modern technology propels itself—

From *Hastings Center Report* 9/1 (1979): 34–43. Copyright © The Hastings Center. Reprinted by permission of the Hastings Center and Eleanor Jonas.

through our agency, to be sure—into ever-succeeding and superceding novelty. The material approach will look at the species of novelties themselves, their taxonomy, as it were, and try to make out how the world furnished with them looks. A third, overarching theme is the *moral* side of technology as a burden on human responsibility, especially its long-term effects on the global condition of man and environment. This—my own main preoccupation over the past years—will only be touched upon.

I   THE FORMAL DYNAMICS OF TECHNOLOGY

First some observations about technology's form as an abstract whole of movement. We are concerned with characteristics of *modern* technology and therefore ask first what distinguishes it *formally* from all previous technology. One major distinction is that modern technology is an enterprise and process, whereas earlier technology was a possession and a state. If we roughly describe technology as comprising the use of artificial implements for the business of life, together with their original invention, improvement, and occasional additions, such a tranquil description will do for most of technology through mankind's career (with which it is coeval), but not for modern technology. In the past, generally speaking, a given inventory of tools and procedures used to be fairly constant, tending toward a mutually adjusting, stable equilibrium of ends and means, which—once established—represented for lengthy periods an unchallenged optimum of technical competence.

To be sure, revolutions occurred, but more by accident than by design. The agricultural revolution, the metallurgical revolution that led from the neolithic to the iron age, the rise of cities, and such developments, *happened* rather than were consciously created. Their pace was so slow that only in the time-contraction of historical retrospect do they appear to be "revolutions" (with the misleading connotation that their contemporaries experienced them as such). Even where the change was sudden, as with the introduction first of the chariot, then of armed horsemen into warfare—a violent, if short-lived, revolution indeed—the innovation did not originate from within the military art of the advanced societies that it affected, but was thrust on it from outside by the (much less civilized) peoples of Central Asia. Instead of spreading through the technological universe of their time, other technical breakthroughs, like Phoenician purple-dying, Byzantine "greek fire," Chinese porcelain and silk, and Damascene steel-tempering, remained jealously guarded monopolies of the inventor communities. Still others, like the hydraulic and steam playthings of Alexandrian mechanics, or compass and gunpowder of the Chinese, passed unnoticed in their serious technological potentials.[1]

On the whole (not counting rare upheavals), the great classical civilizations had comparatively early reached a point of technological saturation—the aforementioned "optimum" in equilibrium of means with acknowledged needs and goals—and had little cause later to go beyond it. From there on, convention reigned supreme. From pottery to monumental architecture, from food growing to shipbuilding, from textiles to engines of war, from time measuring to stargazing: tools, techniques, and objectives remained essentially the same over long times; improvements were sporadic and unplanned. Progress therefore—if it occurred at all[2]—was by inconspicuous increments to a universally high level that still excites our admiration and, in historical fact, was more liable to regression than to surpassing. The former at least was the more noted phenomenon, deplored by the epigones with a nostalgic remembrance of a better past (as in the declining Roman world). More important, there was, even in the best and most vigorous times, no proclaimed *idea* of a future of *constant progress* in the arts. Most important, there was never a deliberate method of going about it like "research," the willingness to undergo the risks of trying unorthodox paths, exchanging information widely about the experience, and so on. Least of all was there a "natural science" as a growing body of theory to guide such semi-theoretical, prepractical activities, plus their social institutionalization. In routines as well as panoply of instruments, accomplished as they were for the purposes they served, the "arts" seemed as settled as those purposes themselves.[3]

Traits of Modern Technology

The exact opposite of this picture holds for modern technology, and this is its first philosophical aspect. Let us begin with some manifest traits.

1. Every new step in whatever direction of whatever technological field tends *not* to approach an equilibrium or saturation point in the process of fitting means to ends (nor is it meant to), but, on the contrary, to give rise, if successful, to further steps in all kinds of direction and with a fluidity of the ends themselves. "Tends to" becomes a compelling "is bound to" with any major or important step (this almost being its criterion); and the innovators themselves expect, beyond the accomplishment, each time, of their immediate task, the constant future repetition of their inventive activity.

2. Every technical innovation is sure to spread quickly through the technological world community, as also do theoretical discoveries in the sciences. The spreading is in terms of knowledge and of practical adoption, the first (and its speed) guaranteed by the universal intercommunication that is itself part of the technological complex, the second enforced by the pressure of competition.

3. The relation of means to ends is not unilinear but circular. Familiar ends of long-standing may find better satisfaction by new technologies whose genesis they had inspired. But equally—and increasingly typical—new technologies may suggest, create, even impose new ends, never before conceived, simply by offering their feasibility. (Who had ever wished to have in his living room the Philharmonic orchestra, or open heart surgery, or a helicopter defoliating a Vietnam forest? or to drink his coffee from a disposable plastic cup? or to have artificial insemination, test-tube babies, and host pregnancies? or to see clones of himself and others walking about?) Technology thus adds to the very objectives of human desires, including objectives for technology itself. The last point indicates the dialectics or circularity of the case: once incorporated into the socioeconomic demand diet, ends first gratuitously (perhaps accidentally) generated by technological invention become necessities of life and set technology the task of further perfecting the means of realizing them.

4. Progress, therefore, is not just an ideological gloss on modern technology, and not at all a mere option offered by it, but an inherent drive which acts willy-nilly in the formal automatics of its *modus operandi* as it interacts with society. "Progress" is here not a value term but purely descriptive. We may resent the fact and despise its fruits and yet must go along with it, for—short of a stop by the fiat of total political power, or by a sustained general strike of its clients or some internal collapse of their societies, or by self-destruction through its works (the last, alas, the least unlikely of these)—the juggernaut moves on relentlessly, spawning its always mutated progeny by coping with the challenges and lures of the now. But while not a value term, "progress" here is not a neutral term either, for which we could simply substitute "change." For it is in the nature of the case, or a law of the series, that a later stage is always, in terms of technology itself, *superior* to the preceding stage.[4] Thus we have here a case of the entropy-defying sort (organic evolution is another), where the internal motion of a system, left

to itself and not interfered with, leads to ever "higher," not "lower" states of itself. Such at least is the present evidence.[5] If Napoleon once said, "Politics is destiny," we may well say today, "Technology is destiny."

These points go some way to explicate the initial statement that modern technology, unlike traditional, is an enterprise and not a possession, a process and not a state, a dynamic thrust and not a set of implements and skills. And they already adumbrate certain "laws of motion" for this restless phenomenon. What we have described, let us remember, were formal traits, which as yet say little about the contents of the enterprise. We ask two questions of this descriptive picture: *why* is this so, that is, what *causes* the restlessness of modern technology; what is the nature of the thrust? And, what is the philosophical import of the facts so explained?

The Nature of Restless Technology

As we would expect in such a complex phenomenon, the motive forces are many, and some causal hints appeared already in the descriptive account. We have mentioned *pressure of competition*—for profit, but also for power, security, and so forth—as one perpetual mover in the universal appropriation of technical improvements. It is equally operative in their origination, that is, in the process of invention itself, nowadays dependent on constant outside subsidy and even goal-setting: potent interests see to both. War, or the threat of it, has proved an especially powerful agent. The less dramatic, but no less compelling, everyday agents are legion. To keep one's head above the water is their common principle (somewhat paradoxical, in view of an abundance already far surpassing what former ages would have lived with happily ever after). Of pressures other than the competitive ones, we must mention those of population growth and of impending exhaustion of natural resources. Since both phenomena are themselves already by-products of technology (the first by way of medical improvements, the second by the voracity of industry), they offer a good example of the more general truth that to a considerable extent technology itself begets the problems which it is then called upon to overcome by a new forward jump. (The Green Revolution and the development of synthetic substitute materials or of alternate sources of energy come under this heading.) These compulsive pressures for progress, then, would operate even for a technology in a noncompetitive, for example, a socialist setting.

A motive force more autonomous and spontaneous than these almost mechanical pushes with their "sink or swim" imperative would be the pull of the quasi-utopian *vision* of an ever better life, whether vulgarly conceived or nobly, once technology had proved the open-ended capacity for procuring the conditions for it: perceived possibility whetting the appetite ("the American dream," "the revolution of rising expectations"). This less palpable factor is more difficult to appraise, but its playing a role is undeniable. Its deliberate fostering and manipulation by the dream merchants of the industrial-mercantile complex is yet another matter and somewhat taints the spontaneity of the motive, as it also degrades the quality of the dream. It is also moot to what extent the vision itself is *post hoc* rather than *ante hoc,* that is, instilled by the dazzling feats of a technological progress already underway and thus more a response to than a motor of it.

Groping in these obscure regions of motivation, one may as well descend, for an explanation of the dynamism as such, into the Spenglerian mystery of a "Faustian soul" innate in Western culture, that drives it, nonrationally, to infinite novelty and unplumbed possibilities for their own sake; or into the Heideggerian depths of a fateful, metaphysical decision of the will for boundless power over the world of things—a decision equally peculiar to the Western mind: speculative intuitions which do strike a resonance in us, but are beyond proof and disproof.

Surfacing once more, we may also look at the very sober, functional facts of industrialism as such, of production and distribution, output maximization, managerial and labor aspects, which even apart from competitive pressure provide their own incentives for technical progress. Similar observations apply to the requirements of *rule* or control in the vast and populous states of our time, those giant territorial super-organisms which for their very cohesion depend on advanced technology (for example, in information, communication, and transportation, not to speak of weaponry) and thus have a stake in its promotion: the more so, the more centralized they are. This holds for socialist systems no less than for free-market societies. May we conclude from this that even a communist world state, freed from external rivals as well as from internal free-market competition, might still have to push technology ahead for purposes of control on this colossal scale? Marxism, in any case, has its own inbuilt commitment to technological progress beyond necessity. But even disregarding all dynamics of these conjectural kinds, the most monolithic case imaginable would, at any rate, still be exposed to those noncompetitive, natural pressures like population growth and dwindling resources that beset industrialism as such. Thus, it seems, the compulsive element of technological progress may not be bound to its original breeding ground, the capitalist system. Perhaps the odds for an eventual stabilization look somewhat better in a socialist system, provided it is worldwide—and possibly totalitarian in the bargain. As it is, the pluralism we are thankful for ensures the constancy of compulsive advance.

We could go on unraveling the causal skein and would be sure to find many more strands. But none nor all of them, much as they explain, would go to the heart of the matter. For all of them have one premise in common without which they could not operate for long: the premise that there *can* be indefinite progress because there *is* always something new and better to find. The, by no means obvious, givenness of this objective condition is also the pragmatic conviction of the performers in the technological drama; but without its being true, the conviction would help as little as the dream of the alchemists. Unlike theirs, it is backed up by an impressive record of past successes, and for many this is sufficient ground for their belief. (Perhaps holding or not holding it does not even greatly matter.) What makes it more than a sanguine belief, however, is an underlying and well-grounded, theoretical view of the nature of things and of human cognition, according to which they do not set a limit to novelty of discovery and invention, indeed, that they of themselves will at each point offer another opening for the as yet unknown and undone. The corollary conviction, then, is that a technology tailored to a nature and to a knowledge of this indefinite potential ensures its indefinitely continued conversion into the practical powers, each step of it begetting the next, with never a cutoff from internal exhaustion of possibilities.

Only habituation dulls our wonder at this wholly unprecedented belief in virtual "infinity." And by all our present comprehension of reality, the belief is most likely true—at least enough of it to keep the road for innovative technology in the wake of advancing science open for a long time ahead. Unless we understand this ontologic-epistemological premise, we have not understood the inmost agent of technological dynamics, on which the working of all the adventitious causal factors is contingent in the long run.

Let us remember that the virtual infinitude of advance we here seek to explain is in essence different from the always avowed perfectibility of every human accomplishment. Even the undisputed master of his craft always had to admit as possible that he might be surpassed in skill or tools or materials; and no excellence of product ever foreclosed that it might still be bettered, just as today's champion runner must know that his time may one day be beaten. But these are improvements within a given genus, not different in kind from what went before, and they must accrue in diminishing fractions. Clearly, the phenomenon of an exponentially growing *generic* innovation is qualitatively different.

Science as a Source of Restlessness

The answer lies in the interaction of *science* and *technology* that is the hallmark of modern progress, and thus ultimately in the kind of nature which modern science progressively discloses. For it is here, in the movement of *knowledge,* where relevant novelty first and constantly occurs. This is itself a novelty. To Newtonian physics, nature appeared simple, almost crude, running its show with a few kinds of basic entities and forces by a few universal laws, and the application of those well-known laws to an ever greater variety of composite phenomena promised ever widening knowledge indeed, but no real surprises. Since the mid-nineteenth century, this minimalistic and somehow finished picture of nature has changed with breathtaking acceleration. In a reciprocal interplay with the growing subtlety of exploration (instrumental and conceptual), nature itself stands forth as ever more subtle. The progress of probing makes the object grow richer in modes of operation, not sparer as classical mechanics had expected. And instead of narrowing the margin of the still-undiscovered, science now surprises itself with unlocking dimension after dimension of new depths. The very essence of matter has turned from a blunt, irreducible ultimate to an always reopened challenge for further penetration. No one can say whether this will go on forever, but a suspicion of intrinsic infinity in the very being of things obtrudes itself and therewith an anticipation of unending inquiry of the sort where succeeding steps will not find the same old story again (Descartes' "matter in motion"), but always add new twists to it. If then the art of technology is correlative to the knowledge of nature, technology too acquires from this source that potential of infinity for its innovative advance.

But it is not just that indefinite scientific progress offers the *option* of indefinite technological progress, to be exercised or not as other interests see fit. Rather the cognitive process itself moves by interaction with the technological, and in the most internally vital sense: for its own *theoretical* purpose, science must generate an increasingly sophisticated and physically formidable technology as its tool. What it finds with this help initiates new departures in the practical sphere, and the latter as a whole, that is, technology at work provides with its experiences a large-scale laboratory for science again, a breeding ground for new questions, and so on in an unending cycle. In brief, a mutual feedback operates between science and technology; each requires and propels the other; and as matters now stand, they can only live together or must die together. For the dynamics of technology, with which we are here concerned, this means that (all external promptings apart) an agent of restlessness is implanted in it by its functionally integral bond with science. As long, therefore, as the cognitive impulse lasts, technology is sure to move ahead with it. The cognitive impulse, in its turn, culturally vulnerable in itself, liable to lag or to grow conservative with a treasured canon—that theoretical eros itself no longer lives on the delicate appetite for truth alone, but is spurred on by its hardier offspring, technology, which communicates to it impulsions from the broadest arena of struggling, insistent life. Intellectual curiosity is seconded by interminably self-renewing practical aim.

I am conscious of the conjectural character of some of these thoughts. The revolutions in science over the last fifty years or so are a fact, and so are the revolutionary style they imparted to technology and the reciprocity between the two concurrent streams (nuclear physics is a good example). But whether those scientific revolutions, which hold primacy in the whole syndrome, will be typical for science henceforth—something like a law of motion for its future—or represent only a singular phase in its longer run, is unsure. To the extent, then, that our forecast of incessant novelty for technology was predicated on a guess concerning the future of science, even concerning the nature of things, it is hypothetical, as such extrapolations are bound to be. But even if the recent past did not usher in a state of permanent revolution for science, and the life of theory settles down again to a more sedate pace, the scope for technological innovation will not easily shrink; and what may no longer be a revolution in science, may still revolutionize our lives in its practical impact through technology. "Infinity" being too large a word anyway, let us say that present signs of potential and of incentives point to an indefinite perpetuation and fertility of the technological momentum.

The Philosophical Implications

It remains to draw philosophical conclusions from our findings, at least to pinpoint aspects of philosophical interest. Some preceding remarks have already been straying into philosophy of science in the technical sense. Of broader issues, two will be ample to provide food for further thought beyond the limitations of this paper. One concerns the status of knowledge in the human scheme, the other the status of technology itself as a human goal, or its tendency to become that from being a means, in a dialectical inversion of the means-end order itself.

Concerning knowledge, it is obvious that the time-honored division of theory and practice has vanished for both sides. The thirst for pure knowledge may persist undiminished, but the involvement of knowing at the heights with doing in the lowlands of life, mediated by technology, has become inextricable; and the aristocratic self-sufficiency of knowing for its own (and the knower's) sake has gone. Nobility has been exchanged for utility. With the possible exception of philosophy, which still can do with paper and pen and tossing thoughts around among peers, all knowledge has become thus tainted, or elevated if you will, whether utility is intended or not. The technological syndrome, in other words, has brought about a thorough *socializing* of the theoretical realm, enlisting it in the service of common need. What used to be the freest of human choices, an extravagance snatched from the pressure of the world—the esoteric life of thought—has become part of the great public play of necessities and a prime necessity in the action of the play.[6] Remotest abstraction has become enmeshed with nearest concreteness. What this pragmatic functionalization of the once highest indulgence in impractical pursuits portends for the image of man, for the restructuring of a hallowed hierarchy of values, for the idea of "wisdom," and so on, is surely a subject for philosophical pondering.

Concerning technology itself, its actual role in modern life (as distinct from the purely instrumental definition of technology as such) has made the relation of means and ends equivocal all the way up from the daily living to the very vocation of man. There could be no question in former technology that its role was that of humble servant—pride of workmanship and aesthetic embellishment of the useful notwithstanding. The Promethean enterprise of modern technology speaks a different language. The word "enterprise" gives the clue, and its unendingness another. We have mentioned that the effect of its innovations is disequilibrating rather than equilibrating with respect to the balance of wants and supply, always breeding its own new wants. This in itself compels the constant attention of the best minds, engaging the full capital of human ingenuity for meeting challenge after challenge and seizing the new chances.

It is psychologically natural for that degree of engagement to be invested with the dignity of dominant purpose. Not only does technology dominate our lives in fact, it nourishes also a belief in its being of predominant worth. The sheer grandeur of the enterprise and its seeming infinity inspire enthusiasm and fire ambition. Thus, in addition to spawning new ends (worthy or frivolous) from the mere invention of means, technology as a grand venture tends to establish *itself* as the transcendent end. At least the suggestion is there and casts its spell on the modern mind. At its most modest, it means elevating *homo faber* to the essential aspect of man; at its most extravagant, it means elevating *power* to the position of his dominant and interminable goal. To become ever more masters of the world, to advance from power to power, even if only collectively and perhaps no longer by choice, *can* now be seen to be the chief vocation of mankind. Surely, this again poses philosophical questions that may well lead unto the uncertain grounds of metaphysics or of faith.

I here break off, arbitrarily, the formal account of the technological movement in general, which as yet has told us little of what the enterprise is about. To this subject I now turn, that is, to the new kinds of powers and objectives that technology opens to modern man and the consequently altered quality of human action itself.

II THE MATERIAL WORKS OF TECHNOLOGY

Technology is a species of power, and we can ask questions about how and on what object any power is exercised. Adopting Aristotle's rule in *De anima* that for understanding a faculty one should begin with its objects, we start from them too—"objects" meaning both the visible *things* technology generates and puts into human use, and the *objectives* they serve. The objects of modern technology are first everything that had always been an object of human artifice and labor: food, clothing, shelter, implements, transportation—all the material necessities and comforts of life. The technological intervention changed at first not the product but its production, in speed, ease, and quantity. However, this is true only of the very first stage of the industrial revolution with which large-scale scientific technology began. For example, the cloth of the steam–driven looms of Lancashire remained the same. Even then, one significant new product was added to the traditional list—the machines themselves, which required an entire new industry with further subsidiary industries to build them. These novel entities, machines—at first capital goods only, not consumer goods—had from the beginning their own impact on man's symbiosis with nature by being consumers themselves. For example: steam-powered water pumps facilitated coal mining, required in turn extra coal for firing their boilers, more coal for the foundries and forges that made those boilers, more for the mining of the requisite iron ore, more for its transportation to the foundries, more—both coal and iron—for the rails and locomotives made in these same foundries, more for the conveyance of the foundries' product to the pitheads and return, and finally more for the distribution of the more abundant coal to the users outside this cycle, among which were increasingly still more machines spawned by the increased availability of coal. Lest it be forgotten over this long chain, we have been speaking of James Watt's modest steam engine for pumping water out of mine shafts. This syndrome of self-proliferation—by no means a linear chain but an intricate web of reciprocity—has been part of modern technology ever since. To generalize, technology exponentially increases man's drain on nature's resources (of substances and of energy), not only through the multiplication of the final goods for consumption, but also, and perhaps more so, through the production and operation of its own mechanical means. And with these means—

machines—it introduced a new category of goods, not for consumption, added to the furniture of our world. That is, among the objects of technology a prominent class is that of technological apparatus itself.

Soon other features also changed the initial picture of a merely mechanized production of familiar commodities. The final products reaching the consumer ceased to be the same, even if still serving the same age-old needs; new needs, or desires, were added by commodities of entirely new kinds which changed the habits of life. Of such commodities, machines themselves became increasingly part of the consumer's daily life to be used directly by himself, as an article not of production but of consumption. My survey can be brief as the facts are familiar.

New Kinds of Commodities

When I said that the cloth of the mechanized looms of Lancashire remained the same, everyone will have thought of today's synthetic fiber textiles for which the statement surely no longer holds. This is fairly recent, but the general phenomenon starts much earlier, in the synthetic dyes and fertilizers with which the chemical industry—the first to be wholly a fruit of science—began. The original rationale of these technological feats was substitution of artificial for natural materials (for reasons of scarcity or cost), with as nearly as possible the same properties for effective use. But we need only think of plastics to realize that art progressed from substitutes to the creation of really new substances with properties not so found in any natural one, raw or processed, thereby also initiating uses not thought of before and giving rise to new classes of objects to serve them. In chemical (molecular) engineering, man does more than in mechanical (molar) engineering which constructs machinery from natural materials; his intervention is deeper, redesigning the infra-patterns of nature, making substances to specification by arbitrary disposition of molecules. And this, be it noted, is done deductively from the bottom, from the thoroughly analyzed last elements, that is, in a real *via compositiva* after the completed *via resolutiva,* very different from the long-known empirical practice of coaxing substances into new properties, as in metal alloys from the bronze age on. Artificiality or creative engineering with abstract construction invades the heart of matter. This, in molecular biology, points to further, awesome potentialities.

With the sophistication of molecular alchemy we are ahead of our story. Even in straightforward hardware engineering, right in the first blush of the mechanical revolution, the objects of use that came out of the factories did not really remain the same, even where the objectives did. Take the old objective of travel. Railroads and ocean liners are relevantly different from the stage coach and from the sailing ship, not merely in construction and efficiency but in the very feel of the user, making travel a different experience altogether, something one may do for its own sake. Airplanes, finally, leave behind any similarity with former conveyances, except the purpose of getting from here to there, with no experience of what lies in between. And these instrumental objects occupy a prominent, even obtrusive place in our world, far beyond anything wagons and boats ever did. Also they are constantly subject to improvement of design, with obsolescence rather than wear determining their life span.

Or take the oldest, most static of artifacts: human habitation. The multistoried office building of steel, concrete, and glass is a qualitatively different entity from the wood, brick, and stone structures of old. With all that goes into it besides the structures as such—the plumbing and wiring, the elevators, the lighting, heating, and cooling systems—it embodies the end products of a whole spectrum of technologies and far-flung industries, where only at the remote sources human hands still meet with primary materials, no longer recognizable in the final result. The ultimate customer inhabiting the product is ensconced in a shell of thoroughly derivative artifacts (perhaps relieved by a nice piece of driftwood). This transformation into utter artificiality is generally, and increasingly, the effect of technology on the human environment, down to the items of daily use. Only in agriculture has the product so far escaped this transformation by the changed modes of its production. We still eat the meat and rice of our ancestors.[7]

Then, speaking of the commodities that technology injects into private use, there are machines themselves, those very devices of its own running, originally confined to the economic sphere. This unprecedented novum in the records of individual living started late in the nineteenth century and has since grown to a pervading mass phenomenon in the Western world. The prime example, of course, is the automobile, but we must add to it the whole gamut of household appliances—refrigerators, washers, dryers, vacuum cleaners—by now more common in the lifestyle of the general population than running water or central heating were one hundred years ago. Add lawn mowers and other power tools for home and garden: we are mechanized in our daily chores and recreations (including the toys of our children) with every expectation that new gadgets will continue to arrive.

These paraphernalia are machines in the precise sense that they perform work and consume energy, and their moving parts are of the familiar magnitudes of our perceptual world. But an additional and profoundly different category of technical apparatus was dropped into the lap of the private citizen, not labor-saving and work-performing, partly not even utilitarian, but—with minimal energy input—catering to the senses and the mind: telephone, radio, television, tape recorders, calculators, record players—all the domestic terminals of the electronics industry, the latest arrival on the technological scene. Not only by their insubstantial, mind-addressed output, also by the subvisible, not literally "mechanical" physics of their functioning do these devices differ in kind from all the macroscopic, bodily moving machinery of the classical type. Before inspecting this momentous turn from power engineering, the hallmark of the first industrial revolution, to communication engineering, which almost amounts to a second industrial-technological revolution, we must take a look at its natural base: electricity.

In the march of technology to ever greater artificiality, abstraction, and subtlety, the unlocking of electricity marks a decisive step. Here is a universal force of nature which yet does not naturally appear to man (except in lightning). It is not a datum of uncontrived experience. Its very "appearance" had to wait for science, which contrived the experience for it. Here, then, a technology depended on science for the mere providing of its "object," the entity itself it would deal with—the first case where theory alone, not ordinary experience, wholly preceded practice (repeated later in the case of nuclear energy). And what sort of entity! Heat and steam are familiar objects of sensuous experience, their force bodily displayed in nature; the matter of chemistry is still the concrete, corporeal stuff mankind had always known. But electricity is an abstract object, disembodied, immaterial, unseen; in its usable form, it is entirely an artifact, generated in a subtle transformation from grosser forms of energy (ultimately from heat via motion). Its theory indeed had to be essentially complete before utilization could begin.

Revolutionary as electrical technology was in itself, its purpose was at first the by now conventional one of the industrial revolution in general: to supply motive power for the propulsion of machines. Its advantages lay in the unique versatility of the new force, the ease of its transmission, transformation and distribution—an unsubstantial commodity, no bulk, no weight, instantaneously delivered at the point of consumption. Nothing like it had ever existed before in man's traffic with matter, space, and time. It made possible the spread of mechanization to every home; this alone was a tremendous boost to the technological tide, at the same time hooking private lives into centralized public networks and thus making them dependent on the functioning of a total system as never before, in fact, for every moment. Remember, you cannot hoard electricity as you can coal and oil, or flour and sugar for that matter.

But something much more unorthodox was to follow. As we all know, the discovery of the universe of electromagnetics caused a revolution in theoretical physics that is still underway. Without it, there would be no relativity theory, no quantum mechanics, no nuclear and subnuclear physics. It also caused a revolution in technology beyond what it contributed, as we noted, to its classical program. The revolution consisted in the passage from electrical to electronic technology which signifies a new level of abstraction in means and ends. It is the difference between power and communication engineering. Its object, the most impalpable of all, is information. Cognitive instruments had been known before—sextant, compass, clock, telescope, microscope, thermometer, all of them for information and not for work. At one time, they were called "philosophical" or "metaphysical" instruments. By the same general criterion, amusing as it may seem, the new electronic information devices, too, could be classed as "philosophical instruments." But those earlier cognitive devices, except the clock, were inert and passive, not generating information actively, as the new instrumentalities do.

Theoretically as well as practically, electronics signifies a genuinely new phase of the scientific-technological revolution. Compared with the sophistication of its theory as well as the delicacy of its apparatus, everything which came before seems crude, almost natural. To appreciate the point, take the man-made satellites now in orbit. In one sense, they are indeed an imitation of celestial mechanics—Newton's laws finally verified by cosmic experiment: astronomy, for millennia the most purely contemplative of the physical sciences, turned into a practical art! Yet, amazing as it is, the astronomic imitation, with all the unleashing of forces and the finesse of techniques that went into it, is the least interesting aspect of those entities. In that respect, they still fall within the terms and feats of classical mechanics (except for the remote-control course corrections).

Their true interest lies in the instruments they carry through the voids of space and in what these do, their measuring, recording, analyzing, computing, their receiving, processing, and transmitting abstract information and even images over cosmic distances. There is nothing in all nature which even remotely foreshadows the kind of things that now ride the heavenly spheres. Man's imitative practical astronomy merely provides the vehicle for something else with which he sovereignly passes beyond all the models and usages of known nature.[8] That the advent of man portended, in its inner secret of mind and will, a cosmic event was known to religion and philosophy: now it manifests itself as such by fact of things and acts in the visible universe. Electronics indeed creates a range of objects imitating nothing and progressively added to by pure invention.

And no less invented are the ends they serve. Power engineering and chemistry for the most part still answered to the natural needs of man: for food, clothing, shelter, locomotion, and so forth. Communication engineering answers to needs of information and control solely created by the civilization that made this technology possible and, once started, imperative. The novelty of the means continues to engender no less novel ends—both becoming as necessary to the functioning of the civilization that spawned them as they would have been pointless for any former one. The world they help to constitute and which needs computers for its very running is no longer nature supplemented, imitated, improved, transformed, the original habitat made more habitable. In the pervasive mentalization of physical relationships it is a *trans-*

nature of human making, but with this inherent paradox: that it threatens the obsolescence of man himself, as increasing automation ousts him from the places of work where he formerly proved his humanhood. And there is a further threat: its strain on nature herself may reach a breaking point.

The Last Stage of the Revolution?

That sentence would make a good dramatic ending. But it is not the end of the story. There may be in the offing another, conceivably the last, stage of the technological revolution, after the mechanical, chemical, electrical, electronic stages we have surveyed, and the nuclear we omitted. All these were based on physics and had to do with what man can put to his use. What about biology? And what about the user himself? Are we, perhaps, on the verge of a technology, based on biological knowledge and wielding an engineering art which, this time, has man himself for its object? This has become a theoretical possibility with the advent of molecular biology and its understanding of genetic programming; and it has been rendered morally possible by the metaphysical neutralizing of man. But the latter, while giving us the license to do as we wish, at the same time denies us the guidance for knowing what to wish. Since the same evolutionary doctrine of which genetics is a cornerstone has deprived us of a valid image of man, the actual techniques, when they are ready, may find us strangely unready for their responsible use. The anti-essentialism of prevailing theory, which knows only of *de facto* outcomes of evolutionary accident and of no valid essences that would give sanction to them, surrenders our being to a freedom without norms. Thus the technological call of the new microbiology is the twofold one of physical feasibility and metaphysical admissibility. Assuming the genetic mechanism to be completely analyzed and its script finally decoded, we can set about rewriting the text. Biologists vary in their estimates of how close we are to the capability; few seem to doubt the right to use it. Judging by the rhetoric of its prophets, the idea of taking our evolution into our own hands is intoxicating even to many scientists.

In any case, the idea of making over man is no longer fantastic, nor interdicted by an inviolable taboo. If and when *that* revolution occurs, if technological power is really going to tinker with the elemental keys on which life will have to play its melody in generations of men to come (perhaps the only such melody in the universe), then a reflection on what is humanly desirable and what should determine the choice—a reflection, in short, on the image of man, becomes an imperative more urgent than any ever inflicted on the understanding of mortal man. Philosophy, it must be confessed, is sadly unprepared for this, its first cosmic task.

III TOWARD AN ETHICS OF TECHNOLOGY

The last topic has moved naturally from the descriptive and analytic plane, on which the objects of technology are displayed for inspection, onto the evaluative plane where their ethical challenge poses itself for decision. The particular case forced the transition so directly because there the (as yet hypothetical) technological object was man directly. But once removed, man is involved in all the other objects of technology, as these singly and jointly remake the worldly frame of his life, in both the narrower and the wider of its senses: that of the artificial frame of civilization in which social man leads his life proximately, and that of the natural terrestrial environment in which this artifact is embedded and on which it ultimately depends.

Again, because of the magnitude of technological effects on both these vital environments in their totality, both the quality of human life and its very preservation in the future are at stake in the rampage of technology. In short, certainly the "image" of man, and possibly the survival of the species (or of much of it), are in jeopardy. This would summon man's duty to his cause even if the jeopardy were not of his own making. But it is, and, in addition to his ageless obligation to meet the threat of things, he bears for the first time the responsibility of prime agent in the threatening disposition of things. Hence nothing is more natural than the passage from the objects to the ethics of technology, from the things made to the duties of their makers and users.

A similar experience of inevitable passage from analysis of fact to ethical significance, let us remember, befell us toward the end of the first section. As in the case of the matter, so also in the case of the form of the technological dynamics, the image of man appeared at stake. In view of the quasi-automatic compulsion of those dynamics, with their perspective of indefinite progression, every existential and moral question that the objects of technology raise assumes the curiously eschatological quality with which we are becoming familiar from the extrapolating guesses of futurology. But apart from thus raising all challenges of present particular matter to the higher powers of future exponential magnification, the despotic dynamics of the technological movement as such, sweeping its captive movers along in its breathless momentum, poses its own questions to man's axiological conception of himself. Thus, form and matter of technology alike enter into the dimension of ethics.

The questions raised for ethics by the objects of technology are defined by the major areas of their impact and thus fall into such fields of knowledge as ecology (with all its biospheric subdivisions of land, sea, and air), demography economics, biomedical and behavioral sciences (even the psychology of mind pollution by television), and so forth. Not even a sketch of the substantive problems, let alone of ethical policies for dealing with them, can here be attempted. Clearly, for a normative rationale of the latter, ethical theory must plumb the very foundations of value, obligation, and the human good.

The same holds of the different kind of questions raised for ethics by the sheer fact of the formal dynamics of technology. But here, a question of another order is added to the straightforward ethical questions of both kinds, subjecting any resolution of them to a pragmatic proviso of harrowing uncertainty. Given the mastery of the creation over its creators, which yet does not abrogate their responsibility nor silence their vital interest, what are the chances and what are the means of gaining *control* of the process, so that the results of any ethical (or even purely prudential) insights can be translated into effective action? How in short can man's freedom prevail against the determinism he has created for himself? On this most clouded question, whereby hangs not only the effectuality or futility of the ethical search which the facts invite (assuming it to be blessed with *theoretical* success!), but perhaps the future of mankind itself, I will make a few concluding, but—alas—inconclusive, remarks. They are intended to touch on the whole ethical enterprise.

Problematic Preconditions of an Effective Ethics

First, a look at the novel state of determinism. Prima facie, it would seem that the greater and more varied powers bequeathed by technology have expanded the range of choices and hence increased human freedom. For economics, for example, the argument has been made[9] that the uniform compulsion which scarcity and subsistence previously imposed on economic behavior with a virtual denial of alternatives (and hence—conjoined with the universal "maximization" motive of capitalist market competition—gave classical economics at least the appearance of a deterministic "science") has given way to a latitude of indeterminacy. The plenty and powers provided by industrial technology allow a pluralism of choosable alternatives (hence disallow scientific prediction). We are not here concerned with the status of economics as a science. But as to the altered state of things alleged in the argument, I submit that the change means rather that one, relatively homogeneous determinism (thus relatively easy to formalize into a law) has been supplanted by another, more complex, multifarious determinism, namely, that exercised by the human artifact itself upon its creator and user. We, abstractly speaking the possessors of those powers, are concretely subject to their emancipated dynamics and the sheer momentum of our own multitude, the vehicle of those dynamics.

I have spoken elsewhere[10] of the "new realm of necessity" set up, like a second nature, by the feedbacks of our achievements. The almighty we, or Man personified is, alas, an abstraction. *Man* may have become more powerful; *men* very probably the opposite, enmeshed as they are in more dependencies than ever before. What ideal Man now can do is not the same as what real men permit or dictate to be done. And here I am thinking not only of the immanent dynamism, almost automatism, of the impersonal technological complex I have invoked so far, but also of the pathology of its client society. Its compulsions, I fear, are at least as great as were those of unconquered nature. Talk of the blind forces of nature! Are those of the sorcerer's creation less blind? They differ indeed in the serial shape of their causality: the action of nature's forces is cyclical, with periodical recurrence of the same, while that of the technological forces is linear, progressive, cumulative, thus replacing the curse of constant toil with the threat of maturing crisis and possible catastrophe. Apart from this significant vector difference, I seriously wonder whether the tyranny of fate has not become greater, the latitude of spontaneity smaller; and whether man has not actually been weakened in his decision-making capacity by his accretion of collective strength.

However, in speaking, as I have just done, of "his" decision-making capacity, I have been guilty of the same abstraction I had earlier criticized in the use of the term "man." Actually, the subject of the statement was no real or representative individual but Hobbes' "Artificial Man," "that great Leviathan, called a Common-Wealth," or the "large horse" to which Socrates likened the city, "which because of its great size tends to be sluggish and needs stirring by a gadfly." Now, the chances of there being such gadflies among the numbers of the commonwealth are today no worse nor better than they have ever been, and in fact they are around and stinging in our field of concern. In that respect, the free spontaneity of personal insight, judgment, and responsible action by speech can be trusted as an ineradicable (if also incalculable) endowment of humanity, and smallness of number is in itself no impediment to shaking public complacency. The problem, however, is not so much complacency or apathy as the counterforces of active, and anything but complacent, interests and the complicity with them of all of us in our daily consumer existence. These interests themselves are factors in the determinism which technology has set up in the space of its sway. The question, then, is that of the possible chances of unselfish insight in the arena of (by nature) selfish *power,* and more particularly; of one long-range, interloping insight against the short-range goals of many incumbent powers. Is there hope that wisdom itself can become power? This renews the thorny old subject of Plato's philosopher-king and—with that inclusion of realism which the utopian Plato did not lack—of the role of myth, not knowledge, in the education of the guardians. Applied to our topic: the *knowledge* of objective dangers and of values endangered, as well as of the technical remedies, is beginning to be there and to be disseminated: but to make it prevail in the marketplace is a matter less of the rational dissemination of truth than of public relations techniques, persuasion, indoctrination, and manipulation, also of unholy alliances, perhaps even conspiracy. The philosopher's descent into the cave may well have to go all the way to "if you can't lick them, join them."

That is so not merely because of the active resistance of special interests but because of the optical illusion of the near and the far which condemns the long-range view to impotence against the enticement and threats of the nearby: it is this incurable shortsightedness of animal-human nature more than ill will that makes it difficult to move even those who have no special axe to grind, but still are in countless ways, as we all are, beneficiaries of the untamed system and so have something dear in the present to lose with the inevitable cost of its taming. The taskmaster, I fear, will have to be actual pain beginning to strike, when the far has moved close to the skin and has vulgar optics on its side. Even then, one may resort to palliatives of the hour. In any event, one should try as much as one can to forestall the advent of emergency with its high tax of suffering or, at the least, prepare for it. This is where the scientist can redeem his role in the technological estate.

The incipient knowledge about technological danger trends must be developed, coordinated, systematized, and the full force of computer-aided projection techniques be deployed to determine priorities of action, so as to inform preventive efforts wherever they can be elicited, to minimize the necessary sacrifices, and at the worst to preplan the saving measures which the terror of beginning calamity will eventually make people willing to accept. Even now, hardly a decade after the first stirrings of "environmental" consciousness, much of the requisite knowledge, plus the rational persuasion, is available inside and outside academia for any well-meaning powerholder to draw upon. To this, we—the growing band of concerned intellectuals—ought persistently to contribute our bit of competence and passion.

But the real problem is to get the well-meaning into power and have that power as little as possible beholden to the interests which the technological colossus generates on its path. It is the problem of the philosopher-king compounded by the greater magnitude and complexity (also sophistication) of the forces to contend with. Ethically, it becomes a problem of playing the game by its impure rules. For the servant of truth to join in it means to sacrifice some of his time-honored role: he may have to turn apostle or agitator or political operator. This raises moral questions beyond those which technology itself poses, that of sanctioning immoral means for a surpassing end, of giving unto Caesar so as to promote what is not Caesar's. It is the grave question of moral casuistry, or of Dostoevsky's Grand Inquisitor, or of regarding cherished liberties as no longer affordable luxuries (which may well bring the anxious friend of mankind into odious political company)—questions one excusably hesitates to touch but in the further tide of things may not be permitted to evade.

What is, prior to joining the fray, the role of philosophy, that is, of a philosophically grounded ethical knowledge, in all this? The somber note of the last remarks responded to the quasi-apocalyptic prospects of the technological tide, where stark issues of planetary survival loom ahead. There, no philosophical ethics is needed to tell us that disaster must be averted. Mainly, this is the case of the ecological dangers. But there are other, noncatastrophic things afoot in technology where not the existence but the image of man is at stake. They are with us now and will accompany us and be joined by others at every new turn technology may take. Mainly, they are in the biomedical, behavioral, and social fields. They lack the stark simplicity of the survival issue, and there is none of the (at least declaratory) unanimity on them which the spectre of extreme crisis commands. It is here where a philosophical ethics or theory of values has its task. Whether its voice will be listened to in the dispute on policies is not for it to ask; perhaps it cannot even muster an authoritative voice with which to speak—a house divided, as philosophy is. But the philosopher must try for normative knowledge, and if his labors fall predictably short of producing a compelling axiomatics, at least his clarifications can counteract rashness and make people pause for a thoughtful view.

Where not existence but "quality" of life is in question, there is room for honest dissent on goals, time for theory to ponder them, and freedom from the tyranny of the lifeboat situation. Here, philosophy can have its try and its say. Not so on the extremity of the survival issue. The philosopher, to be sure, will also strive for a theoretical grounding of the very proposition that there ought to be men on earth, and that present generations are obligated to the existence of future ones. But such esoteric, ultimate validation of the perpetuity imperative for the species—whether obtainable or not to the satisfaction of reason—is happily not needed for consensus in the face of ultimate threat. Agreement in favor of life is pretheoretical, instinctive, and universal. Averting disaster takes precedence over everything else, including pursuit of the good, and suspends otherwise inviolable prohibitions and rules. All moral standards for individual or group behavior, even demands for individual sacrifice of life, are premised on the continued existence of human life. As I have said elsewhere,[11] "No rules can be devised for the waiving of rules in extremities. As with the famous shipwreck examples of ethical theory, the less said about it, the better."

Never before was there cause for considering the contingency that all mankind may find itself in a lifeboat, but this is exactly what we face when the viability of the planet is at stake. Once the situation becomes desperate, then what there is to do for salvaging it must be done, so that there be life—which "then," after the storm has been weathered, can again be adorned by ethical conduct. The moral inference to be drawn from this lurid eventuality of a moral pause is that we must never allow a lifeboat situation for humanity to arise.[12] One part of the ethics of technology is precisely to guard the space in which any ethics can operate. For the rest, it must grapple with the cross-currents of value in the complexity of life.

A final word on the question of determinism versus freedom which our presentation of the technological syndrome has raised. The best hope of man rests in his most troublesome gift: the spontaneity of human acting which confounds all prediction. As the late Hannah Arendt never tired of stressing: the continuing arrival of newborn individuals in the world assures ever-new beginnings. We should expect to be surprised and to see our predictions come to naught. But those predictions themselves, with their warning voice, can have a vital share in provoking and informing the spontaneity that is going to confound them.

NOTES

1. But as serious an actuality as the Chinese plough "wandered" slowly westward with little traces of its route and finally caused a major, highly beneficial revolution in medieval European agriculture, which almost no one deemed worth recording when it happened (cf. Paul Leser, *Entstehung und Verbreitung des Pfluges*, Münster, 1931; reprint: The International Secretariate for Research on the History of Agricultural Implements, Brede-Lingby, Denmark, 1971).

2. Progress did, in fact, occur even at the heights of classical civilizations. The Roman arch and vault, for example, were distinct engineering advances over the horizontal entablature and flat ceiling of Greek (and Egyptian) architecture, permitting spanning feats and thereby construction objectives not contemplated before (stone bridges, aqueducts, the vast baths and other public halls of Imperial Rome). But materials, tools, and techniques were still the same, the role of human labor and crafts remained unaltered, stone-cutting and brick-baking went on as before. An existing technology was enlarged in its scope of performance, but none of its means or even goals made obsolete. [Ed.]

3. One meaning of "classical" is that those civilizations had somehow implicitly "defined" themselves

and neither encouraged nor even allowed themselves to pass beyond their innate terms. The—more or less—achieved "equilibrium" was their very pride. [Ed.]

4. This only seems to be but is not a value statement, as the reflection on, for example, an ever more destructive atom bomb shows. [Ed.]

5. There may conceivably be internal degenerative factors—such as the overloading of finite information-processing capacity—that may bring the (exponential) movement to a halt or even make the system fall apart. We don't know yet. [Ed.]

6. There is a paradoxical side effect to this change of roles. That very science which forfeited its place in the domain of leisure to become a busy toiler in the field of common needs, creates by its toils a growing domain of leisure for the masses, who reap this with the other fruits of technology as an additional (and no less novel) article of forced consumption. Hence leisure, from a privilege of the few, has become a problem for the many to cope with. Science, not idle, provides for the needs of this idleness too: no small part of technology is spent on filling the leisure-time gap which technology itself has made a fact of life. [Ed.]

7. Not so, objects my colleague Robert Heilbroner in a letter to me; "I'm sorry to tell you that meat and rice are both *profoundly* influenced by technology. Not even they are left untouched." Correct, but they are at least generically the same (their really profound changes lie far back in the original breeding of domesticated strains from wild ones—as in the case of all cereal plants under cultivation). I am speaking here of an order of transformation in which the results bear no resemblance to the natural materials at their source, nor to any naturally occurring state of them. [Ed.]

8. Note also that in radio technology, the medium of action is nothing material, like wires conducting currents, but the entirely immaterial electromagnetic "field," i.e., space itself. The symbolic picture of "waves" is the last remaining link to the forms of our perceptual world. [Ed.]

9. I here loosely refer to Adolph Lowe, "The Normative Roots of Economic Values," in Sidney Hook, ed., *Human Values and Economic Policy* (New York: New York University Press, 1967) and, more perhaps, to the many discussions I had with Lowe over the years. For my side of the argument, see "Economic Knowledge and the Critique of Goals," in R. L. Heibroner, ed., *Economic Means and Social Ends* (Englewood Cliffs, N.J.: Prentice-Hall, 1969), reprinted in Hans Jonas, *Philosophical Essays* (Englewood Cliffs, N.J.: Prentice-Hall, 1969), reprinted in Hans Jonas, *Philosophical Essays* (Englewood Cliffs, N.J.: Prentice-Hall, 1974).

10. "The Practical Uses of Theory," *Social Research* 26 (1959), reprinted in Hans Jonas, *The Phenomenon of Life* (New York, 1966). The reference is to pp. 209–10 in the latter edition.

11. "Philosophical Reflections on Experimenting with Human Subjects," in Paul A. Freund, ed., *Experimentation with Human Subjects* (New York: George Braziller, 1970), reprinted in Hans Jonas, *Philosophical Essays.* The reference is to pp. 124–25 in the latter edition.

12. For a comprehensive view of the demands which such a situation or even its approach would make on our social and political values, see Geoffrey Vickers, *Freedom in a Rocking Boat* (London, 1970).

3

Question Concerning Technology

Martin Heidegger

In what follows we shall be *questioning* concerning technology. Questioning builds a way. We would be advised, therefore, above all to pay heed to the way, and not to fix our attention on isolated sentences and topics. The way is a way of thinking. All ways of thinking, more or less perceptibly, lead through language in a manner that is extraordinary. We shall be questioning concerning *technology,* and in so doing we should like to prepare a free relationship to it. The relationship will be free if it opens our human existence to the essence of technology. When we can respond to this essence, we shall be able to experience the technological within its own bounds.

Technology is not equivalent to the essence of technology. When we are seeking the essence of "tree," we have to become aware that that which pervades every tree, as tree, is not itself a tree that can be encountered among all the other trees.

Likewise, the essence of technology is by no means anything technological. Thus we shall never experience our relationship to the essence of technology so long as we merely conceive and push forward the technological, put up with it, or evade it. Everywhere we remain unfree and chained to technology, whether we passionately affirm or deny it. But we are delivered over to it in the worst possible way when we regard it as something neutral; for this conception of it, to which today we particularly like to do homage, makes us utterly blind to the essence of technology.

According to ancient doctrine, the essence of a thing is considered to be *what* the thing is. We ask the question concerning technology when we ask what it is. Everyone knows the two statements that answer our question. One says: Technology is a means to an end. The other says: Technology is a human activity. The two definitions of technology belong together. For to posit ends and procure and utilize the means to them is a human activity. The manufacture and utilization of equipment, tools, and machines, the manufactured and used things themselves, and the needs and ends that they serve, all belong to what technology is. The whole complex of these contrivances is technology. Technology itself is a contrivance, or, in Latin, an *instrumentum.*

The current conception of technology, according to which it is a means and a human activity, can therefore be called the instrumental and anthropological definition of technology.

Who would ever deny that it is correct? It is in obvious conformity with what we are envisioning when we talk about technology. The instrumental definition of technology is indeed so uncannily correct that it even holds for modern technology, of which, in other respects, we maintain with some justification that it is, in contrast to the older handwork technology, something completely different and therefore new. Even the power plant with its turbines and generators is a man-made means to an end established by man. Even the jet aircraft and the high-frequency apparatus are means to ends. A radar station is of course less simple than a weather vane. To be sure, the construction of a high-frequency apparatus requires the interlocking of various processes of technical-industrial production. And certainly a sawmill in a secluded valley of the Black Forest is a primitive means compared with the hydroelectric plant in the Rhine River.

But this much remains correct: modern technology too is a means to an end. That is why the instrumental conception of technology conditions every attempt to bring man into the right relation to technology. Everything depends on our manipulating technology in the proper manner as a means. We will, as we say, "get" technology "spiritually in hand." We will master it. The will to mastery becomes all the more urgent the more technology threatens to slip from human control.

But suppose now that technology were no mere means, how would it stand with the will to master it? Yet we said, did we not, that the instrumental definition of technology is correct? To be sure. The correct always fixes upon something pertinent in whatever is under consideration. However, in order to be correct, this fixing by no means needs to uncover the thing in question in its essence. Only at the point where such an uncovering happens does the true come to pass. For that reason the merely correct is not yet the true. Only the true brings us into a free relationship with that which concerns us from out of its essence. Accordingly, the correct instrumental definition of technology still does not show us technology's essence. In order that we may arrive at this, or at least come close to it, we must seek the true by way of the correct. We must ask: What is the instrumental itself? Within what do such things as means and end belong? A means is that whereby something is effected and thus attained. Whatever has an effect as its consequence is called a cause. But not only that by means of which something else is effected is a cause. The end in keeping with which the kind of means to be used is determined is also considered a cause. Wherever ends are pursued and means are employed, wherever instrumentality reigns, there reigns causality.

For centuries philosophy has taught that there are four causes: (1) the *causa materialis,* the material, the matter out of which, for example, a silver chalice is made; (2) the *causa formalis,* the form, the shape into which the material enters; (3) the *causa finalis,* the end, for example, the sacrificial rite in relation to which the chalice required is determined as to its form and matter; (4) the *causa efficiens,* which brings about the effect that is the finished, actual chalice, in this instance, the silversmith. What technology is, when represented as a means, discloses itself when we trace instrumentality back to fourfold causality.

But suppose that causality, for its part, is veiled in darkness with respect to what it is? Certainly for centuries we have acted as though the doctrine of the four causes had fallen from heaven as a truth as clear as daylight. But it might be that the time has come to ask, Why are there just four causes? In relation to the aforementioned four, what does "cause" really mean?

From whence does it come that the causal *character* of the four causes is so unifiedly determined that they belong together?

So long as we do not allow ourselves to go into these questions, causality, and with it instrumentality, and with the latter the accepted definition of technology, remain obscure and groundless.

For a long time we have been accustomed to representing cause as that which brings something about. In this connection, to bring about means to obtain results, effects. The *causa efficiens,* but one among the four causes, sets the standard for all causality. This goes so far that we no longer even count the *causa finalis,* telic finality, as causality. *Causa, casus,* belongs to the verb *cadere,* "to fall," and means that which brings it about that something falls out as a result in such and such a way. The doctrine of the four causes goes back to Aristotle. But everything that later ages seek in Greek thought under the conception and rubric "causality," in the realm of Greek thought and for Greek thought per se has simply nothing at all to do with bringing about and effecting. What we call cause [*Ursache*] and the Romans call *causa* is called *aition* by the Greeks, that to which something else is indebted [*das, was ein anderes verschuldet*]. The four causes are the ways, all belonging at once to each other, of being responsible for something else. An example can clarify this.

Silver is that out of which the silver chalice is made. As this matter (*hyle*), it is co-responsible for the chalice. The chalice is indebted to, i.e., owes thanks to, the silver for that out of which it consists. But the sacrificial vessel is indebted not only to the silver. As a chalice, that which is indebted to the silver appears in the aspect of a chalice and not in that of a brooch or a ring. Thus the sacrificial vessel is at the same time indebted to the aspect (*eidos*) of chaliceness. Both the silver into which the aspect is admitted as chalice and the aspect in which the silver appears are in their respective ways co-responsible for the sacrificial vessel.

But there remains yet a third that is above all responsible for the sacrificial vessel. It is that which in advance confines the chalice within the realm of consecration and bestowal. Through this the chalice is circumscribed as sacrificial vessel. Circumscribing gives bounds to the thing. With the bounds the thing does not stop; rather from out of them it begins to be what, after production, it will be. That which gives bounds, that which completes, in this sense is called in Greek *telos,* which is all too often translated as "aim" or "purpose," and so misinterpreted. The *telos* is responsible for what as matter and for what as aspect are together co-responsible for the sacrificial vessel.

Finally there is a fourth participant in the responsibility for the finished sacrificial vessel's lying before us ready for use, i.e., the silversmith—but not at all because he, in working, brings about the finished sacrificial chalice as if it were the effect of a making; the silversmith is not a *causa efficiens.*

The Aristotelian doctrine neither knows the cause that is named by this term nor uses a Greek word that would correspond to it.

The silversmith considers carefully and gathers together the three aforementioned ways of being responsible and indebted. To consider carefully [*überlegen*] is in Greek *legein, logos. Legein* is rooted in *apophainesthai,* to bring forward into appearance. The silversmith is co-responsible as that from whence the sacrificial vessel's bringing forth and resting-in-self take and retain their first departure. The three previously mentioned ways of being responsible owe thanks to the pondering of the silversmith for the "that" and the "how" of their coming into appearance and into play for the production of the sacrificial vessel.

Thus four ways of being responsible hold sway in the sacrificial vessel that lies ready before us. They differ from one another, yet they belong together. What unites them from the beginning? In what does this playing in unison of the four ways of being responsible play? What is the source of the unity of the four causes? What, after all, does this owing and being responsible mean, thought as the Greeks thought it?

Today we are too easily inclined either to understand being responsible and being indebted moralistically as a lapse, or else to construe them in terms of effecting. In either case we bar to ourselves the way to the primal meaning of that which is later called causality. So long as this way is not opened up to us we shall also fail to see what instrumentality, which is based on causality, actually is.

In order to guard against such misinterpretations of being responsible and being indebted, let us clarify the four ways of being responsible in terms of that for which they are responsible. According to our example, they are responsible for the silver chalice's lying ready before us as a sacrificial vessel. Lying before and lying ready (*hypokeisthai*) characterize the presencing of something that presences. The four ways of being responsible bring something into appearance. They let it come forth into presencing [*An-wesen*]. They set it free to that place and so start it on its way, namely, into its complete arrival. The principal characteristic of being responsible is this starting something on its way into arrival. It is in the sense of such a starting something on its way into arrival that being responsible is an occasioning or an inducing to go forward [*Ver-an-lassen*]. On the basis of a look at what the Greeks experienced in being responsible, in *aitia*, we now give this verb "to occasion" a more inclusive meaning, so that it now is the name for the essence of causality thought as the Greeks thought it. The common and narrower meaning of "occasion" in contrast is nothing more than striking against and releasing, and means a kind of secondary cause within the whole of causality.

But in what, then, does the playing in unison of the four ways of occasioning play? They let what is not yet present arrive into presencing. Accordingly, they are unifiedly ruled over by a bringing that brings what presences into appearance. Plato tells us what this bringing is in a sentence from the *Symposium* (205b): *hē gar toi ek tou mē onton eis to on ionti hotōioun aitia pasa esti poiēsis.* "Every occasion for whatever passes over and goes forward into presencing from that which is not presencing is *poiēsis*, is bringing-forth [*Her-vor-bringen*]."

It is of utmost importance that we think bringing-forth in its full scope and at the same time in the sense in which the Greeks thought it. Not only handcraft manufacture, not only artistic and poetical bringing into appearance and concrete imagery, is a bringing-forth, *poiēsis*. *Physis* also, the arising of something from out of itself, is a bringing-forth, *poiēsis*. *Physis* is indeed *poiēsis* in the highest sense. For what presences by means of *physis* has the bursting open belonging to bringing-forth, e.g., the bursting of a blossom into bloom, in itself (*en heautōi*). In contrast, what is brought forth by the artisan or the artist, e.g., the silver chalice, has the bursting open belonging to bringing-forth not in itself, but in another (*en allōi*), in the craftsman or artist.

The modes of occasioning, the four causes, are at play, then, within bringing-forth. Through bringing-forth, the growing things of nature as well as whatever is completed through the crafts and the arts come at any given time to their appearance.

But how does bringing-forth happen, be it in nature or in handwork and art? What is the bringing-forth in which the fourfold way of occasioning plays? Occasioning has to do with the presencing [*Anwesen*] of that which at any given time comes to appearance in bringing-forth. Bringing-forth brings hither out of concealment forth into unconcealment. Bringing-forth comes to pass only insofar as something concealed comes into unconcealment. This coming rests and moves freely within what we call revealing [*das Entbergen*]. The Greeks have the word *alētheia* for revealing. The Romans translate this with *veritas.* We say "truth" and usually understand it as the correctness of an idea.

But where have we strayed to? We are questioning concerning technology, and we have arrived now at *alētheia,* at revealing. What has the essence of technology to do with revealing? The answer: everything. For every bringing-forth is grounded in revealing. Bringing-forth, indeed, gathers within itself the four modes of occasioning—causality—and rules them throughout. Within its domain belong end and means, belongs instrumentality. Instrumentality is considered to be the fundamental characteristic of technology. If we inquire, step by step, into what technology, represented as means, actually is, then we shall arrive at revealing. The possibility of all productive manufacturing lies in revealing.

Technology is therefore no mere means. Technology is a way of revealing. If we give heed to this, then another whole realm for the essence of technology will open itself up to us. It is the realm of revealing, i.e., of truth.

This prospect strikes us as strange. Indeed, it should do so, should do so as persistently as possible and with so much urgency that we will finally take seriously the simple question of what the name "technology" means. The word stems from the Greek. *Technikon* means that which belongs to *technē.* We must observe two things with respect to the meaning of this word. One is that *technē* is the name not only for the activities and skills of the craftsman, but also for the arts of the mind and the fine arts. *Technē* belongs to bringing-forth, to *poiēsis;* it is something poietic.

The other point that we should observe with regard to *technē* is even more important. From earliest times until Plato the word *technē* is linked with the word *epistēmē.* Both words are names for knowing in the widest sense. They mean to be entirely at home in something, to understand and be expert in it. Such knowing provides an opening up. As an opening up it is a revealing. Aristotle, in a discussion of special importance (*Nicomachean Ethics,* Bk. VI, chaps. 3 and 4), distinguishes between *epistēmē* and *technē* and indeed with respect to what and how they reveal. *Technē* is a mode of *alētheuein.* It reveals whatever does not bring itself forth and does not yet lie here before us, whatever can look and turn out now one way and now another. Whoever builds a house or a ship or forges a sacrificial chalice reveals what is to be brought forth, according to the perspectives of the four modes of occasioning. This revealing gathers together in advance the aspect and the matter of ship or house, with a view to the finished thing envisioned as completed, and from this gathering determines the manner of its construction. Thus what is decisive in *technē* does not lie at all in making and manipulating nor in the using of means, but rather in the aforementioned revealing. It is as revealing, and not as manufacturing, that *technē* is a bringing-forth.

Thus the clue to what the word *technē* means and to how the Greeks defined it leads us into the same context that opened itself to us when we pursued the question of what instrumentality as such in truth might be.

Technology is a mode of revealing. Technology comes to presence in the realm where revealing and unconcealment take place, where *alētheia,* truth, happens.

In opposition to this definition of the essential domain of technology, one can object that it indeed holds for Greek thought and that at best it might apply to the techniques of the handcraftsman, but that it simply does not fit modern machine-powered technology. And it is precisely the latter and it alone that is the disturbing thing, that moves us to ask the question concerning technology per se. It is said that modern technology is something incomparably different from all earlier technologies because it is based on modern physics as an exact science. Meanwhile we have come to understand more clearly that the reverse holds true as well:

Modern physics, as experimental, is dependent upon technical apparatus and upon progress in the building of apparatus. The establishing of this mutual relationship between technology and physics is correct. But it remains a merely historiographical establishing of facts and says nothing about that in which this mutual relationship is grounded. The decisive question still remains: Of what essence is modern technology that it happens to think of putting exact science to use?

What is modern technology? It too is a revealing. Only when we allow our attention to rest on this fundamental characteristic does that which is new in modern technology show itself to us.

And yet the revealing that holds sway throughout modern technology does not unfold into a bringing-forth in the sense of *poiēsis*. The revealing that rules in modern technology is a challenging [*Herausfordern*], which puts to nature the unreasonable demand that it supply energy that can be extracted and stored as such. But does this not hold true for the old windmill as well? No. Its sails do indeed turn in the wind; they are left entirely to the wind's blowing. But the windmill does not unlock energy from the air currents in order to store it.

In contrast, a tract of land is challenged into the putting out of coal and ore. The earth now reveals itself as a coal mining district, the soil as a mineral deposit. The field that the peasant formerly cultivated and set in order appears differently than it did when to set in order still meant to take care of and to maintain. The work of the peasant does not challenge the soil of the field. In the sowing of the grain it places the seed in the keeping of the forces of growth and watches over its increase. But meanwhile even the cultivation of the field has come under the grip of another kind of setting-in-order, which *sets* upon nature. It sets upon it in the sense of challenging it. Agriculture is now the mechanized food industry. Air is now set upon to yield nitrogen, the earth to yield ore, ore to yield uranium, for example; uranium is set upon to yield atomic energy, which can be released either for destruction or for peaceful use.

This setting-upon that challenges forth the energies of nature is an expediting, and in two ways. It expedites in that it unlocks and exposes. Yet that expediting is always itself directed from the beginning toward furthering something else, i.e., toward driving on to the maximum yield at the minimum expense. The coal that has been hauled out in some mining district has not been supplied in order that it may simply be present somewhere or other. It is stockpiled; that is, it is on call, ready to deliver the sun's warmth that is stored in it. The sun's warmth is challenged forth for heat, which in turn is ordered to deliver steam whose pressure turns the wheels that keep a factory running.

The hydroelectric plant is set into the current of the Rhine. It sets the Rhine to supplying its hydraulic pressure, which then sets the turbines turning. This turning sets those machines in motion whose thrust sets going the electric current for which the long-distance power station and its network of cables are set up to dispatch electricity. In the context of the interlocking processes pertaining to the orderly disposition of electrical energy, even the Rhine itself appears as something at our command. The hydroelectric plant is not built into the Rhine River as was the old wooden bridge that joined bank with bank for hundreds of years. Rather the river is dammed up into the power plant. What the river is now, namely, a water power supplier, derives from out of the essence of the power station. In order that we may even remotely consider the monstrousness that reigns here, let us ponder for a moment the contrast that speaks out of the two titles, "The Rhine" as dammed up into the *power* works, and "The Rhine" as uttered out of the *art* work, in Hölderlin's hymn by that name. But, it will be replied, the Rhine is still a river in the landscape, is it not? Perhaps. But how? In no other way than as an object on call for inspection by a tour group ordered there by the vacation industry.

The revealing that rules throughout modern technology has the character of a setting-upon, in the sense of a challenging-forth. That challenging happens in that the energy concealed in nature is unlocked, what is unlocked is transformed, what is transformed is stored up, what is stored up is, in turn, distributed, and what is distributed is switched about ever anew. Unlocking, transforming, storing, distributing, and switching about are ways of revealing. But the revealing never simply comes to an end. Neither does it run off into the indeterminate. The revealing reveals to itself its own manifoldly interlocking paths, through regulating their course. This regulating itself is, for its part, everywhere secured. Regulating and securing even become the chief characteristics of the challenging revealing.

What kind of unconcealment is it, then, that is peculiar to that which comes to stand forth through this setting-upon that challenges? Everywhere everything is ordered to stand by, to be immediately at hand, indeed to stand there just so that it may be on call for a further ordering. Whatever is ordered about in this way has its own standing. We call it the standing-reserve [*Bestand*]. The word expresses here something more, and something more essential, than mere "stock." The name "standing-reserve" assumes the rank of an inclusive rubric. It designates nothing less than the way in which everything presences that is wrought upon by the challenging revealing. Whatever stands by in the sense of standing-reserve no longer stands over against us as object.

Yet an airliner that stands on the runway is surely an object. Certainly. We can represent the machine so. But then it conceals itself as to what and how it is. Revealed, it stands on the taxi strip only as standing-reserve, inasmuch as it is ordered to ensure the possibility of transportation. For this it must be in its whole structure and in every one of its constituent parts, on call for duty, i.e., ready for takeoff. (Here it would be appropriate to discuss Hegel's definition of the machine as an autonomous tool. When applied to the tools of the craftsman, his characterization is correct. Characterized in this way, however, the machine is not thought at all from out of the essence of technology within which it belongs. Seen in terms of the standing-reserve, the machine is completely unautonomous, for it has its standing only from the ordering of the orderable.)

The fact that now, wherever we try to point to modern technology as the challenging revealing, the words "setting-upon," "ordering," "standing-reserve," obtrude and accumulate in a dry, monotonous, and therefore oppressive way, has its basis in what is now coming to utterance.

Who accomplishes the challenging setting-upon through which what we call the real is revealed as standing-reserve? Obviously, man. To what extent is man capable of such a revealing? Man can indeed conceive, fashion, and carry through this or that in one way or another. But man does not have control over unconcealment itself, in which at any given time the real shows itself or withdraws. The fact that the real has been showing itself in the light of Ideas ever since the time of Plato, Plato did not bring about. The thinker only responded to what addressed itself to him.

Only to the extent that man for his part is already challenged to exploit the energies of nature can this ordering revealing happen. If man is challenged, ordered, to do this, then does not man himself belong even more originally than nature within the standing-reserve? The current talk about human resources, about the supply of patients for a clinic, gives evidence of this. The forester who, in the wood, measures the felled timber and to all appearances walks the same forest path in the same way as did his grandfather is today commanded by profit-making in the lumber industry, whether he knows it or not. He is made subordinate to the orderability of cellulose, which for its part is challenged forth by the need for paper, which is then delivered to newspapers and illustrated magazines. The latter, in their turn, set public opinion to swallowing what is printed, so that a set configuration of opinion becomes available on demand. Yet precisely because man is challenged more originally than are the energies of nature, i.e., into the process of ordering, he never is transformed into mere standing-reserve. Since man drives technology forward, he takes part in ordering as a way of revealing. But the unconcealment itself, within which ordering unfolds, is never a human handiwork, any more than is the realm through which man is already passing every time he as a subject relates to an object.

Where and how does this revealing happen if it is no mere handiwork of man? We need not look far. We need only apprehend in an unbiased way that which has already claimed man and has done so, so decisively that he can only be man at any given time as the one so claimed. Wherever man opens his eyes and ears, unlocks his heart, and gives himself over to meditating and striving, shaping and working, entreating and thanking, he finds himself everywhere already brought into the unconcealed. The unconcealment of the unconcealed has already come to pass whenever it calls man forth into the modes of revealing allotted to him. When man, in his way, from within unconcealment reveals that which presences, he merely responds to the call of unconcealment even when he contradicts it. Thus when man, investigating, observing, ensnares nature as an area of his own conceiving, he has already been claimed by a way of revealing that challenges him to approach nature as an object of research, until even the object disappears into the objectlessness of standing-reserve.

Modern technology as an ordering revealing is, then, no merely human doing. Therefore we must take that challenging that sets upon man to order the real as standing-reserve in accordance with the way in which it shows itself. That challenging gathers man into ordering. This gathering concentrates man upon ordering the real as standing-reserve.

That which primordially unfolds the mountains into mountain ranges and courses through them in their folded togetherness is the gathering that we call *"Ge*birg" [mountain chain].

That original gathering from which unfold the ways in which we have feelings of one kind or another we name *"Gemüt"* [disposition].

We now name that challenging claim which gathers man thither to order the self-revealing as standing-reserve: *"Ge-stell"* [Enframing].

We dare to use this word in a sense that has been thoroughly unfamiliar up to now.

According to ordinary usage, the word *Gestell* [frame] means some kind of apparatus, e.g., a bookrack. *Gestell* is also the name for a skeleton. And the employment of the word *Ge-stell* [Enframing] that is now required of us seems equally eerie, not to speak of the arbitrariness with which words of a mature language are thus misused. Can anything be more strange? Surely not. Yet this strangeness is an old usage of thinking. And indeed thinkers accord with this usage precisely at the point where it is a matter of thinking that which is highest. We, late born, are no longer in a position to appreciate the significance of Plato's daring to use the word *eidos* for that which in everything and in each particular thing endures as present. For *eidos,* in the common speech, meant the outward aspect [*Ansicht*] that a visible thing offers to the physical eye. Plato exacts of this word, however, something utterly extraordinary: that it name what precisely is not and never will be perceivable with physical eyes. But even this is by no means the full extent of what is extraordinary here. For *idea* names not only the nonsensuous aspect of what is physically visible. Aspect (*idea*) names and is, also, that which constitutes the essence in the audible, the tasteable, the tactile, in everything that is in any way accessible. Compared with the demands that Plato makes on language and thought in this and other instances, the use of the word *Gestell* as the name for the essence of modern technology, which we now venture here, is almost harmless. Even so, the usage now required remains something exacting and is open to misinterpretation.

Enframing means the gathering together of that setting-upon which sets upon man, i.e., challenges him forth, to reveal the real, in the mode of ordering, as standing-reserve. Enframing means that way of revealing which holds sway in the essence of modern technology and which is itself nothing technological. On the other hand, all those things that are so familiar to us and are standard parts of an assembly, such as rods, pistons, and chassis, belong to the technological. The assembly itself, however, together with the aforementioned stockparts, falls within the sphere of technological activity; and this activity always merely responds to the challenge of Enframing, but it never comprises Enframing itself or brings it about.

The word *stellen* [to set upon] in the name *Ge-stell* [Enframing] not only means challenging. At the same time it should preserve the suggestion of another *Stellen* from which it stems, namely, that producing and presenting [*Her- und Dar-stellen*] which, in the sense of *poiēsis*, lets what presences come forth into unconcealment. This producing that brings forth—e.g., the erecting of a statue in the temple precinct—and the challenging ordering now under consideration are indeed fundamentally different, and yet they remain related in their essence. Both are ways of revealing, of *alētheia*. In Enframing, that unconcealment comes to pass in conformity with which the work of modern technology reveals the real as standing-reserve. This work is therefore neither only a human activity nor a mere means within such activity. The merely instrumental, merely anthropological definition of technology is therefore in principle untenable. And it cannot be rounded out by being referred back to some metaphysical or religious explanation that undergirds it.

It remains true, nonetheless, that man in the technological age is, in a particularly striking way, challenged forth into revealing. That revealing concerns nature, above all, as the chief storehouse of the standing energy reserve. Accordingly, man's ordering attitude and behavior display themselves first in the rise of modern physics as an exact science. Modern science's way of representing pursues and entraps nature as a calculable coherence of forces. Modern physics is not experimental physics because it applies apparatus to the questioning of nature. Rather the reverse is true. Because physics, indeed already as pure theory, sets nature up to exhibit itself as a coherence of forces calculable in advance, it therefore orders its experiments precisely for the purpose of asking whether and how nature reports itself when set up in this way.

But after all, mathematical physics arose almost two centuries before technology. How, then, could it have already been set upon by modern technology and placed in its service? The facts testify to the contrary. Surely technology got under way only when it could be supported by exact physical science. Reckoned chronologically, this is correct. Thought historically, it does not hit upon the truth.

The modern physical theory of nature prepares the way first not simply for technology but for the essence of modern technology. For already in physics the challenging gathering-together into ordering revealing holds sway. But in it that gathering does not yet come expressly to appearance. Modern physics is the herald of Enframing, a herald whose origin is still unknown. The essence of modern technology has for a long time been concealing itself, even where power machinery has been invented, where electrical technology is in full swing, and where atomic technology is well under way.

All coming to presence, not only modern technology, keeps itself everywhere concealed to the last. Nevertheless, it remains, with respect to its holding sway, that which precedes all: the earliest. The Greek thinkers already knew of this when they said: That which is earlier with regard to the arising that holds sway becomes manifest to us men only later. That which is primally early shows itself only ultimately to men. Therefore, in the realm of thinking, a painstaking effort to think through still more primally what was primally thought is not the absurd wish to revive what is past, but rather the sober readiness to be astounded before the coming of what is early.

Chronologically speaking, modern physical science begins in the seventeenth century. In contrast, machine-power technology develops only in the second half of the eighteenth century. But modern technology, which for chronological reckoning is the later, is, from the point of view of the essence holding sway within it, the historically earlier.

If modern physics must resign itself ever increasingly to the fact that its realm of representation remains inscrutable and incapable of being visualized, this resignation is not dictated by any committee of researchers. It is challenged forth by the rule of Enframing, which demands that nature be orderable as standing-reserve. Hence physics, in all its retreating from the representation turned only toward objects that has alone been standard till recently, will never be able to renounce this one thing: that nature reports itself in some way or other that is identifiable through calculation and that it remains orderable as a system of information. This system is determined, then, out of a causality that has changed once again. Causality now displays neither the character of the occasioning that brings forth nor the nature of the *causa efficiens,* let alone that of the *causa formalis.* It seems as though causality is shrinking into a reporting—a reporting challenged forth—of standing-reserves that must be guaranteed either simultaneously or in sequence. To this shrinking would correspond the process of growing resignation that Heisenberg's lecture depicts in so impressive a manner.*

Because the essence of modern technology lies in Enframing, modern technology must employ exact physical science. Through its so doing, the deceptive illusion arises that modern technology is applied physical science. This illusion can maintain itself only so long as neither the essential origin of modern science nor indeed the essence of modern technology is adequately found out through questioning.

We are questioning concerning technology in order to bring to light our relationship to its essence. The essence of modern technology shows itself in what we call Enframing. But simply to point to this is still in no way to answer the question concerning technology, if to answer means to respond, in the sense of correspond, to the essence of what is being asked about.

Where do we find ourselves brought to, if now we think one step further regarding what Enframing itself actually is? It is nothing technological, nothing on the order of a machine. It is the way in which the real reveals itself as standing-reserve. Again we ask: Does this revealing happen somewhere beyond all human doing? No. But neither does it happen exclusively *in* man, or decisively *through* man.

Enframing is the gathering together that belongs to that setting-upon which sets upon man and puts him in position to reveal the real, in the mode of ordering, as standing-reserve. As the one who is challenged forth in this way, man stands within the essential realm of Enframing. He can never take up a relationship to it only subsequently. Thus the question as

---

*W. Heisenberg, "Das Naturbild in der heutigen Physik," in *Die Künste im technischen Zeitalter* (Munich, 1954), pp. 43 ff.

to how we are to arrive at a relationship to the essence of technology, asked in this way, always comes too late. But never too late comes the question as to whether we actually experience ourselves as the ones whose activities everywhere, public and private, are challenged forth by Enframing. Above all, never too late comes the question as to whether and how we actually admit ourselves into that wherein Enframing itself comes to presence.

The essence of modern technology starts man upon the way of that revealing through which the real everywhere, more or less distinctly, becomes standing-reserve. "To start upon a way" means "to send" in our ordinary language. We shall call that sending-that-gathers [*versammelde Schicken*] which first starts man upon a way of revealing, *destining* [*Geschick*]. It is from out of this destining that the essence of all history [*Geschichte*] is determined. History is neither simply the object of written chronicle nor simply the fulfillment of human activity. That activity first becomes history as something destined.* And it is only the destining into objectifying representation that makes the historical accessible as an object for historiography, i.e., for a science, and on this basis makes possible the current equating of the historical with that which is chronicled.

Enframing, as a challenging-forth into ordering, sends into a way of revealing. Enframing is an ordaining of destining, as is every way of revealing. Bringing-forth, *poiēsis,* is also a destining in this sense.

Always the unconcealment of that which is goes upon a way of revealing. Always the destining of revealing holds complete sway over man. But that destining is never a fate that compels. For man becomes truly free only insofar as he belongs to the realm of destining and so becomes one who listens and hears, and not one who is simply constrained to obey.

The essence of freedom is *originally* not connected with the will or even with the causality of human willing.

Freedom governs the open in the sense of the cleared and lighted up, i.e., of the revealed. It is to the happening of revealing, i.e., of truth, that freedom stands in the closest and most intimate kinship. All revealing belongs within a harboring and a concealing. But that which frees—the mystery—is concealed and always concealing itself. All revealing comes out of the open, goes into the open, and brings into the open. The freedom of the open consists neither in unfettered arbitrariness nor in the constraint of mere laws. Freedom is that which conceals in a way that opens to light, in whose clearing there shimmers that veil that covers what comes to presence of all truth and lets the veil appear as what veils. Freedom is the realm of the destining that at any given time starts a revealing upon its way.

The essence of modern technology lies in Enframing. Enframing belongs within the destining of revealing. These sentences express something different from the talk that we hear more frequently, to the effect that technology is the fate of our age, where "fate" means the inevitableness of an unalterable course.

But when we consider the essence of technology, then we experience Enframing as a destining of revealing. In this way we are already sojourning within the open space of destining, a destining that in no way confines us to a stultified compulsion to push on blindly with technology or, what comes to the same thing, to rebel helplessly against it and curse it as the work of the devil. Quite to the contrary, when we once open ourselves expressly to the *essence* of technology, we find ourselves unexpectedly taken into a freeing claim.

The essence of technology lies in Enframing. Its holding sway belongs within destining.

---

*See *Vom Wesen der Wahrheit,* 1930; 1st ed., 1943, pp. 16 ff. [English translation, "On the Essence of Truth," in *Existence and Being,* ed. Werner Brock (Chicago: Regnery, 1949), pp. 308 ff.]

Since destining at any given time starts man on a way of revealing, man, thus under way, is continually approaching the brink of the possibility of pursuing and pushing forward nothing but what is revealed in ordering, and of deriving all his standards on this basis. Through this the other possibility is blocked, that man might be admitted more and sooner and ever more primally to the essence of that which is unconcealed and to its unconcealment, in order that he might experience as his essence his needed belonging to revealing.

Placed between these possibilities, man is endangered from out of destining. The destining of revealing is as such, in every one of its modes, and therefore necessarily, *danger.*

In whatever way the destining of revealing may hold sway, the unconcealment in which everything that is shows itself at any given time harbors the danger that man may misconstrue the unconcealed and misinterpret it. Thus where everything that presences exhibits itself in the light of a cause-effect coherence, even God can, for representational thinking, lose all that is exalted and holy, the mysteriousness of his distance. In the light of causality, God can sink to the level of a cause, of *causa efficiens.* He then becomes, even in theology, the god of the philosophers, namely, of those who define the unconcealed and the concealed in terms of the causality of making, without ever considering the essential origin of this causality.

In a similar way the unconcealment in accordance with which nature presents itself as a calculable complex of the effects of forces can indeed permit correct determinations; but precisely through these successes the danger can remain that in the midst of all that is correct the true will withdraw.

The destining of revealing is in itself not just any danger, but *the* danger.

Yet when destining reigns in the mode of Enframing, it is the supreme danger. This danger attests itself to us in two ways. As soon as what is unconcealed no longer concerns man even as object, but does so, rather, exclusively as standing-reserve, and man in the midst of objectlessness is nothing but the orderer of the standing-reserve, then he comes to the very brink of a precipitous fall; that is, he comes to the point where he himself will have to be taken as standing-reserve. Meanwhile man, precisely as the one so threatened, exalts himself to the posture of lord of the earth. In this way the impression comes to prevail that everything man encounters exists only insofar as it is his construct. This illusion gives rise in turn to one final delusion: It seems as though man everywhere and always encounters only himself. Heisenberg has with complete correctness pointed out that the real must present itself to contemporary man in this way.* *In truth, however, precisely nowhere does man today any longer encounter himself, i.e., his essence.* Man stands so decisively in attendance on the challenging-forth of Enframing that he does not apprehend Enframing as a claim, that he fails to see himself as the one spoken to, and hence also fails in every way to hear in what respect he ek-sists, from out of his essence, in the realm of an exhortation or address, and thus *can never* encounter only himself.

But Enframing does not simply endanger man in his relationship to himself and to everything that is. As a destining, it banishes man into that kind of revealing which is an ordering. Where this ordering holds sway, it drives out every other possibility of revealing. Above all, Enframing conceals that revealing which, in the sense of *poiēsis,* lets what presences come forth into appearance. As compared with that other revealing, the setting-upon that challenges forth thrusts man into a relation to that which is, that is at once antithetical and rigorously ordered. Where Enframing holds sway, regulating and securing of the standing-reserve mark all revealing. They no longer even let their own fundamental characteristic appear, namely, this revealing as such.

---

*"Das Naturbild," pp. 60 ff.

Thus the challenging Enframing not only conceals a former way of revealing, bringing-forth, but it conceals revealing itself and with it That wherein unconcealment, i.e., truth, comes to pass.

Enframing blocks the shining-forth and holding-sway of truth. The destining that sends into ordering is consequently the extreme danger. What is dangerous is not technology. There is no demonry of technology, but rather there is the mystery of its essence. The essence of technology, as a destining of revealing, is the danger. The transformed meaning of the word "Enframing" will perhaps become somewhat more familiar to us now if we think Enframing in the sense of destining and danger.

The threat to man does not come in the first instance from the potentially lethal machines and apparatus of technology. The actual threat has already affected man in his essence. The rule of Enframing threatens man with the possibility that it could be denied to him to enter into a more original revealing and hence to experience the call of a more primal truth.

Thus, where Enframing reigns, there is *danger* in the highest sense.

> But where danger is, grows
> The saving power also.

Let us think carefully about these words of Hölderlin. What does it mean "to save"? Usually we think that it means only to seize hold of a thing threatened by ruin, in order to secure it in its former continuance. But the verb "to save" says more. "To save" is to fetch something home into its essence, in order to bring the essence for the first time into its genuine appearing. If the essence of technology, Enframing, is the extreme danger, and if there is truth in Hölderlin's words, then the rule of Enframing cannot exhaust itself solely in blocking all lighting-up of every revealing, all appearing of truth. Rather, precisely the essence of technology must harbor in itself the growth of the saving power. But in that case, might not an adequate look into what Enframing is as a destining of revealing bring into appearance the saving power in its arising?

In what respect does the saving power grow there also where the danger is? Where something grows, there it takes root, from thence it thrives. Both happen concealedly and quietly and in their own time. But according to the words of the poet we have no right whatsoever to expect that there where the danger is we should be able to lay hold of the saving power immediately and without preparation. Therefore we must consider now, in advance, in what respect the saving power does most profoundly take root and thence thrive even in that wherein the extreme danger lies, in the holding sway of Enframing. In order to consider this, it is necessary, as a last step upon our way, to look with yet clearer eyes into the danger. Accordingly, we must once more question concerning technology. For we have said that in technology's essence roots and thrives the saving power.

But how shall we behold the saving power in the essence of technology so long as we do not consider in what sense of "essence" it is that Enframing is actually the essence of technology?

Thus far we have understood "essence" in its current meaning. In the academic language of philosophy, "essence" means *what* something is; in Latin, *quid. Quidditas,* whatness, provides the answer to the question concerning essence. For example, what pertains to all kinds of trees—oaks, beeches, birches, firs—is the same "treeness." Under this inclusive genus—the "universal"—fall all real and possible trees. Is then the essence of technology, Enframing, the common genus for everything technological? If that were the case then the steam turbine, the radio transmitter, and the cyclotron would each be an Enframing. But the word "Enframing" does not mean here a tool or any kind of apparatus. Still less does it mean the general concept of such resources. The machines and apparatus are no more cases and kinds of Enframing than are the man at the switchboard and the engineer in the drafting room. Each of these in its own way indeed belongs as stockpart, available resource, or executer, within Enframing; but Enframing is never the essence of technology in the sense of a genus. Enframing is a way of revealing having the character of destining, namely, the way that challenges forth. The revealing that brings forth (*poiēsis*) is also a way that has the character of destining. But these ways are not kinds that, arrayed beside one another, fall under the concept of revealing. Revealing is that destining which, ever suddenly and inexplicably to all thinking, apportions itself into the revealing that brings forth and that also challenges, and which allots itself to man. The challenging revealing has its origin as a destining in bringing-forth. But at the same time Enframing, in a way characteristic of a destining, blocks *poiēsis*.

Thus Enframing, as a destining of revealing, is indeed the essence of technology, but never in the sense of genus and *essentia*. If we pay heed to this, something astounding strikes us: It is technology itself that makes the demand on us to think in another way what is usually understood by "essence." But in what way?

If we speak of the "essence of a house" and the "essence of a state," we do not mean a generic type; rather we mean the ways in which house and state hold sway, administer themselves, develop and decay—the way in which they "essence" [*Wesen*]. Johann Peter Hebel in a poem, "Ghost on Kanderer Street," for which Goethe had a special fondness, uses the old word *die Weserei*. It means the city hall inasmuch as there the life of the community gathers and village existence is constantly in play, i.e., comes to presence. It is from the verb *wesen* that the noun is derived. *Wesen* understood as a verb is the same as *währen* [to last or endure], not only in terms of meaning, but also in terms of the phonetic formation of the word. Socrates and Plato already think the essence of something as what essences, what comes to presence, in the sense of what endures. But they think what endures as what remains permanently (*aei on*). And they find what endures permanently in what, as that which remains, tenaciously persists throughout all that happens. That which remains they discover, in turn, in the aspect (*eidos, idea*), for example, the Idea "house."

The Idea "house" displays what anything is that is fashioned as a house. Particular, real, and possible houses, in contrast, are changing and transitory derivatives of the Idea and thus belong to what does not endure.

But it can never in any way be established that enduring is based solely on what Plato thinks as *idea* and Aristotle thinks as *to ti ēn einai* (that which any particular thing has always been), or what metaphysics in its most varied interpretations thinks as *essentia*.

All essencing endures. But is enduring only permanent enduring? Does the essence of technology endure in the sense of the permanent enduring of an Idea that hovers over everything technological, thus making it seem that by technology we mean some mythological abstraction? The way in which technology essences lets itself be seen only from out of that permanent enduring in which Enframing comes to pass as a destining of revealing. Goethe once uses the mysterious word *fortgewähren* [to grant permanently] in place of *fortwähren* [to endure permanently].* He hears *währen* [to endure] and *gewähren* [to grant] here in one unarticulated accord. And if we now ponder more carefully than we did before what it is that

---

*"Die Wahlverwandtschaften" [Congeniality], pt. II, chap. 10, in the novelette *Die wunderlichen Nachbarskinder* [The strange neighbor's children].

actually endures and perhaps alone endures, we may venture to say: *Only what is granted endures. That which endures primally out of the earliest beginning is what grants.*

As the essencing of technology, Enframing is that which endures. Does Enframing hold sway at all in the sense of granting? No doubt the question seems a horrendous blunder. For according to everything that has been said, Enframing is, rather, a destining that gathers together into the revealing that challenges forth. Challenging is anything but a granting. So it seems, so long as we do not notice that the challenging-forth into the ordering of the real as standing-reserve still remains a destining that starts man upon a way of revealing. As this destining, the coming to presence of technology gives man entry into That which, of himself, he can neither invent nor in any way make. For there is no such thing as a man who, solely of himself, is only man.

But if this destining, Enframing, is the extreme danger, not only for man's coming to presence, but for all revealing as such, should this destining still be called a granting? Yes, most emphatically, if in this destining the saving power is said to grow. Every destining of revealing comes to pass from out of a granting and as such a granting. For it is granting that first conveys to man that share in revealing which the coming-to-pass of revealing needs. As the one so needed and used, man is given to belong to the coming-to-pass of truth. The granting that sends in one way or another into revealing is as such the saving power. For the saving power lets man see and enter into the highest dignity of his essence. This dignity lies in keeping watch over the unconcealment—and with it, from the first, the concealment—of all coming to presence on this earth. It is precisely in Enframing, which threatens to sweep man away into ordering as the supposed single way of revealing, and so thrusts man into the danger of the surrender of his free essence—it is precisely in this extreme danger that the innermost indestructible belongingness of man within granting may come to light, provided that we, for our part, begin to pay heed to the coming to presence of technology.

Thus the coming to presence of technology harbors in itself what we least suspect, the possible arising of the saving power.

Everything, then, depends upon this: that we ponder this arising and that, recollecting, we watch over it. How can this happen? Above all through our catching sight of what comes to presence in technology, instead of merely staring at the technological. So long as we represent technology as an instrument, we remain held fast in the will to master it. We press on past the essence of technology.

When, however, we ask how the instrumental comes to presence as a kind of causality, then we experience this coming to presence as the destining of a revealing.

When we consider, finally, that the coming to presence of the essence of technology comes to pass in the granting that needs and uses man so that he may share in revealing, then the following becomes clear:

The essence of technology is in a lofty sense ambiguous. Such ambiguity points to the mystery of all revealing, i.e., of truth.

On the one hand, Enframing challenges forth into the frenziedness of ordering that blocks every view into the coming-to-pass of revealing and so radically endangers the relation to the essence of truth.

On the other hand, Enframing comes to pass for its part in the granting that lets man endure—as yet unexperienced, but perhaps more experienced in the future—that he may be the one who is needed and used for the safekeeping of the coming to presence of truth. Thus does the arising of the saving power appear.

The irresistibility of ordering and the restraint of the saving power draw past each other like the paths of two stars in the course of the heavens. But precisely this, their passing by, is the hidden side of their nearness.

When we look into the ambiguous essence of technology, we behold the constellation, the stellar course of the mystery.

The question concerning technology is the question concerning the constellation in which revealing and concealing, in which the coming to presence of truth, comes to pass.

But what help is it to us to look into the constellation of truth? We look into the danger and see the growth of the saving power.

Through this we are not yet saved. But we are thereupon summoned to hope in the growing light of the saving power. How can this happen? Here and now and in little things, that we may foster the saving power in its increase. This includes holding always before our eyes the extreme danger.

The coming to presence of technology threatens revealing, threatens it with the possibility that all revealing will be consumed in ordering and that everything will present itself only in the unconcealedness of standing-reserve. Human activity can never directly counter this danger. Human achievement alone can never banish it. But human reflection can ponder the fact that all saving power must be of a higher essence than what is endangered, though at the same time kindred to it.

But might there not perhaps be a more primally granted revealing that could bring the saving power into its first shining forth in the midst of the danger, a revealing that in the technological age rather conceals than shows itself?

There was a time when it was not technology alone that bore the name *technē*. Once that revealing that brings forth truth into the splendor of radiant appearing also was called *technē*.

Once there was a time when the bringing-forth of the true into the beautiful was called *technē*. And the *poiēsis* of the fine arts also was called *technē*.

In Greece, at the outset of the destining of the West, the arts soared to the supreme height of the revealing granted them. They brought the presence [*Gegenwart*] of the gods, brought the dialogue of divine and human destinings, to radiance. And art was simply called *technē*. It was a single, manifold revealing. It was pious, *promos,* i.e., yielding to the holding-sway and the safekeeping of truth.

The arts were not derived from the artistic. Art works were not enjoyed aesthetically. Art was not a sector of cultural activity.

What, then, was art—perhaps only for that brief but magnificent time? Why did art bear the modest name *technē?* Because it was a revealing that brought forth and hither, and therefore belonged within *poiēsis*. It was finally that revealing which holds complete sway in all the fine arts, in poetry, and in everything poetical that obtained *poiēsis* as its proper name.

The same poet from whom we heard the words

> *But where danger is, grows*
> *The saving power also.*

says to us:

> *. . . poetically dwells man upon this earth.*

The poetical brings the true into the splendor of what Plato in the *Phaedrus* calls to *ekphanestaton,* that which shines forth most purely. The poetical thoroughly pervades every art, every revealing of coming to presence into the beautiful.

Could it be that the fine arts are called to poetic revealing? Could it be that revealing lays claim to the arts most primally, so that they for their part may expressly foster the growth of the saving power, may awaken and found anew our look into that which grants and our trust in it?

Whether art may be granted this highest possibility of its essence in the midst of the extreme danger, no one can tell. Yet we can be astounded. Before what? Before this other possibility: that the frenziedness of technology may entrench itself everywhere to such an extent that someday, throughout everything technological, the essence of technology may come to presence in the coming-to-pass of truth.

Because the essence of technology is nothing technological, essential reflection upon technology and decisive confrontation with it must happen in a realm that is, on the one hand, akin to the essence of technology and, on the other, fundamentally different from it.

Such a realm is art. But certainly only if reflection on art, for its part, does not shut its eyes to the constellation of truth after which we are *questioning*.

Thus questioning, we bear witness to the crisis that in our sheer preoccupation with technology we do not yet experience the coming to presence of technology, that in our sheer aesthetic-mindedness we no longer guard and preserve the coming to presence of art. Yet the more questioningly we ponder the essence of technology, the more mysterious the essence of art becomes.

The closer we come to the danger, the more brightly do the ways into the saving power begin to shine and the more questioning we become. For questioning is the piety of thought.

4

Heidegger on Gaining a Free Relation to Technology

Hubert Dreyfus

INTRODUCTION: WHAT HEIDEGGER IS NOT SAYING

In *The Question Concerning Technology* Heidegger describes his aim:

"We shall be questioning concerning technology, and in so doing we should like to prepare a free relationship to it."

He wants to reveal the essence of technology in such a way that "in no way confines us to a stultified compulsion to push on blindly with technology or, what comes to the same thing, to rebel helplessly against it."[1] Indeed, he claims that "When we once open ourselves expressly to the *essence* of technology, we find ourselves unexpectedly taken into a freeing claim."[2]

We will need to explain essence, opening, and freeing before we can understand Heidegger here. But already Heidegger's project should alert us to the fact that he is not announcing one more reactionary rebellion against technology, although many respectable philosophers, including Jürgen Habermas, take him to be doing just that; nor is he doing what progressive thinkers such as Habermas want him to do, proposing a way to get technology under control so that it can serve our rationally chosen ends.

The difficulty in locating just where Heidegger stands on technology is no accident. Heidegger has not always been clear about what distinguishes his approach from a romantic reaction to the domination of nature, and when he does finally arrive at a clear formulation of his own original view, it is so radical that everyone is tempted to translate it into conventional platitudes about the evils of technology. Thus Heidegger's ontological concerns are mistakenly assimilated to humanistic worries about the devastation of nature.

Those who want to make Heidegger intelligible in terms of current anti-technological banalities can find support in his texts. During the war he attacks consumerism:

> The circularity of consumption for the sake of consumption is the sole procedure which distinctively characterizes the history of a world which has become an unworld.[3]

And as late as 1955 he holds that:

> The world now appears as an object open to the attacks of calculative thought. . . . Nature
> becomes a gigantic gasoline station, an energy source for modern technology and industry.[4]

In this address to the Schwartzwald peasants he also laments the appearance of television
antennae on their dwellings.

> Hourly and daily they are chained to radio and television. . . . All that with which modern tech-
> niques of communication stimulate, assail, and drive man—all that is already much closer to
> man today than his fields around his farmstead, closer than the sky over the earth, closer than the
> change from night to day, closer than the conventions and customs of his village, than the tradi-
> tion of his native world.[5]

Such statements suggest that Heidegger is a Luddite who would like to return from the exploitation of the earth, consumerism, and mass media to the world of the pre-Socratic Greeks or
the good old Schwartzwald peasants.

HEIDEGGER'S ONTOLOGICAL APPROACH TO TECHNOLOGY

As his thinking develops, however, Heidegger does not deny these are serious problems, but
he comes to the surprising and provocative conclusion that focusing on loss and destruction is
still technological.

> All attempts to reckon existing reality . . . in terms of decline and loss, in terms of fate, catastro-
> phe, and destruction, are merely technological behavior.[6]

Seeing our situation as posing a problem that must be solved by appropriate action turns out
to be technological too:

> [T]he instrumental conception of technology conditions every attempt to bring man into the right
> relation to technology. . . . The will to mastery becomes all the more urgent the more technology
> threatens to slip from human control.[7]

Heidegger is clear this approach cannot work:

> No single man, no group of men, no commission of prominent statesmen, scientists, and techni-
> cians, no conference of leaders of commerce and industry, can brake or direct the progress of
> history in the atomic age.[8]

His view is both darker and more hopeful. He thinks there is a more dangerous situation facing
modern man than the technological destruction of nature and civilization, yet a situation about
which something *can* be done—at least indirectly. The threat is not a *problem* for which there
can be a *solution* but an ontological *condition* from which we can be *saved.*

Heidegger's concern is the human distress caused by the *technological understanding of
being,* rather than the destruction caused by specific technologies. Consequently, Heidegger
distinguishes the current problems caused by technology—ecological destruction, nuclear danger, consumerism, etc.—from the devastation that would result if technology solved all our
problems.

> What threatens man in his very nature is the . . . view that man, by the peaceful release, transformation, storage, and channeling of the energies of physical nature, could render the human condition . . . tolerable for everybody and happy in all respects.[9]

The "greatest danger" is that

> the approaching tide of technological revolution in the atomic age could so captivate, bewitch, dazzle, and beguile man that calculative thinking may someday come to be accepted and practiced as *the only* way of thinking.[10]

The danger, then, is not the destruction of nature or culture but a restriction in our way of thinking—a leveling of our understanding of being.

To evaluate this claim we must give content to what Heidegger means by an understanding of being. Let us take an example. Normally we deal with things, and even sometimes people, as resources to be used until no longer needed and then put aside. A styrofoam cup is a perfect example. When we want a hot or cold drink it does its job, and when we are through with it we throw it away. How different this understanding of an object is from what we can suppose to be the everyday Japanese understanding of a delicate teacup. The teacup does not preserve temperature as well as its plastic replacement, and it has to be washed and protected, but it is preserved from generation to generation for its beauty and its social meaning. It is hard to picture a tea ceremony around a styrofoam cup.

Note that the traditional Japanese understanding of what it is to be human (passive, contented, gentle, social, etc.) fits with their understanding of what it is to be a thing (delicate, beautiful, traditional, etc.). It would make no sense for us, who are active, independent, and aggressive—constantly striving to cultivate and satisfy our desires—to relate to things the way the Japanese do; or for the Japanese (before their understanding of being was interfered with by ours) to invent and prefer styrofoam teacups. In the same vein *we* tend to think of politics as the negotiation of individual desires while the Japanese seek consensus. In sum the social practices containing an understanding of what it is to be a human self, those containing an interpretation of what it is to be a thing, and those defining society fit together. They add up to an understanding of being.

The shared practices into which we are socialized, then, provide a background understanding of what counts as things, what counts as human beings, and ultimately what counts as real, on the basis of which we can direct our actions toward particular things and people. Thus the understanding of being creates what Heidegger calls a *clearing* in which things and people can show up for us. We do not produce the clearing. It produces us as the kind of human beings that we are. Heidegger describes the clearing as follows:

> [B]eyond what is, not away from it but before it, there is still something else that happens. In the midst of beings as a whole an open place occurs. There is a clearing, a lighting. . . . This open center is . . . not surrounded by what is; rather, the lighting center itself encircles all that is. . . . Only this clearing grants and guarantees to human beings a passage to those entities that we ourselves are not, and access to the being that we ourselves are.[11]

What, then, is the essence of technology, i.e., the technological understanding of being, i.e., the technological clearing, and how does opening ourselves to it give us a free relation to technological devices? To begin with, when we ask about the essence of technology we are able to see that Heidegger's question cannot be answered by defining technology. Technology is as old as civilization. Heidegger notes that it can be correctly defined as "a means and a human activity." He calls this "the instrumental and anthropological definition of technology."[12] But if we ask about the *essence* of technology (the technological understanding of being) we find that modern technology is "something completely different and . . . new."[13] Even different from using styrofoam cups to serve our desires. The essence of modern technology, Heidegger tells us, is to seek more and more flexibility and efficiency *simply for its own sake.* "[E]xpediting is always itself directed from the beginning . . . towards driving on to the maximum yield at the minimum expense."[14] That is, our only goal is optimization:

> Everywhere everything is ordered to stand by, to be immediately at hand, indeed to stand there just so that it may be on call for a further ordering. Whatever is ordered about in this way has its own standing. We call it standing-reserve. . . .[15]

No longer are we subjects turning nature into an object of exploitation:

> The subject-object relation thus reaches, for the first time, its pure "relational," i.e., ordering, character in which both the subject and the object are sucked up as standing-reserves.[16]

A modern airliner is not an object at all, but just a flexible and efficient cog in the transportation system.[17] (And passengers are presumably not subjects but merely resources to fill the planes.) Heidegger concludes: "Whatever stands by in the sense of standing-reserve no longer stands over against us as object."[18]

All ideas of serving God, society, our fellow men, or even our own calling disappear. Human beings, on this view, become a resource to be used, but more important to be enhanced—like any other.

> Man, who no longer conceals his character of being the most important raw material, is also drawn into this process.[19]

In the film *2001,* the robot HAL, when asked if he is happy on the mission, answers: "I'm using all my capacities to the maximum. What more could a rational entity desire?" This is a brilliant expression of what anyone would say who is in touch with our current understanding of being. We pursue the growth or development of our potential simply for its own sake—it is our only goal. The human potential movement perfectly expresses this technological understanding of being, as does the attempt to better organize the future use of our natural resources. We thus become part of a system which no one directs but which moves toward the total mobilization of all beings, even us. This is why Heidegger thinks the perfectly ordered society dedicated to the welfare of all is not the solution of our problems but the distressing culmination of the technological understanding of being.

WHAT THEN CAN WE DO?

But, of course, Heidegger uses and depends upon modern technological devices. He is no Luddite and he does not advocate a return to the pre-technological world.

> It would be foolish to attack technology blindly. It would be shortsighted to condemn it as the work of the devil. We depend on technical devices; they even challenge us to ever greater advances.[20]

Instead, Heidegger suggests that there is a way we can keep our technological devices and yet remain true to ourselves:

> We can affirm the unavoidable use of technical devices, and also deny them the right to dominate us, and so to warp, confuse, and lay waste our nature.[21]

To understand how this might be possible we need an illustration of Heidegger's important distinction between technology and the technological understanding of being. Again we can turn to Japan. In contemporary Japan a traditional, non-technological understanding of being still exists alongside the most advanced high-tech production and consumption. The TV set and the household gods share the same shelf—the styrofoam cup co-exists with the porcelain one. We can thus see that one can have technology without the technological understanding of being, so it becomes clear that the technological understanding of being can be dissociated from technological devices.

To make this dissociation, Heidegger holds, one must rethink the history of being in the West. Then one will see that although a technological understanding of being is our destiny, it is not our fate. That is, although our understanding of things and ourselves as resources to be ordered, enhanced, and used efficiently has been building up since Plato and dominates our practices, we are not stuck with it. It is not the way things have to be, but nothing more or less than our current cultural clearing.

Only those who think of Heidegger as opposing technology will be surprised at his next point. Once we see that technology is our latest understanding of being, we will be grateful for it. We did not make this clearing nor do we control it, but if it were not given to us to encounter things and ourselves as resources, nothing would show up *as* anything at all and no possibilities for action would make sense. And once we realize—in our practices, of course, not just in our heads—that we *receive* our technological understanding of being, we have stepped out of the technological understanding of being, for we then see that what is most important in our lives is not subject to efficient enhancement. This transformation in our sense of reality—this overcoming of calculative thinking—is precisely what Heideggerian thinking seeks to bring about. Heidegger seeks to show how we can recognize and thereby overcome our restricted, willful modern clearing precisely by recognizing our essential receptivity to it.

> [M]odern man must first and above all find his way back into the full breadth of the space proper to his essence. That essential space of man's essential being receives the dimension that unites it to something beyond itself . . . that is the way in which the safekeeping of being itself is given to belong to the essence of man as the one who is needed and used by being.[22]

But precisely how can we experience the technological understanding of being as a gift to which we are receptive? What is the phenomenon Heidegger is getting at? We can break out of the technological understanding of being whenever we find ourselves gathered by things rather than controlling them. When a thing like a celebratory meal, to take Heidegger's example, pulls our practices together and draws us in, we experience a focusing and a nearness that resists technological ordering. Even a technological object like a highway bridge, when experienced as a gathering and focusing of our practices, can help us resist the very technological ordering it furthers. Heidegger describes the bridge so as to bring out both its technological ordering function and its continuity with pre-technological things.

> The old stone bridge's humble brook crossing gives to the harvest wagon its passage from the
> fields into the village and carries the lumber cart from the field path to the road. The highway
> bridge is tied into the network of long-distance traffic, paced as calculated for maximum yield.
> Always and ever differently the bridge escorts the lingering and hastening ways of men to and
> fro. . . . The bridge *gathers* to itself in *its own way* earth and sky, divinities and mortals.[23]

Getting in sync with the highway bridge in its technological functioning can make us sensitive
to the technological understanding of being as the way our current clearing works, so that we
experience our role as receivers, and the importance of receptivity, thereby freeing us from
our compulsion to force all things into one efficient order.

This transformation in our understanding of being, unlike the slow process of cleaning
up the environment which is, of course, also necessary, would take place in a sudden Gestalt
switch.

> The turning of the danger comes to pass suddenly. In this turning, the clearing belonging to the
> essence of being suddenly clears itself and lights up.[24]

The danger, when grasped as the danger, becomes that which saves us. "The self-same danger
is, when it is *as* the danger, the saving power."[25]

This remarkable claim gives rise to two opposed ways of understanding Heidegger's
response to technology. Both interpretations agree that once one recognizes the technological
understanding of being for what it is—a historical understanding—one gains a free relation to
it. We neither push forward technological efficiency as our only goal nor always resist it. If we
are free of the technological imperative we can, in each case, discuss the pros and cons. As
Heidegger puts it:

> We let technical devices enter our daily life, and at the same time leave them outside . . . as things
> which are nothing absolute but remain dependent upon something higher [the clearing]. I would
> call this comportment toward technology which expresses "yes" and at the same time "no", by
> an old word, *releasement towards things.*[26]

One way of understanding this proposal—represented here by Richard Rorty—holds that
once we get in the right relation to technology, viz. recognize it as a clearing, it is revealed as
just as good as any other clearing. Efficiency—getting the most out of ourselves and everything else—is fine, so long as we do not think that efficiency for its own sake is the *only* end
for man, dictated by reality itself, to which all others must be subordinated. Heidegger seems
to support this acceptance of the technological understanding of being when he says:

> That which shows itself and at the same time withdraws [i.e., the clearing] is the essential trait
> of what we call the mystery. I call the comportment which enables us to keep open to the mean-
> ing hidden in technology, *openness to the mystery.* Releasement toward things and openness to
> the mystery belong together. They grant us the possibility of dwelling in the world in a totally
> different way. They promise us a new ground and foundation upon which we can stand and
> endure in the world of technology without being imperiled by it.[27]

But acceptance of the mystery of the gift of understandings of being cannot be Heidegger's whole story, for he immediately adds:

> Releasement toward things and openness to the mystery give us a vision of a new rootedness
> which *someday* might even be fit to recapture the old and now rapidly disappearing rootedness
> in a changed form.[28]

We then look back at the preceding remark and realize *releasement* gives only a "possibility"
and a "promise" of "dwelling in the world in a totally different way."

Mere openness to technology, it seems, leaves out much that Heidegger finds essential
to human being: embeddedness in nature, nearness or localness, shared meaningful differences
such as noble and ignoble, justice and injustice, salvation and damnation, mature and immature—to name those that have played important roles in our history. *Releasement,* while giving
us a free relation to technology and protecting our nature from being distorted and distressed,
cannot give us any of these.

For Heidegger, there are, then, two issues. One issue is clear:

> The issue is the saving of man's essential nature. Therefore, the issue is keeping meditative think-
> ing alive.[29]

But that is not enough:

> If releasement toward things and openness to the mystery awaken within us, then we should
> arrive at a path that will lead to a new ground and foundation.[30]

Releasement, it turns out, is only a stage, a kind of holding pattern, awaiting a new understanding of being, which would give some content to our openness—what Heidegger calls a new
rootedness. That is why each time Heidegger talks of *releasement* and the saving power of
understanding technology as a gift he then goes on to talk of the divine.

> Only when man, in the disclosing coming-to-pass of the insight by which he himself is beheld
> . . . renounces human self-will . . . does he correspond in his essence to the claim of that insight.
> In thus corresponding man is gathered into his own, that he . . . may, as the mortal, look out
> toward the divine.[31]

The need for a new centeredness is reflected in Heidegger's famous remark in his last interview: "Only a god can save us now."[32] But what does this mean?

THE NEED FOR A GOD

Just preserving pre-technical practices, even if we could do it, would not give us what we need.
The pre-technological practices no longer add up to a shared sense of reality and one cannot
legislate a new understanding of being. For such practices to give meaning to our lives, and
unite us in a community, they would have to be focused and held up to the practitioners. This
function, which later Heidegger calls "truth setting itself to work," can be performed by what
he calls a work of art. Heidegger takes the Greek temple as his illustration of an artwork working. The temple held up to the Greeks what was important, and so let there be heroes and
slaves, victory and disgrace, disaster and blessing, and so on. People whose practices were
manifested and focused by the temple had guidelines for leading good lives and avoiding bad
ones. In the same way, the medieval cathedral made it possible to be a saint or a sinner by

showing people the dimensions of salvation and damnation. In either case, one knew where one stood and what one had to do. Heidegger holds that "there must always be some being in the open [the clearing], something that is, in which the openness takes its stand and attains its constancy."[33]

We could call such special objects cultural paradigms. A cultural paradigm focuses and collects the scattered practices of a culture, unifies them into coherent possibilities for action, and holds them up to the people who can then act and relate to each other in terms of the shared exemplar.

When we see that for later Heidegger only those practices focused in a paradigm can establish what things can show up as and what it makes sense to do, we can see why he was pessimistic about salvaging aspects of the Enlightenment or reviving practices focused in the past. Heidegger would say that we should, indeed, try to preserve such practices, but they can save us only if they are radically transformed and integrated into a new understanding of reality. In addition we must learn to appreciate marginal practices—what Heidegger calls the saving power of insignificant things—practices such as friendship, back-packing into the wilderness, and drinking the local wine with friends. All these practices are marginal precisely because they are not efficient. They can, of course, be engaged in for the sake of health and greater efficiency. This expanding of technological efficiency is the greatest danger. But these saving practices could come together in a new cultural paradigm that held up to us a new way of doing things, thereby focusing a world in which formerly marginal practices were central and efficiency marginal. Such a new object or event that grounded a new understanding of reality Heidegger would call a new god. This is why he holds that "only another god can save us."[34]

Once one sees what is needed, one also sees that there is not much we can do to bring it about. A new sense of reality is not something that can be made the goal of a crash program like the moon flight—a paradigm of modern technological power. A hint of what such a new god might look like is offered by the music of the sixties. The Beatles, Bob Dylan, and other rock groups became for many the articulation of new understanding of what really mattered. This new understanding almost coalesced into a cultural paradigm in the Woodstock Music Festival, where people actually lived for a few days in an understanding of being in which mainline contemporary concern with rationality, sobriety, willful activity, and flexible, efficient control were made marginal and subservient to Greek virtues such as openness, enjoyment of nature, dancing, and Dionysian ecstasy along with a neglected Christian concern with peace, tolerance, and love of one's neighbor without desire and exclusivity. Technology was not smashed or denigrated but all the power of the electronic media was put at the service of the music which focused all the above concerns.

If enough people had found in Woodstock what they most cared about, and recognized that all the others shared this recognition, a new understanding of being might have coalesced and been stabilized. Of course, in retrospect we see that the concerns of the Woodstock generation were not broad and deep enough to resist technology and to sustain a culture. Still we are left with a hint of how a new cultural paradigm would work, and the realization that we must foster human receptivity and preserve the endangered species of pre-technological practices that remain in our culture, in the hope that one day they will be pulled together into a new paradigm, rich enough and resistant enough to give new meaningful directions to our lives.

To many, however, the idea of *a* god which will give us a unified but open community—one set of concerns which everyone shares if only as a focus of disagreement—sounds either unrealistic or dangerous. Heidegger would probably agree that its open democratic version looks increasingly unobtainable and that we have certainly seen that its closed totalitarian form can be disastrous. But Heidegger holds that given our historical essence—the kind of beings we have become during the history of our culture—such a community is necessary to us. This raises the question of whether our need for one community is, indeed, dictated by our historical essence, or whether the claim that we can't live without a centered and rooted culture is simply romantic nostalgia.

It is hard to know how one could decide such a question, but Heidegger has a message even for those who hold that we, in this pluralized modern world, should not expect and do not need one all-embracing community. Those who, from Dostoievsky, to the hippies, to Richard Rorty, think of communities as local enclaves in an otherwise impersonal society still owe us an account of what holds these local communities together. If Dostoievsky and Heidegger are right, each local community still needs its local god—its particular incarnation of what the community is up to. In that case we are again led to the view that releasement is not enough, and to the modified Heideggerian slogan that only some new *gods* can save us.

NOTES

1. Martin Heidegger, "The Question Concerning Technology," *The Question Concerning Technology* (New York: Harper Colophon, 1977), pp. 25–26.

2. Ibid.

3. Heidegger, "Overcoming Metaphysics," *The End of Philosophy* (New York: Harper and Row, 1973), p. 107.

4. Heidegger, *Discourse on Thinking* (New York: Harper and Row, 1966), p. 50.

5. Ibid., p. 48.

6. Heidegger, "The Turning," *The Question Concerning Technology,* p. 48.

7. Heidegger, "The Question Concerning Technology," *The Question Concerning Technology,* p. 5.

8. Heidegger, *Discourse on Thinking,* p. 52.

9. Martin Heidegger, "What Are Poets For?" *Poetry, Language, Thought* (New York: Harper and Row, 1971), p. 116.

10. Heidegger, *Discourse on Thinking,* p. 56.

11. Heidegger, "The Origin of the Work of Art," *Poetry, Language, Thought,* p. 53.

12. Heidegger, "The Question Concerning Technology," p. 5.

13. Ibid.

14. Ibid., p. 15.

15. Ibid., p. 17.

16. Heidegger, "Science and Reflection," *The Question Concerning Technology,* p. 173.

17. Heidegger, "The Question Concerning Technology," p. 17.

18. Ibid.

19. Heidegger, "Overcoming Metaphysics," *The End of Philosophy,* p. 104.

20. Heidegger, *Discourse on Thinking,* p. 53.

21. Ibid., p. 54.

22. Heidegger, "The Turning," *The Question Concerning Technology,* p. 39.

23. Heidegger, *Poetry, Language, Thought,* pp. 152–53.

24. Ibid., p. 44.

25. Heidegger, "The Turning," *The Question Concerning Technology,* p. 39.

26. Heidegger, *Discourse on Thinking,* p. 54.

27. Ibid., p. 55.

28. Ibid. (My italics.)

29. Ibid., p. 56.

30. Ibid.
31. Heidegger, "The Turning," *The Question Concerning Technology,* p. 47.
32. "Nur noch ein Gott kann uns retten," *Der Spiegel,* May 31, 1976.
33. Heidegger, "The Origin of the Work of Art," *Poetry, Language, Thought,* p. 61.
34. This is an equally possible translation of the famous phrase from *Der Spiegel.*

5

Social Implications of Technology

Herbert Marcuse

In this chapter, technology is taken as a social process in which technics proper (that is, the technical apparatus of industry, transportation, communication) is but a partial factor. We do not ask for the influence or effect of technology on the human individuals. For they are themselves an integral part and factor of technology, not only as the men who invent or attend to machinery but also as the social groups which direct its application and utilization. Technology, as a mode of production, as the totality of instruments, devices and contrivances which characterize the machine age is thus at the same time a mode of organizing and perpetuating (or changing) social relationships, a manifestation of prevalent thought and behavior patterns, an instrument for control and domination.[1]

Technics by itself can promote authoritarianism as well as liberty, scarcity as well as abundance, the extension as well as the abolition of toil. National Socialism is a striking example of the ways in which a highly rationalized and mechanized economy with the utmost efficiency in production can operate in the interest of totalitarian oppression and continued scarcity. The Third Reich is indeed a form of "technocracy": the technical considerations of imperialistic efficiency and rationality supersede the traditional standards of profitability and general welfare. In National Socialist Germany, the reign of terror is sustained not only by brute force which is foreign to technology but also by the ingenious manipulation of the power inherent in technology: the intensification of labor, propaganda, the training of youths and workers, the organization of the governmental, industrial and party bureaucracy—all of which constitute the daily implements of terror—follow the lines of greatest technological efficiency. This terroristic technocracy cannot be attributed to the exceptional requirements of "war economy"; war economy is rather the normal state of the National Socialist ordering of the social and economic process, and technology is one of the chief stimuli of this ordering.[2]

In the course of the technological process a new rationality and new standards of individuality have spread over society, different from and even opposed to those which initiated the

From Herbert Marcuse, "Some Social Implications of Modern Technology," in *The Essential Frankfurt School Reader*, ed. Andrew Arato and Eike Gebhardt, pp. 138–162. Copyright © 1982 by Continuum Publishing. Reprinted by permission of The Continuum International Publishing Group.

march of technology. These changes are not the (direct or derivative) effect of machinery on its users or of mass production on its consumers; they are rather themselves determining factors in the development of machinery and mass production. In order to understand their full import, it is necessary to survey briefly the traditional rationality and standards of individuality which are being dissolved by the present stage of the machine age.

The human individual whom the exponents of the middle class revolution had made the ultimate unit as well as the end of society stood for values which strikingly contradict those holding sway over society today. If we try to assemble in one guiding concept the various religious, political and economic tendencies which shaped the idea of the individual in the sixteenth and seventeenth century, we may define the individual as the subject of certain fundamental standards and values which no external authority was supposed to encroach upon. These standards and values pertained to the forms of life, social as well as personal, which were most adequate to the full development of man's faculties and abilities. By the same token, they were the "truth" of his individual and social existence. The individual, as a rational being, was deemed capable of finding these forms by his own thinking and, once he had acquired freedom of thought, of pursuing the course of action which would actualize them. Society's task was to grant him such freedom and to remove all restrictions upon his rational course of action.

The principle of individualism, the pursuit of self-interest, was conditioned upon the proposition that self-interest was rational, that is to say, that it resulted from and was constantly guided and controlled by autonomous thinking. The rational self-interest did not coincide with the individual's immediate self-interest, for the latter depended upon the standards and requirements of the prevailing social order, placed there not by his autonomous thought and conscience but by external authorities. In the context of radical Puritanism, the principle of individualism thus set the individual against his society. Men had to break through the whole system of ideas and values imposed upon them, and to find and seize the ideas and values that conformed to their rational interest. They had to live in a state of constant vigilance, apprehension, and criticism, to reject everything that was not true, not justified by free reason. This, in a society which was not yet rational, constituted a principle of permanent unrest and opposition. For false standards still governed the life of men, and the free individual was therefore he who criticised realization. The theme has nowhere been more fittingly expressed than in Milton's image of a "wicked race of deceivers, who . . . took the virgin Truth, hewd her lovely form into a thousand peeces, and scatter'd them to the four winds. From that time ever since, the sad friends of Truth, such as durst appear, imitating the careful search that Isis made for the mangl'd body of Osiris, went up and down gathering up limb by limb still as they could find them. We have not yet found them all, . . . nor ever shall do, till her Master's second coming . . .—To be still searching what we know not, by what we know, still closing up truth to truth as we find it (for all her body is homogeneal and proportionall)," this was the principle of individualistic rationality.[3]

To fulfill this rationality presupposed an adequate social and economic setting, one that would appeal to individuals whose social performance was, at least to a large extent, their own work. Liberalist society was held to be the adequate setting for individualistic rationality. In the sphere of free competition, the tangible achievements of the individual which made his products and performances a part of society's need, were the marks of his individuality. In the course of time, however, the process of commodity production undermined the economic basis on which individualistic rationality was built. Mechanization and rationalization forced the weaker competitor under the dominion of the giant enterprises of machine industry which, in establishing society's dominion over nature, abolished the free economic subject.

The principle of competitive efficiency favors the enterprises with the most highly mechanized and rationalized industrial equipment. Technological power tends to the concentration of economic power, to "large units of production, of vast corporate enterprises producing large quantities and often a striking variety of goods, of industrial empires owning and controlling materials, equipment, and processes from the extraction of raw materials to the distribution of finished products, of dominance over an entire industry by a small number of giant concerns. . . ." And technology "steadily increases the power at the command of giant concerns by creating new tools, processes and products."[4] Efficiency here called for integral unification and simplification, for the removal of all "waste," the avoidance of all detours, it called for radical coordination. A contradiction exists, however, between the profit incentive that keeps the apparatus moving and the rise of the standard of living which this same apparatus has made possible. "Since control of production is in the hands of enterprisers working for profit, they will have at their disposal whatever emerges as surplus after rent, interest, labor, and other costs are met. These costs will be kept at the lowest possible minimum as a matter of course."[5] Under these circumstances, profitable employment of the apparatus dictates to a great extent the quantity, form and kind of commodities to be produced, and through this mode of production and distribution, the technological power of the apparatus affects the entire rationality of those whom it serves.

Under the impact of this apparatus,[6] individualistic rationality has been transformed into technological rationality. It is by no means confined to the subjects and objects of large scale enterprises but characterizes the pervasive mode of thought and even the manifold forms of protest and rebellion. This rationality establishes standards of judgment and fosters attitudes which make men ready to accept and even to introcept the dictates of the apparatus.

Lewis Mumford has characterized man in the machine age as an "objective personality," one who has learned to transfer all subjective spontaneity to the machinery which he serves, to subordinate his life to the "matter-of-factness" of a world in which the machine is the factor and he the factum.[7] Individual distinctions in the aptitude, insight and knowledge are transformed into different quanta of skill and training, to be coordinated at any time within the common framework of standardized performances.

Individuality, however, has not disappeared. The free economic subject rather has developed into the object of large-scale organization and coordination, and individual achievement has been transformed into standardized efficiency. The latter is characterized by the fact that the individual's performance is motivated, guided and measured by standards external to him, standards pertaining to predetermined tasks and functions. The efficient individual is the one whose performance is an action only insofar as it is the proper reaction to the objective requirements of the apparatus, and his liberty is confined to the selection of the most adequate means for reaching a goal which he did not set. Whereas individual achievement is independent of recognition and consummated in the work itself, efficiency is a rewarded performance and consummated only in its value for the apparatus.

With the majority of the population, the former freedom of the economic subject was gradually submerged in the efficiency with which he performed services assigned to him. The world had been rationalized to such an extent, and this rationality had become such a social power that the individual could do no better than adjust himself without reservation. Veblen was among the first to derive the new matter-of-factness from the machine process, from which it spread over the whole society: "The share of the operative workman in the machine industry is (typically) that of an attendant, an assistant, whose duty it is to keep pace with the machine process and to help out with workmanlike manipulation at points where the machine process engaged is incomplete. His work supplements the machine process rather than makes use of it. On the contrary the machine process makes use of the workman. The ideal mechanical contrivance in this technological system is the automatic machine."[8] The machine process requires a knowledge oriented to "a ready apprehension of opaque facts, in passably exact quantitative terms. This class of knowledge presumes a certain intellectual or spiritual attitude on the part of the workman, such an attitude as will readily apprehend and appreciate matter of fact and will guard against the suffusion of this knowledge with putative animistic or anthropomorphic subtleties, quasi-personal interpretations of the observed phenomena and of their relations to one another."[9]

As an attitude, matter-of-factness is not bound to the machine process. Under all forms of social production men have taken and justified their motives and goals from the facts that made up their reality, and in doing so they have arrived at the most diverging philosophies. Matter-of-factness animated ancient materialism and hedonism, it was responsible in the struggle of modern physical science against spiritual oppression, and in the revolutionary rationalism of the Enlightenment. The new attitude differs from all these in the highly rational compliance which typifies it. The facts directing man's thought and action are not those of nature which must be accepted in order to be mastered, or those of society which must be changed because they no longer correspond to human needs and potentialities. Rather are they those of the machine process, which itself appears as the embodiment of rationality and expediency.

Let us take a simple example. A man who travels by automobile to a distant place chooses his route from the highway maps. Towns, lakes and mountains appear as obstacles to be bypassed. The countryside is shaped and organized by the highway. Numerous signs and posters tell the traveler what to do and think; they even request his attention to the beauties of nature or the hallmarks of history. Others have done the thinking for him, and perhaps for the better. Convenient parking spaces have been constructed where the broadest and most surprising view is open. Giant advertisements tell him when to stop and find the pause that refreshes. And all this is indeed for his benefit, safety and comfort; he receives what he wants. Business, technics, human needs and nature are welded together into one rational and expedient mechanism. He will fare best who follows its directions, subordinating his spontaneity to the anonymous wisdom which ordered everything for him.

The decisive point is that this attitude—which dissolves all actions into a sequence of semi-spontaneous reactions to prescribed mechanical norms—is not only perfectly rational but also perfectly reasonable. All protest is senseless, and the individual who would insist on his freedom of action would become a crank. There is no personal escape from the apparatus which has mechanized and standardized the world. It is a rational apparatus, combining utmost expediency with utmost convenience, saving time and energy, removing waste, adapting all means to the end, anticipating consequences, sustaining calculability and security.

In manipulating the machine, man learns that obedience to the directions is the only way to obtain desired results. Getting along is identical with adjustment to the apparatus. There is no room for autonomy. Individualistic rationality has developed into efficient compliance with the pregiven continuum of means and ends. The latter absorbs the liberating efforts of thought, and the various functions of reason converge upon the unconditional maintenance of the apparatus. It has been frequently stressed that scientific discoveries and inventions are shelved as soon as they seem to interfere with the requirements of profitable marketing.[10] The necessity which is the mother inventions is to a great extent the necessity of maintaining and expanding the apparatus. Inventions have "their chief use . . . in the service of business, not of industry, and their great further use is in the furtherance, or rather the acceleration, of obligatory social amenities." They are mostly of a competitive nature, and "any technological advantage gained by one competitor forthwith becomes a necessity to all the rest, on pain of defeat," so that one might as well say that, in the monopolistic system, "invention is the mother of necessity."[11]

Everything cooperates to turn human instincts, desires and thoughts into channels that feed the apparatus. Dominant economic and social organizations "do not maintain their power by force. . . . They do it by identifying themselves with the faiths and loyalties of the people,"[12] and the people have been trained to identify their faiths and loyalties with them. The relationships among men are increasingly mediated by the machine process. But the mechanical contrivances which facilitate intercourse among individuals also intercept and absorb their libido, thereby diverting it from the all too dangerous realm in which the individual is free of society. The average man hardly cares for any living being with the intensity and persistence he shows for his automobile. The machine that is adored is no longer dead matter but becomes something like a human being. And it gives back to man what it possesses: the life of the social apparatus to which it belongs. Human behavior is outfitted with the rationality of the machine process, and this rationality has a definite social content. The machine process operates according to the laws of mass production. Expediency in terms of technological reason is, at the same time, expediency in terms of profitable efficiency, and rationalization is, at the same time, monopolistic standardization and concentration. The more rationally the individual behaves and the more lovingly he attends to his rationalized work, the more he succumbs to the frustrating aspects of this rationality. He is losing his ability to abstract from the special form in which rationalization is carried through and is losing his faith in its unfulfilled potentialities. His matter-of-factness, his distrust of all values which transcend the facts of observation, his resentment against all "quasi-personal" and metaphysical interpretations, his suspicion of all standards which relate the observable order of things, the rationality of the apparatus, to the rationality of freedom—this whole attitude serves all too well those who are interested in perpetuating the prevailing form of matters of fact. The machine process requires a "consistent training in the mechanical apprehension of things," and this training, in turn, promotes "conformity to the schedule of living," a "degree of trained insight and a facile strategy in all manner of quantitative adjustments and adaptations."[13] The "mechanics of conformity" spread from the technological to the social order; they govern performance not only in the factories and shops, but also in the offices, schools, assemblies and, finally, in the realm of relaxation and entertainment.

Individuals are stripped of their individuality, not by external compulsion, but by the very rationality under which they live. Industrial psychology correctly assumes that "the dispositions of men are fixed emotional habits and as such they are quite dependable reaction patterns."[14] True, the force which transforms human performance into a series of dependable reactions is an external force: the machine process imposes upon men the patterns of mechanical behavior, and the standards of competitive efficiency are the more enforced from outside the less independent the individual competitor becomes. But man does not experience this loss of his freedom as the work of some hostile and foreign force; he relinquishes his liberty to the dictum of reason itself. The point is that today the apparatus to which the individual is to adjust and adopt himself is so rational that individual protest and liberation appear not only as hopeless but as utterly irrational. The system of life created by modern industry is one of the highest expediency, convenience and efficiency. Reason, once defined in these terms, becomes equivalent to an activity which perpetuates this world. Rational behavior becomes identical with a matter-of-factness which teaches reasonable submissiveness and thus guarantees getting along in the prevailing order.

At first glance, the technological attitude rather seems to imply the opposite of resignation. Teleological and theological dogmas no longer interfere with man's struggle with matter; he develops his experimental energies without inhibition. There is no constellation of matter which he does not try to break up, to manipulate and to change according to his will and interest. This experimentalism, however, frequently serves the effort to develop a higher efficiency of hierarchical control over men. Technological rationality may easily be placed into the service of such control: in the form of "scientific management," it has become one of the most profitable means for streamlined autocracy. F. W. Taylor's exposition of scientific management shows within it the union of exact science, matter-of-factness and big industry: "Scientific management attempts to substitute, in the relation between employers and workers, the government of fact and law for the rule of force and opinion. It substitutes exact knowledge for guesswork, and seeks to establish a code of natural laws equally binding upon employers and workmen. Scientific management thus seeks to substitute in the shop discipline, natural law in place of a code of discipline based upon the caprice and arbitrary power of men. No such democracy has ever existed in industry before. Every protest of every workman must be handled by those on the management side and the right and wrong of the complaint must be settled, not by the opinion either of the management or the workman but by the great code of laws which has been developed and which must satisfy both sides."[15] The scientific effort aims at eliminating waste, intensifying production and standardizing the product. And this whole scheme to increase profitable efficiency poses as the final fulfillment of individualism, ending up with a demand to "develop the individuality of the workers."[16]

The idea of compliant efficiency perfectly illustrates the structure of technological rationality. Rationality is being transformed from a critical force into one of adjustment and compliance. Autonomy of reason loses its meaning in the same measure as the thoughts, feelings and actions of men are shaped by the technical requirements of the apparatus which they have themselves created. Reason has found its resting place in the system of standardized control, production and consumption. There it reigns through the laws and mechanisms which insure the efficiency, expediency and coherence of this system.

As the laws and mechanisms of technological rationality spread over the whole society, they develop a set of truth values of their own which hold good for the functioning of the apparatus—and for that alone. Propositions concerning competitive or collusive behavior, business methods, principles of effective organization and control, fair play, the use of science and technics are true or false in terms of this value system, that is to say, in terms of instrumentalities that dictate their own ends. These truth values are tested and perpetuated by experience and must guide the thoughts and actions of all who wish to survive. Rationality here calls for unconditional compliance and coordination, and consequently, the truth values related to this rationality imply the subordination of thought to pre-given external standards. We may call this set of truth values the technological truth, technological in the twofold sense that it is an instrument of expediency rather than an end in itself, and that it follows the pattern of technological behavior.

By virtue of its subordination to external standards, the technological truth comes into striking contradiction with the form in which individualistic society had established its supreme values. The pursuit of self-interest now appears to be conditioned upon heteronomy, and autonomy as an obstacle rather than stimulus for rational action. The originally identical and "homogenous" truth seems to be split into two different sets of truth values and two different patterns of behavior: the one assimilated to the apparatus, the other antagonistic to it; the one making up the prevailing technological rationality and governing the behavior required by it, the other pertaining to a critical rationality whose values can be fulfilled only if it has itself shaped all personal and social relationships. The critical rationality derives from the principles of autonomy which individualistic society itself had declared to be its self-evident truths. Measuring these principles against the form in which individualistic society has actualized them, critical rationality accuses social injustice in the name of individualistic society's own ideology.[17] The relationship between technological and critical truth is a difficult problem which cannot be dealt with here, but two points must be mentioned. (1) The two sets of truth values are neither wholly contradictory nor complementary to each other; many truths of technological rationality are preserved or transformed in critical rationality. (2) The distinction between the two sets is not rigid; the content of each set changes in the social process so that what were once critical truth values become technological values. For example, the proposition that every individual is equipped with certain inalienable rights is a critical proposition but it was frequently interpreted in favor of efficiency and concentration of power.[18]

The standardization of thought under the sway of technological rationality also affects the critical truth values. The latter are torn from the context to which they originally belonged and, in their new form, are given wide, even official publicity. For example, propositions which, in Europe, were the exclusive domain of the labor movement are today adopted by the very forces which these propositions denounced. In the fascist countries, they serve as ideological instruments for the attack on "Jewish capitalism" and "Western plutocracy," thereby concealing the actual front in the struggle. The materialistic analysis of present-day economy is employed to justify fascism to the German industrialists in whose interest it operates, as the regime of last resort for imperialistic expansion.[19] In other countries, the critique of political economy functions in the struggle among conflicting business groups and as governmental weapon for unmasking monopolistic practices; it is propagated by the columnists of the big press syndicates and finds its way even into the popular magazines and the addresses to manufacturers associations. As these propositions become part and parcel of the established culture, however, they seem to lose their edge and to merge with the old and the familiar. This familiarity with the truth illuminates the extent to which society has become indifferent and insusceptible to the impact of critical thought. For the categories of critical thought preserve their truth value only if they direct the full realization of the social potentialities which they envision, and they lose their vigor if they determine an attitude of fatalistic compliance or competitive assimilation.

Several influences have conspired to bring about the social impotence of critical thought. The foremost among them is the growth of the industrial apparatus and of its all-embracing control over all spheres of life. The technological rationality inculcated in those who attend to this apparatus has transformed numerous modes of external compulsion and authority into modes of self-discipline and self-control. Safety and order are, to a large extent, guaranteed by the fact that man has learned to adjust his behavior to the other fellow's down to the most minute detail. All men act equally rationally, that is to say, according to the standards which insure the functioning of the apparatus and thereby the maintenance of their own life. But this "introversion" of compulsion and authority has strengthened rather than attenuated the mechanisms of social control. Men, in following their own reason, follow those who put their reason to profitable use. In Europe, these mechanisms helped to prevent the individual from acting in accordance with the conspicuous truth, and they were efficiently supplemented by the physical control mechanisms of the apparatus. At this point, the otherwise diverging interests and their agencies are synchronized and adjusted in such a manner that they efficiently counteract any serious threat to their dominion.

The ever growing strength of the apparatus, however, is not the only influence responsible. The social impotence of critical thought has been further facilitated by the fact that important strata of the opposition have for long been incorporated into the apparatus itself—without losing the title of the opposition. The history of this process is well known and is illustrated in the development of the labor movement. Shortly after the first World War, Veblen declared that "the A.F. of L. is itself one of the Vested Interests, as ready as any other to do battle for its own margin of privilege and profit. . . . The A.F. of L. is a business organization with a vested interest of its own; for keeping up prices and keeping down the supply, quite after the usual fashion of management by the other Vested Interests."[20] The same holds true for the labor bureaucracy in leading European countries. The question here pertains not to the political expediency and the consequences of such a development, but to the changing function of the truth values which labor had represented and carried forward.

These truth values belonged, to a large extent, to the critical rationality which interpreted the social process in terms of its restrained potentialities. Such a rationality can fully develop only in social groups whose organization is not patterned on the apparatus in its prevailing forms or on its agencies and institutions. For the latter are pervaded by the technological rationality which shapes the attitude and interests of those dependent on them, so that all transcending aims and values are cut off. A harmony prevails between the "spirit" and its material embodiment such that the spirit cannot be supplanted without disrupting the functioning of the whole. The critical truth values borne by an oppositional social movement change their significance when this movement incorporates itself into the apparatus. Ideas such as liberty, productive industry, planned economy, satisfaction of needs are then fused with the interests of control and competition. Tangible organizational success thus outweighs the exigencies of critical rationality.

Its tendency to assimilate itself to the organizational and psychological pattern of the apparatus caused a change in the very structure of the social opposition in Europe. The critical rationality of its aims was subordinated to the technological rationality of its organization and thereby "purged" of the elements which transcended the established pattern of thought and action. This process was the apparently inevitable result of the growth of large-scale industry and of its army of dependents. The latter could hope effectively to assert their interests only if these were effectively coordinated in large-scale organizations. The oppositional groups were being transformed into mass parties, and their leadership into mass bureaucracies. This transformation, however, far from dissolving the structure of individualistic society into a new system, sustained and strengthened its basic tendencies.

It seems to be self-evident that mass and individual are contradictory concepts and incompatible facts. The crowd "is, to be sure, composed of individuals—but of individuals who cease to be isolated, who cease thinking. The isolated individual within the crowd cannot help thinking, criticizing the emotions. The others, on the other hand, cease to think: they are moved, they are carried away, they are elated; they feel united with their fellow members in the crowd, released from all inhibitions; they are changed and feel no connection with their former state of mind."[21] This analysis, although it correctly describes certain features of the masses, contains one wrong assumption, that in the crowd the individuals "cease to be isolated," are changed and "feel no connection with their former state of mind." Under authoritarianism, the function of the masses rather consists in consummating the isolation of the individual and in realizing his "former state of mind." The crowd is an association of individuals who have been stripped of all "natural" and personal distinctions and reduced to the standardized expression of their abstract individuality, namely, the pursuit of self-interest. As member of a crowd, man has become the standardized subject of brute self-preservation. In the crowd, the restraint placed by society upon the competitive pursuit of self-interest tends to become ineffective and aggressive impulses are easily released. These impulses have been developed under the exigencies of scarcity and frustration, and their release rather accentuates the "former state of mind." True, the crowd "unites," but it unites the atomic subjects of self-preservation who are detached from everything that transcends their selfish interests and impulses. The crowd is thus the antithesis of the "community," and the perverted realization of individuality.

The weight and import of the masses grow with the growth of rationalization, but at the same time they are transformed into a conservative force which itself perpetuates the existence of the apparatus. As there is a decrease in the number of those who have the freedom of individual performance, there is an increase in the number of those whose individuality is reduced to self-preservation by standardization. They can pursue their self-interest only by developing "dependable reaction patterns" and by performing pre-arranged functions. Even the highly differentiated professional requirements of modern industry promote standardization. Vocational training is chiefly training in various kinds of skill, psychological and physiological adaptation to a "job" which has to be done. The job, a pregiven "type of work . . . requires a particular combination of abilities,"[22] and those who create the job also shape the human material to fill it. The abilities developed by such training make the "personality" a means for attaining ends which perpetuate man's existence as an instrumentality, replaceable at short notice by other instrumentalities of the same brand. The psychological and "personal" aspects of vocational training are the more emphasized the more they are subjected to regimentation and the less they are left to free and complete development. The "human side" of the employee and the concern for his personal aptitudes and habits play an important part in the total mobilization of the private sphere for mass production and mass culture. Psychology and individualization serve to consolidate stereotyped dependability, for they give the human object the feeling that he unfolds himself by discharging functions which dissolve his self into a series of required actions and responses. Within this range, individuality is not only preserved but also fostered and rewarded, but such individuality is only the special form in which a man introcepts and discharges, within a general pattern, certain duties allocated to him. Specialization fixates the prevailing scheme of standardization. Almost everyone has become a potential member of the crowd, and the masses belong to the daily implements of the social process. As such, they can easily be handled, for the thoughts, feelings and interests of their members have been assimilated to the pattern of the apparatus. To be sure, their outbursts are terrifying and violent but these are readily directed against the weaker competitors and the conspicuous "outsiders" (Jews, foreigners, national minorities). The coordinated masses do not crave a new order but a larger share in the prevailing one. Through their action, they strive to rectify, in an anarchic way, the injustice of competition. Their uniformity is in the competitive self-interest they all manifest, in the equalized expressions of self-preservation. The members of the masses are individuals.

The individual in the crowd is certainly not the one whom the individualist principle exhorted to develop his self, nor is his self-interest the same as the rational interest urged by this principle. Where the daily social performance of the individual has become antagonistic to his "true interest," the individualist principle has changed its meaning. The protagonists of individualism were aware of the fact that "individuals can be developed only by being trusted with somewhat more than they can, at the moment, do well";[23] today, the individual is trusted with precisely what he can, at the moment, do well. The philosophy of individualism has seen the "essential freedom" of the self to be "that it stands for a fateful moment outside of all belongings, and determines for itself alone whether its primary attachments shall be with actual earthly interests or with those of an ideal and potential 'Kingdom of God.'"[24] This ideal and potential kingdom has been defined in different ways, but it has always been characterized by contents which were opposed and transcendent to the prevailing kingdom. Today, the prevailing type of individual is no longer capable of seizing the fateful moment which constitutes his freedom. He has changed his function; from a unit of resistance and autonomy, he has passed to one of ductility and adjustment. It is this function which associates individuals in masses.

The emergence of the modern masses, far from endangering the efficiency and coherence of the apparatus, has facilitated the progressing coordination of society and the growth of authoritarian bureaucracy, thus refuting the social theory of individualism at a decisive point. The technological process seemed to tend to the conquest of scarcity and thus to the slow transformation of competition into cooperation. The philosophy of individualism viewed this process as the gradual differentiation and liberation of human potentialities, as the abolition of the "crowd." Even in the Marxian conception, the masses are not the spearhead of freedom. The Marxian proletariat is not a crowd but a class, defined by its determinate position in the productive process, the maturity of its "consciousness," and the rationality of its common interest. Critical rationality, in the most accentuated form, is the prerequisite for its liberating function. In one aspect at least, this conception is in line with the philosophy of individualism: it envisions the rational form of human association as brought about and sustained by the autonomous decision and action of free men.

This is the one point at which the technological and the critical rationality seem to converge, for the technological process implies a democratization of functions. The system of production and distribution has been rationalized to such an extent that the hierarchical distinction between executive and subordinate performances is to an ever smaller degree based upon essential distinctions in aptitude and insight, and to an ever greater degree upon inherited power and a vocational training to which everyone could be subjected. Even experts and "engineers" are no exception. To be sure, the gap between the underlying population and those who design the blueprints for rationalization, who lay out production, who make the inventions and discoveries which accelerate technological progress, becomes daily more conspicuous, particularly in a period of war economy. At the same time, however, this gap is maintained more by the division of power than by the division of work. The hierarchical distinction of the experts and engineers results from the fact that their ability and knowledge is utilized in the interest of autocratic power. The "technological leader" is also a "social leader"; his "social leadership overshadows and conditions his function as a scientist, for it gives him institutional power within the group . . . ," and the "captain of industry" acts in "perfect accordance with the traditional dependence of the expert's function."[25] Were it not for this fact, the task of the expert and engineer would not be an obstacle to the general democratization of functions. Technological rationalization has created a common framework of experience for the various professions and occupations. This experience excludes or restrains those elements that transcend the technical control over matters of fact and thus extends the scope of rationalization from the objective to the subjective world. Underneath the complicated web of stratified control is an array of more or less standardized techniques, tending to one general pattern, which insure the material reproduction of society. The "persons engaged in a practical occupation" seem to be convinced that "any situation which appears in the performance of their role can be fitted into some general pattern with which the best, if not all, of them are familiar."[26] Moreover, the instrumentalistic conception of technological rationality is spreading over almost the whole realm of thought and gives the various intellectual activities a common denominator. They too become a kind of technique,[27] a matter of training rather than individuality, requiring the expert rather than the complete human personality.

The standardization of production and consumption, the mechanization of labor, the improved facilities of transportation and communication, the extension of training, the general dissemination of knowledge—all these factors seem to facilitate the exchangeability of functions. It is as if the basis were shrinking on which the pervasive distinction between "specialized (technical)" and "common" knowledge[28] has been built, and as if the authoritarian control of functions would prove increasingly foreign to the technological process. The special form, however, in which the technological process is organized, counteracts this trend. The same development that created the modern masses as the standardized attendants and dependents of large-scale industry also created the hierarchical organization of private bureaucracies. Max Weber has already stressed the connection between mass-democracy and bureaucracy: "In contrast to the democratic self-administration of small homogeneous units," the bureaucracy is "the universal concomitant of modern mass democracy."[29]

The bureaucracy becomes the concomitant of the modern masses by virtue of the fact that standardization proceeds along the lines of specialization. The latter by itself, provided that it is not arrested at the point where it interferes with the domain of vested control, is quite compatible with the democratization of functions. Fixated specialization, however, tends to atomize the masses and to insulate the subordinate from the executive functions. We have mentioned that specialized vocational training implies fitting a man to a particular job or a particular line of jobs, thus directing his "personality," spontaneity and experience to the special situations he may meet in filling the job. In this manner, the various professions and occupations, notwithstanding their convergence upon one general pattern, tend to become atomic units which require coordination and management from above. The technical democratization of functions is counteracted by their atomization, and the bureaucracy appears as the agency which guarantees their rational course and order.

The bureaucracy thus emerges on an apparently objective and impersonal ground, provided by the rational specialization of functions, and this rationality in turn serves to increase the rationality of submission. For, the more the individual functions are divided, fixated and synchronized according to objective and impersonal patterns, the less reasonable is it for the individual to withdraw or withstand. "The material fate of the masses becomes increasingly dependent upon the continuous and correct functioning of the increasingly bureaucratic order of private capitalistic organizations."[30] The objective and impersonal character of technological rationality bestows upon the bureaucratic groups the universal dignity of reason. The rationality embodied in the giant enterprises makes it appear as if men, in obeying them, obey the dictum of an objective rationality. The private bureaucracy fosters a delusive harmony between the special and the common interest. Private power relationships appear not only as relationships between objective things but also as the rule of rationality itself.

In the fascist countries, this mechanism facilitated the merger between private, semiprivate (party) and public (governmental) bureaucracies. The efficient realization of the interests of large-scale enterprise was one of the strongest motives for the transformation of economic into totalitarian political control, and efficiency is one of the main reasons for the fascist regime's hold over its regimented population. At the same time, however, it is also the force which may break this hold. Fascism can maintain its rule only by aggravating the restraint which it is compelled to impose upon society. It will ever more conspicuously manifest its inability to develop the productive forces, and it will fall before that power which proves to be more efficient than fascism.

In the democratic countries, the growth of the private bureaucracy can be balanced by the strengthening of the public bureaucracy. The rationality inherent in the specialization of functions tends to enlarge the scope and weight of bureaucratization. In the private bureau-cracy, however, such an expansion will intensify rather than alleviate the irrational elements of the social process, for it will widen the discrepancy between the technical character of the division of functions and the autocratic character of control over them. In contrast, the public bureaucracy, if democratically constituted and controlled, will overcome this discrepancy to the extent that it undertakes the "conservation of those human and material resources which technology and corporations have tended to misuse and waste."[31] In the age of mass society, the power of the public bureaucracy can be the weapon which protects the people from the encroachment of special interests upon the general welfare. As long as the will of the people can effectively assert itself, the public bureaucracy can be a lever of democratization. Large-scale industry tends to organize on a national scale, and fascism has transformed economic expansion into the military conquest of whole continents. In this situation, the restoration of society to its own right, and the maintenance of individual freedom have become directly polit-ical questions, their solution depending upon the outcome of the international struggle.

The social character of bureaucratization is largely determined by the extent to which it allows for a democratization of functions that tends to close the gap between the governing bureaucracy and the governed population. If everyone has become a potential member of the public bureaucracy (as he has become a potential member of the masses), society will have passed from the stage of hierarchical bureaucratization to the stage of technical self-administration. Insofar as technocracy implies a deepening of the gap between specialized and common knowledge, between the controlling and coordinating experts and the controlled and coordinated people, the technocratic abolition of the "price system" would stabilize rather than shatter the forces which stand in the way of progress. The same holds true for the so-called managerial revolution. According to the theory of the managerial revolution,[32] the growth of the apparatus entails the rise of a new social class, the "managers," to take over social domination and to establish a new economic and political order. Nobody will deny the increasing importance of management and the simultaneous shift in the function of control. But these facts do not make the managers a new social class or the spearhead of a revolution. Their "source of income" is the same as that of the already existing classes: they either draw salaries, or, insofar as they possess a share in the capital, are themselves capitalists. Moreover, their specific function in the prevailing division of labor does not warrant the expectation that they are predestined to inaugurate a new and more rational division of labor. This function is either determined by the requirement of profitable utilization of capital, and, in this case, the managers are simply capitalists or deputy-capitalists (comprising the "executives" and the corporation-managers[33]); or it is determined by the material process of production (engineers, technicians, production managers, plant superintendents). In the latter case, the managers would belong to the vast army of the "immediate producers" and share its "class interest," were it not for the fact that, even in this function, they work as deputy-capitalists and thus form a segregated and privileged group between capital and labor. Their power, and the awe which it inspires, are derived not from their actual "technological" performance but from their social position, and this they owe to the prevailing organization of production. "The leading managerial and directorial figures within the inner business sancta . . . are drawn from, or have been absorbed into, the upper layers of wealth and income whose stakes it is their function to defend."[34] To sum up, as a separate social group, the managers are thoroughly tied up with the vested interests, and as performers of necessary productive functions they do not constitute a separate "class" at all.

The spreading hierarchy of large-scale enterprise and the precipitation of individuals into masses determine the trends of technological rationality today. What results is the mature form of that individualistic rationality which characterized the free economic subject of the industrial revolution. Individualistic rationality was born as a critical and oppositional attitude that derived freedom of action from the unrestricted liberty of thought and conscience and measured all social standards and relations by the individual's rational self-interest. It grew into the rationality of competition in which the rational interest was superseded by the interest of the market, and individual achievement absorbed by efficiency. It ended with standardized submission to the all-embracing apparatus which it had itself created. This apparatus is the embodiment and resting place of individualistic rationality, but the latter now requires that individuality must go. He is rational who most efficiently accepts and executes what is allocated to him, who entrusts his fate to the large-scale enterprises and organizations which administer the apparatus.

Such was the logical outcome of a social process which measured individual performance in terms of competitive efficiency. The philosophers of individualism have always had an inkling of this outcome and they expressed their anxiety in many different forms, in the skeptical conformism of Hume, in the idealistic introversion of individual freedom, in the frequent attacks of the Transcendentalists against the rule of money and power. But the social forces were stronger than the philosophic protests, and the philosophic justification of individualism took on more and more of the overtones of resignation. Toward the end of the nineteenth century, the idea of the individual became increasingly ambiguous: it combined insistence upon free social performance and competitive efficiency with glorification of smallness, privacy and self-limitation. The rights and liberties of the individual in society were interpreted as the rights and liberties of privacy and withdrawal from society. William James, faithful to the individualistic principle, asserted that, in the "rivalry between real organizable goods," the "world's trial is better than the closest solution," provided that the victorious keep "the vanquished somehow represented."[35] His doubt, however, as to whether this trial is really a fair one seems to motivate his hatred of "bigness and greatness in all their forms,"[36] his declaration that "the smaller and more intimate is the truer,—the man more than the home, the home more than the state or the church."[37] The counterposition of individual and society, originally meant to provide the ground for a militant reformation of society in the interest of the individual, comes to prepare and justify the individual's withdrawal from society. The free and self-reliant "soul," which originally nourished the individual's critique of external authority, now becomes a refuge from external authority. Tocqueville had already defined individualism in terms of acquiescence and peaceful resignation: "a mature and calm feeling, which disposes each member of the community to sever himself from the mass of his fellow-creatures; and to draw apart with his family and his friends; so that, after he has thus formed a little circle of his own, he willingly leaves society at large to itself."[38] Autonomy of the individual came to be regarded as a private rather than a public affair, an element of retreat rather than aggression. All these factors of resignation are comprehended in Benjamin Constant's statement that "our liberty should be composed of the peaceful enjoyment of private independence."[39]

The elements of restraint and resignation which became increasingly strong in the individualist philosophy of the nineteenth century elucidate the connection between individualism and scarcity. Individualism is the form liberty assumes in a society wherein the acquisition and utilization of wealth is dependent on competitive toil. Individuality is a distinct possession of "pioneers"; it presupposes the open and empty spaces, the freedom of "hewing out a home" as well as the need to do so. The individual's world is a "world of labor and the march," as Walt Whitman says, one in which the available intellectual and material resources must be conquered and appropriated through incessant struggle with man and nature, and in which human forces are released to distribute and administer scarcity.

In the period of large-scale industry, however, the existential conditions making for individuality give way to conditions which render individuality unnecessary. In clearing the ground for the conquest of scarcity, the technological process not only levels individuality but also tends to transcend it where it is concurrent with scarcity. Mechanized mass production is filling the empty spaces in which individuality could assert itself. The cultural standardization points, paradoxically enough, to potential abundance as well as actual poverty. This standardization may indicate the extent to which individual creativeness and originality have been rendered unnecessary. With the decline of the liberalistic era, these qualities were vanishing from the domain of material production and becoming the ever more exclusive property of the highest intellectual activities. Now, they seem to disappear from this sphere too: mass culture is dissolving the traditional forms of art, literature and philosophy together with the "personality" which unfolded itself in producing and consuming them. The striking impoverishment which characterizes the dissolution of these forms may involve a new source of enrichment. They derived their truth from the fact that they represented the potentialities of man and nature which were excluded or distorted in the reality. So far were those potentialities from their actualization in the social consciousness that much cried out for unique expression. But today, *humanitas,* wisdom, beauty, freedom and happiness can no longer be represented as the realm of the "harmonious personality" nor as the remote heaven of art nor as metaphysical systems. The "ideal" has become so concrete and so universal that it grips the life of every human being, and the whole of mankind is drawn into the struggle for its realization. Under the terror that now threatens the world the ideal constricts itself to one single and at the same time common issue. Faced with fascist barbarism, everyone knows what freedom means, and everyone is aware of the irrationality in the prevailing rationality.

Modern mass society quantifies the qualitative features of individual labor and standardizes the individualistic elements in the activities of intellectual culture. This process may bring to the fore the tendencies which make individuality a historical form of human existence, to be surpassed by further social development. This does not mean that society is bound to enter a stage of "collectivism." The collectivistic traits which characterize the development today may still belong to the phase of individualism. Masses and mass culture are manifestations of scarcity and frustration, and the authoritarian assertion of the common interest is but another form of the rule of particular interests over the whole. The fallacy of collectivism consists in that it equips the whole (society) with the traditional properties of the individual. Collectivism abolishes the free pursuit of competing individual interests but retains the idea of the common interest as a separate entity. Historically, however, the latter is but the counterpart of the former. Men experience their society as the objective embodiment of the collectivity as long as the individual interests are antagonistic to and competing with each other for a share in the social wealth. To such individuals, society appears as an objective entity, consisting of numerous things, institutions and agencies: plants and shops, business, police and law, government, schools and churches, prisons and hospitals, theaters and organizations, etc. Society is almost everything the individual is not, everything that determines his habits, thoughts and behavior patterns, that affects him from "outside." Accordingly, society is noticed chiefly as a power of restraint and control, providing the framework which integrates the goals, faculties and aspirations of men. It is this power which collectivism retains in its picture of society, thus perpetuating the rule of things and men over men.

The technological process itself furnishes no justification for such a collectivism. Technics hamper individual development only insofar as they are tied to a social apparatus which perpetuates scarcity, and this same apparatus has released forces which may shatter the special historical form in which technics is utilized. For this reason, all programs of an anti-technological character, all propaganda for an anti-industrial revolution[40] serve only those who regard human needs as a by-product of the utilization of technics. The enemies of technics readily join forces with a terroristic technocracy.[41] The philosophy of the simple life, the struggle against big cities and their culture frequently serves to teach men distrust of the potential instruments that could liberate them. We have pointed to the possible democratization of functions which technics may promote and which may facilitate complete human development in all branches of work and administration. Moreover, mechanization and standardization may one day help to shift the center of gravity from the necessities of material production to the arena of free human realization. The less individuality is required to assert itself in standardized social performances, the more it could retreat to a free "natural" ground. These tendencies, far from engendering collectivism, may lead to new forms of individualization. The machine individualizes men by following the physiological lines of individuality: it allocates the work to finger, hand, arm, foot, classifying and occupying men according to the dexterity of these organs.[42] The external mechanisms which govern standardization here meet a "natural" individuality; they lay bare the ground on which a hitherto suppressed individualization might develop. On this ground, man is an individual by virtue of the uniqueness of his body and its unique position in the space-time continuum. He is an individual insofar as this natural uniqueness molds his thoughts, instincts, emotions, passions and desires. This is the "natural" *principium individuationis.* Under the system of scarcity, men developed their senses and organs chiefly as implements of labor and competitive orientation: skill, taste, proficiency, tact, refinement and endurance were qualities molded and perpetuated by the hard struggle for life, business and power. Consequently, man's thoughts, appetites and the ways of their fulfillment were not "his;" they showed the oppressive and inhibitive features which this struggle imposed upon him. His senses, organs and appetites became acquisitive, exclusive and antagonistic. The technological process has reduced the variety of individual qualities down to this natural basis of individualization, but this same basis may become the foundation for a new form of human development.

The philosophy of individualism established an intrinsic connection between individuality and property.[43] According to this philosophy, man could not develop a self without conquering and cultivating a domain of his own, to be shaped exclusively by his free will and reason. The domain thus conquered and cultivated had become part and parcel of his own "nature." Man removed the objects in this domain from the state in which he found them, and made them the tangible manifestation of his individual labor and interest. They were his property because they were fused with the very essence of his personality. This construction did not correspond to the facts and lost its meaning in the era of mechanized commodity production, but it contained the truth that individual development, far from being an inner value only, required an external sphere of manifestation and an autonomous concern for men and things.

The process of production has long dissolved the link between individual labor and property and now tends to dissolve the link between the traditional form of property and social control, but the tightening of this control counteracts a tendency which may give the individualistic theory a new content. Technological progress would make it possible to decrease the time and energy spent in the production of the necessities of life, and a gradual reduction of scarcity and abolition of competitive pursuits could permit the self to develop from its natural roots. The less time and energy man has to expend in maintaining his life and that of society, the greater the possibility that he can "individualize" the sphere of his human realization. Beyond the realm of necessity, the essential differences between men could unfold themselves: everyone could think and act by himself, speak his own language, have his own emotions and follow his own passions. No longer chained to competitive efficiency, the self could grow in the realm of satisfaction. Man could come into his own in his passions. The objects of his desires would be the less exchangeable the more they were seized and shaped by his free self. They would "belong" to him more than ever before, and such ownership would not be injurious, for it would not have to defend its own against a hostile society.

Such a Utopia would not be a state of perennial happiness. The "natural" individuality of man is also the source of his natural sorrow. If the human relations are nothing but human, if they are freed from all foreign standards, they will be permeated with the sadness of their singular content. They are transitory and irreplaceable, and their transitory character will be accentuated when concern for the human being is no longer mingled with fear for his material existence and overshadowed by the threat of poverty, hunger, and ostracism.

The conflicts, however, which may arise from the natural individuality of men may not bear the violent and aggressive features which were so frequently attributed to the "state of nature." These features may be the marks of coercion and privation. "Appetite is never excessive, never furious, save when it has been starved. The frantic hunger we see it so often exhibiting under every variety of criminal form, marks only the hideous starvation to which society subjects it. It is not a normal but a morbid state of the appetite, growing exclusively out of the unnatural compression which is imposed upon it by the exigencies of our immature society. Every appetite and passion of man's nature is good and beautiful, and destined to be fully enjoyed . . . Remove, then, the existing bondage of humanity, remove those factitious restraints which keep appetite and passion on the perpetual lookout for escape, like steam from an overcharged boiler, and their force would instantly become conservative instead of destructive."[44]

NOTES

1. Cf. Lewis Mumford, *Technics and Civilization,* New York 1936, p. 364: The motive in back of "mechanical discipline and many of the primary inventions . . . was not technical efficiency but holiness, or power over other men. In the course of their development machines have extended these aims and provided a vehicle for their fulfillment."

2. Cf. A.R.L. Gurland, "Technological Trends and Economic Structure under National Socialism," in *Studies in Philosophy and Social Science,* IX (1941), No. 2, pp. 226ff.

3. Areopagitica.

4. *Temporary National Committee,* Monograph No. 22, "Technology in Our Economy," Washington, 1941, p. 195.

5. *Temporary National Economic Committee, Final Report of the Executive Secretary,* Washington 1941, p. 140.

6. The term "apparatus" denotes the institutions, devices and organizations of industry in their prevailing social setting.

7. L. Mumford, *op. cit.,* pp. 361ff.

8. *The Instinct of Workmanship,* New York, 1922, p. 306f.

9. *Ibid.,* p. 310. This training in "matter of factness" applies not only to the factory worker but also to those who direct rather than attend the machine.

10. Florian Znaniecki, *The Social Role of the Man of Knowledge,* New York 1940, p. 54f. Bernard J. Stern, *Society and Medical Progress,* Princeton 1941, Chapter IX, and the same author's contribution to *Technological Trends and National Policy,* U. S. National Resources Committee, Washington 1937.

11. Thorstein Veblen, *op. cit.,* p. 315f.

12. Thurmann Arnold, *The Folklore of Capitalism,* New York 1941, p. 193f.

13. Thorstein Veblen, *op. cit.,* p. 314.

14. Albert Walton, *Fundamentals of Industrial Psychology,* New York 1941, p. 24.

15. Robert F. Hoxie, *Scientific Management and Labor,* New York 1916, p. 140f.

16. *Ibid.,* p. 149.

17. Cf. Max Horkheimer and Herbert Marcuse, "Traditionelle und kritische Theorie," in *Zeitschrift für Socialforschung,* VI (1937), pp. 245ff.

18. Cf. the discussion on the law Le Chapelier in the National Assembly of the French Revolution.

19. Hitler's speech before the Industry Club in Düsseldorf, January 27, 1932, in *My New Order,* New York 1941, pp. 93ff.

20. *The Engineers and The Price System,* New York 1940, pp. 88ff.

21. E. Lederer, *State of the Masses,* New York 1940, p. 32f.

22. Albert Walton, *op. cit.,* p. 27.

23. W. E. Hocking, *The Lasting Elements of Individualism,* New Haven 1937, p. 5.

24. *Ibid.,* p. 23.

25. Florian Znaniecki, *op. cit.,* pp. 40, 55.

26. *Op. cit.,* p. 31—Znaniecki's description refers to a historical state of affairs in which "no demand for a scientist can arise," but it appears to refer to a basic tendency of the prevailing state of affairs.

27. Cf. Max Horkheimer, "The End of Reason," above.

28. Florian Znaniecki, *op. cit.,* p. 25.

29. *Wirtschaft und Gesellschaft,* Tübingen 1922, p. 666.

30. Max Weber, *op. cit.,* p. 669.

31. Henry A. Wallace, *Technology, Corporations, and the General Welfare,* Chapel Hill 1937, p. 56.

32. J. Burnham, *The Managerial Revolution,* New York 1941, pp. 78ff.

33. *Ibid.,* p. 83f.

34. Robert A. Brady, "Policies of National Manufacturing Spitzenverbände," in *Political Science Quarterly,* LVI, p. 537.

35. *The Thought and Character of William James,* ed. R. B. Perry, Boston 1935, II, p. 265.

36. *Ibid.,* p. 315.

37. *Ibid.,* p. 383.

38. *Democracy in America,* trans. H. Reeve, New York 1904, p. 584.

39. Quoted in E. Mims, *The Majority of the People,* New York 1941, p. 152.

40. See for example Oswald Spengler, *Man and Technics,* New York 1932, p. 96f., and Roy Helton, "The Anti-Industrial Revolution," in *Harpers,* December 1941, pp. 65ff.

41. In National Socialist Germany, the ideology of blood and soil and the glorification of the peasant is an integral part of the imperialistic mobilization of industry and labor.

42. For examples of the degree to which this physiological individualization has been utilized see *Changes in Machinery and Job Requirements in Minnesota Manufacturing 1931–36,* Works Projects Administration, National Research Project, Report No. 1–6. Philadelphia, p. 19.

43. See Max Horkheimer, "The End of Reason," above.

44. Henry James, "Democracy and Its Issues," in *Lectures and Miscellanies,* New York 1852, p. 47f.

6

Technical Progress and the Social Life-World

Jürgen Habermas

When C. P. Snow published *The Two Cultures* in 1959, he initiated a discussion of the relation of science and literature which has been going on in other countries as well as in England. Science in this connection has meant the strictly empirical sciences, while literature has been taken more broadly to include methods of interpretation in the cultural sciences. The treatise with which Aldous Huxley entered the controversy, however, *Literature and Science,* does limit itself to confronting the natural sciences with the belles-lettres.

Huxley distinguishes the two cultures primarily according to the specific experiences with which they deal: literature makes statements mainly about private experiences, the sciences about intersubjectively accessible experiences. The latter can be expressed in a formalized language, which can be made universally valid by means of general definitions. In contrast, the language of literature must verbalize what is in principle unrepeatable and must generate an intersubjectivity of mutual understanding in each concrete case. But this distinction between private and public experience allows only a first approximation to the problem. The element of ineffability that literary expression must overcome derives less from a private experience encased in subjectivity than from the constitution of these experiences within the horizon of a life-historical environment. The events whose connection is the object of the lawlike hypotheses of the sciences can be described in a spatio-temporal coordinate system, but they do not make up a world:

> The world with which literature deals is the world in which human beings are born and live and finally die; the world in which they love and hate, in which they experience triumph and humiliation, hope and despair; the world of sufferings and enjoyments, of madness and common sense, of silliness, cunning and wisdom; the world of social pressures and individual impulses, of reason against passion, of instincts and conventions, of shared language and unsharable feelings and sensations . . . [1]

In contrast, science does not concern itself with the contents of a life-world of this sort, which is culture-bound, ego-centered, and pre-interpreted in the ordinary language of social groups and socialized individuals:

> . . . As a professional chemist, say, a professional physicist or physiologist, [the scientist] is the inhabitant of a radically different universe—not the universe of given appearances, but the world of inferred fine structures, not the experienced world of unique events and diverse qualities, but the world of quantified regularities. (LS, p. 8)

Huxley juxtaposes the *social life-world* and the *worldless universe of facts.* He also sees precisely the way in which the sciences transpose their information about this worldless universe into the life-world of social groups:

> Knowledge is power and, by a seeming paradox, it is through their knowledge of what happens in this unexperienced world of abstractions and inferences that scientists have acquired their enormous and growing power to control, direct, and modify the world of manifold appearances in which human beings are privileged and condemned to live. (LS, p. 10)

But Huxley does not take up the question of the relation of the two cultures at this juncture, where the sciences enter the social life-world through the technical exploitation of their information. Instead he postulates an immediate relation. Literature should assimilate scientific statements as such, so that science can take on "flesh and blood."

> . . . Until some great artist comes along and tells us what to do, we shall not know how the muddled words of the tribe and the too precise words of the textbooks should be poetically purified, so as to make them capable of harmonizing our private and unsharable experiences with the scientific hypotheses in terms of which they are explained. (LS, p. 107)

This postulate is based, I think, on a misunderstanding. Information provided by the strictly empirical sciences can be incorporated in the social life-world only through its technical utilization, as technological knowledge, serving the expansion of our power of technical control. Thus, such information is not on the same level as the action-orienting self-understanding of social groups. Hence, without mediation, the information content of the sciences cannot be relevant to that part of practical knowledge which gains expression in literature. It can only attain significance through the detour marked by the practical results of technical progress. Taken for itself, knowledge of atomic physics remains without consequence for the interpretation of our life-world, and to this extent the cleavage between the two cultures is inevitable. Only when with the aid of physical theories we can carry out nuclear fission, only when information is exploited for the development of productive or destructive forces, can its revolutionary practical results penetrate the literary consciousness of the life-world: poems arise from consideration of Hiroshima and not from the elaboration of hypotheses about the transformation of mass into energy.

The idea of an atomic poetry that would elaborate on hypotheses follows from false premises. In fact, the problematic relation of literature and science is only one segment of a much broader problem: *How is it possible to translate technically exploitable knowledge into the practical consciousness of a social life-world?* This question obviously sets a new task, not only or even primarily for literature. The skewed relation of the two cultures is so disquieting only because, in the seeming conflict between the two competing cultural traditions, a true life-problem of scientific civilization becomes apparent: namely, how can the relation between technical progress and the social life-world, which today is still clothed in a primitive, traditional, and unchosen form, be reflected upon and brought under the control of rational discussion?

To a certain extent practical questions of government, strategy, and administration had to be dealt with through the application of technical knowledge even at an earlier period. Yet today's problem of transposing technical knowledge into practical consciousness has changed not merely its order of magnitude. The mass of technical knowledge is no longer restricted to pragmatically acquired techniques of the classical crafts. It has taken the form of scientific information that can be exploited for technology. On the other hand, behavior-controlling traditions no longer naively define the self-understanding of modern societies. Historicism has broken the natural-traditional validity of action-orienting value systems. Today, the self-understanding of social groups and their worldview as articulated in ordinary language is mediated by the hermeneutic appropriation of traditions as traditions. In this situation questions of life conduct demand a rational discussion that is not focused exclusively either on technical means or on the application of traditional behavioral norms. The reflection that is required extends beyond the production of technical knowledge and the hermeneutical clarification of traditions to the employment of technical means in historical situations whose objective conditions (potentials, institutions, interests) have to be interpreted anew each time in the framework of a self-understanding determined by tradition.

This problem-complex has only entered consciousness within the last two or three generations. In the nineteenth century one could still maintain that the sciences entered the conduct of life through two separate channels: through the technical exploitation of scientific information and through the processes of individual education and culture during academic study. Indeed, in the German university system, which goes back to Humboldt's reform, we still maintain the fiction that the sciences develop their action-orienting power through educational processes within the life history of the individual student. I should like to show that the intention designated by Fichte as a "transformation of knowledge into works" can no longer be carried out in the private sphere of education, but rather can be realized only on the politically relevant level at which technically exploitable knowledge is translatable into the context of our life-world. Though literature participates in this, it is primarily a problem of the sciences themselves.

At the beginning of the nineteenth century, in Humboldt's time, it was still impossible, looking at Germany, to conceive of the scientific transformation of social life. Thus, the university reformers did not have to break seriously with the tradition of practical philosophy. Despite the profound ramifications of revolutions in the political order, the structures of the preindustrial work world persisted, permitting for the last time, as it were, the classical view of the relation of theory to practice. In this tradition, the technical capabilities employed in the sphere of social labor are not capable of immediate direction by theory. They must be pragmatically practiced according to traditional patterns of skill. Theory, which is concerned with the immutable essence of things beyond the mutable region human affairs, can obtain practical validity only by molding the manner of life of men engaged in theory. Understanding the cosmos as a whole yields norms of individual human behavior, and it is through the actions of the philosophically educated that theory assumes a positive form. This was the only relation of theory to practice incorporated in the traditional idea of university education. Even where Schelling attempts to provide the physician's practice with a scientific basis in natural philosophy, the medical *craft* is unexpectedly transformed into a medical *praxiology*. The physician must orient himself to Ideas derived from natural philosophy in the same way that the subject of moral action orients itself through the Ideas of practical reason.

Since then it has become common knowledge that the scientific transformation of medicine succeeds only to the extent that the pragmatic doctrine of the medical art can be transformed into the control of isolated natural processes, checked by scientific method. The same holds for other areas of social labor. Whether it is a matter of rationalizing the production of goods, management and administration, construction of machine tools, roads, or airplanes, or the manipulation of electoral, consumer, or leisure-time behavior, the professional practice in question will always have to assume the form of technical control of objectified processes.

In the early nineteenth century, the maxim that scientific knowledge is a source of culture required a strict separation between the university and the technical school because the preindustrial forms of professional practice were impervious to theoretical guidance. Today, research processes are coupled with technical conversion and economic exploitation, and production and administration in the industrial system of labor generate feedback for science. The application of science in technology and the feedback of technical progress to research have become the substance of the world of work. In these circumstances, unyielding opposition to the decomposition of the university into specialized schools can no longer invoke the old argument. Today, the reason given for delimiting study on the university model from the professional sphere is not that the latter is still foreign to science, but conversely, that science—to the very extent that it has penetrated professional practice—has estranged itself from humanistic culture. The philosophical conviction of German idealism that scientific knowledge is a source of culture no longer holds for the strictly empirical scientist. It was once possible for theory, via humanistic culture, to become a practical force. Today, theories can become technical power while remaining unpractical, that is, without being expressly oriented to the interaction of a community of human beings. Of course, the sciences now transmit a specific capacity: but the capacity for control, which they teach, is not the same capacity for life and action that was to be the expected of the scientifically educated and cultivated.

The cultured possessed orientation in action. Their culture was universal only in the sense of the universality of a culture-bound horizon of a world in which scientific experiences could be interpreted and turned into practical abilities, namely, into a reflected consciousness of the practically necessary. The only type of experience which is admitted as scientific today according to positivistic criteria is not capable of this transposition into practice. The capacity for *control* made possible by the empirical sciences is not to be confused with the capacity for *enlightened action*. But is science, therefore, completely discharged of this task of action-orientation, or does the question of academic education in the framework of a civilization transformed by scientific means arise again today as a problem of the sciences themselves?

First, production processes were revolutionized by scientific methods. Then expectations of technically correct functioning were also transferred to those areas of society that had become independent in the course of the industrialization of labor and thus supported planned organization. The power of technical control over nature made possible by science is extended today directly to society: for every isolatable social system, for every cultural area that has become a separate, closed system whose relations can be analyzed immanently in terms of presupposed system goals, a new discipline emerges in the social sciences. In the same measure, however, the problems of technical control solved by science are transformed into life problems. For the scientific control of natural and social processes—in a word, technology—does not release men from action. Just as before, conflicts must be decided, interests realized, interpretations found—through both action and transaction structured by ordinary language.

Today, however, these practical problems are themselves in large measure determined by the system of our technical achievements.

But if technology proceeds from science, and I mean the technique of influencing human behavior no less than that of dominating nature, then the assimilation of this technology into the practical life-world, bringing the technical control of particular areas within the reaches of the communication of acting men, really requires scientific reflection. The prescientific horizon of experience becomes infantile when it naively incorporates contact with the products of the most intensive rationality.

Culture and education can then no longer indeed be restricted to the ethical dimension of personal attitude. Instead, in the political dimension at issue, the theoretical guidance of action must proceed from a scientifically explicated understanding of the world.

The relation of technical progress and social life-world and the translation of scientific information into practical consciousness is not an affair of private cultivation.

I should like to reformulate this problem with reference to political decision-making. In what follows we shall understand "technology" to mean scientifically rationalized control of objectified processes. It refers to the system in which research and technology are coupled with feedback from the economy and administration. We shall understand "democracy" to mean the institutionally secured forms of general and public communication that deal with the practical question of how men can and want to live under the objective conditions of their ever-expanding power of control. Our problem can then be stated as one of the relation of technology and democracy: how can the power of technical control be brought within the range of the consensus of acting and transacting citizens?

I should like first to discuss two antithetical answers. The first, stated in rough outline, is that of Marxian theory. Marx criticizes the system of capitalist production as a power that has taken on its own life in opposition to the interests of productive freedom, of the producers. Through the private form of appropriating socially produced goods, the technical process of producing use values falls under the alien law of an economic process that produces exchange values. Once we trace this self-regulating character of the accumulation of capital back to its origins in private property in the means of production, it becomes possible for mankind to comprehend economic compulsion as an alienated result of its own free productive activity and then abolish it. Finally, the reproduction of social life can be rationally planned as a process of producing use values; society places this process under its technical control. The latter is exercised democratically in accordance with the will and insight of the associated individuals. Here Marx equates the practical insight of a political public with successful technical control. Meanwhile we have learned that even a well-functioning planning bureaucracy with scientific control of the production of goods and services is not a sufficient condition for realizing the associated material and intellectual productive forces in the interest of the enjoyment and freedom of an emancipated society. For Marx did not reckon with the possible emergence at every level of a discrepancy between scientific control of the material conditions of life and a democratic decision-making process. This is the philosophical reason why socialists never anticipated the authoritarian welfare state, where social wealth is relatively guaranteed while political freedom is excluded.

Even if technical control of physical and social conditions for preserving life and making it less burdensome had attained the level that Marx expected would characterize a communist stage of development, it does not follow that they would be linked automatically with social emancipation of the sort intended by the thinkers of the Enlightenment in the eighteenth century and the Young Hegelians in the nineteenth. For the techniques with which the development of a highly industrialized society could be brought under control can no longer be interpreted according to an instrumental model, as though appropriate means were being organized for the realization of goals that are either presupposed without discussion or clarified through communication.

Hans Freyer and Helmut Schelsky have outlined a counter-model which recognizes technology as an independent force. In contrast to the primitive state of technical development, the relation of the organization of means to given or preestablished goals today seems to have been reversed. The process of research and technology—which obeys immanent laws—precipitates in an unplanned fashion new methods for which we then have to find purposeful application. Through progress that has become automatic, Freyer argues, abstract potential continually accrues to us in renewed thrusts. Subsequently, both life interests and fantasy that generates meaning have to take this potential in hand and expend it on concrete goals. Schelsky refines and simplifies this thesis to the point of asserting that technical progress produces not only unforeseen methods but the unplanned goals and applications themselves: technical potentialities command their own practical realization. In particular, he puts forth this thesis with regard to the highly complicated objective exigencies that in political situations allegedly prescribe solutions without alternatives.

> Political norms and laws are replaced by objective exigencies of scientific-technical civilization, which are not posited as political decisions and cannot be understood as norms of conviction or weltanschauung. Hence, the idea of democracy loses its classical substance, so to speak. In place of the political will of the people emerges an objective exigency, which man himself produces as science and labor.

In the face of research, technology, the economy, and administration—integrated as a system that has become autonomous—the question prompted by the neohumanistic ideal of culture, namely, how can society possible exercise sovereignty over the technical conditions of life and integrate them into the practice of the life-world, seems hopelessly obsolete. In the technical state such ideas are suited at best for "the manipulation of motives to help bring about what must happen anyway from the point of view of objective necessity."

It is clear that this thesis of the autonomous character of technical development is not correct. The pace and *direction* of technical development today depend to a great extent on public investments: in the United States the defense and space administrations are the largest sources of research contracts. I suspect that the situation is similar in the Soviet Union. The assertion that politically consequential decisions are reduced to carrying out the immanent exigencies of disposable techniques and that therefore they can no longer be made the theme of practical considerations, serves in the end merely to conceal preexisting, unreflected social interests and prescientific decisions. As little as we can accept the optimistic convergence of technology and democracy, the pessimistic assertion that technology excludes democracy is just as untenable.

These two answers to the question of how the force of technical control can be made subject to the consensus of acting and transacting citizens are inadequate. Neither of them can deal appropriately with the problem with which we are objectively confronted in the West and East, namely, how we can actually bring under control the preexisting, unplanned relations of technical progress and the social life-world. The tensions between productive forces and social intentions that Marx diagnosed and whose explosive character has intensified in an unforeseen manner in the age of thermonuclear weapons are the consequence of an ironic relation of theory to practice. The direction of technical progress is still largely determined today by social interests that arise autochthonously out of the compulsion of the reproduction of social life without being reflected upon and confronted with the declared political self-understanding of social groups. In consequence, new technical capacities erupt without preparation into existing forms of life-activity and conduct. New potentials for expanded power of technical control make obvious the disproportion between the results of the most organized rationality and unreflected goals, rigidified value systems, and obsolete ideologies.

Today, in the industrially most advanced systems, an energetic attempt must be made consciously to take in hand the mediation between technical progress and the conduct of life in the major industrial societies, a mediation that has previously taken place without direction, as a mere continuation of natural history. This is not the place to discuss the social, economic, and political conditions on which a long-term central research policy would have to depend. It is not enough for a social system to fulfill the conditions of technical rationality. Even if the cybernetic dream of a virtually instinctive self-stabilization could be realized, the value system would have contracted in the meantime to a set of rules for the maximization of power and comfort; it would be equivalent to the biological base value of survival at any cost, that is, ultrastability. Through the unplanned sociocultural consequences of technological progress, the human species has challenged itself to learn not merely to affect its social destiny, but to control it. This challenge of technology cannot be met with technology alone. It is rather a question of setting into motion a politically effective discussion that rationally brings the social potential constituted by technical knowledge and ability into a defined and controlled relation to our practical knowledge and will. On the one hand, such discussion could enlighten those who act politically about the tradition-bound self-understanding of their interests in relation to what is technically possible and feasible. On the other hand, they would be able to judge practically, in the light of their now articulated and newly interpreted needs, the direction and the extent to which they want to develop technical knowledge for the future.

This *dialectic of potential and will* takes place today without reflection in accordance with interests for which public justification is neither demanded nor permitted. Only if we could elaborate this dialectic with political consciousness could we succeed in directing the mediation of technical progress and the conduct of social life, which until now has occurred as an extention of natural history; its conditions being left outside the framework of discussion and planning. The fact that this is a matter for reflection means that it does not belong to the professional competence of specialists. The substance of domination is not dissolved by the power of technical control. To the contrary, the former can simply hide behind the latter. The irrationality of domination, which today has become a collective peril to life, could be mastered only by the development of a political decision-making process tied to the principle of general discussion free from domination. Our only hope for the rationalization of the power structure lies in conditions that favor political power for thought developing through dialogue. The redeeming power of reflection cannot be supplanted by the extension of technically exploitable knowledge.

NOTE

1. Aldous Huxley, *Literature and Science*. New York: Harper & Row, 1963, p. 8 (henceforth LS).

Part II

RECENT PHILOSOPHY OF TECHNOLOGY

Recent philosophy of technology has taken an "empirical turn" away from the transcendental orientation of early philosophy of technology toward a more practical, contextual interpretation. As opposed to the early pessimistic assessments of a singular technological rationality, philosophers since the 1980s tend to view technology empirically and historically—in terms of its actual uses in particular situations. This approach treats technology as a social construction that interacts with other social forces rather than as an autonomous entity with its own unique rationality. Technology is now seen as interdependent in relation to society rather than independent of it. Technology and society form an inseparable pair; neither is intelligible without reference to the other. This new generation of philosophers reinterprets the relationship between technology and society to explore all of the different ways that our devices and systems mediate our lives. Recent philosophy of technology complicates our understanding of what is natural and artificial, social and technical.

In "The Culture of Technology," Arnold Pacey argues that an adequate understanding of technology has to include its technical, organizational, and cultural aspects. The narrowly construed (instrumental) meaning of technology understands it only in terms of technical properties, mechanisms, techniques, and precise knowledge, all of which are universally valid and may be applied anywhere in the world. The broader, and more accurate in Pacey's estimation, understanding of technology includes cultural values; organizational activities; economic, political, and legal regulations; as well as a host of other social factors that form the technological nature of our lives. Pacey defines technology as "the application of scientific and other knowledge to practical tasks by ordered systems that involve people and organizations, living things and machines." The virtue of this definition is that it broadens our understanding of technology to include neglected cultural aspects and hidden background values that not only form the context in which a technology is used but also affect its design, administration, use, and meaning. The study of technology practice must consider a comprehensive network of researchers, technicians, administrators, regulators, policy planners, users, affected nonusers, social use contexts, and belief systems. Through case histories of snowmobiles and water pumps, we see that neither the technical, organizational, nor cultural aspect of technology should be privileged. Each is equally fundamental.

In "Technologies as Forms of Life," Langdon Winner wants to awaken us from our "technological somnambulism." Typically we think about technology in terms of two categories: making and use. Only some people are concerned with making, maintaining, and understanding the workings of technologies (e.g., technicians, engineers, and repair specialists). The rest of us use technologies, presumably in limited ways and without any detailed understanding of the technologies in use. I may use a device correctly or incorrectly or for good or bad purposes, but this had more to do with my activity than with the technology itself. The implicit, commonsense assumption in the use of a device is that technologies are neutral with respect to ends. If we assume that making technology only concerns practitioners and the use of technology is unproblematic, we fail to see how technology creates the underlying structures of social-political life. Unfortunately, too many analyses of technology (by social scientists, historians, and politicians) look only at the side effects, impacts, costs, and benefits of technology without examining the extent to which technological things are involved in establishing the very conditions of social life, shaping who we are and how we live. Winner asserts that a philosophy of technology should critically examine the nature and significance of artificial aids to human activity in order to uncover the crucial role they play in transforming our lives.

Winner claims a better model for understanding technology can be found in Wittgenstein and Marx. From Wittgenstein we get the concept of a "form of life," or the background and context of meaningful activity. Technology is a part of the cultural practices that shape our understanding and pattern our actions. From Marx we get the concept of "historical materialism," the insight that changes in material conditions change human activities. The philosophies of Wittgenstein and Marx direct our attention to ways that technologies are woven into the fabric of our lives. They remind us, Winner states, that social activity is "an on-going process of world-making." The relevant questions to ask about technology are, What kind of world are we making? What are the forms of life that emerge? How are they interrelated? What kind of social relations are made through technologies? Ultimately Winner's concerns are political. He reminds us that technical choices matter as much as political choices in designing and shaping a world.

Albert Borgmann's "Focal Things and Practices," excerpted from his *Technology and the Character of Contemporary Life,* develops Heidegger's analysis by specifying in greater detail what it would mean to launch a "reform of technology." Borgmann's key distinction is between "things" and "devices." To encounter a thing is to engage it fully and to participate in its "world"—all of the social dimensions of using and experiencing something. A device is merely an instrument for producing a commodity and what the device is for. The device functions inconspicuously by disburdening us and making a commodity available. The "promise of modern technology" Borgmann explains is that the use of devices will free us from the misery and work imposed on us by nature and social pressures, and will make our lives better by liberating and enriching our experience. However, Borgmann points out that technology has failed to live up to its promise of liberation, because it is silent as to the ends, purposes, and goods that we desire. To reform technology, we need to revive *focal things* and *focal practices.* Like a hearth, a focal thing gathers us: it centers and engages people, technologies, and places. Above all, focal things and practices contain within them a vision of the good life missing from the device paradigm.

Focal things also direct our attention to devices by making their role in our lives conspicuous. Borgmann's examples of focal practices that revise our relationship to technology include performing music (rather than listening to it on a device), running (rather than working out at a gym), and enjoying a traditional home-cooked meal with friends and family (rather than consuming fast food alone). These focal activities combine humans, technologies, ends, and means in a more engaging and satisfying way than the mere use of devices. Borgmann agrees with Heidegger that we need to question our traditional notion of technology and recognize its true essence in order to appreciate focal things. Like Heidegger, Borgmann invites us to see through the pervasiveness and self-reinforcing patterns of technology. Nonfocal things and devices are perfectly fine, Borgmann states, but as a complete lifestyle it is an empty, shallow, pointless life. But he moves beyond Heidegger by suggesting more practical and contemporary ways to reform technology.

Don Ihde argues in "A Phenomenology of Technics," excerpted from *Technology and the Lifeworld: From Garden to Earth,* that human life has always been suffused with technology. He makes no sweeping claims about technology as such. Instead, Ihde provides a perspective and framework to analyze our experience of technology. The method of analysis is phenomenology, a descriptive method premised on the idea that experience is always relational. The "intentionality of consciousness" of which Ihde writes means that every instance of experience has its reference or direction toward what is experienced. The aim of a phenomenological description is to identify the essential or invariant features of experienced phenomena. Ihde undertakes a phenomenological description of several sets of human–technology relations in order to analyze how technologies often mediate and transform our experience. A phenomenology of human–technology relations shows that the structural dimensions of technological mediation produce a range of possible experiences. Technological objects are ambiguous; their meaning is relative to use and context. We cannot understand how technology affects us unless we treat technological objects as being in relation to us, and not just objects in themselves.

According to Ihde, when we consider the ways our experience is mediated by technological objects, we find several unique sets of human–technology relations, each positioning us in a slightly different relation to technology. One set of relations he calls "embodiment relations" with devices we use to experience the world and that, at the same time, alter and modify our perception of the world. (These include devices like reading glasses, hearing aids, writing implements, and handheld tools.) A second set he calls "hermeneutic relations" that involve instruments that we read rather than use as tools. (These include devices like clocks, thermometers, altimeters, spectrographic devices, and other technologies with visual displays that must be interpreted to be understood.) A third set is "alterity relations," in which technologies appear as "other" to us, possessing a kind of independence from humans as creators and users. (These include things like toys, robots, ATM machines, video games, and visual technologies that we interact with as if they were autonomous beings.) The final set is "background relations," in which technologies form the context of experience in a way that is seldom consciously perceived. (This set of devices includes things like the lighting, air conditioning, clothing, shelter, and automatic machines that operate in the background, subtly affecting our experience.) The virtue of Ihde's phenomenological approach is its attention to detail and its emphasis on an active sense of the human–technology relationship.

Donna Haraway, in "A Cyborg Manifesto: Science, Technology, and Socialist-Feminism in the Late Twentieth Century," champions human–machine hybrids as an ironic model for overcoming the dualisms that haunt our thinking about technology, politics, and human identity—for example, man and woman, human and machine, nature and culture, natural and artificial, and reason and emotion. Because cyborg imagery includes both human and machine, it suggests ways of overcoming theories that propose false dichotomies and privilege one part of the dyad over the other. The *myth of the cyborg* also reminds us to avoid demonizing science and technology as agents of dehumanization. Haraway exhorts us to overcome our desire for organic wholeness, original unity, and a nostalgic return to a pure, uncorrupted state of nature.

The cyborg myth helps us recognize that the very ideas of humanity, nature, and technology are socially constructed realities. There is nothing essential or necessary about them. Once we are liberated from dualistic thinking and nostalgic longing for unity, Haraway writes, we are free to take "pleasure in the confusion of boundaries and responsibility in their construction." The political consequence of cyborg ontology is a model of solidarity—not unity—among diverse groups. Instead of trying to make our world more humane—modeled after a single notion of what a human is—we should instead try to become more like cyborgs. Haraway calls into question the very distinction between natural and artificial. In so doing, she affirms a positive role for technological mediation in all aspects of lives—right down to our bodies and identities.

Bruno Latour, in "A Collective of Humans and Nonhumans," also seeks to overcome the dualistic paradigm that defines modernity. His theory of a "sociotechnical collective" implies networks of relations among humans and machines that overcome the traditional distinction between subjectivity and objectivity. Writing in response to the "science wars" that have occurred between sociologists (who claim that scientific facts are mere social constructions) and philosophers of science (who claim that scientific facts at very least aspire to objectivity), Latour argues that both sides miss the point. There has never been such a thing as humanity without technology nor technology without humanity. Consequently, the very idea of the social construction of objectivity in science and technology is incoherent. The idea of social construction presupposes a very conventional modern myth in which scientific and social progress is a function of the increasing differentiation between the domains of objectivity and subjectivity. The modernist narrative of progress involves the differentiation, as Kant and Weber say, of the natural world (including science, technology, and rationality) and the social world (including values, politics, and individuality). But Latour argues that the exact opposite is happening: science, technologies, and society are becoming more closely linked. In place of the modern "master narrative" that separates subjects and object, he proposes a "servant narrative" that shows how "imbroglios of humans and non-humans" occur on an ever increasing scale. The science wars will end when we realize they are based on a false dichotomy.

The alternative *pragmatogony* (mythic origine of machines and devices) describes eleven distinctive layers through which humans and nonhumans exchange their respective properties. Each even-numbered step in the mythical pragmatogony shows how skills and properties relevant to maintaining social relations become relevant for establishing relations with and among nonhumans. The odd-numbered steps go in the opposite direction and show how nonhuman relations are incorporated into human relations. By following this pragmatogony, we come to see how we are *sociotechnical animals*. There is no pure humanity and no pure technology. Once we recognize how intertwined humans and nonhumans are, a number of modernist dichotomies disappear—for example, objectivity and subjectivity, realism and relativism, nature and society, and scientific fact and social understanding. If Latour is right, then we must give up the illusion that humans and technology ever were free and independent from each other.

Andrew Light, in "Ecological Restoration and the Culture of Nature," takes a rather different approach in examining the nature–culture distinction. Light examines from the perspective of environmental philosophy the theoretical and practical importance of the practice of restoring damaged ecological systems. He disagrees with philosophers who view restorations of nature as misguided attempts to replicate nature and restore its original value. These philosophers maintain that restorations are merely human creations like any other artifact. Eric Katz, for example, argues that interventions that aim to recreate and restore nature in fact manipulate, dominate, and ultimately diminish the value of nature. From this perspective, restorations of nature are attempts at control that serve human ends rather than let nature pursue its own development.

Addressing these criticisms, Light grants that a restoration of nature is never the same as the original but suggests that the practice be seen more like a benevolent art restoration that seeks to remedy a past harm than a malicious forgery that attempts to pass off a fake as the real thing. Benevolent restorations neither harm nor dominate nature, nor are they attempts to trick us. Rather, Light argues, they are earnest attempts to correct harm done to the environment by human intervention, to rectify the balance of nature, and to let nature pursue its own interests or natural course. Furthermore, ecological restoration not only restores nature but also restores our moral relationship with the environment. The more people participate in restorations, the more connected we become to nature and the less likely to allow it to be further harmed. The environmental pragmatism that Light endorses sidesteps the nature–culture distinction by granting it and nonetheless affirming the importance of participating together in restoring nature. Nothing but good, Light claims, comes from restoring nature.

In "Democratic Rationalization: Technology, Power, and Freedom," Andrew Feenberg makes a strong case against understanding technological rationality as singular and dystopic. His proposed alternative to a theory of technological rationalization (the spread of calculating reason to every aspect of social life) is a version of "constructivism" that maintains that facts and artifacts are socially fabricated. People make decisions about technologies based on a number of contingent factors, not just according to scientific and technical criteria. Technology is flexible; it adapts to a variety of social demands. It is also a social object open to multiple interpretations, not a single technically explicable function. Yet technology is still partially composed by technological rationality incorporated into the very structure of machines. The "double aspect theory" that Feenberg proposes explains that social meaning and functional (i.e., technological) rationality are intertwined as two inextricable aspects of the same object.

In order to understand technology, we should endeavor to uncover the social horizon in which a device is produced, remove any illusion of its necessity, and expose the ever present relativity of technical choices. The danger of the apparent neutrality of functional rationality is that it is often enlisted in support of a hegemony (i.e., a specific mode of social and political control). A critical theory of technology such as Feenberg's shows how the "technical code" of a device or system is the sedimentation of social values and interests in the form of technical rules and procedures that secure power and advantage for a hegemony. Feenberg concludes with an appeal for a "democratic rationalization" that would incorporate democratic values into industrial design. We have the potential to change the values designed into our technologies once we recognize the indeterminate character of technical design, use, and transformation as well as the social and political stakes of technical design. Then we can begin to criticize our society's culture of technology and imagine alternatives that would foster a more democratic, meaningful, and livable environment.

7

The Culture of Technology

Arnold Pacey

QUESTIONS OF NEUTRALITY

Winter sports in North America gained a new dimension during the 1960s with the introduction of the snowmobile. Ridden like a motorcycle, and having handlebars for steering, this little machine on skis gave people in Canada and the northern United States extra mobility during their long winters. Snowmobile sales doubled annually for awhile, and in the boom year of 1970–71 almost half a million were sold. Subsequently the market dropped back, but snowmobiling had established itself, and organized trails branched out from many newly prosperous winter holiday resorts. By 1978, there were several thousand miles of public trails, marked and maintained for snowmobiling, about half in the province of Quebec.

Although other firms had produced small motorized toboggans, the type of snowmobile which achieved this enormous popularity was only really born in 1959, chiefly on the initiative of Joseph-Armand Bombardier of Valcourt, Quebec.[1] He had experimented with vehicles for travel over snow since the 1920s, and had patented a rubber-and-steel crawler track to drive them. His first commercial success, which enabled his motor repair business to grow into a substantial manufacturing firm, was a machine capable of carrying seven passengers which was on the market from 1936. He had other successes later, but nothing that caught the popular imagination like the little snowmobile of 1959, which other manufacturers were quick to follow up.

However, the use of snowmobiles was not confined to the North American tourist centres. In Sweden, Greenland and the Canadian Arctic, snowmobiles have now become part of the equipment on which many communities depend for their livelihood. In Swedish Lapland they are used for reindeer herding. On Canada's Banks Island they have enabled Eskimo trappers to continue providing their families' cash income from the traditional winter harvest of fox furs.

Such use of the snowmobile by people with markedly different cultures may seem to

From Arnold Pacey, *The Culture of Technology*, Cambridge, Mass.: MIT Press, 1983, pp. 1–8, 10–12.

illustrate an argument very widely advanced in discussions of problems associated with technology. This is the argument which states that technology is culturally, morally and politically neutral—that it provides tools independent of local value-systems which can be used impartially to support quite different kinds of lifestyle.

Thus in the world at large, it is argued that technology is 'essentially amoral, a thing apart from values, an instrument which can be used for good or ill'.[2] So if people in distant countries starve; if infant mortality within the inner cities is persistently high; if we feel threatened by nuclear destruction or more insidiously by the effects of chemical pollution, then all that, it is said, should not be blamed on technology, but on its misuse by politicians, the military, big business and others.

The snowmobile seems the perfect illustration of this argument. Whether used for reindeer herding or for recreation, for ecologically destructive sport, or to earn a basic living, it is the same machine. The engineering principles involved in its operation are universally valid, whether its users are Lapps or Eskimos, Dene (Indian) hunters, Wisconsin sportsmen, Quebecois vacationists, or prospectors from multinational oil companies. And whereas the snowmobile has certainly had a social impact, altering the organization of work in Lapp communities, for example, it has not necessarily influenced basic cultural values. The technology of the snowmobile may thus appear to be something quite independent of the lifestyles of Lapps or Eskimos or Americans.

One look at a modern snowmobile with its fake streamlining and flashy colours suggests another point of view. So does the advertising which portrays virile young men riding the machines with sexy companions, usually blonde and usually riding pillion. The Eskimo who takes a snowmobile on a long expedition in the Arctic quickly discovers more significant discrepancies. With his traditional means of transport, the dog-team and sledge, he could refuel as he went along by hunting for his dogs' food. With the snowmobile he must take an ample supply of fuel and spare parts; he must be skilled at doing his own repairs and even then he may take a few dogs with him for emergency use if the machine breaks down. A vehicle designed for leisure trips between well-equipped tourist centres presents a completely different set of servicing problems when used for heavier work in more remote areas. One Eskimo 'kept his machine in his tent so it could be warmed up before starting in the morning, and even then was plagued by mechanical failures'.[3] There are stories of other Eskimos, whose mechanical aptitude is well known, modifying their machines to adapt them better to local use.

So is technology culturally neutral? If we look at the construction of a basic machine and its working principles, the answer seems to be yes. But if we look at the web of human activities surrounding the machine, which include its practical uses, its role as a status symbol, the supply of fuel and spare parts, the organized tourist trails, and the skills of its owners, the answer is clearly no. Looked at in this second way, technology is seen as a part of life, not something that can be kept in a separate compartment. If it is to be of any use, the snowmobile must fit into a pattern of activity which belongs to a particular lifestyle and set of values.

The problem here, as in much public discussion, is that 'technology' has become a catchword with a confusion of different meanings. Correct usage of the word in its original sense seems almost beyond recovery, but consistent distinction between different levels of meaning is both possible and necessary. In medicine, a distinction of the kind required is often made by talking about 'medical practice' when a general term is required, and employing the phrase 'medical science' for the more strictly technical aspects of the subject. Sometimes, references to 'medical practice' only denote the organization necessary to use medical knowledge and skill for treating patients. Sometimes, however, and more usefully, the term refers to the whole activity of medicine, including its basis in technical knowledge, its organization, and its cultural aspects. The latter comprise the doctor's sense of vocation, his personal values and satisfactions, and the ethical code of his profession. Thus 'practice' may be a broad and inclusive concept.

Once this distinction is established, it is clear that although medical practice differs quite markedly from one country to another, medical science consists of knowledge and techniques which are likely to be useful in many countries. It is true that medical science in many western countries is biased by the way that most research is centred on large hospitals. Even so, most of the basic knowledge is widely applicable and relatively independent of local cultures. Similarly, the design of snowmobiles reflects the way technology is practised in an industrialized country—standardized machines are produced which neglect some of the special needs of Eskimos and Lapps. But one can still point to a substratum of knowledge, technique and underlying principle in engineering which has universal validity, and which may be applied anywhere in the world.

We would understand much of this more clearly, I suggest, if the concept of practice were to be used in all branches of technology as it has traditionally been used in medicine. We might then be better able to see which aspects of technology are tied up with cultural values, and which aspects are, in some respects, value-free. We would be better able to appreciate technology as a human activity and as part of life. We might then see it not only as comprising machines, techniques and crisply precise knowledge, but also as involving characteristic patterns of organization and imprecise values.

Medical practice may seem a strange exemplar for the other technologies, distorted as it so often seems to be by the lofty status of the doctor as an expert. But what is striking to anybody more used to engineering is that medicine has at least got concepts and vocabulary which allow vigorous discussion to take place about different ways of serving the community. For example, there are phrases such as 'primary health care' and 'community medicine' which are sometimes emphasized as the kind of medical practice to be encouraged wherever the emphasis on hospital medicine has been pushed too far. There are also some interesting adaptations of the language of medical practice. In parts of Asia, para-medical workers, or paramedics, are now paralleled by 'para-agros' in agriculture, and the Chinese barefoot doctors have inspired the suggestion that barefoot technicians could be recruited to deal with urgent problems in village water supply. But despite these occasional borrowings, discussion about practice in most branches of technology has not progressed very far.

PROBLEMS OF DEFINITION

In defining the concept of technology-practice more precisely, it is necessary to think with some care about its human and social aspect. Those who write about the social relations and social control of technology tend to focus particularly on organization. In particular, their emphasis is on planning and administration, the management of research, systems for regulation of pollution and other abuses, and professional organization among scientists and technologists. These are important topics, but there is a wide range of other human content in technology-practice which such studies often neglect, including personal values and individual experience of technical work.

To bring all these things into a study of technology-practice may seem likely to make it bewilderingly comprehensive. However, by remembering the way in which medical practice has a technical and ethical as well as an organizational element, we can obtain a more orderly view of what technology-practice entails. To many politically minded people, the *organizational aspect* may seem most crucial. It represents many facets of administration and public policy; it relates to the activities of designers, engineers, technicians, and production workers, and also concerns the users and consumers of whatever is produced. Many other people, however, identify technology with its *technical aspect,* because that has to do with machines, techniques, knowledge and the essential activity of making things work.

Beyond that, though, there are values which influence the creativity of designers and inventors. These, together with the various beliefs and habits of thinking which are characteristic of technical and scientific activity, can be indicated by talking about an ideological or *cultural aspect* of technology-practice. There is some risk of ambiguity here, because strictly speaking, ideology, organization, technique and tools are all aspects of the culture of a society. But in common speech, culture refers to values, ideas and creative activity, and it is convenient to use the term with this meaning. It is in this sense that the title of this chapter refers to the cultural aspect of technology-practice.

All these ideas are summarized by Figure 7.1, in which the whole triangle stands for the concept of technology-practice and the corners represent its organizational, technical and cultural aspects. This diagram is also intended to illustrate how the word technology is sometimes used by people in a restricted sense, and sometimes with a more general meaning. When technology is discussed in the more restricted way, cultural values and organizational factors are regarded as external to it. Technology is then identified entirely with its technical aspects, and the words 'technics' or simply 'technique' might often be more appropriately used. The more general meaning of the word, however, can be equated with technology-practice, which clearly is not value-free and politically neutral, as some people say it should be.

Some formal definitions of technology hover uncertainly between the very general and

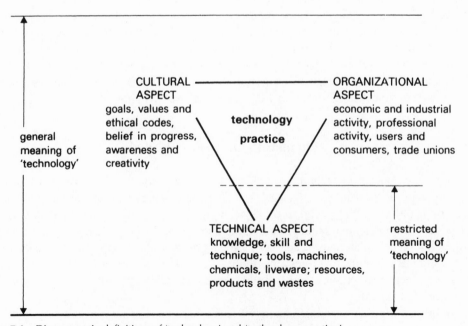

Figure 7.1   Diagrammatic definitions of 'technology' and 'technology practice'

the more restricted usage. Thus J. K. Galbraith defines technology as 'the systematic application of scientific or other organized knowledge to practical tasks'.[4] This sounds like a fairly narrow definition, but on reading further one finds that Galbraith thinks of technology as an activity involving complex organization and value-systems. In view of this, other authors have extended Galbraith's wording.

For them a definition which makes explicit the role of people and organizations as well as hardware is one which describes technology as 'the application of scientific and other organized knowledge to practical tasks by . . . ordered systems that involve people and machines'.[5] In most respects, this sums up technology-practice very well. But some branches of technology deal with processes dependent on living organisms. Brewing, sewage treatment and the new biotechnologies are examples. Many people also include aspects of agriculture, nutrition and medicine in their concept of technology. Thus our definition needs to be enlarged further to include 'liveware' as well as hardware; technology-practice is thus *the application of scientific and other knowledge to practical tasks by ordered systems that involve people and organizations, living things and machines.*

This is a definition which to some extent includes science within technology. That is not, of course, the same as saying that science is merely one facet of technology with no purpose of its own. The physicist working on magnetic materials or semiconductors may have an entirely abstract interest in the structure of matter, or in the behaviour of electrons in solids. In that sense, he may think of himself as a pure scientist, with no concern at all for industry and technology. But it is no coincidence that the magnetic materials he works on are precisely those that are used in transformer cores and computer memory devices, and that the semiconductors investigated may be used in microprocessors. The scientist's choice of research subject is inevitably influenced by technological requirements, both through material pressures and also via a climate of opinion about what subjects are worth pursuing. And a great deal of science is like this, with goals that are definitely outside technology-practice, but with a practical function within it.

Given the confusion that surrounds usage of the word 'technology', it is not surprising that there is also confusion about the two adjectives 'technical' and 'technological'. Economists make their own distinction, defining change of technique as a development based on choice from a range of known methods, and technological change as involving fundamentally new discovery or invention. This can lead to a distinctive use of the word 'technical'. However, I shall employ this adjective when I am referring solely to the technical aspects of practice as defined by Figure 7.1. For example, the application of a chemical water treatment to counteract river pollution is described here as a 'technical fix' (not a 'technological fix'). It represents an attempt to solve a problem by means of technique alone, and ignores possible changes in practice that might prevent the dumping of pollutants in the river in the first place.

By contrast, when I discuss developments in the practice of technology which include its organizational aspects, I shall describe these as 'technological developments', indicating that they are not restricted to technical form. The terminology that results from this is usually consistent with everyday usage, though not always with the language of economics.

EXPOSING BACKGROUND VALUES

One problem arising from habitual use of the word technology in its more restricted sense is that some of the wider aspects of technology-practice have come to be entirely forgotten. Thus behind the public debates about resources and the environment, or about world food supplies, there is a tangle of unexamined beliefs and values, and a basic confusion about what technology is for. Even on a practical level, some projects fail to get more than half way to solving the problems they address, and end up as unsatisfactory technical fixes, because important organizational factors have been ignored. Very often the users of equipment and their patterns of organization are largely forgotten.

My aim is to strip away some of the attitudes that restrict our view of technology in order to expose these neglected cultural aspects. With the snowmobile, a first step was to look at different ways in which the use and maintenance of the machine is organized in different communities. This made it clear that a machine designed in response to the values of one culture needed a good deal of effort to make it suit the purposes of another.

A further example concerns the apparently simple hand-pumps used at village wells in India. During a period of drought in the 1960s, large power-driven drilling rigs were brought in to reach water at considerable depths in the ground by means of bore-holes. It was at these new wells that most of the hand-pumps were installed. By 1975 there were some 150,000 of them, but surveys showed that at any one time as many as two-thirds had broken down. New pumps sometimes failed within three or four weeks of installation. Engineers identified several faults, both in the design of the pumps and in standards of manufacture. But although these defects were corrected, pumps continued to go wrong. Eventually it was realized that the breakdowns were not solely an engineering problem. They were also partly an administrative or management issue, in that arrangements for servicing the pumps were not very effective. There was another difficulty, too, because in many villages, nobody felt any personal responsibility for looking after the pumps. It was only when these things were tackled together that pump performance began to improve.

This episode and the way it was handled illustrate very well the importance of an integrated appreciation of technology-practice. A breakthrough only came when all aspects of the administration, maintenance and technical design of the pump were thought out in relation to one another. What at first held up solution of the problem was a view of technology which began and ended with the machine—a view which, in another similar context, has been referred to as tunnel vision in engineering.

Any professional in such a situation is likely to experience his own form of tunnel vision. If a management consultant had been asked about the hand-pumps, he would have seen the administrative failings of the maintenance system very quickly, but might not have recognized that mechanical improvements to the pumps were required. Specialist training inevitably restricts people's approach to problems. But tunnel vision in attitudes to technology extends far beyond those who have had specialized training; it also affects policy-making, and influences popular expectations. People in many walks of life tend to focus on the tangible, technical aspect of any practical problem, and then to think that the extraordinary capabilities of modern technology ought to lead to an appropriate 'fix'. This attitude seems to apply to almost everything from inner city decay to military security, and from pollution to a cure for cancer. But all these issues have a social component. To hope for a technical fix for any of them that does not also involve social and cultural measures is to pursue an illusion.

So it was with the hand-pumps. The technical aspect of the problem was exemplified by poor design and manufacture. There was the organizational difficulty about maintenance. Also important, though, was the cultural aspect of technology as it was practised by the engineers involved. This refers, firstly, to the engineers' way of thinking, and the tunnel vision it led to; secondly, it indicates conflicts of values between highly trained engineers and the relatively uneducated people of the Indian countryside whom the pumps were meant to benefit. The local people probably had exaggerated expectations of the pumps as the products of an all-powerful, alien technology, and did not see them as vulnerable bits of equipment needing care in use and protection from damage; in addition, the local people would have their own views about hygiene and water use.

Many professionals in technology are well aware that the problems they deal with have social implications, but feel uncertainty about how these should be handled. To deal only with the technical detail and leave other aspects on one side is the easier option, and after all, is what they are trained for. With the hand-pump problem, an important step forward came when one of the staff of a local water development unit started looking at the case-histories of individual pump breakdowns. It was then relatively easy for him to pass from a technical review of components which were worn or broken to looking at the social context of each pump. He was struck by the way some pumps had deteriorated but others had not. One well-cared-for pump was locked up during certain hours; another was used by the family of a local official; others in good condition were in places where villagers had mechanical skills and were persistent with improvised repairs. It was these specific details that enabled suggestions to be made about the reorganization of pump maintenance.[6]

A first thought prompted by this is that a training in science and technology tends to focus on general principles, and does not prepare one to look for specifics in quite this way. But the human aspect of technology—its organization and culture—is not easily reduced to general principles, and the investigator with an eye for significant detail may sometimes learn more than the professional with a highly systematic approach.

A second point concerns the way in which the cultural aspect of technology-practice tends to be hidden beneath more obvious and more practical issues. Behind the tangible aspect of the broken handpumps lies an administrative problem concerned with maintenance. Behind that lies a problem of political will—the official whose family depended on one of the pumps was somehow well served. Behind that again were a variety of questions concerning cultural values regarding hygiene, attitudes to technology, and the outlook of the professionals involved.

This need to strip away the more obvious features of technology-practice to expose the background values is just as evident with new technology in western countries. Very often concern will be expressed about the health risk of a new device when people are worried about more intangible issues, because health risk is partly a technical question that is easy to discuss openly. A relatively minor technical problem affecting health may thus become a proxy for deeper worries about the way technology is practised which are more difficult to discuss.

An instance of this is the alleged health risks associated with visual display units (VDUs) in computer installations. Careful research has failed to find any real hazard except that operators may suffer eyestrain and fatigue. Yet complaints about more serious problems continue, apparently because they can be discussed seriously with employers while misgivings about the overall systems are more difficult to raise. Thus a negative reaction to new equipment may be expressed in terms of a fear of 'blindness, sterility, etc.', because in our society, this is regarded as a legitimate reason for rejecting it. But to take such fears at face value will often be to ignore deeper, unspoken anxieties about 'deskilling, inability to handle new procedures, loss of control over work'.[7]

Here, then, is another instance where, beneath the overt technical difficulty there are questions about the organizational aspect of technology—especially the organization of specific tasks. These have political connotations, in that an issue about control over work raises

Arnold Pacey

questions about where power lies in the work-place, and perhaps ultimately, where it lies within industrial society. But beyond arguments of that sort, there are even more basic values about creativity in work and the relationship of technology and human need.

In much the same way as concern about health sometimes disguises work-place issues, so the more widely publicized environmental problems may also hide underlying organizational and political questions. C. S. Lewis once remarked that 'Man's power over Nature often turns out to be a power exerted by some men over other men with Nature as its instrument', and a commentator notes that this, 'and not the environmental dilemma as it is usually conceived', is the central issue for technology.[8] As such, it is an issue whose political and social ramifications have been ably analysed by a wide range of authors.[9]

Even this essentially political level of argument can be stripped away to reveal another cultural aspect of technology. If we look at the case made out in favour of almost any major project—a nuclear energy plant, for example—there are nearly always issues concerning political power behind the explicit arguments about tangible benefits and costs. In a nuclear project, these may relate to the power of management over trade unions in electricity utilities; or to prestige of governments and the power of their technical advisers. Yet those who operate these levers of power are able to do so partly because they can exploit deeper values relating to the so-called technological imperative, and to the basic creativity that makes innovation possible. This, I argue, is a central part of the culture of technology, and its analysis occupies several chapters in this book. If these values underlying the technological imperative are understood, we may be able to see that here is a stream of feeling which politicians can certainly manipulate at times, but which is stronger than their short-term purposes, and often runs away beyond their control.

NOTES

1. M. B. Doyle, *An Assessment of the Snowmobile Industry and Sport,* Washington DC: International Snowmobile Industry Association, 1978, pp. 14, 47; on Joseph-Armand Bombardier, see Alexander Ross, *The Risk Takers,* Toronto: Macmillan and the Financial Post, 1978, p. 155.

2. R. A. Buchanan, *Technology and Social Progress,* Oxford: Pergamon Press, 1965, p. 163.

3. P. J. Usher, 'The use of snowmobiles for trapping on Banks Island', *Arctic* (Arctic Institute of North America), 25, 1972, p. 173.

4. J. K. Galbraith, *The New Industrial State,* 2nd British edition, London: André Deutsch, 1972, chapter 2.

5. John Naughton, 'Introduction: technology and human values', in *Living with Technology: a Foundation Course,* Milton Keynes: The Open University Press, 1979.

6. Charles Heineman, 'Survey of hand-pumps in Vellakovil . . .', unpublished report, January 1975, quoted by Arnold Pacey, *Hand-pump Maintenance,* London: Intermediate Technology Publications, 1977.

7. Leela Damodaran, 'Health hazards of VDUs?—Chairman's introduction', conference at Loughborough University of Technology, 11 December 1980.

8. Quoted by Peter Hartley, 'Educating engineers', *The Ecologist,* 10 (10), December 1980, p. 353.

9. E.g. David Elliott and Ruth Elliott, *The Control of Technology,* London and Winchester: Wykeham, 1976.

8

Technologies as Forms of Life

Langdon Winner

It is reasonable to suppose that a society thoroughly committed to making artificial realities would have given a great deal of thought to the nature of that commitment. One might expect, for example, that the philosophy of technology would be a topic widely discussed by scholars and technical professionals, a lively field of inquiry often chosen by students at our universities and technical institutes. One might even think that the basic issues in this field would be well defined, its central controversies well worn. However, such is not the case. At this late date in the development of our industrial/technological civilization the most accurate observation to be made about the philosophy of technology is that there really isn't one.

The basic task for a philosophy of technology is to examine critically the nature and significance of artificial aids to human activity. That is its appropriate domain of inquiry, one that sets it apart from, say, the philosophy of science. Yet if one turns to the writings of twentieth-century philosophers, one finds astonishingly little attention given to questions of that kind. The eight-volume *Encyclopedia of Philosophy,* a recent compendium of major themes in various traditions of philosophical discourse, contains no entry under the category "technology."[1] Neither does that work contain enough material under possible alternative headings to enable anyone to piece together an idea of what a philosophy of technology might be.

True, there are some writers who have taken up the topic. The standard bibliography in the philosophy of technology lists well over a thousand books and articles in several languages by nineteenth- and twentieth-century authors.[2] But reading through the material listed shows, in my view, little of enduring substance. The best writing on this theme comes to us from a few powerful thinkers who have encountered the subject in the midst of much broader and ambitious investigations—for example, Karl Marx in the development of his theory of historical materialism or Martin Heidegger as an aspect of his theory of ontology. It may be, in fact, that the philosophy is best seen as a derivative of more fundamental questions. For despite the fact that nobody would deny its importance to an adequate understanding of the human condi-

From *Epistemology, Methodology and the Social Sciences*, pp. 249–263, eds. Cohen and Wartofsky. Kluwer Academic Publishers. Used with kind permission from Kluwer Academic Publishers.

tion, technology has never joined epistemology, metaphysics, esthetics, law, science, and politics as a fully respectable topic for philosophical inquiry.

Engineers have shown little interest in filling this void. Except for airy pronouncements in yearly presidential addresses at various engineering societies, typically ones that celebrate the contributions of a particular technical vocation to the betterment of humankind, engineers appear unaware of any philosophical questions their work might entail. As a way of starting a conversation with my friends in engineering, I sometimes ask, "What are the founding principles of your discipline?" The question is always greeted with puzzlement. Even when I explain what I am after, namely, a coherent account of the nature and significance of the branch of engineering in which they are involved, the question still means nothing to them. The scant few who raise important first questions about their technical professions are usually seen by their colleagues as dangerous cranks and radicals. If Socrates' suggestion that the "unexamined life is not worth living" still holds, it is news to most engineers.[3]

TECHNOLOGICAL SOMNAMBULISM

Why is it that the philosophy of technology has never really gotten under way? Why has a culture so firmly based upon countless sophisticated instruments, techniques, and systems remained so steadfast in its reluctance to examine its own foundations? Much of the answer can be found in the astonishing hold the idea of "progress" has exercised on social thought during the industrial age. In the twentieth century it is usually taken for granted that the only reliable sources for improving the human condition stem from new machines, techniques, and chemicals. Even the recurring environmental and social ills that have accompanied technological advancement have rarely dented this faith. It is still a prerequisite that the person running for public office swear his or her unflinching confidence in a positive link between technical development and human well-being and affirm that the next wave of innovations will surely be our salvation.

There is, however, another reason why the philosophy of technology has never gathered much steam. According to conventional views, the human relationship to technical things is too obvious to merit serious reflection. The deceptively reasonable notion that we have inherited from much earlier and less complicated times divides the range of possible concerns about technology into two basic categories: *making* and *use*. In the first of these our attention is drawn to the matter of "how things work" and of "making things work." We tend to think that this is a fascination of certain people in certain occupations, but not for anyone else. "How things work" is the domain of inventors, technicians, engineers, repairmen, and the like who prepare artificial aids to human activity and keep them in good working order. Those not directly involved in the various spheres of "making" are thought to have little interest in or need to know about the materials, principles, or procedures found in those spheres.

What the others do care about, however, are tools and uses. This is understood to be a straightforward matter. Once things have been made, we interact with them on occasion to achieve specific purposes. One picks up a tool, uses it, and puts it down. One picks up a telephone, talks on it, and then does not use it for a time. A person gets on an airplane, flies from point A to point B, and then gets off. The proper interpretation of the meaning of technology in the mode of use seems to be nothing more complicated than an occasional, limited, and nonproblematic interaction.

The language of the notion of "use" also includes standard terms that enable us to interpret technologies in a range of moral contexts. Tools can be "used well or poorly" and for "good or bad purposes"; I can use my knife to slice a loaf of bread or to stab the next person that walks by. Because technological objects and processes have a promiscuous utility, they are taken to be fundamentally neutral as regards their moral standing.

The conventional idea of what technology is and what it means, an idea powerfully reinforced by familiar terms used in everyday language, needs to be overcome if a critical philosophy of technology is to move ahead. The crucial weakness of the conventional idea is that it disregards the many ways in which technologies provide structure for human activity. Since, according to accepted wisdom, patterns that take shape in the sphere of "making" are of interest to practitioners alone, and since the very essence of "use" is its occasional, innocuous, nonstructuring occurrence, any further questioning seems irrelevant.[4]

If the experience of modern society shows us anything, however, it is that technologies are not merely aids to human activity, but also powerful forces acting to reshape that activity and its meaning. The introduction of a robot to an industrial workplace not only increases productivity, but often radically changes the process of production, redefining what "work" means in that setting. When a sophisticated new technique or instrument is adopted in medical practice, it transforms not only what doctors do, but also the ways people think about health, sickness, and medical care. Widespread alterations of this kind in techniques of communication, transportation, manufacturing, agriculture, and the like are largely what distinguish our times from early periods of human history. The kinds of things we are apt to see as "mere" technological entities become much more interesting and problematic if we begin to observe how broadly they are involved in conditions of social and moral life.

It is true that recurring patterns of life's activity (whatever their origins) tend to become unconscious processes taken for granted. Thus, we do not pause to reflect upon how we speak a language as we are doing so or the motions we go through in taking a shower. There is, however, one point at which we may become aware of a pattern taking shape—the very first time we encounter it. An opportunity of that sort occurred several years ago at the conclusion of a class I was teaching. A student came to my office on the day term papers were due and told me his essay would be late. "It crashed this morning," he explained. I immediately interpreted this as a "crash" of the conceptual variety, a flimsy array of arguments and observations that eventually collapses under the weight of its own ponderous absurdity. Indeed, some of my own papers have "crashed" in exactly that manner. But this was not the kind of mishap that had befallen this particular fellow. He went on to explain that his paper had been composed on a computer terminal and that it had been stored in a time-sharing minicomputer. It sometimes happens that the machine "goes down" or "crashes," making everything that happens in and around it stop until the computer can be "brought up," that is, restored to full functioning.

As I listened to the student's explanation, I realized that he was telling me about the facts of a particular form of activity in modern life in which he and others similarly situated were already involved and that I had better get ready for. I remembered J. L. Austin's little essay "A Plea for Excuses" and noticed that the student and I were negotiating one of the boundaries of contemporary moral life—where and how one gives and accepts an excuse in a particular technology-mediated situation.[5] He was, in effect, asking me to recognize a new world of parts and pieces and to acknowledge appropriate practices and expectations that hold in that world. From then on, a knowledge of this situation would be included in my understanding of not only "how things work" in that generation of computers, but also how we do things as consequence, including which rules to follow when the machines break down. Shortly thereafter I got used to computers crashing, disrupting hotel reservations, banking, and other everyday transactions; eventually, my own papers began crashing in this new way.

Some of the moral negotiations that accompany technological change eventually become matters of law. In recent times, for example, a number of activities that employ computers as their operating medium have been legally defined as "crimes." Is unauthorized access to a computerized data base a criminal offense? Given the fact that electronic information is in the strictest sense intangible, under what conditions is it "property" subject to theft? The law has had to stretch and reorient its traditional categories to encompass such problems, creating whole new classes of offenses and offenders.

The ways in which technical devices tend to engender distinctive worlds of their own can be seen in a more familiar case. Picture two men traveling in the same direction along a street on a peaceful, sunny day, one of them afoot and the other driving an automobile. The pedestrian has a certain flexibility of movement: he can pause to look in a shop window, speak to passersby, and reach out to pick a flower from a sidewalk garden. The driver, although he has the potential to move much faster, is constrained by the enclosed space of the automobile, the physical dimensions of the highway, and the rules of the road. His realm is spatially structured by his intended destination, by a periphery of more-or-less irrelevant objects (scenes for occasional side glances), and by more important objects of various kinds—moving and parked cars, bicycles, pedestrians, street signs, etc., that stand in his way. Since the first rule of good driving is to avoid hitting things, the immediate environment of the motorist becomes a field of obstacles.

Imagine a situation in which the two persons are next-door neighbors. The man in the automobile observes his friend strolling along the street and wishes to say hello. He slows down, honks his horn, rolls down the window, sticks out his head, and shouts across the street. More likely than not the pedestrian will be startled or annoyed by the sound of the horn. He looks around to see what's the matter and tries to recognize who can be yelling at him across the way. "Can you come to dinner Saturday night?" the driver calls out over the street noise. "What?" the pedestrian replies, straining to understand. At that moment another car to the rear begins honking to break up the temporary traffic jam. Unable to say anything more, the driver moves on.

What we see here is an automobile collision of sorts, although not one that causes bodily injury. It is a collision between the *world* of the driver and that of the pedestrian. The attempt to extend a greeting and invitation, ordinarily a simple gesture, is complicated by the presence of a technological device and its standard operating conditions. The communication between the two men is shaped by an incompatibility of the form of locomotion known as walking and a much newer one, automobile driving. In cities such as Los Angeles, where the physical landscape and prevailing social habits assume everyone drives a car, the simple act of walking can be cause for alarm. The U.S. Supreme Court decided one case involving a young man who enjoyed taking long walks late at night through the streets of San Diego and was repeatedly arrested by police as a suspicious character. The Court decided in favor of the pedestrian, noting that he had not been engaged in burglary or any other illegal act. Merely traveling by foot is not yet a crime.[6]

Knowing how automobiles are made, how they operate, and how they are used and knowing about traffic laws and urban transportation policies does little to help us understand how automobiles affect the texture of modern life. In such cases a strictly instrumental/functional understanding fails us badly. What is needed is an interpretation of the ways, both obvious and subtle, in which everyday life is transformed by the mediating role of technical devices. In hindsight the situation is clear to everyone. Individual habits, perceptions, concepts of self, ideas of space and time, social relationships, and moral and political boundaries have all been powerfully restructured in the course of modern technological development. What is fascinating about this process is that societies involved in it have quickly altered some of the fundamental terms of human life without appearing to do so. Vast transformations in the structure of our common world have been undertaken with little attention to what those alterations mean. Judgments about technology have been made on narrow grounds, paying attention to such matters as whether a new device serves a particular need, performs more efficiently than its predecessor, makes a profit, or provides a convenient service. Only later does the broader significance of the choice become clear, typically as a series of surprising "side effects" or "secondary consequences." But it seems characteristic of our culture's involvement with technology that we are seldom inclined to examine, discuss, or judge pending innovations with broad, keen awareness of what those changes mean. In the technical realm we repeatedly enter into a series of social contracts, the terms of which are revealed only after the signing.

It may seem that the view I am suggesting is that of technological determinism: the idea that technological innovation is the basic cause of changes in society and that human beings have little choice other than to sit back and watch this ineluctable process unfold. But the concept of determinism is much too strong, far too sweeping in its implications to provide an adequate theory. It does little justice to the genuine choices that arise, in both principle and practice, in the course of technical and social transformation. Being saddled with it is like attempting to describe all instances of sexual intercourse based only on the concept of rape. A more revealing notion, in my view, is that of technological somnambulism. For the interesting puzzle in our times is that we so willingly sleepwalk through the process of reconstituting the conditions of human existence.

BEYOND IMPACTS AND SIDE EFFECTS

Social scientists have tried to awaken the sleeper by developing methods of technology assessment. The strength of these methods is that they shed light on phenomena that were previously overlooked. But an unfortunate shortcoming of technology assessment is that it tends to see technological change as a "cause" and everything that follows as an "effect" or "impact." The role of the researcher is to identify, observe, and explain these effects. This approach assumes that the causes have already occurred or are bound to do so in the normal course of events. Social research boldly enters the scene to study the "consequences" of the change. After the bulldozer has rolled over us, we can pick ourselves up and carefully measure the treadmarks. Such is the impotent mission of technological "impact" assessment.

A somewhat more farsighted version of technology assessment is sometimes used to predict which changes are likely to happen, the "social impacts of computerization" for example. With these forecasts at its disposal, society is, presumably, better able to chart its course. But, once again, the attitude in which the predictions are offered usually suggests that the "impacts" are going to happen in any case. Assertions of the sort "Computerization will bring about a revolution in the way we educate our children" carry the strong implication that those who will experience the change are obliged simply to endure it. Humans must adapt. That is their destiny. There is no tampering with the source of change, and only minor modifications are possible at the point of impact (perhaps some slight changes in the fashion contour of this year's treadmarks).

But we have already begun to notice another view of technological development, one that transcends the empirical and moral shortcomings of cause-and-effect models. It begins with the recognition that as technologies are being built and put to use, significant alterations in patterns of human activity and human institutions are already taking place. New worlds are being made. There is nothing "secondary" about this phenomenon. It is, in fact, the most important accomplishment of any new technology. The construction of a technical system that involves human beings as operating parts brings a reconstruction of social roles and relationships. Often this is a result of a new system's own operating requirements: it simply will not work unless human behavior changes to suit its form and process. Hence, the very act of using the kinds of machines, techniques, and systems available to us generates patterns of activities and expectations that soon become "second nature." We do indeed "use" telephones, automobiles, electric lights, and computers in the conventional sense of picking them up and putting them down. But our world soon becomes one in which telephony, automobility, electric lighting, and computing are forms of life in the most powerful sense: life would scarcely be thinkable without them.

My choice of the term "forms of life" in this context derives from Ludwig Wittgenstein's elaboration of that concept in *Philosophical Investigations.* In his later writing Wittgenstein sought to overcome an extremely narrow view of the structure of language then popular among philosophers, a view that held language to be primarily a matter of naming things and events. Pointing to the richness and multiplicity of the kinds of expression or "language games" that are a part of everyday speech, Wittgenstein argued that "the speaking of language is a part of an activity, or of a form of life."[7] He gave a variety of examples—the giving of orders, speculating about events, guessing riddles, making up stories, forming and testing hypotheses, and so forth—to indicate the wide range of language games involved in various "forms of life." Whether he meant to suggest that these are patterns that occur naturally to all human beings or that they are primarily cultural conventions that can change with time and setting is a question open to dispute.[8] For the purposes here, what matters is not the ultimate philosophical status of Wittgenstein's concept but its suggestiveness in helping us to overcome another widespread and extremely narrow conception: our normal understanding of the meaning of technology in human life.

As they become woven into the texture of everyday existence, the devices, techniques, and systems we adopt shed their tool-like qualities to become part of our very humanity. In an important sense we become the beings who work on assembly lines, who talk on telephones, who do our figuring on pocket calculators, who eat processed foods, who clean our homes with powerful chemicals. Of course, working, talking, figuring, eating, cleaning, and such things have been parts of human activity for a very long time. But technological innovations can radically alter these common patterns and on occasion generate entirely new ones, often with surprising results. The role television plays in our society offers some poignant examples. None of those who worked to perfect the technology of television in its early years and few of those who brought television sets into their homes ever intended the device to be employed as the universal babysitter. That, however, has become one of television's most common functions in the modern home. Similarly, if anyone in the 1930s had predicted people would eventually be watching seven hours of television each day, the forecast would have been laughed away as absurd. But recent surveys indicate that we Americans do spend that much time, roughly one-third of our lives, staring at the tube. Those who wish to reassert freedom of choice in the matter sometimes observe, "You can always turn off your TV." In a trivial sense that is true. At least for the time being the on/off button is still included as standard equipment on most sets (perhaps someday it will become optional). But given how central television has become to the content of everyday life, how it has become the accustomed topic of conversation in workplaces, schools, and other social gatherings, it is apparent that television is a phenomenon that, in the larger sense, cannot be "turned off" at all. Deeply insinuated into people's perceptions, thoughts, and behavior, it has become an indelible part of modern culture.

Most changes in the content of everyday life brought on by technology can be recognized as versions of earlier patterns. Parents have always had to entertain and instruct children and to find ways of keeping the little ones out of their hair. Having youngsters watch several hours of television cartoons is, in one way of looking at the matter, merely a new method for handling this age-old task, although the "merely" is of no small significance. It is important to ask, Where, if at all, have modern technologies added *fundamentally new* activities to the range of things human beings do? Where and how have innovations in science and technology begun to alter the very *conditions of life* itself? Is computer programming only a powerful recombination of forms of life known for ages—doing mathematics, listing, sorting, planning, organizing, etc.—or is it something unprecedented? Is industrialized agribusiness simply a renovation of older ways of farming, or does it amount to an entirely new phenomenon?

Certainly, there are some accomplishments of modern technology, manned air flight, for example, that are clearly altogether novel. Flying in airplanes is not just another version of modes of travel previously known; it is something new. Although the hope of humans flying is as old as the myth of Daedalus and Icarus or the angels of the *Old Testament,* it took a certain kind of modern machinery to realize the dream in practice. Even beyond the numerous breakthroughs that have pushed the boundaries of human action, however, lie certain kinds of changes now on the horizon that would amount to a fundamental change in the conditions of human life itself. One such prospect is that of altering human biology through genetic engineering. Another is the founding of permanent settlements in outer space. Both of these possibilities call into question what it means to be human and what constitutes "the human condition."[9] Speculation about such matters is now largely the work of science fiction, whose notorious perversity as a literary genre signals the troubles that lie in wait when we begin thinking about becoming creatures fundamentally different from any the earth has seen. A great many futuristic novels are blatantly technopornographic.

But, on the whole, most of the transformations that occur in the wake of technological innovation are actually variations of very old patterns. Wittgenstein's philosophically conservative maxim "What has to be accepted, the given, is—so one could say—*forms of life"* could well be the guiding rule of a phenomenology of technical practice.[10] For instance, asking a question and awaiting an answer, a form of interaction we all know well, is much the same activity whether it is a person we are confronting or a computer. There are, of course, significant differences between persons and computers (although it is fashionable in some circles to ignore them). Forms of life that we mastered before the coming of the computer shape our expectations as we begin to use the instrument. One strategy of software design, therefore, tries to "humanize" the computers by having them say "Hello" when the user logs in or having them respond with witty remarks when a person makes an error. We carry with us highly structured anticipations about entities that appear to participate, if only minimally, in forms of life and associated language games that are parts of human culture. Those anticipations provide much of the persuasive power of those who prematurely claim great advances in "artificial intelligence" based on narrow but impressive demonstrations of computer performance. But then children have always fantasized that their dolls were alive and talking.

The view of technologies as forms of life I am proposing has its clearest beginnings in the writings of Karl Marx. In Part I of *The German Ideology,* Marx and Engels explain the relationship of human individuality and material conditions of production as follows: "The way in which men produce their means of subsistence depends first of all on the nature of the means of subsistence they actually find in existence and have to reproduce. This mode of production must not be considered simply as being the reproduction of the physical existence of the individuals. Rather it is a definite form of activity of these individuals, a definite form of expressing their life, a definite *mode of life* on their part. As individuals express their life, so they are."[11]

Marx's concept of production here is a very broad and suggestive one. It reveals the total inadequacy of any interpretation that finds social change a mere "side effect" or "impact" of technological innovation. While he clearly points to means of production that sustain life in an immediate, physical sense, Marx's view extends to a general understanding of human development in a world of diverse natural resources, tools, machines, products, and social relations. The notion is clearly not one of occasional human interaction with devices and material conditions that leave individuals unaffected. By changing the shape of material things, Marx observes, we also change ourselves. In this process human beings do not stand at the mercy of a great deterministic punch press that cranks out precisely tailored persons at a certain rate during a given historical period. Instead, the situation Marx describes is one in which individuals are actively involved in the daily creation and recreation, production and reproduction of the world in which they live. Thus, as they employ tools and techniques, work in social labor arrangements, make and consume products, and adapt their behavior to the material conditions they encounter in their natural and artificial environment, individuals realize possibilities for human existence that are inaccessible in more primitive modes of production.

Marx expands upon this idea in "The Chapter on Capital" in the *Grundrisse.* The development of forces of production in history, he argues, holds the promise of the development of a many-sided individuality in all human beings. Capital's unlimited pursuit of wealth leads it to develop the productive powers of labor to a state "where the possession and preservation of general wealth require a lesser labour time of society as a whole, and where the labouring society relates scientifically to the process of its progressive reproduction, its reproduction in constantly greater abundance." This movement toward a general form of wealth "creates the material elements for the development of the rich individuality which is all-sided in its production as in its consumption, and whose labour also therefore appears no longer as labour, but as the full development of activity itself."[12]

If one has access to tools and materials of woodworking, a person can develop the human qualities found in the activities of carpentry. If one is able to employ the instruments and techniques of music making, one can become (in that aspect of one's life) a musician. Marx's ideal here, a variety of materialist humanism, anticipates that in a properly structured society under modern conditions of production, people would engage in a very wide range of activities that enrich their individuality along many dimensions. It is that promise which, he argues, the institutions of capitalism thwart and cripple.[13]

As applied to an understanding of technology, the philosophies of Marx and Wittgenstein direct our attention to the fabric of everyday existence. Wittgenstein points to a vast multiplicity of cultural practices that comprise our common world. Asking us to notice "what we say when," his approach can help us recognize the way language reflects the content of technical practice. It makes sense to ask, for example, how the adoption of digital computers might alter the way people think of their own faculties and activities. If Wittgenstein is correct, we would expect that changes of this kind would appear, sooner or later, in the language people use to talk about themselves. Indeed, it has now become commonplace to hear people say "I need to access your data." "I'm not programmed for that." "We must improve our interface." "The mind is the best computer we have."

Marx, on the other hand, recommends that we see the actions and interactions of every-day life within an enormous tapestry of historical developments. On occasion, as in the chapter on "Machinery and Large-Scale Industry" in *Capital,* his mode of interpretation also includes a place for a more microscopic treatment of specific technologies in human experience.[14] But on the whole his theory seeks to explain very large patterns, especially relationships between different social classes, that unfold at each stage in the history of material production. These developments set the stage for people's ability to survive and express themselves, for their ways of being human.

RETURN TO MAKING

To invoke Wittgenstein and Marx in this context, however, is not to suggest that either one or both provide a sufficient basis for a critical philosophy of technology. Proposing an attitude in which forms of life must be accepted as "the given," Wittgenstein decides that philosophy "leaves everything as it is."[15] Although some Wittgensteinians are eager to point out that this position does not necessarily commit the philosopher to conservatism in an economic or politi-cal sense, it does seem that as applied to the study of forms of life in the realm of technology, Wittgenstein leaves us with little more than a passive traditionalism. If one hopes to interpret technological phenomena in a way that suggests positive judgments and actions, Wittgenstei-nian philosophy leaves much to be desired.

In a much different way Marx and Marxism contain the potential for an equally woeful passivity. This mode of understanding places its hopes in historical tendencies that promise human emancipation at some point. As forces of production and social relations of production develop and as the proletariat makes its way toward revolution, Marx and his orthodox follow-ers are willing to allow capitalist technology, for example, the factory system, to develop to its farthest extent. Marx and Engels scoffed at the utopians, anarchists, and romantic critics of industrialism who thought it possible to make moral and political judgments about the course a technological society ought to take and to influence that path through the application of phil-osophical principles. Following this lead, most Marxists have believed that while capitalism is a target to be attacked, technological expansion is entirely good in itself, something to be encouraged without reservation. In its own way, then, Marxist theory upholds an attitude as nearly lethargic as the Wittgensteinian decision to "leave everything as it is." The famous eleventh thesis on Feuerbach—"The philosophers have only interpreted the world in various ways; the point, however, is to change it"—conceals an important qualification: that judgment, action, and change are ultimately products of history. In its view of technological development Marxism anticipates a history of rapidly evolving material productivity, an inevitable course of events in which attempts to propose moral and political limits have no place. When social-ism replaces capitalism, so the promise goes, the machine will finally move into high gear, presumably releasing humankind from its age-old miseries.

Whatever their shortcomings, however, the philosophies of Marx and Wittgenstein share a fruitful insight: the observation that social activity is an ongoing process of world-making. Throughout their lives people come together to renew the fabric of relationships, transactions, and meanings that sustain their common existence. Indeed, if they did not engage in this continuing activity of material and social production, the human world would literally fall apart. All social roles and frameworks—from the most rewarding to the most oppressive—must somehow be restored and reproduced with the rise of the sun each day.

From this point of view, the important question about technology becomes, As we "make things work," what kind of *world* are we making? This suggests that we pay attention not only to the making of physical instruments and processes, although that certainly remains important, but also to the production of psychological, social, and political conditions as a part of any significant technical change. Are we going to design and build circumstances that enlarge possibilities for growth in human freedom, sociability, intelligence, creativity, and self-government? Or are we headed in an altogether different direction?

It is true that not every technological innovation embodies choices of great significance. Some developments are more-or-less innocuous; many create only trivial modifications in how we live. But in general, where there are substantial changes being made in what people are doing and at a substantial investment of social resources, then it always pays to ask in advance about the qualities of the artifacts, institutions, and human experiences currently on the drawing board.

Inquiries of this kind present an important challenge to all disciplines in the social sciences and humanities. Indeed, there are many historians, anthropologists, sociologists, psychologists, and artists whose work sheds light on long-overlooked human dimensions of technology. Even engineers and other technical professionals have much to contribute here when they find courage to go beyond the narrow-gauge categories of their training.

The study of politics offers its own characteristic route into this territory. As the political imagination confronts technologies as forms of life, it should be able to say something about the choices (implicit or explicit) made in the course of technological innovation and the grounds for making those choices wisely. That is a task I take up elsewhere. Through technological creation and many other ways as well, we make a world for each other to live in. Much more than we have acknowledged in the past, we must admit our responsibility for what we are making.

NOTES

1. *The Encyclopedia of Philosophy,* 8 vols., Paul Edwards (editor-in-chief) (New York: Macmillan, 1967).

2. *Bibliography of the Philosophy of Technology,* Carl Mitcham and Robert Mackey (eds.) (Chicago: University of Chicago Press, 1973).

3. There are, of course, exceptions to this general attitude. See Stephen H. Unger, *Controlling Technology: Ethics and the Responsible Engineer* (New York: Holt, Rinehart, and Winston, 1982).

4. An excellent corrective to the general thoughtfulness about "making" and "use" is to be found in Carl Mitcham, "Types of Technology," in *Research in Philosophy and Technology,* Paul Durbin (ed.) (Greenwich, Conn.: JAI Press, 1978), 229–294.

5. J. L. Austin, *Philosophical Papers* (Oxford: Oxford University Press, 1961), 123–153.

6. See William Kolender et al., "Petition v. Edward Lawson," *Supreme Court Reporter* 103: 1855–1867, 1983. Edward Lawson had been arrested approximately fifteen times on his long walks and refused to provide identification when stopped by the police. Lawson cited his rights guaranteed by the Fourth and Fifth Amendments of the U.S. Constitution. The Court found the California vagrancy statute requiring "credible and reliable" identification to be unconstitutionally vague. See also Jim Mann, "State Vagrancy Law Voided as Overly Vague," *Los Angeles Times,* May 3, 1983, 1, 19.

7. Ludwig Wittgenstein, *Philosophical Investigations,* ed. 3, translated by G. E. M. Anscombe, with English and German indexes (New York: Macmillan, 1958), 11e.

8. Hanna Pitkin, *Wittgenstein and Justice: On the Significance of Ludwig Wittgenstein for Social and Political Thought* (Berkeley: University of California Press, 1972), 293.

9. For a thorough discussion of this idea, see Hannah Arendt, *The Human Condition* (Chicago: University of Chicago Press, 1958); and Hannah Arendt, *Willing,* vol. II of *The Life of the Mind* (New York: Harcourt Brace Jovanovich, 1978).

10. *Philosophical Investigations,* 226e.

11. Karl Marx and Friedrich Engels, "The German Ideology," in *Collected Works,* vol. 5 (New York: International Publishers, 1976), 31.

12. Karl Marx, *Grundrisse,* translated with a foreword by Martin Nicolaus (Harmondsworth, England: Penguin Books, 1973), 325.

13. An interesting discussion of Marx in this respect is Kostas Axelos' *Alienation, Praxis and Techne in the Thought of Karl Marx,* translated by Ronald Bruzina (Austin: University of Texas Press, 1976).

14. Karl Marx, *Capital,* vol. 1, translated by Ben Fowkes, with an introduction by Ernest Mandel (Harmondsworth, England: Penguin Books, 1976), chap. 15.

15. *Philosophical Investigations,* 49e.

9

Focal Things and Practices

Albert Borgmann

THE DEVICE PARADIGM

We must now provide an explicit account of the pattern or paradigm of technology. I begin with two clear cases and analyze them in an intuitive way to bring out the major features of the paradigm. And I attempt to raise those features into sharper relief against the sketch of a pretechnological setting and through the consideration of objections that may be advanced against the distinctiveness of the pattern.

Technology, as we have seen, promises to bring the forces of nature and culture under control, to liberate us from misery and toil, and to enrich our lives. To speak of technology making promises suggests a substantive view of technology and is misleading. But the parlance is convenient and can always be reconstructed to mean that implied in the technological mode of taking up with the world there is a promise that this approach to reality will, by way of the domination of nature, yield liberation and enrichment. Who issues the promise to whom is a question of political responsibility; and who the beneficiaries of the promise are is a question of social justice. These questions are taken up in later chapters. What we must answer first is the question of how the promise of liberty and prosperity was specified and given a definite pattern of implementation.

As a first let us note that the notions of liberation and enrichment are joined in that of availability. Goods that are available to us enrich our lives and, if they are technologically available, they do so without imposing burdens on us. Something is available in this sense if it has been rendered instantaneous, ubiquitous, safe, and easy.[1] Warmth, e.g., is now available. We get a first glimpse of the distinctiveness of availability when we remind ourselves that warmth was not available, e.g., in Montana a hundred years ago. It was not instantaneous because in the morning a fire first had to be built in the stove or the fireplace. And before it could be built, trees had to be felled, logs had to be sawed and split, the wood had to be hauled

Albert Borgmann, *Technology and the Character of Contemporary Life: A Philosophical Inquiry*, Chicago: University of Chicago Press, 1984, pp. 40–44, 196–210, 221–226. Copyright © 1984 by The University of Chicago Press. Reprinted by permission of The University of Chicago Press.

and stacked. Warmth was not ubiquitous because some rooms remained unheated, and none was heated evenly. The coaches and sleighs were not heated, nor were the boardwalks or all of the shops and stores. It was not entirely safe because one could get burned or set the house on fire. It was not easy because work, some skills, and attention were constantly required to build and sustain a fire.

Such observations, however, are not sufficient to establish the distinctiveness of avail-ability. In the common view, technological progress is seen as a more or less gradual and straightforward succession of lesser by better implements.[2] The wood-burning stove yields to the coal-fired central plant with heat distribution by convection, which in turn gives way to a plant fueled by natural gas and heating through forced air, and so on.[3] To bring the distinctive-ness of availability into relief we must turn to the distinction between things and devices. A thing, in the sense in which I want to use the word here, is inseparable from its context, namely, its world, and from our commerce with the thing and its world, namely, engagement. The experience of a thing is always and also a bodily and social engagement with the thing's world. In calling forth a manifold engagement, a thing necessarily provides more than one commod-ity. Thus a stove used to furnish more than mere warmth. It was a *focus,* a hearth, a place that gathered the work and leisure of a family and gave the house a center. Its coldness marked the morning, and the spreading of its warmth the beginning of the day. It assigned to the different family members tasks that defined their place in the household. The mother built the fire, the children kept the firebox filled, and the father cut the firewood. It provided for the entire family a regular and bodily engagement with the rhythm of the seasons that was woven together of the threat of cold and the solace of warmth, the smell of wood smoke, the exertion of sawing and of carrying, the teaching of skills, and the fidelity to daily tasks. These features of physical engagement and of family relations are only first indications of the full dimensions of a thing's world. Physical engagement is not simply physical contact but the experience of the world through the manifold sensibility of the body. That sensibility is sharpened and strengthened in skill. Skill is intensive and refined world engagement. Skill, in turn, is bound up with social engagement. It molds the person and gives the person character.[4] Limitations of skill confine any one person's primary engagement with the world to a small area. With the other areas one is immediately engaged through one's acquaintance with the characteristic demeanor and hab-its of the practitioners of the other skills. That acquaintance is importantly enriched through one's use of their products and the observation of their working. Work again is only one exam-ple of the social context that sustains and comes to be focused in a thing. If we broaden our focus to include other practices, we can see similar social contexts in entertainment, in meals, in the celebration of the great events of birth, marriage, and death. And in these wider horizons of social engagement we can see how the cultural and natural dimensions of the world open up.

We have now sketched a background against which we can outline a specific notion of the device. We have seen that a thing such as a fireplace provides warmth, but it inevitably provides those many other elements that compose the world of the fireplace. We are inclined to think of these additional elements as burdensome, and they were undoubtedly often so expe-rienced. A device such as a central heating plant procures mere warmth and disburdens us of all other elements. These are taken over by the machinery of the device. The machinery makes no demands on our skill, strength, or attention, and it is less demanding the less it makes its presence felt. In the progress of technology, the machinery of a device has therefore a tendency to become concealed or to shrink. Of all the physical properties of a device, those alone are crucial and prominent which constitute the commodity that the device procures. Informally speaking, the commodity of a device is "what a device is there for." In the case of a central heating plant it is warmth, with a telephone it is communication, a car provides transportation, frozen food makes up a meal, a stereo set furnishes music. "Commodity" for the time being is to be taken flexibly. The emphasis lies on the commodious way in which devices make goods and services available. There are at first unavoidable ambiguities in the notion of the device and the commodity; they can gradually be resolved through substantive analyses and methodological reflections.[5] Tentatively, then, those aspects or properties of a device that provide the answer to "What is the device for?" constitute its commodity, and they remain relatively fixed. The other properties are changeable and are changed, normally on the basis of scientific insight and engineering ingenuity, to make the commodity still more available. Hence every device has functional equivalents, and equivalent devices may be physically and structurally very dissimilar from one another.

The development of television provides an illustration of these points. The bulky machinery of the first sets was obtrusive in relation to the commodity it procured, namely, the moving two-dimensional picture which appeared in fuzzy black and white on a screen with the size and shape of a bull's-eye. Gradually the screens became larger, more rectangular; the picture became sharper and eventually colored. The sets became relatively smaller and less conspicuous in their machinery. And this development continues and has its limit in match-box-sized sets which provide arbitrarily large and most finely grained moving and colored pictures. The example also shows how radical changes in the machinery amounted to continuous improvements of the function as tubes gave way to transistors and these yielded to silicon chips. Cables and satellites were introduced as communication links. Pictures could be had in recorded rather than transmitted form, and recordings can be had on tapes or discs. These considerations in turn show how the technical development of a device increases availability. Increasingly, video programs can be seen nearly everywhere—in bars, cars, in every room of a home. Every conceivable film can be had. A program broadcast at an inconvenient time can be recorded and played later. The constraints of time and place are more and more dissolved. It is an instructive exercise to see how in the implements that surround us daily the machinery becomes less conspicuous, the function more prominent, how radical technical changes in the machinery are but degrees of advancement in the commodity, and how the availability of the commodities increases all the while.

The distinction in the device between its machinery and its function is a specific instance of the means-ends distinction. In agreement with the general distinction, the machinery or the means is subservient to and validated by the function or the end. The technological distinction of means and ends differs from the general notion in two respects. In the general case, it is very questionable how clearly and radically means and ends can be distinguished without doing violence to the phenomena.[6] In the case of the technological device, however, the machinery can be changed radically without threat to the identity and familiarity of the function of the device. No one is confused when one is invited to replace one's watch, powered by a spring, regulated by a balance wheel, displaying time with a dial and pointers, with a watch that is powered electrically, is regulated by a quartz crystal, and displays time digitally. This concomitance of radical variability of means and relative stability of ends is the first distinguishing feature. The second, closely tied to the first, is the concealment and unfamiliarity of the means and the simultaneous prominence and availability of the ends.[7]

The concealment of the machinery and the disburdening character of the device go hand in hand. If the machinery were forcefully present, it would eo ipso make claims on our faculties. If claims are felt to be onerous and are therefore removed, then so is the machinery. A commodity is truly available when it can be enjoyed as a mere end, unencumbered by means. It must be noted that the disburdenment resting on a feudal household is ever incomplete. The lord and the lady must always reckon with the moods, the insubordination, and the frailty of the servants.[8] The device provides social disburdenment, i.e., anonymity. The absence of the master-servant relation is of course only one instance of social anonymity. The starkness of social anonymity in the technological universe can be gauged only against a picture of the social relations in a world of things. Such a picture will also show that social anonymity necessarily shades off into one of nature, culture, and history.

FOCAL THINGS AND PRACTICES

To see that the force of nature can be encountered analogously in many other places, we must develop the general notions of focal things and practices. This is the first point of this chapter. The Latin word *focus,* its meaning and etymology, are our best guides to this task. But once we have learned tentatively to recognize the instances of focal things and practices in our midst, we must acknowledge their scattered and inconspicuous character too. Their hidden splendor comes to light when we consider Heidegger's reflections on simple and eminent things. But an inappropriate nostalgia clings to Heidegger's account. It can be dispelled, so I will argue, when we remember and realize more fully that the technological environment heightens rather than denies the radiance of genuine focal things and when we learn to understand that focal things require a practice to prosper within. These points I will try to give substance in the subsequent parts of this chapter by calling attention to the focal concerns of running and of the culture of the table.

The Latin word *focus* means hearth. We came upon it earlier where the device paradigm was first delineated and where the hearth or fireplace, a thing, was seen as the counterpart to the central heating plant, a device. It was pointed out that in a pretechnological house the fireplace constituted a center of warmth, of light, and of daily practices. For the Romans the *focus* was holy, the place where the housegods resided. In ancient Greece, a baby was truly joined to the family and household when it was carried about the hearth and placed before it. The union of a Roman marriage was sanctified at the hearth. And at least in the early periods the dead were buried by the hearth. The family ate by the hearth and made sacrifices to the housegods before and after the meal. The hearth sustained, ordered, and centered house and family.[9] Reflections of the hearth's significance can yet be seen in the fireplace of many American homes. The fireplace often has a central location in the house. Its fire is now symbolical since it rarely furnishes sufficient warmth. But the radiance, the sounds, and the fragrance of living fire consuming logs that are split, stacked, and felt in their grain have retained their force. There are no longer images of the ancestral gods placed by the fire; but there often are pictures of loved ones on or above the mantel, precious things of the family's history, or a clock, measuring time.[10]

The symbolical center of the house, the living room with the fireplace, often seems forbidding in comparison with the real center, the kitchen with its inviting smells and sounds. Accordingly, the architect Jeremiah Eck has rearranged homes to give them back a hearth, "a place of warmth and activity" that encompasses cooking, eating, and living and so is central to the house whether it literally has a fireplace or not.[11] Thus we can satisfy, he says, "the need for a place of focus in our family lives."[12]

"Focus," in English, is now a technical term of geometry and optics. Johannes Kepler was the first so to use it, and he probably drew on the then already current sense of focus as the "burning point of lens or mirror."[13] Correspondingly, an optic or geometric focus is a point where lines or rays converge or from which they diverge in a regular or lawful way. Hence "focus" is used as a verb in optics to denote moving an object in relation to a lens or modifying a combination of lenses in relation to an object so that a clear and well-defined image is produced.

These technical senses of "focus" have happily converged with the original one in ordinary language. Figuratively they suggest that a focus gathers the relations of its context and radiates into its surroundings and informs them. To focus on something or to bring it into focus is to make it central, clear, and articulate. It is in the context of these historical and living senses of "focus" that I want to speak of focal things and practices. Wilderness on this continent, it now appears, is a focal thing. It provides a center of orientation; when we bring the surrounding technology into it, our relations to technology become clarified and well-defined. But just how strong its gathering and radiating force is requires further reflection. And surely there will be other focal things and practices: music, gardening, the culture of the table, or running.

We might in a tentative way be able to see these things as focal; what we see more clearly and readily is how inconspicuous, homely, and dispersed they are. This is in stark contrast to the focal things of pretechnological times, the Greek temple or the medieval cathedral that we have mentioned before. Martin Heidegger was deeply impressed by the orienting force of the Greek temple. For him, the temple not only gave a center of meaning to its world but had orienting power in the strong sense of first originating or establishing the world, of disclosing the world's essential dimensions and criteria.[14] Whether the thesis so extremely put is defensible or not, the Greek temple was certainly more than a self-sufficient architectural sculpture, more than a jewel of well-articulated and harmoniously balanced elements, more, even, than a shrine for the image of the goddess or the god. As Vincent Scully has shown, a temple or a temple precinct gathered and disclosed the land in which it was situated. The divinity of land and sea was focused in the temple.[15]

To see the work of art as the focus and origin of the world's meaning was a pivotal discovery for Heidegger. He had begun in the modern tradition of Western philosophy where, the sense of reality is to be grasped by determining the antecedent and controlling conditions of all there is (the *Bedingungen der Möglichkeit* as Immanuel Kant has it). Heidegger wanted to outdo this tradition in the radicality of his search for the fundamental conditions of being. Perhaps it was the relentlessness of his pursuit that disclosed the ultimate futility of it. At any rate, when the universal conditions are explicated in a suitably general and encompassing way, what truly matters still hangs in the balance because everything depends on how the conditions come to be actualized and instantiated.[16] The preoccupation with antecedent conditions not only leaves this question unanswered; it may even make it inaccessible by leaving the impression that, once the general and fundamental matters are determined, nothing of consequence remains to be considered. Heidegger's early work, however, already contained the seeds of its overcoming. In his determination to grasp reality in its concreteness, Heidegger had found and stressed the inexorable and unsurpassable givenness of human existence, and he had provided analyses of its pretechnological wholeness and its technological distraction though the significance of these descriptions for technology had remained concealed to him.[17] And then he discovered that the unique event of significance in the singular work of art, in the prophet's proclamation, and in the political deed was crucial. This insight was worked out in detail with regard to the artwork. But in an epilogue to the essay that develops this point, Heidegger recognized that the insight comes too late. To be sure, our time has brought forth admirable works of art. "But," Heidegger insists, "the question remains: is art still an essential and necessary way in which that truth happens which is decisive for historical existence, or is art no longer of this character?"[18]

Heidegger began to see technology (in his more or less substantive sense) as the force that has eclipsed the focusing powers of pretechnological times. Technology becomes for him the final phase of a long metaphysical development. The philosophical concern with the conditions of the possibility of whatever is now itself seen as a move into the oblivion of what finally matters. But how are we to recover orientation in the oblivious and distracted era of technology when the great embodiments of meaning, the works of art, have lost their focusing power? Amidst the complication of conditions, of the *Bedingungen,* we must uncover the simplicity of things, of the *Dinge.*[19] A jug, an earthen vessel from which we pour wine, is such a thing. It teaches us what it is to hold, to offer, to pour, and to give. In its clay, it gathers for us the earth as it does in containing the wine that has grown from the soil. It gathers the sky whose rain and sun are present in the wine. It refreshes and animates us in our mortality. And in the libation it acknowledges and calls on the divinities. In these ways the thing (in agreement with its etymologically original meaning) gathers and discloses what Heidegger calls the fourfold, the interplay of the crucial dimensions of earth and sky, mortals and divinities.[20] A thing, in Heidegger's eminent sense, is a focus; to speak of focal things is to emphasize the central point twice.

Still, Heidegger's account is but a suggestion fraught with difficulties. When Heidegger described the focusing power of the jug, he might have been thinking of a rural setting where wine jugs embody in their material, form, and craft a long and local tradition; where at noon one goes down to the cellar to draw a jug of table wine whose vintage one knows well; where at the noon meal the wine is thoughtfully poured and gratefully received.[21] Under such circumstances, there might be a gathering and disclosure of the fourfold, one that is for the most part understood and in the background and may come to the fore on festive occasions. But all of this seems as remote to most of us and as muted in its focusing power as the Parthenon or the Cathedral of Chartres. How can so simple a thing as a jug provide that turning point in our relation to technology to which Heidegger is looking forward? Heidegger's proposal for a reform of technology is even more programmatic and terse than his analysis of technology.[22] Both, however, are capable of fruitful development.[23] Two points in Heidegger's consideration of the turn of technology must particularly be noted. The first serves to remind us of arguments already developed which must be kept in mind if we are to make room for focal things and practices. Heidegger says, broadly paraphrased, that the orienting force of simple things will come to the fore only as the rule of technology is raised from its anonymity, is disclosed as the orthodoxy that heretofore has been taken for granted and allowed to remain invisible.[24] As long as we overlook the tightly patterned character of technology and believe that we live in a world of endlessly open and rich opportunities, as long as we ignore the definite ways in which we, acting technologically, have worked out the promise of technology and remain vaguely enthralled by that promise, so long simple things and practices will seem burdensome, confining, and drab. But if we recognize the central vacuity of advanced technology, that emptiness can become the opening for focal things. It works both ways, of course. When we see a focal concern of ours threatened by technology, our sight for the liabilities of mature technology is sharpened.

A second point of Heidegger's is one that we must develop now. The things that gather the fourfold, Heidegger says, are inconspicuous and humble. And when we look at his litany of things, we also see that they are scattered and of yesterday: jug and bench, footbridge and plow, tree and pond, brook and hill, heron and deer, horse and bull, mirror and clasp, book and picture, crown and cross.[25] That focal things and practices are inconspicuous is certainly true; they flourish at the margins of public attention. And they have suffered a diaspora; this too must be accepted, at least for now. That is not to say that a hidden center of these dispersed focuses may not emerge some day to unite them and bring them home. But it would clearly be a forced growth to proclaim such a unity now. A reform of technology that issues from focal concerns will be radical not in imposing a new and unified master plan on the technological universe but in discovering those sources of strength that will nourish principled and confident beginnings, measures, i.e., which will neither rival nor deny technology.

But there are two ways in which we must go beyond Heidegger. One step in the first direction has already been taken. It led us to see that the simple things of yesterday attain a new splendor in today's technological context. The suggestion in Heidegger's reflections that we have to seek out pretechnological enclaves to encounter focal things is misleading and dispiriting. Rather we must see any such enclave itself as a focal thing heightened by its technological context. The turn to things cannot be a setting aside and even less an escape from technology but a kind of affirmation of it. The second move beyond Heidegger is in the direction of practice, into the social and, later, the political situation of focal things.[26] Though Heidegger assigns humans their place in the fourfold when he depicts the jug in which the fourfold is focused, we scarcely see the hand that holds the jug, and far less do we see of the social setting in which the pouring of the wine comes to pass. In his consideration of another thing, a bridge, Heidegger notes the human ways and works that are gathered and directed by the bridge.[27] But these remarks too present practices from the viewpoint of the focal thing. What must be shown is that focal things can prosper in human practices only. Before we can build a bridge, Heidegger suggests, we must be able to dwell.[28] But what does that mean concretely?

The consideration of the wilderness has disclosed a center that stands in a fruitful counterposition to technology. The wilderness is beyond the procurement of technology, and our response to it takes us past consumption. But it also teaches us to accept and to appropriate technology. We must now try to discover if such centers of orientation can be found in greater proximity and intimacy to the technological everyday life. And I believe they can be found if we follow up the hints that we have gathered from and against Heidegger, the suggestions that focal things seem humble and scattered but attain splendor in technology if we grasp technology properly, and that focal things require a practice for their welfare. Running and the culture of the table are such focal things and practices. We have all been touched by them in one way or another. If we have not participated in a vigorous or competitive run, we have certainly taken walks; we have felt with surprise, perhaps, the pleasure of touching the earth, of feeling the wind, smelling the rain, of having the blood course through our bodies more steadily. In the preparation of a meal we have enjoyed the simple tasks of washing leaves and cutting bread; we have felt the force and generosity of being served a good wine and homemade bread. Such experiences have been particularly vivid when we came upon them after much sitting and watching indoors, after a surfeit of readily available snacks and drinks. To encounter a few simple things was liberating and invigorating. The normal clutter and distraction fall away when, as the poet says,

> there, in limpid brightness shine,
> on the table, bread and wine.[29]

If such experiences are deeply touching, they are fleeting as well. There seems to be no thought or discourse that would shelter and nurture such events; not in politics certainly, nor in philosophy where the prevailing idiom sanctions and applies equally to lounging and walking, to Twinkies, and to bread, the staff of life. But the reflective care of the good life has not withered away. It has left the profession of philosophy and sprung up among practical people. In fact, there is a tradition in this country of persons who are engaged by life in its concreteness and simplicity and who are so filled with this engagement that they have reached for the pen to become witnesses and teachers, speakers of deictic discourse. Melville and Thoreau are among the great prophets of this tradition. Its present health and extent are evident from the fact that it now has no overpowering heroes but many and various more or less eminent practitioners. Their work embraces a spectrum between down-to-earth instruction and soaring speculation. The span and center of their concerns vary greatly. But they all have their mooring in the attention to tangible and bodily things and practices, and they speak with an enthusiasm that is nourished by these focal concerns. Pirsig's book is an impressive and troubling monument in this tradition, impressive in the freshness of its observations and its pedagogical skill, troubling in its ambitious and failing efforts to deal with the large philosophical issues. Norman Maclean's *A River Runs through It* can be taken as a fly-fishing manual, a virtue that pleases its author.[30] But it is a literary work of art most of all and a reflection on technology inasmuch as it presents the engaging life, both dark and bright, from which we have so recently emerged. Colin Fletcher's treatise of *The Complete Walker* is most narrowly a book of instruction about hiking and backpacking.[31] The focal significance of these things is found in the interstices of equipment and technique; and when the author explicitly engages in deictic discourse he has "an unholy awful time" with it.[32] Roger B. Swain's contemplation of gardening in *Earthly Pleasures* enlightens us in cool and graceful prose about the scientific basis and background of what we witness and undertake in our gardens.[33] Philosophical significance enters unbidden and easily in the reflections on time, purposiveness, and the familiar. Looking at these books, I see a stretch of water that extends beyond my vision, disappearing in the distance. But I can see that it is a strong and steady stream, and it may well have parts that are more magnificent than the ones I know.[34]

To discover more clearly the currents and features of this, the other and more concealed, American mainstream, I take as witnesses two books where enthusiasm suffuses instruction vigorously, Robert Farrar Capon's *The Supper of the Lamb* and George Sheehan's *Running and Being*.[35] Both are centered on focal events, the great run and the great meal. The great run, where one exults in the strength of one's body, in the ease and the length of the stride, where nature speaks powerfully in the hills, the wind, the heat, where one takes endurance to the breaking point, and where one is finally engulfed by the good will of the spectators and the fellow runners.[36] The great meal, the long session as Capon calls it, where the guests are thoughtfully invited, the table has been carefully set, where the food is the culmination of tradition, patience, and skill and the presence of the earth's most delectable textures and tastes, where there is an invocation of divinity at the beginning and memorable conversation throughout.[37]

Such focal events are compact, and if seen only in their immediate temporal and spatial extent they are easily mistaken. They are more mistakable still when they are thought of as experiences in the subjective sense, events that have their real meaning in transporting a person into a certain mental or emotional state. Focal events, so conceived, fall under the rule of technology. For when a subjective state becomes decisive, the search for a machinery that is functionally equivalent to the traditional enactment of that state begins, and it is spurred by endeavors to find machineries that will procure the state more instantaneously, ubiquitously, more assuredly and easily. If, on the other hand, we guard focal things in their depth and integrity, then, to see them fully and truly, we must see them in context. Things that are deprived of their context become ambiguous.[38] The letter "a" by itself means nothing in particular. In the context of "table" it conveys or helps to convey a more definite meaning. But "table" in turn can mean many things. It means something more powerful in the text of Capon's book where he speaks of "The Vesting of the Table."[39] But that text must finally be seen in the context and texture of the world. To say that something becomes ambiguous is to say that it is made to say less, little, or nothing. Thus to elaborate the context of focal events is to grant them their proper eloquence.

"The distance runner," Sheehan says, "is the least of all athletes. His sport the least of all sports."[40] Running is simply to move through time and space, step-by-step. But there is splendor in that simplicity. In a car we move of course much faster, farther, and more comfortably. But we are not moving on our own power and in our own right. We cash in prior labor for present motion. Being beneficiaries of science and engineering and having worked to be able to pay for a car, gasoline, and roads, we now release what has been earned and stored and use it for transportation. But when these past efforts are consumed and consummated in my driving, I can at best take credit for what I have done. What I am doing now, driving, requires no effort, and little or no skill or discipline. I am a divided person; my achievement lies in the past, my enjoyment in the present. But in the runner, effort and joy are one; the split between means and ends, labor and leisure is healed.[41] To be sure, if I have trained conscientiously, my past efforts will bear fruit in a race. But they are not just cashed in. My strength must be risked and enacted in the race which is itself a supreme effort and an occasion to expand my skill.

This unity of achievement and enjoyment, of competence and consummation, is just one aspect of a central wholeness to which running restores us. Good running engages mind and body. Here the mind is more than an intelligence that happens to be housed in a body. Rather the mind is the sensitivity and the endurance of the body.[42] Hence running in its fullness, as Sheehan stresses over and over again, is in principle different from exercise designed to procure physical health. The difference between running and physical exercise is strikingly exhibited in one and the same issue of the *New York Times Magazine*. It contains an account by Peter Wood of how, running the New York City Marathon, he took in the city with body and mind, and it has an account by Alexandra Penney of corporate fitness programs where executives, concerned about their Coronary Risk Factor Profile, run nowhere on treadmills or ride stationary bicycles.[43] In another issue, the *Magazine* shows executives exercising their bodies while busying their dissociated minds with reading.[44] To be sure, unless a runner concentrates on bodily performance, often in an effort to run the best possible race, the mind wanders as the body runs. But as in free association we range about the future and the past, the actual and the possible, our mind, like our breathing, rhythmically gathers itself to the here and now, having spread itself to distant times and faraway places.

It is clear from these reflections that the runner is mindful of the body because the body is intimate with the world. The mind becomes relatively disembodied when the body is severed from the depth of the world, i.e., when the world is split into commodious surfaces and inaccessible machineries. Thus the unity of ends and means, of mind and body, and of body and world is one and the same. It makes itself felt in the vividness with which the runner experiences reality. "Somehow you feel more in touch," Wood says, "with the realities of a massive inner-city housing problem when you are running through it slowly enough to take in the grim details, and, surprisingly, cheered on by the remaining occupants."[45] As this last remark suggests, the wholeness that running establishes embraces the human family too. The experience of that simple event releases an equally simple and profound sympathy. It is a natural goodwill, not in need of drugs nor dependent on a common enemy. It wells up from depths that have been forgotten, and it overwhelms the runners ever and again.[46] As Wood recounts his running through streets normally besieged by crime and violence, he remarks: "But we can only be amazed today at the warmth that emanates from streets usually better known for violent crime." And his response to the spectators' enthusiasm is this: "I feel a great proximity to the crowd, rushing past at all of nine miles per hour; a great affection for them individually; a commitment to run as well as I possibly can, to acknowledge their support."[47] For George Sheehan, finally, running discloses the divine. When he runs, he wrestles with God.[48] Serious running takes us to the limits of our being. We run into threatening and seemingly unbearable pain. Sometimes, of course, the plunge into that experience gets arrested in ambition and vanity. But it can take us further to the point where in suffering our limits we experience our greatness too. This, surely, is a hopeful place to escape technology, metaphysics, and the God of the philosophers and reach out to the God of Abraham, Isaac, and Jacob.[49]

If running allows us to center our lives by taking in the world through vigor and simplicity, the culture of the table does so by joining simplicity with cosmic wealth. Humans are such complex and capable beings that they can fairly comprehend the world and, containing it, constitute a cosmos in their own right. Because we are standing so eminently over against the world, to come in touch with the world becomes for us a challenge and a momentous event. In one sense, of course, we are always already in the world, breathing the air, touching the ground, feeling the sun. But as we can in another sense withdraw from the actual and present world, contemplating what is past and to come, what is possible and remote, we celebrate correspondingly our intimacy with the world. This we do most fundamentally when in eating we take in the world in its palpable, colorful, nourishing immediacy. Truly human eating is the union of the primal and the cosmic. In the simplicity of bread and wine, of meat and vegetable, the world is gathered.

The great meal of the day, be it at noon or in the evening, is a focal event par excellence. It gathers the scattered family around the table. And on the table it gathers the most delectable things nature has brought forth. But it also recollects and presents a tradition, the immemorial experiences of the race in identifying and cultivating edible plants, in domesticating and butchering animals; it brings into focus closer relations of national or regional customs, and more intimate traditions still of family recipes and dishes. This living texture is being rent through the procurement of food as a commodity and the replacement of the culture of the table by the food industry. Once food has become freely available, it is only consistent that the gathering of the meal is shattered and disintegrates into snacks, T.V. dinners, bites that are grabbed to be eaten; and eating itself is scattered around television shows, late and early meetings, activities, overtime work, and other business. This is increasingly the normal condition of technological eating. But it is within our power to clear a central space amid the clutter and distraction. We can begin with the simplicity of a meal that has a beginning, a middle, and an end and that breaks through the superficiality of convenience food in the simple steps of beginning with raw ingredients, preparing and transforming them, and bringing them to the table. In this way we can again become freeholders of our culture. We are disfranchised from world citizenship when the foods we eat are mere commodities. Being essentially opaque surfaces, they repel all efforts at extending our sensibility and competence into the deeper reaches of the world. A Big Mac and a Coke can overwhelm our tastebuds and accommodate our hunger. Technology is not, after all, a children's crusade but a principled and skillful enterprise of defining and satisfying human needs. Through the diversion and busyness of consumption we may have unlearned to feel constrained by the shallowness of commodities. But having gotten along for a time and quite well, it seemed, on institutional or convenience food, scales fall from our eyes when we step up to a festively set family table. The foods stand out more clearly, the fragrances are stronger, eating has once more become an occasion that engages and accepts us fully.

To understand the radiance and wealth of a festive meal we must be alive to the interplay of things and humans, of ends and means. At first a meal, once it is on the table, appears to have commodity character since it is now available before us, ready to be consumed without effort or merit. But though there is of course in any eating a moment of mere consuming, in a festive meal eating is one with an order and discipline that challenges and ennobles the participants. The great meal has its structure. It begins with a moment of reflection in which we place ourselves in the presence of the first and last things. It has a sequence of courses; it requires and sponsors memorable conversation; and all this is enacted in the discipline called table manners. They are warranted when they constitute the respectful and skilled response to the great things that are coming to pass in the meal. We can see how order and discipline have collapsed when we eat a Big Mac. In consumption there is the pointlike and inconsequential conflation of a sharply delimited human need with an equally contextless and closely fitting commodity. In a Big Mac the sequence of courses has been compacted into one object and the discipline of table manners has been reduced to grabbing and eating. The social context reaches no further than the pleasant faces and quick hands of the people who run the fast-food outlet. In a festive meal, however, the food is served, one of the most generous gestures human beings are capable of. The serving is of a piece with garnishing; garnishing is the final phase of cooking, and cooking is one with preparing the food. And if we are blessed with rural circumstances, the preparation of food draws near the harvesting and the raising of the vegetables in the garden close by. This context of activities is embodied in persons. The dish and the cook, the vegetable and the gardener tell of one another. Especially when we are guests, much of the meal's deeper context is socially and conversationally mediated. But that mediation has translucence and intelligibility because it extends into the farther and deeper recesses without break and with a bodily immediacy that we too have enacted or at least witnessed firsthand. And what seems to be a mere receiving and consuming of food is in fact the enactment of generosity and gratitude, the affirmation of mutual and perhaps religious obligations. Thus eating in a focal setting differs sharply from the social and cultural anonymity of a fast-food outlet.

The pretechnological world was engaging through and through, and not always positively. There also was ignorance, to be sure, of the final workings of God and king; but even the unknown engaged one through mystery and awe. In this web of engagement, meals already had focal character, certainly as soon as there was anything like a culture of the table.[50] Today, however, the great meal does not gather and order a web of thoroughgoing relations of engagement; within the technological setting it stands out as a place of profound calm, one in which we can leave behind the narrow concentration and one-sided strain of labor and the tiring and elusive diversity of consumption. In the technological setting, the culture of the table not only focuses our life; it is also distinguished as a place of healing, one that restores us to the depth of the world and to the wholeness of our being.

As said before, we all have had occasion to experience the profound pleasure of an invigorating walk or a festive meal. And on such occasions we may have regretted the scarcity of such events; we might have been ready to allow such events a more regular and central place in our lives. But for the most part these events remain occasional, and indeed the ones that still grace us may be slipping from our grasp. We have seen various aspects of this malaise, especially its connection with television. But why are we acting against our better insights and aspirations? This at first seems all the more puzzling as the engagement in a focal activity is for most citizens of the technological society an instantaneous and ubiquitous possibility. On any day I can decide to run or to prepare a meal after work. Everyone has some sort of suitable equipment. At worst one has to stop on the way home to pick up this or that. It is of course technology that has opened up these very possibilities. But why are they lying fallow for the most part? There is a convergence of several factors. Labor is exhausting, especially when it is divided. When we come home, we often feel drained and crippled. Diversion and pleasurable consumption appear to be consonant with this sort of disability. They promise to untie the knots and to soothe the aches. And so they do at a shallow level of our existence. At any rate, the call for exertion and engagement seems like a cruel and unjust demand. We have sat in the easy chair, beer at hand and television before us; when we felt stirrings of ambition, we found it easy to ignore our superego.[51] But we also may have had our alibi refuted on occasion when someone to whom we could not say no prevailed on us to put on our coat and to step out into cold and windy weather to take a walk. At first our indignation grew. The discomfort was worse than we had thought. But gradually a transformation set in. Our gait became steady, our blood began to flow vigorously and wash away our tension, we smelled the rain, began thoughtfully to speak with our companion, and finally returned home settled, alert, and with a fatigue that was capable of restful sleep.

But why did such occurrences remain episodes also? The reason lies in the mistaken assumption that the shaping of our lives can be left to a series of individual decisions. Whatever goal in life we entrust to this kind of implementation we in fact surrender to erosion. Such a policy ignores both the frailty and strength of human nature. On the spur of the moment, we normally act out what has been nurtured in our daily practices as they have been shaped by the norms of our time. When we sit in our easy chair and contemplate what to do, we are firmly enmeshed in the framework of technology with our labor behind us and the blessings of our labor about us, the diversions and enrichments of consumption. This arrangement has had our lifelong allegiance, and we know it to have the approval and support of our fellows. It would take superhuman strength to stand up to this order ever and again. If we are to challenge *the rule of technology,* we can do so only through *the practice of engagement.*

The human ability to establish and commit oneself to a practice reflects our capacity to comprehend the world, to harbor it in its expanse as a context that is oriented by its focal points. To found a practice is to guard a focal concern, to shelter it against the vicissitudes of fate and our frailty. John Rawls has pointed out that there is decisive difference between the justification of a practice and of a particular action falling under it.[52] Analogously, it is one thing to decide for a focal practice and quite another to decide for a particular action that appears to have focal character.[53] Putting the matter more clearly, we must say that without a practice an engaging action or event can momentarily light up our life, but it cannot order and orient it focally. Competence, excellence, or virtue, as Aristotle first saw, come into being as an *éthos,* a settled disposition and a way of life.[54] Through a practice, Alasdaire MacIntyre says accordingly, "human powers to achieve excellence, and human conceptions of the ends and goods involved, are systematically extended."[55] Through a practice we are able to accomplish what remains unattainable when aimed at in a series of individual decisions and acts.

How can a practice be established today? Here, as in the case of focal things, it is helpful to consider the foundation of pretechnological practices. In mythic times the latter were often established through the founding and consecrating act of a divine power or mythic ancestor. Such an act set up a sacred precinct and center that gave order to a violent and hostile world. A sacred practice, then, consisted in the regular reenactment of the founding act, and so it renewed and sustained the order of the world. Christianity came into being this way; the eucharistic meal, the Supper of the Lamb, is its central event, established with the instruction that it be reenacted. Clearly a focal practice today should have centering and orienting force as well. But it differs in important regards from its grand precursors. A mythic focal practice derived much force from the power of its opposition. The alternative to the preservation of the cosmos was chaos, social and physical disorder and collapse. It is a reduction to see mythic practices merely as coping behavior of high survival value. A myth does not just aid survival; it defines what truly human life is. Still, as in the case of pretechnological morality, economic and social factors were interwoven with mythic practices. Thus the force of brute necessity supported, though it did not define, mythic focal practices. Since a mythic focal practice united in itself the social, the economic, and the cosmic, it was naturally a prominent and public affair. It rested securely in collective memory and in the mutual expectations of the people.

This sketch, of course, fails to consider many other kinds of pretechnological practices. But it does present one important aspect of them and more particularly one that serves well as a backdrop for focal practices in a technological setting. It is evident that technology is itself a sort of practice, and it procures its own kind of order and security. Its history contains great moments of innovation, but it did not arise out of a founding event that would have focal character; nor has it produced focal things. Thus it is not a *focal* practice, and it has indeed, so I have urged, a debilitating tendency to scatter our attention and to clutter our surroundings. A focal practice today, then, meets no tangible or overtly hostile opposition from its context and is so deprived of the wholesome vigor that derives from such opposition. But there is of course an opposition at a more profound and more subtle level. To feel the support of that opposing force one must have experienced the subtly debilitating character of technology, and above all one must understand, explicitly or implicitly, that the peril of technology lies not in this or that of its manifestations but in *the pervasiveness and consistency of its pattern.* There are always occasions where a Big Mac, an exercycle, or a television program are unobjectionable and truly helpful answers to human needs. This makes a case-by-case appraisal of technology so inconclusive. It is when we attempt to take the measure of technological life in its normal totality that we are distressed by its shallowness. And I believe that the more strongly we sense and the more clearly we understand the coherence and the character of technology, the more evident it becomes to us that technology must be countered by an equally patterned and social commitment, i.e., by a practice.

At this level the opposition of technology does become fruitful to focal practices. They can now be seen as restoring a depth and integrity to our lives that are in principle excluded within the paradigm of technology. MacIntyre, though his foil is the Enlightenment more than technology, captures this point by including in his definition of practice the notion of "goods internal to a practice."[56] These are one with the practice and can only be obtained through that practice. The split between means and ends is healed. In contrast "there are those goods externally and contingently attached" to a practice; and in that case there "are always alternative ways for achieving such goods, and their achievement is never to be had *only* by engaging in some particular kind of practice."[57] Thus practices (in a looser sense) that serve external goods are subvertible by technology. But MacIntyre's point needs to be clarified and extended to include or emphasize not only the essential unity of human being and a particular sort of doing but also the tangible things in which the world comes to be focused. The importance of this point has been suggested by the consideration of running and the culture of the table. There are objections to this suggestion. Here I want to advance the thesis by considering Rawls's contention that a practice is defined by rules. We can take a rule as an instruction for a particular domain of life to act in a certain way under specified circumstances. How important is the particular character of the tangible setting of the rules? Though Rawls does not address this question directly he suggests in using baseball for illustration that "a peculiarly shaped piece of wood" and a kind of bag become a bat and base only within the confines defined by the rules of baseball.[58] Rules and the practice they define, we might argue in analogy to what Rawls says about their relation to particular cases, are logically prior to their tangible setting. But the opposite contention seems stronger to me. Clearly the possibilities and challenges of baseball are crucially determined by the layout and the surface of the field, the weight and resilience of the ball, the shape and size of the bat, etc. One might of course reply that there are rules that define the physical circumstances of the game. But this is to take "rule" in broader sense. Moreover it would be more accurate to say that the rules of this latter sort reflect and protect the identity of the original tangible circumstances in which the game grew up. The rules, too, that circumscribe the actions of the players can be taken as ways of securing and ordering the playful challenges that arise in the human interplay with reality. To be sure there are developments and innovations in sporting equipment. But either they quite change the nature of the sport as in pole vaulting, or they are restrained to preserve the identity of the game as in baseball.

It is certainly the purpose of a focal practice to guard in its undiminished depth and identity the thing that is central to the practice, to shield it against the technological diremption into means and end. Like values, rules and practices are recollections, anticipations, and, we can now say, guardians of the concrete things and events that finally matter. Practices protect focal things not only from technological subversion but also against human frailty. It was emphasized that the ultimately significant things to which we respond in deictic discourse cannot be possessed or controlled. Hence when we reach out for them, we miss them occasionally and sometimes for quite some time. Running becomes unrelieved pain and cooking a thankless chore. If in the technological mode we insisted on assured results or if more generally we estimated the value of future efforts on the basis of recent experience, focal things would vanish from our lives. A practice keeps faith with focal things and saves for them an opening in our lives. To be sure, eventually the practice needs to be empowered again by the reemergence of the great thing in its splendor. A practice that is not so revived degenerates into an empty and perhaps deadening ritual.

We can now summarize the significance of a focal practice and say that such a practice is required to counter technology in its patterned pervasiveness and to guard focal things in their depth and integrity. Countering technology through a practice is to take account of our susceptibility to technological distraction, and it is also to engage the peculiarly human strength of comprehension, i.e., the power to take in the world in its extent and significance and to respond through an enduring commitment. Practically a focal practice comes into being through resoluteness, either an explicit resolution where one vows regularly to engage in a focal activity from this day on or in a more implicit resolve that is nurtured by a focal thing in favorable circumstances and matures into a settled custom.

In considering these practical circumstances we must acknowledge final difference between focal practices today and their eminent pretechnological predecessors. The latter, being public and prominent, commanded elaborate social and physical settings: hierarchies, offices, ceremonies, and choirs; edifices, altars, implements, and vestments. In comparison our focal practices are humble and scattered. Sometimes they can hardly be called practices, being private and limited. Often they begin as a personal regimen and mature into a routine without ever attaining the social richness that distinguishes a practice. Given the often precarious and inchoate nature of focal practices, evidently focal things and practices, for all the splendor of their simplicity and their fruitful opposition to technology, must be further clarified in their relation to our everyday world if they are to be seen as a foundation for the reform of technology.

WEALTH AND THE GOOD LIFE

Strong claims have been made for focal things and practices. Focal concerns supposedly allow us to center our lives and to launch a reform of technology and so to usher in the good life that has eluded technology. We have seen that focal practices today tend to be isolated and rudimentary. But these are marginal deficiencies, due to unfavorable circumstances. Surely there are central problems as well that pertain to focal practices no matter how well developed. Before we can proceed to suggestions about how technology may be reformed to make room for the good life, the most important objections regarding focal practices, the pivots of that reform, must be considered and, if possible, refuted. These disputations are not intended to furnish the impregnable defense of focal concerns which is neither possible nor to be wished for. The deliberations of this chapter are rather efforts to connect the notion of a focal practice more closely with the prevailing conceptual and social situation and so to advance the standing of focal concerns in our midst. To make the technological universe hospitable to focal things turns out to be the heart of the reform of technology.

Let me now draw out the concrete consequences of this kind of reform. I begin with particular illustrations and proceed to broader observations. Sheehan's focal concern is running, but he does not run everywhere he wants to go. To get to work he drives a car. He depends on that technological device and its entire associated machinery of production, service, resources, and roads. Clearly, one in Sheehan's position would want the car to be as perfect a technological device as possible: safe, reliable, easy to operate, free of maintenance. Since runners deeply enjoy the air, the trees, and the open spaces that grace their running, and since human vigor and health are essential to their enterprise, it would be consistent of them to want an environmentally benign car, one that is free of pollution and requires a minimum of resources for its production and operation. Since runners express themselves through running, they would not need to do so through the glitter, size, and newness of their vehicles.[59]

At the threshold of their focal concern, runners leave technology behind, technology, i.e., as a way of taking up with the world. The products of technology remain ubiquitous, of course: clothing, shoes, watches, and the roads. But technology can produce instruments as well as devices, objects that call forth engagement and allow for a more skilled and intimate contact with the world.[60] Runners appreciate shoes that are light, firm, and shock absorbing. They allow one to move faster, farther, and more fluidly. But runners would not want to have such movement procured by a motorcycle, nor would they, on the other side, want to obtain merely the physiological benefit of such bodily movement from a treadmill.

A focal practice engenders an intelligent and selective attitude toward technology. It leads to a simplification and perfection of technology in the background of one's focal concern and to a discerning use of technological products at the center of one's practice. I am not, of course, describing an evident development or state of affairs. It does appear from what little we know statistically of the runners in this country, for instance, that they lead a more engaged, discriminating, and a socially more profound life.[61] I am rather concerned to draw out the consequences that naturally follow for technology from a focal commitment and from a recognition of the device pattern. There is much diffidence, I suspect, among people whose life is centered, even in their work, around a great concern. Music is surely one of these. But at times, it seems to me, musicians confine the radiance, the rhythm, and the order of music and the ennobling competence that it requires to the hours and places of performance. The entrenchment of technology may make it seem quixotic to want to lead a fully musical life or to change the larger technological setting so that it would be more hospitable and attentive to music. Moreover, as social creatures we seek the approval of our fellows according to the prevailing standards. One may be a runner first and most of all; but one wants to prove too that one has been successful in the received sense. Proof requires at least the display, if not the consumption, of expensive commodities. Such inconsistency is regrettable, not because we just have to have reform of technology but because it is a partial disavowal of one's central concern. To have a focal thing radiate transformatively into its environment is not to exact some kind of service from it but to grant it its proper eloquence.

There is of course intuitive evidence for the thesis that a focal commitment leads to an intelligent limitation of technology. There are people who, struck by a focal concern, remove much technological clutter from their lives. In happy situations, the personal and private reforms take three directions. The first is of course to clear a central space for the focal thing, to establish an inviolate time for running, or to establish a hearth in one's home for the culture of the table. And this central clearing goes hand in hand, as just suggested, with a newly discriminating use of technology.[62] The second direction of reform is the simplification of the context that surrounds and supports the focal area. And then there is a third endeavor, that of extending the sphere of engagement as far as possible. Having experienced the depth of things and the pleasure of full-bodied competence at the center, one seeks to extend such excellence to the margins of life. "Do it yourself" is the maxim of this tendency and "self-sufficiency" its goal. But the tendencies for which these titles stand also exhibit the dangers of this third direction of reform. Engagement, however skilled and disciplined, becomes disoriented when it exhausts itself in the building, rebuilding, refinement, and maintenance of stages on which nothing is ever enacted. People finish their basements, fertilize their lawns, fix their cars. What for? The peripheral engagement suffocates the center, and festivity, joy, and humor disappear. Similarly, the striving for self-sufficiency may open up a world of close and intimate relations with things and people. But the demands of the goal draw a narrow and impermeable boundary about that world. There is no time to be a citizen of the cultural and political world at large and no possibility of assuming one's responsibility in it. The antidote to such disorientation and constriction is the appropriate acceptance of technology. In one or another area of one's life one should gratefully accept the disburdenment from daily and time-consuming chores and allow celebration and world citizenship to prosper in the time that has been gained.

What emerges here is a distinct notion of the good life or more precisely the private or personal side of one. Clearly, it will remain crippled if it cannot unfold into the world of labor and the public realm. To begin on the side of leisure and privacy is to acknowledge the presently dispersed and limited standing of focal powers. It is also to avail oneself of the immediate and undeniably large discretion one has in shaping one's free time and private sphere.[63] Even within these boundaries the good life that is centered on focal concerns is distinctive enough. Evidently, it is a favored and prosperous life. It possesses the time and the implements that are needed to devote oneself to a great calling. Technology provides us with the leisure, the space, the books, the instruments, the equipment, and the instruction that allow us to become equal to some great thing that has beckoned us from afar or that has come to us through a tradition. The citizen of the technological society has been spared the abysmal bitterness of knowing himself or herself to be capable of some excellence or achievement and of being at the same time worn-out by poor and endless work, with no time to spare and no possibility of acquiring the implements of one's desire. That bitterness is aggravated when one has a gifted child that is similarly deprived, and is exacerbated further through class distinctions where one sees richer but less gifted and dedicated persons showered with opportunities of excellence. There is prosperity also in knowing that one is able to engage in a focal practice with a great certainty of physical health and economic security. One can be relatively sure that the joy that one receives from a focal thing will not be overshadowed by the sudden loss of a loved one with whom that joy is shared. And one prospers not only in being engaged in a profound and living center but also in having a view of the world at large in its essential political, cultural, and scientific dimensions. Such a life is centrally prosperous, of course, in opening up a familiar world where things stand out clearly and steadily, where life has a rhythm and depth, where we encounter our fellow human beings in the fullness of their capacities, and where we know ourselves to be equal to that world in depth and strength.

This kind of prosperity is made possible by technology, and it is centered in a focal concern. Let us call it wealth to distinguish it from the prosperity that is confined to technology and that I want to call affluence. Affluence consists in the possession and consumption of the most numerous, refined, and varied commodities. This superlative formulation betrays its relative character. "Really" to be affluent is to live now and to rank close to the top of the hierarchy of inequality. All of the citizens of a typical technological society are more affluent than anyone in the Middle Ages. But this affluence, astounding when seen over time, is dimmed or even insensible at any one time for all but those who have a disproportionately large share of it. Affluence, strictly defined, has an undeniable glamour. It is the embodiment of the free, rich, and imperial life that technology has promised. So at least it appears from below whence it is seen by most people. Wealth in comparison is homely, homely in the sense of being plain and simple but homely also in allowing us to be at home in our world, intimate with its great things, and familiar with our fellow human beings. This simplicity, as said before, has its own splendor that is more sustaining than the glamour of affluence which leaves its beneficiaries, so we hear, sad and bored.[64] Wealth is a romantic notion also in that it continues and develops a tradition of concerns and of excellence that is rooted on the other side of the modern divide, i.e., of the Enlightenment. A life of wealth is certainly not romantic in the sense of constituting an uncomprehending rejection of the modern era and a utopian reform proposal.[65]

I will conclude by considering the narrower sphere of wealth and by connecting it with the traditional notions of excellence and of the family. The virtues of world citizenship, of gallantry, musicianship, and charity still command an uneasy sort of allegiance and it is natural, therefore, to measure the technological culture by these standards. Perhaps people are ready to accept the distressing results of such measurement with a rueful sort of agreement. But obviously the acceptance of the standards, if there is one, is not strong enough to engender the reforms that the pursuit of traditional excellence would demand. This, I believe, is due to the fact that the traditional virtues have for too long been uprooted from the soil that used to nourish them. Values, standards, and rules, I have urged repeatedly, are recollections and anticipations of great things and events. They provide bonds of continuity with past greatness and allow us to ready ourselves and our children for the great things we look forward to. Rules and values inform and are acted out in practices. A virtue is the practiced and accomplished faculty that makes one equal to a great event. From such considerations it is evident that the real circumstances and forces to which the traditional values, virtues, and rules used to answer are all but beyond recollection, and there is little in the technological universe that they can anticipate and ready us for. The peculiar character of technological reality has escaped the attention of the modern students of ethics.

To sketch a notion of excellence that is appropriate to technology is, in one sense, simply to present another version of the reform of technology that has been developed so far. But it is also to uncover and to strengthen ties to a tradition that the modern era has neglected to its peril. As regards world citizenship today, the problem is not confinement but the proliferation of channels of communication and of information. From the mass of available information we select by the criteria of utility and entertainment. We pay attention to information that is useful to the maintenance and advancement of technology, and we consume those news items that divert us. In the latter case the world is shredded into colorful bits of entertainment, and the distracted kind of knowledge that corresponds to that sort of information is the very opposite of the principled appropriation of the world that is meant by world citizenship.[66] The realm of technically useful information does not provide access to world citizenship either. Technical information is taken up primarily in one's work. Since most work in technology is unskilled, the demands on technical knowledge are low, and most people know little of science, engineering, economics, and politics. The people at the leading edge of technology have difficulty in absorbing and integrating the information that pertains to their field.[67] But even if the flood of technical information is appropriately channeled, as I think it can be, its mastery still constitutes knowledge of the social machinery, of the means rather than the ends of life. What is needed if we are to make the world truly and finally ours again is the recovery of a center and a standpoint from which one can tell what matters in the world and what merely clutters it up. A focal concern is that center of orientation. What is at issue here comes to the fore when we compare the simple and authentic world appropriation of someone like Mother Teresa with the shallow and vagrant omniscience of a technocrat.

Gallantry in a life of wealth is the fitness of the human body for the greatness and the playfulness of the world. Thus it has a grounding and a dignity that are lost in traditional gallantry, a loss that leaves the latter open to the technological concept of the perfect body where the body is narcissistically stylized into a glamorous something by whatever scientific means and according to the prevailing fashion. In the case of musicianship the tradition of excellence is unbroken and has expanded into jazz and popular music. What the notion of wealth can contribute to the central splendor and competence of music is to make us sensible to the confinement and the procurement of music. Confinement and procurement are aspects of the same phenomenon. The discipline and the rhythmic grace and order that characterize music are often confined, as said above, to the performance proper and are not allowed to inform the broader environment. This is because the unreformed structure of the technological universe leaves no room for such forces. Accordingly, music is allowed to conform to technology and is procured as a commodity that is widely and inconsequentially consumed. A focal concern for musicianship, then, will curtail the consumption of music and secure a more influential position for the authentic devotion to music.

Finally, one may hope that focal practices will lead to a deepening of charity and compassion. Focal practices provide a profounder commerce with reality and bring us closer to that intensity of experience where the world engages one painfully in hunger, disease, and confinement. A focal practice also discloses fellow human beings more fully and may make us more sensitive to the plight of those persons whose integrity is violated or suppressed. In short, a life of engagement may dispel the astounding callousness that insulates the citizens of the technological societies from the well-known misery in much of the world. The crucial point has been well made by Duane Elgin:

> When people deliberately choose to live closer to the level of material sufficiency, they are brought closer to the reality of material existence for a majority of persons on this planet. There is not the day-to-day insulation from material poverty that accompanies the hypnosis of a culture of affluence.[68]

The plight of the family, finally, consists in the absorption of its tasks and substance by technology. The reduction of the household to the family and the growing emptiness of family life leave the parents bewildered and the children without guidance. Since less and less of vital significance remains entrusted to the family, the parents have ceased to embody rightful authority and a tradition of competence, and correspondingly there is less and less legitimate reason to hold children to any kind of discipline. Parental love is deprived of tangible and serious circumstances in which to realize itself. Focal practices naturally reside in the family, and the parents are the ones who should initiate and train their children in them. Surely parental love is one of the deepest forms of sympathy. But sympathy needs enthusiasm to have substance. Families, I have found, that we are willing to call healthy, close, or warm turn out, on closer inspection, to be centered on a focal concern. And even in families that exhibit the typical looseness of structure, the diffidence of parents, and the impertinence of children, we can often discover a bond of respect and deep affection between parent and youngster, one that is secured in a common concern such as a sport and keeps the family from being scattered to the winds.

NOTES

1. Earlier versions of this notion of technology can be found in "Technology and Reality," *Man and World* 4 (1971): 59–69; "Orientation in Technology," *Philosophy Today* 16 (1972): 135–47; "The Explanation of Technology," *Research in Philosophy and Technology* 1 (1978): 99–118. Daniel J. Boorstin similarly describes the character of everyday America in terms of availability and its constituents. See his *Democracy and Its Discontents* (New York, 1975).

2. See Emmanuel G. Mesthene, *Technological Change* (New York, 1970), p. 28.

3. See Melvin M. Rotsch, "The Home Environment," in *Technology in Western Civilization,* ed. Melvin Kranzberg and Carroll W. Pursell, Jr. (New York, 1967), 2: 226–28. For the development of the kitchen stove (the other branch into which the original fireplace or stove developed), see Siegfried Giedion, *Mechanization Takes Command* (New York, 1969 [first published in 1948]), pp. 527–47.

4. See George Sturt's description of the sawyers in *The Wheelwright's Shop* (Cambridge, 1974 [first published in 1923]), pp. 32–40.

5. In economics, "commodity" is a technical term for a tradable (and usually movable) economic good. In social science, it has become a technical term as a translation of Marx's *Ware* (merchandise). Marx's use and the use here suggested and to be developed agree inasmuch as both are intended to capture a novel and ultimately detrimental transformation of a traditional (pretechnological) phenomenon. For Marx, a commodity of the negative sort is the result of the reification of social relations, in particular of the reification of the workers' labor power, into something tradable and exchangeable which is then wrongfully appropriated by the capitalists and used against the workers. This constitutes the exploitation of the workers and their alienation from their work. It finally leads to their pauperization. I disagree that this transformation is at the center of gravity of the modern social order. The crucial change is rather the splitting of the pretechnological fabric of life into machinery and commodity according to the device paradigm. Though I concede and stress the

tradable and exchangeable character of commodities, as I use the term, their primary character, here intended, is their commodious and consumable availability with the technological machinery as their basis and with disengagement and distraction as their recent consequences. On Marx's notion of commodity and commodity fetishism, see Paul M. Sweezy, *The Theory of Capitalist Development* (New York, 1968), pp. 34–40.

6. See Morton Kaplan, "Means/Ends Rationality," *Ethics* 87 (1976): 61–65.

7. Martin Heidegger gives a careful account of the interpenetration of means and ends in the pretechnological disclosure of reality. But when he turns to the technological disclosure of being (*das Gestell*) and to the device in particular (*das Gerät*), he never points out the peculiar technological diremption of means and ends though he does mention the instability of the machine within technology. Heidegger's emphasis is perhaps due to his concern to show that technology as a whole is not a means or an instrument. See his "The Question Concerning Technology," in *The Question Concerning Technology and Other Essays,* trans. William Lovitt (New York, 1977).

8. It also turns out that a generally rising standard of living makes personal services disproportionately expensive. See Staffan B. Linder, *The Harried Leisure Class* (New York, 1970), pp. 34–37.

9. See *Paulys Realencyclopädie der classischen Altertumswissenschaft* (Stuttgart, 1893–1963), 15: 615–17; See also Fustel de Coulanges, "The Sacred Fire," in *The Ancient City,* trans. Willard Small (Garden City, N.Y., n.d. [first published in 1864]), pp. 25–33.

10. See Kent C. Bloomer and Charles W. Moore, *Body, Memory, and Architecture* (New Haven, 1977), pp. 2–3 and 50–51.

11. See Jeremiah Eck, "Home Is Where the Hearth Is," *Quest* 3 (April 1979): 12.

12. Ibid., p. 11.

13. See *The Oxford English Dictionary.*

14. See Martin Heidegger, "The Origin of the Work of Art," in *Poetry, Language, Thought,* trans. Albert Hofstadter (New York, 1971), pp. 15–87.

15. See Vincent Scully, *The Earth, the Temple, and the Gods* (New Haven, 1962).

16. See my *The Philosophy of Language* (The Hague, 1974), pp. 126–31.

17. See Heidegger, *Being and Time,* trans. John Macquarrie and Edward Robinson (New York, 1962), pp. 95–107, 163–68, 210–24.

18. See Heidegger, "The Origin of the Work of Art," p. 80.

19. See Heidegger, "The Thing," in *Poetry, Language, Thought,* pp. 163–82. Heidegger alludes to the turn from the *Bedingungen to the Dinge* on p. 179 of the original, "Das Ding," in *Vorträge und Aufsätze* (Pfullingen, 1959). He alludes to the turn from technology to (focal) things in "The Question Concerning Technology."

20. See Heidegger, "The Thing."

21. See M. F. K. Fisher, *The Cooking of Provincial France* (New York, 1968), p. 50.

22. Though there are seeds for a reform of technology to be found in Heidegger as I want to show, Heidegger insists that "philosophy will not be able to effect an immediate transformation of the present condition of the world. Only a god can save us." See "Only a God Can Save Us: Der Spiegel's Interview with Martin Heidegger," trans. Maria P. Alter and John D. Caputo, *Philosophy Today* 20 (1976): 277.

23. I am not concerned to establish or defend the claim that my account of Heidegger or my development of his views is authoritative. It is merely a matter here of acknowledging a debt.

24. See Heidegger, "The Question Concerning Technology"; Langdon Winner makes a similar point in "The Political Philosophy of Alternative Technology," in *Technology and Man's Future,* ed. Albert H. Teich, 3d ed. (New York, 1981), pp. 369–73.

25. See Heidegger, "The Thing," pp. 180–82.

26. The need of complementing Heidegger's notion of the thing with the notion of practice was brought home to me by Hubert L. Dreyfus's essay, "Holism and Hermeneutics," *Review of Metaphysics* 34 (1980): 22–23.

27. See Heidegger, "Building Dwelling Thinking," in *Poetry, Language, Thought,* pp. 152–53.

28. Ibid., pp. 148–49.

29. Georg Trakl, quoted by Heidegger in "Language," in *Poetry, Language, Thought,* pp. 194–95 (I have taken some liberty with Hofstadter's translation).

30. See Normal Maclean, *A River Runs through It and Other Stories* (Chicago, 1976). Only the first of the three stories instructs the reader about fly fishing.

31. See Colin Fletcher, *The Complete Walker* (New York, 1971).

32. Ibid., p. 9.

33. See Roger B. Swain, *Earthly Pleasures: Tales from a Biologist's Garden* (New York, 1981).

34. Here are a few more: Wendell Berry, *Farming: A Handbook* (New York, 1970); Stephen Kiesling, *The Shell Game: Reflections on Rowing and the Pursuit of Excellence* (New York, 1982); John Richard Young, *Schooling for Young Riders* (Norman, Okla., 1970); W. Timothy Gallwey, *The Inner Game of Tennis* (New York, 1974); Ruedi Bear, *Pianta Su: Ski Like the Best* (Boston, 1976). Such books must be sharply distinguished from those that promise to teach accomplishments without effort and in no time. The latter kind of book is technolgical in intent and fraudulent in fact.

35. See Robert Farrar Capon, *The Supper of the Lamb: A Culinary Reflection* (Garden City, N.Y., 1969); and George Sheehan, *Running and Being: The Total Experience* (New York, 1978).

36. See Sheehan, pp. 211–20 and elsewhere.

37. See Capon, pp. 167–181.

38. See my "Mind, Body, and World," *Philosophical Forum* 8 (1976): 76–79.

39. See Capon, pp. 176–77.

40. See Sheehan, p. 127.

41. On the unity of achievement and enjoyment, see Alasdair MacIntyre, *After Virtue* (Notre Dame, Ind., 1981), p. 184.

42. See my "Mind, Body, and World," pp. 68–86.

43. See Peter Wood, "Seeing New York on the Run," *New York Times Magazine,* 7 October 1979; Alexandra Penney, "Health and Grooming: Shaping Up the Corporate Image," ibid.

44. See *New York Times Magazine,* 3 August 1980, pp. 20–21.

45. See Wood, p. 112.

46. See Sheehan, pp. 211–17.

47. See Wood, p. 116.

48. See Sheehan, pp. 221–31 and passim.

49. There is substantial anthropological evidence to show that running has been a profound focal practice in certain pretechnological cultures. I am unable to discuss it here. Nor have I discussed the problem, here and elsewhere touched upon, of technology and religion. The present study, I believe, has important implications for that issue, but to draw them out would require more space and circumspection than are available now. I have made attempts to provide an explication in "Christianity and the Cultural Center of Gravity," *Listening* 18 (1983): 93–102; and in "Prospects for the Theology of Technology," *Theology and Technology,* ed. Carl Mitcham and Jim Grote (Lanham, Md., 1984), pp. 305–22.

50. See M. F. K. Fisher, pp. 9–31.

51. Some therapists advise lying down till these stirrings go away.

52. See John Rawls, "Two Concepts of Rules," *Philosophical Review* 64 (1955): 3–32.

53. Conversely, it is one thing to break a practice and quite another to omit a particular action. For we define ourselves and our lives in our practices; hence to break a practice is to jeopardize one's identity while omitting a particular action is relatively inconsequential.

54. See Aristotle's *Nicomachean Ethics,* the beginning of Book Two in particular.

55. See MacIntyre, p. 175.

56. Ibid., pp. 175–77.

57. Ibid., p. 176.

58. See Rawls, p. 25.

59. On the general rise and decline of the car as a symbol of success, see Daniel Yankelovich, *New Rules: Searching for Self-Fulfillment in a World Turned Upside Down* (Toronto, 1982), pp. 36–39.

60. Although these technological instruments are translucent relative to the world and so permit engagement with the world, they still possess an opaque machinery that mediates engagement but is not itself experienced either directly or through social mediation. See also the remarks in n. 12 above.

61. See "Who Is the American Runner?" *Runner's World* 15 (December 1980): 36–42.

62. Capon's book is the most impressive document of such discriminating use of technology.

63. A point that is emphatically made by E. F. Schumacher in *Small Is Beautiful* (New York, 1973) and in *Good Work* (New York, 1979); by Duane Elgin in *Voluntary Simplicity* (New York, 1981); and by Yankelovich in *New Rules.*

64. See Roger Rosenblatt, "The Sad Truth about Big Spenders," *Time,* 8 December 1980, pp. 84 and 89.

65. On the confusions that beset romanticism in its opposition to technology, see Lewis Mumford, *Technics and Civilization* (New York, 1963), pp. 285–303.

66. See Daniel J. Boorstin, *Democracy and Its Discontents* (New York, 1975), pp. 12–25.

67. See Elgin, pp. 251–71. In believing that the mass of complex technical information poses a mortal threat to bureaucracies, Elgin, it seems to me, indulges in the unwarranted pessimism of the optimists.

68. Ibid., p. 71.

10

A Phenomenology of Technics

Don Ihde

The task of a phenomenology of human-technology relations is to discover the various structural features of those ambiguous relations. In taking up this task, I shall begin with a focus upon experientially recognizable features that are centered upon the ways we are bodily engaged with technologies. The beginning will be within the various ways in which I-as-body interact with my environment by means of technologies.

A. TECHNICS EMBODIED

If much of early modern science gained its new vision of the world through optical technologies, the process of embodiment itself is both much older and more pervasive. To embody one's praxis *through* technologies is ultimately an *existential* relation with the world. It is something humans have always—since they left the naked perceptions of the Garden—done.

I have previously and in a more suggestive fashion already noted some features of the visual embodiment of optical technologies. Vision is technologically transformed through such optics. But while the fact *that* optics transform vision may be clear, the variants and invariants of such a transformation are not yet precise. That becomes the task for a more rigorous and structural phenomenology of embodiment. I shall begin by drawing from some of the previous features mentioned in the preliminary phenomenology of visual technics.

Within the framework of phenomenological relativity, visual technics first may be located within the intentionality of seeing.

I see—through the optical artifact—the world

This seeing is, in however small a degree, at least minimally distinct from a direct or naked seeing.

I see—the world

I call this first set of existential technological relations with the world *embodiment relations,* because in this use context I take the technologies *into* my experiencing in a particular way by way of perceiving *through* such technologies and through the reflexive transformation of my perceptual and body sense.

In Galileo's use of the telescope, he embodies his seeing through the telescope thusly:

Galileo—telescope—Moon

Equivalently, the wearer of eyeglasses embodies eyeglass technology:

I—glasses—world

The technology is actually *between* the seer and the seen, in a *position of mediation.* But the referent of the seeing, that towards which sight is directed, is "on the other side" of the optics. One sees *through* the optics. This, however, is not enough to specify this relation as an embodiment one. This is because one first has to determine *where* and *how,* along what will be described as a continuum of relations, the technology is experienced.

There is an initial sense in which this positioning is doubly ambiguous. First, the technology must be *technically* capable of being seen through; it must be transparent. I shall use the term *technical* to refer to the physical characteristics of the technology. Such characteristics may be designed or they may be discovered. Here the disciplines that deal with such characteristics are informative, although indirectly so for the philosophical analysis per se. If the glass is not transparent enough, seeing-through is not possible. If it is transparent enough, approximating whatever "pure" transparency could be empirically attainable, then it becomes possible to embody the technology. This is a material condition for embodiment.

Embodying as an activity, too, has an initial ambiguity. It must be learned or, in phenomenological terms, constituted. If the technology is good, this is usually easy. The very first time I put on my glasses, I see the now-corrected world. The adjustments I have to make are not usually focal irritations but fringe ones (such as the adjustment to backglare and the slight changes in spatial motility). But once learned, the embodiment relation can be more precisely described as one in which the technology becomes maximally "transparent." It is, as it were, taken into my own perceptual-bodily self experience thus:

(I-glasses)-world

My glasses become part of the way I ordinarily experience my surroundings; they "withdraw" and are barely noticed, if at all. I have then actively embodied the technics of vision. Technics is the symbiosis of artifact and user within a human action.

Embodiment relations, however, are not at all restricted to visual relations. They may occur for any sensory or microperceptual dimension. A hearing aid does this for hearing, and the blind man's cane for tactile motility. Note that in these corrective technologies *the same structural features of embodiment* obtain as with the visual example. Once learned, cane and hearing aid "withdraw" (if the technology is good—and here we have an experiential clue for the perfecting of technologies). I hear the world through the hearing aid and feel (and hear) it through the cane. The juncture (I-artifact)-world is through the technology and brought close by it.

Such relations *through* technologies are not limited to either simple or complex technologies. Glasses, insofar as they are engineered systems, are much simpler than hearing aids. More complex than either of these monosensory devices are those that entail whole-body motility. One such common technology is automobile driving. Although driving an automobile encompasses more than embodiment relations, its pleasurability is frequently that associated with embodiment relations.

One experiences the road and surroundings *through* driving the car, and motion is the focal activity. In a finely engineered sports car, for example, one has a more precise feeling of the road and of the traction upon it than in the older, softer-riding, large cars of the fifties. One embodies the car, too, in such activities as parallel parking: when well embodied, one feels rather than sees the distance between car and curb—one's bodily sense is "extended" to the parameters of the driver-car "body." And although these embodiment relations entail larger, more complex artifacts and entail a somewhat longer, more complex learning process, the bodily tacit knowledge that is acquired is perceptual-bodily.

Here is a first clue to the polymorphous sense of bodily extension. The experience of one's "body image" is not fixed but malleably extendable and/or reducible in terms of the material or technological mediations that may be embodied. I shall restrict the term embodiment, however, to those types of mediation that can be so experienced. The same dynamic polymorphousness can also be located in non-mediational or direct experience. Persons trained in the martial arts, such as karate, learn to feel the vectors and trajectories of the opponent's moves within the space of the combat. The near space around one's material body is charged.

Embodiment relations are a particular kind of use-context. They are technologically relative in a double sense. First, the technology must "fit" the use. Indeed, within the realm of embodiment relations one can develop a quite specific set of qualities for design relating to attaining the requisite technological "withdrawal." For example, in handling highly radioactive materials at a distance, the mechanical arms and hands which are designed to pick up and pour glass tubes inside the shielded enclosure have to "feed back" a delicate sense of touch to the operator. The closer to invisibility, transparency, and the extension of one's own bodily sense this technology allows, the better. Note that the design perfection is not one related to the machine alone but to the combination of machine and human. The machine is perfected along a bodily vector, molded to the perceptions and actions of humans.

And when such developments are most successful, there may arise a certain romanticizing of technology. In much anti-technological literature there are nostalgic calls for returns to simple tool technologies. In part, this may be because long-developed tools are excellent examples of bodily expressivity. They are both direct in actional terms and immediately experienced; but what is missed is that such embodiment relations may take any number of directions. Both the sports car driver within the constraints of the racing route and the bulldozer driver destroying a rainforest may have the satisfactions of powerful embodiment relations.

There is also a deeper desire which can arise from the experience of embodiment relations. It is the doubled desire that, on one side, is a wish for *total transparency,* total embodiment, for technology to truly "become me." Were this possible, it would be equivalent to there being no technology, for total transparency would *be* my body and senses; I desire the face-to-face that I would experience without the technology. But that is only one side of the desire. The other side is the desire to have the power, the transformation that the technology makes available. Only by using the technology is my bodily power enhanced and magnified by speed, through distance, or by any of the other ways in which technologies change my capacities. These capacities are always *different* from my naked capacities. The desire is, at best, contradictory. I want the transformation that the technology allows, but I want it in such a way that

I am basically unaware of its presence. I want it in such a way that it becomes me. Such a desire both secretly *rejects* what technologies are and overlooks the transformational effects which are necessarily tied to human-technology relations. This illusory desire belongs equally to pro- and anti-technology interpretations of technology.

The desire is the source of both utopian and dystopian dreams. The actual, or material, technology always carries with it only a partial or quasi-transparency, which is the price for the extension of magnification that technologies give. In extending bodily capacities, the technology also transforms them. In that sense, all technologies in use are non-neutral. They change the basic situation, however subtly, however minimally; but this is the other side of the desire. The desire is simultaneously a desire for a change in situation—to inhabit the earth, or even to go beyond the earth—while sometimes inconsistently and secretly wishing that this movement could be without the mediation of the technology.

The direction of desire opened by embodied technologies also has its positive and negative thrusts. Instrumentation in the knowledge activities, notably science, is the gradual extension of perception into new realms. The desire is to see, but seeing is seeing through instrumentation. Negatively, the desire for pure transparency is the wish to escape the limitations of the material technology. It is a platonism returned in a new form, the desire to escape the newly extended body of technological engagement. In the wish there remains the contradiction: the user both wants and does not want the technology. The user wants what the technology gives but does not want the limits, the transformations that a technologically extended body implies. There is a fundamental ambivalence toward the very human creation of our own earthly tools.

The ambivalence that can arise concerning technics is a reflection of one kind upon the *essential ambiguity* that belongs to technologies in use. But this ambiguity, I shall argue, has its own distinctive shape. Embodiment relations display an essential magnification/reduction structure which has been suggested in the instrumentation examples. Embodiment relations simultaneously magnify or amplify and reduce or place aside what is experienced through them.

The sight of the mountains of the moon, through all the transformational power of the telescope, removes the moon from its setting in the expanse of the heavens. But if our technologies were only to replicate our immediate and bodily experience, they would be of little use and ultimately of little interest. A few absurd examples might show this:

In a humorous story, a professor bursts into his club with the announcement that he has just invented a reading machine. The machine scans the pages, reads them, and perfectly reproduces them. (The story apparently was written before the invention of photocopying. Such machines might be said to be "perfect reading machines" in actuality.) The problem, as the innocent could see, was that this machine leaves us with precisely the problem we had prior to its invention. To have reproduced through mechanical "reading" all the books in the world leaves us merely in the library.

A variant upon the emperor's invisible clothing might work as well. Imagine the invention of perfectly transparent clothing through which we might technologically experience the world. We could see through it, breathe through it, smell and hear through it, touch through it. Indeed, it effects no changes of any kind, since it is *perfectly* invisible. Who would bother to pick up such clothing (even if the presumptive wearer could find it)? Only by losing some invisibility—say, with translucent coloring—would the garment begin to be usable and interesting. For here, at least, fashion would have been invented—but at the price of losing total transparency—by becoming that through which we relate to an environment.

Such stories belong to the extrapolated imagination of fiction, which stands in contrast to even the most minimal actual embodiment relations, which in their material dimensions simultaneously extend and reduce, reveal and conceal.

In actual human-technology relations of the embodiment sort, the transformational structures may also be exemplified by variations: In optical technologies, I have already pointed out how spatial significations change in observations through lenses. The entire gestalt changes. When the apparent size of the moon changes, along with it the apparent position of the observer changes. Relativistically, the moon is brought "close"; and equivalently, this optical near-distance applies to both the moon's appearance and my bodily sense of position. More subtly, every dimension of spatial signification also changes. For example, with higher and higher magnification, the well-known phenomenon of depth, instrumentally mediated as a "focal plane," also changes. Depth diminishes in optical near-distance.

A related phenomenon in the use of an optical instrument is that it transforms the spatial significations of vision in an instrumentally focal way. But my seeing without instrumentation is a full bodily seeing—I see not just with my eyes but with my whole body in a unified sensory experience of things. In part, this is why there is a noticeable irreality to the apparent position of the observer, which only diminishes with the habits acquired through practice with the instrument. But the optical instrument cannot so easily transform the entire sensory gestalt. The focal sense that is magnified through the instrument is monodimensioned.

Here may be the occasion (although I am not claiming a cause) for a certain interpretation of the senses. Historians of perception have noted that, in medieval times, not only was vision not the supreme sense but sound and smell may have had greatly enhanced roles so far as the interpretation of the senses went. Yet in the Renaissance and even more exaggeratedly in the Enlightenment, there occurred the reduction to sight as the favored sense, and within sight, a certain reduction *of* sight. This favoritism, however, also carried implications for the other senses.

One of these implications was that each of the senses was interpreted to be clear and distinct from the others, with only certain features recognizable through a given sense. Such an interpretation impeded early studies in echo location.

In 1799 Lazzaro Spallanzani was experimenting with bats. He noticed not only that they could locate food targets in the dark but also that they could do so blindfolded. Spallanzani wondered if bats could guide themselves by their ears rather than by their eyes. Further experimentation, in which the bats' ears were filled with wax, showed that indeed they could not guide themselves without their ears. Spallanzani surmised that either bats locate objects through hearing or they had some sense of which humans knew nothing. Given the doctrine of separate senses and the identification of shapes and objects through vision alone, George Montagu and Georges Cuvier virtually laughed Spallanzani out of the profession.

This is not to suggest that such an interpretation of sensory distinction was due simply to familiarity with optical technologies, but the common experience of enhanced vision through such technologies was at least the standard practice of the time. Auditory technologies were to come later. When auditory technologies did become common, it was possible to detect the same amplification/reduction structure of the human-technology experience.

The telephone in use falls into an auditory embodiment relation. If the technology is good, I hear *you* through the telephone and the apparatus "withdraws" into the enabling background:

(I-telephone)-you

But as a monosensory instrument, your phenomenal presence is that of a voice. The ordinary multidimensioned presence of a face-to-face encounter does not occur, and I must at best imagine those dimensions through your vocal gestures. Also, as with the telescope, the spatial significations are changed. There is here an auditory version of visual near-distance. It makes little difference whether you are geographically near or far, none at all whether you are north or south, and none with respect to anything but your bodily relation to the instrument. Your voice retains its partly irreal near-distance, reduced from the full dimensionality of direct perceptual situations. This telephonic distance is different both from immediate face-to-face encounters and from visual or geographical distance as normally taken. Its distance is a mediated distance with its own identifiable significations.

While my primary set of variations is to locate and demonstrate the invariance of a magnification/reduction structure to any embodiment relation, there are also secondary and important effects noted in the histories of technology. In the very first use of the telephone, the users were fascinated and intrigued by its auditory transparency. Watson heard and recognized Bell's *voice,* even though the instrument had a high ratio of noise to message. In short, the fascination attaches to magnification, amplification, enhancement. But, contrarily, there can be a kind of forgetfulness that equally attaches to the reduction. What is *revealed* is what excites; what is concealed may be forgotten. Here lies one secret for technological trajectories with respect to development. There are *latent telics* that occur through inventions.

Such telics are clear enough in the history of optics. Magnification provided the fascination. Although there were stretches of time with little technical progress, this fascination emerged from time to time to have led to compound lenses by Galileo's day. If some magnification shows the new, opens to what was poorly or not at all previously detected, what can greater magnification do? In our own time, the explosion of such variants upon magnification is dramatic. Electron enhancement, computer image enhancement, CAT and NMR internal scanning, "big-eye" telescopes—the list of contemporary magnificational and visual instruments is very long.

I am here restricting myself to what may be called a *horizontal* trajectory, that is, optical technologies that bring various micro- or macro-phenomena to vision through embodiment relations. By restricting examples to such phenomena, one structural aspect of embodiment relations may be pointed to concerning the relation to microperception and its Adamic context. While *what* can be seen has changed dramatically—Galileo's New World has now been enhanced by astronomical phenomena never suspected and by micro-phenomena still being discovered—there remains a strong phenomenological constant in *how* things are seen. All lenses and optical technologies of the sort being described bring what is to be seen into a normal bodily space and distance. Both the macroscopic and the microscopic appear within the same near-distance. The "image size" of galaxy or amoeba is the *same.* Such is the existential condition for visibility, the counterpart to the technical condition, that the instrument makes things visually present.

The mediated presence, however, must fit, be made close to my actual bodily position and sight. Thus there is a reference within the instrumental context to my face-to-face capacities. These remain primitive and central within the new mediational context. Phenomenological theory claims that for every change in what is seen (the object correlate), there is a noticeable change in how (the experiential correlate) the thing is seen.

In embodiment relations, such changes retain both an equivalence and a difference from non-mediated situations. What remains constant is the bodily focus, the reflexive reference back to my bodily capacities. What is seen must be seen from or within my visual field, from the apparent distance in which discrimination can occur regarding depth, etc., just as in face-to-face relations. But the range of what can be brought into this proximity is transformed by means of the instrument.

Let us imagine for a moment what was never in fact a problem for the history of instrumentation: If the "image size" of both a galaxy and an amoeba is the "same" for the observer using the instrument, how can we tell that one is macrocosmic and the other microcosmic? The "distance" between us and these two magnitudes, Pascal noted, was the same in that humans were interpreted to be between the infinitely large and the infinitely small.

What occurs through the mediation is not a problem *because our construction of the observation presupposes ordinary praxical spatiality.* We handle the paramecium, placing it on the slide and then under the microscope. We aim the telescope at the indicated place in the sky and, before looking through it, note that the distance is at least that of the heavenly dome. But in our imagination experiment, what if our human were *totally immersed* in a technologically mediated world? What if, from birth, all vision occurred only through lens systems? Here the problem would become more difficult. But in our distance from Adam, it is precisely the presumed difference that makes it possible for us to see both nakedly *and* mediately—and thus to be able to locate the difference—that places us even more distantly from any Garden. It is because we retain this ordinary spatiality that we have a reflexive point of reference from which to make our judgments.

The noetic or bodily reflexivity implied in all vision also may be noticed in a magnified way in the learning period of embodiment. Galileo's telescope had a small field, which, combined with early hand-held positioning, made it very difficult to locate any particular phenomenon. What must have been noted, however, even if not commented upon, was the exaggerated sense of bodily motion experienced through trying to fix upon a heavenly body—and more, one quickly learns something about the earth's very motion in the attempt to use such primitive telescopes. Despite the apparent fixity of the stars, the hand-held telescope shows the earth-sky motion dramatically. This magnification effect is within the experience of one's own bodily viewing.

This bodily and actional point of reference retains a certain privilege. All experience refers to it in a taken-for-granted and recoverable way. The bodily condition of the possibility for seeing is now twice indicated by the very situation in which mediated experience occurs. Embodiment relations continue to locate that privilege of my being here. The partial symbiosis that occurs in well-designed embodied technologies retains that motility which can be called expressive. Embodiment relations constitute one existential form of the full range of the human-technology field.

B. HERMENEUTIC TECHNICS

Heidegger's hammer in use displays an embodiment relation. Bodily action through it occurs within the environment. But broken, missing, or malfunctioning, it ceases to be the means of praxis and becomes an obtruding *object* defeating the work project. Unfortunately, that negative derivation of objectness by Heidegger carries with it a block against understanding a second existential human-technology relation, the type of relation I shall term *hermeneutic.*

The term hermeneutic has a long history. In its broadest and simplest sense it means "interpretation," but in a more specialized sense it refers to *textual* interpretation and thus entails *reading.* I shall retain both these senses and take hermeneutic to mean a special interpretive action within the technological context. That kind of activity calls for special modes of action and perception, modes analogous to the reading process.

Reading is, of course, a reading of _____; and in its ordinary context, what fills the intentional blank is a text, something *written*. But all writing entails technologies. Writing has a product. Historically, and more ancient than the revolution brought about by such crucial technologies as the clock or the compass, the invention and development of writing was surely even more revolutionary than clock or compass with respect to human experience. Writing transformed the very perception and understanding we have of language. Writing is a technologically embedded form of language.

There is a currently fashionable debate about the relationship between speech and writing, particularly within current Continental philosophy. The one side argues that speech is primary, both historically and ontologically, and the other—the French School—inverts this relation and argues for the primacy of writing. I need not enter this debate here in order to note the *technological difference* that obtains between oral speech and the materially connected process of writing, at least in its ancient forms.

Writing is inscription and calls for both a process of writing itself, employing a wide range of technologies (from stylus for cuneiform to word processors for the contemporary academic), and other material entities upon which the writing is recorded (from clay tablet to computer printout). Writing is technologically mediated language. From it, several features of hermeneutic technics may be highlighted. I shall take what may at first appear as a detour into a distinctive set of human-technology relations by way of a phenomenology of reading and writing.

Reading is a specialized perceptual activity and praxis. It implicates my body, but in certain distinctive ways. In an ordinary act of reading, particularly of the extended sort, what is read is placed before or somewhat under one's eyes. We read in the immediate context from some miniaturized bird's-eye perspective. What is read occupies an expanse within the focal center of vision, and I am ordinarily in a somewhat rested position. If the object-correlate, the "text" in the broadest sense, is a chart, as in the navigational examples, what is represented retains a representational isomorphism with the natural features of the landscape. The chart represents the land- (or sea)scape and insofar as the features are isomorphic, there is a kind of representational "transparency." The chart in a peculiar way "refers" beyond itself to what it represents.

Now, with respect to the embodiment relations previously traced, such an isomorphic representation is both similar and dissimilar to what would be seen on a larger scale from some observation position (at bird's-eye level). It is similar in that the shapes on the chart are reduced representations of distinctive features that can be directly or technologically mediated in face-to-face or embodied perceptions. The reader can compare these similarities. But chart reading is also different in that, during the act of reading, the perceptual focus is the chart itself, a substitute for the landscape.

I have deliberately used the chart-reading example for several purposes. First, the "textual" isomorphism of a representation allows this first example of hermeneutic technics to remain close to yet differentiated from the perceptual isomorphism that occurs in the optical examples. The difference is at least perceptual in that one sees *through* the optical technology, but now one *sees* the chart as the visual terminus, the "textual" artifact itself.

Something much more dramatic occurs, however, when the representational isomorphism disappears in a printed text. There is no isomorphism between the printed word and what it "represents," although there is some kind of *referential* "transparency" that belongs to this new technologically embodied form of language. It is apparent from the chart example that the chart itself becomes the *object of perception* while simultaneously referring beyond itself to what is not immediately seen. In the case of the printed text, however, the referential transparency is distinctively different from technologically embodied perceptions. *Textual transparency is hermeneutic transparency, not perceptual transparency.*

Once attained, like any other acquisition of the lifeworld, writing could be read and understood in terms of its unique linguistic transparency. Writing becomes an embodied hermeneutic technics. Now the descriptions may take a different shape. What is referred to is referred by the text and is referred to *through* the text. What now presents itself is the "world" of the text.

This is not to deny that all language has its unique kind of transparency. Reference beyond itself, the capacity to let something become present through language, belongs to speech as well. But here the phenomenon being centered upon is the new embodiment of language in writing. Even more thematically, the concern is for the ways in which writing as a "technology" transforms experiential structures.

Linguistic transparency is what makes present the *world* of the text. Thus, when I read Plato, Plato's "world" is made present. But this presence is a *hermeneutic* presence. Not only does it occur *through* reading, but it takes its shape in the interpretative context of my language abilities. His world is linguistically mediated, and while the words may elicit all sorts of imaginative and perceptual phenomena, it is through language that such phenomena occur. And while such phenomena may be strikingly rich, they do not appear as word-like.

We take this phenomenon of reading for granted. It is a sedimented acquisition of the literate lifeworld and thus goes unnoticed until critical reflection isolates its salient features. It is the same with the wide variety of hermeneutic technics we employ.

The movement from embodiment relations to hermeneutic ones can be very gradual, as in the history of writing, with little-noticed differentiations along the human-technology continuum. A series of wide-ranging variants upon readable technologies will establish the point. First, a fairly explicit example of a readable technology: Imagine sitting inside on a cold day. You look out the window and notice that the snow is blowing, but you are toasty warm in front of the fire. You can clearly "see" the cold in Merleau-Ponty's pregnant sense of perception—but you do not actually *feel* it. Of course, you could, were you to go outside. You would then have a full face-to-face verification of what you had seen.

But you might also see the thermometer nailed to the grape arbor post and *read* that it is 28°F. You would now "know" how cold it was, but you still would not feel it. To retain the full sense of an embodiment relation, there must also be retained some isomorphism with the felt sense of the cold—in this case, tactile—that one would get through face-to-face experience. One could invent such a technology; for example, some conductive material could be placed through the wall so that the negative "heat," which is cold, could be felt by hand. But this is not what the thermometer does.

Instead, you read the thermometer, and in the immediacy of your reading you *hermeneutically* know that it is cold. There is an instantaneity to such reading, as it is an already constituted intuition (in phenomenological terms). But you should not fail to note that *perceptually* what you have seen is the dial and the numbers, the thermometer "text." And that text has hermeneutically delivered its "world" reference, the cold.[1]

Such constituted immediacy is not always available. For instance, although I have often enough lived in countries where Centigrade replaces Fahrenheit, I still must translate from my intuitive familiar language to the less familiar one in a deliberate and self-conscious hermeneu-

tic act. Immediacy, however, is not the test for whether the relation is hermeneutic. A hermeneutic relation mimics sensory perception insofar as it is also a kind of seeing as _____ ; but
it is a referential seeing, which has as its immediate perceptual focus seeing the thermometer.

Now let us make the case more complex. In the example cited, the experiencer had both
embodiment (seeing the cold) and hermeneutic access to the phenomenon (reading the thermometer). Suppose the house were hermetically sealed, with no windows, and the only access
to the weather were through the thermometer (and any other instruments we might include).
The hermeneutic character of the relation becomes more obvious. I now clearly have to know
how to read the instrumentation and from this reading knowledge get hold of the "world"
being referred to.

This example has taken actual shape in nuclear power plants. In the Three Mile Island
incident, the nuclear power system was observed only through instrumentation. Part of the
delay that caused a near meltdown was *misreadings* of the instruments. There was no face-to-
face, independent access to the pile or to much of the machinery involved, nor could there be.

An intentionality analysis of this situation retains the mediational position of the technology:

I-technology-world
(engineer-instruments-pile)

The operator has instruments between him or her and the nuclear pile. But—and here, an
essential difference emerges between embodiment and hermeneutic relations—what is immediately perceived is the instrument panel itself. It becomes the object of my microperception,
although in the special sense of a hermeneutic transparency, I *read* the pile through it. This
situation calls for a different formalization:

I-(technology-world)

The parenthesis now indicates that the immediate *perceptual* focus of my experience *is* the
control panel. I read through it, but this reading is now dependent upon the semi-opaque connection between the instruments and the referent object (the pile). This *connection* may now
become enigmatic.

In embodiment relations, what allows the partial symbiosis of myself and the technology
is the capacity of the technology to become perceptually transparent. In the optical examples,
the glass-maker's and lens-grinder's arts must have accomplished this end if the embodied use
is to become possible. Enigmas which may occur regarding embodiment-use transparency thus
may occur within the parenthesis of the embodiment relation:

(I-technology) → World
‖_____‖
enigma position

(This is not to deny that once the transparency is established, thus making microperception
clear, the observer may still fail, particularly at the macroperceptual level. For the moment,
however, I shall postpone this type of interpretive problem.) It would be an oversimplification
of the history of lens-making were not problems of this sort recognized. Galileo's instrument
not only was hard to look through but was good only for certain "middle range" sightings in
astronomical terms (it did deliver the planets and even some of their satellites). As telescopes

became more powerful, levels, problems with chromatic effects, diffraction effects, etc., occurred. As Ian Hacking has noted,

> Magnification is worthless if it magnifies two distinct dots into one big blur. One needs to resolve the dots into two distinct images. . . . It is a matter of diffraction. The most familiar example of diffraction is the fact that shadows of objects with sharp boundaries are fuzzy. This is a consequence of the wave character of light.[2]

Many such examples may be found in the history of optics, technical problems that had to be solved before there could be any extended reach within embodiment relations. Indeed, many of the barriers in the development of experimental science can be located in just such limitations in instrumental capacity.

Here, however, the task is to locate a parallel difficulty in the emerging new human-technology relation, hermeneutic relations. The location of the technical problem in hermeneutic relations lies in the *connector* between the instrument and the referent. Perceptually, the user's visual (or other) terminus is *upon* the instrumentation itself. To read an instrument is an analogue to reading a text. But if the text does not correctly refer, its reference object or its world cannot be present. Here is a new location for an enigma:

$$I \rightarrow (\text{technology-world})$$

enigma position

While breakdown may occur at any part of the relation, in order to bring out the graded distinction emerging between embodiment and hermeneutic relations, a short pathology of connectors might be noted.

If there is nothing that impedes my direct perceptual situation with respect to the instrumentation (in the Three Mile Island example, the lights remain on, etc.), interpretive problems in reading a strangely behaving "text" at least occur in the open; but the technical enigma may also occur within the text-referent relation. How could the operator tell if the instrument was malfunctioning or that to which the instrument refers? Some form of *opacity* can occur within the technology-referent pole of the relation. If there is some independent way of verifying which aspect is malfunctioning (a return to unmediated face-to-face relations), such a breakdown can be easily detected. Both such occurrences are reasons for instrumental redundancy. But in examples where such independent verification is not possible or untimely, the opacity would remain.

Let us take a simple mechanical connection as a borderline case. In shifting gears on my boat, there is a lever in the cockpit that, when pushed forward, engages the forward gear; upward, neutral; and backwards, reverse. Through it, I can ordinarily feel the gear change in the transmission (embodiment) and recognize the simple hermeneutic signification (forward for forward) as immediately intuitive. Once, however, on coming in to the dock at the end of the season, I disengaged the forward gear—and the propeller continued to drive the boat forward. I quickly reversed—and again the boat continued. The hermeneutic significance had failed; and while I also felt a difference in the way the gear lever felt, I did not discover until later that the clasp that retained the lever itself had corroded, thus preventing any actual shifting at all. But even at this level there can be opacity within the technology-object relation.

The purpose of this somewhat premature pathology of human-technology relations is not to cast a negative light upon hermeneutic relations in contrast to embodiment ones but rather to indicate that there are different locations where perceptual and human-technology relations interact. Normally, when the technologies work, the technology-world relation would retain its unique hermeneutic transparency. But if the I-(technology-world) relation is far enough along the continuum to identify the relation as a hermeneutic one, the intersection of perceptual-bodily relations with the technology changes.

Readable technologies call for the extension of my hermeneutic and "linguistic" capacities *through* the instruments, while the reading itself retains its bodily perceptual location as a relation *with* or *towards* the technology. What is emerging here is the first suggestion of an emergence of the technology as "object" but without its negative Heideggerian connotation. Indeed, the type of special capacity as a "text" is a condition for hermeneutic transparency.

The transformation made possible by the hermeneutic relation is a transformation that occurs precisely through *differences* between the text and what is referred to. What is needed is a particular set of textually clear perceptions that "reduce" to that which is immediately readable. To return to the Three Mile Island example, one problem uncovered was that the instrument panel design was itself faulty. It did not incorporate its dials and gauges in an easily readable way. For example, in airplane instrument panel design, much thought has been given to pattern recognition, which occurs as a perceptual gestalt. Thus, in a four-engined aircraft, the four dials indicating r.p.m. will be coordinated so that a single glance will indicate which, if any, engine is out of synchronization. Such technical design accounts for perceptual structures.

There is a second caution concerning the focus upon connectors and pathology. In all the examples I have used to this point, the hermeneutic technics have involved material connections. (The thermometer employs a physical property of a bimetallic spring or mercury in a column; the instrument panel at TMI employs mechanical, electrical, or other material connections; the shift lever, a simple mechanical connection.) If reading does not employ any such material connections, it might seem that its referentiality is essentially different, yet not even all technological connections are strictly material. Photography retains representational isomorphism with the object, yet does not "materially" connect with its object; it is a minimal beginning of action at a distance.

I have been using contemporary or post-scientific examples, but non-material hermeneutic relations do not obtain only for contemporary humans. As existential relations, they are as "old" as post-Garden humanity. Anthropology and the history of religions have long been familiar with a wide variety of shamanistic praxes which fall into the pattern of hermeneutic technics. In what may at first seem a somewhat outrageous set of examples, note the various "reading" techniques employed in shamanism. The reading of animal entrails, of thrown bones, of bodily marks—all are hermeneutic techniques. The patterns of the entrails, bones, or whatever are taken to *refer* to some state of affairs, instrumentally or textually.

Not only are we here close to a familiar association between magic and the origins of technology suggested by many writers, but we are, in fact, closer to a wider hermeneutic praxis in an intercultural setting. For that reason, the very strangeness of the practice must be critically examined. If the throwing of bones is taken as a "primitive" form of medical diagnosis—which does play a role in shamanism—we might conclude that it is indeed a poor form of hermeneutic relations. What we might miss, however, is that the entire gestalt of what is being diagnosed may differ radically from the other culture and ours.

It may well be that as a focused form of diagnosis upon some particular bodily ailment (appendicitis, for example), the diagnosis will fail. But since one important element in shamanism is a wider diagnosis, used particularly as the occasion of locating certain communal or social problems, it may work better. The sometimes socially contextless emphasis of Western medicine upon a presumably "mechanical" body may overlook precisely the context which the shaman so clearly recognizes. The entire gestalt is different and differently focused, but in both cases there are examples of hermeneutic relations.

In our case, the very success of Western medicine in certain diseases is due to the introduction of technologies into the hermeneutic relation (fever/thermometer; blood pressure/manometer, etc.) The point is that hermeneutic relations are as commonplace in traditional and ancient social groups as in ours, even if they are differently arranged and practiced.

By continuing the intentionality analysis I have been following, one can now see that hermeneutic relations vary the continuum of human-technology-world relations. Hermeneutic relations maintain the general mediation position of technologies within the context of human praxis towards a world, but they also change the variables within the human-technology-world relation. A comparative formalism may be suggestive:

General intentionality relations
Human-technology-world

Variant A: embodiment relations
(I-technology) → world

Variant B: hermeneutic relations
I → (technology-world)

While each component of the relation changes within the correlation, the overall shapes of the variants are distinguishable. Nor are these matters of simply how technologies are experienced.

Another set of examples from the set of optical instruments may illustrate yet another way in which instrumental intentionalities can follow new trajectories. Strictly embodiment relations can be said to work best when there is both a transparency and an isomorphism between perceptual and bodily action within the relation. I have suggested that a trajectory for development in such cases may often be a horizontal one. Such a trajectory not only follows greater and greater degrees of magnification but also entails all the difficulties of a technical nature that go into allowing what is to be seen as though by direct vision. But not all optical technologies follow this strategy. The introduction of hermeneutic possibilities opens the trajectory into what I shall call *vertical* directions, possibilities that rely upon quite deliberate hermeneutic transformations.

It might be said that the telescope and microscope, by extending vision while transforming it, remained *analogue* technologies. The enhancement and magnification made possible by such technologies remain visual and transparent to ordinary vision. The moon remains recognizably the moon, and the microbe—even if its existence was not previously suspected—remains under the microscope a beastie recognized as belonging to the animate continuum. Here, just as the capacity to magnify becomes the foreground phenomenon to the background phenomenon of the reduction necessarily accompanying the magnification, so the similitude of what is seen with ordinary vision remains central to embodiment relations.

Not all optical technologies mediate such perceptions. In gradually moving towards the visual "alphabet" of a hermeneutic relation, deliberate variations may occur which enhance previously undiscernible *differences*:

1) Imagine using spectacles to correct vision, as previously noted. What is wanted is to *return* vision as closely as possible to ordinary perception, not to distort or modify it in any extreme micro- or macroperceptual direction. But now, for snowscapes or sun on the water or desert, we modify the lenses by coloring or polarizing them to cut glare. Such a variation

transforms *what* is seen in some degree. Whether we say the polarized lens removes glare or "darkens" the landscape, what is seen is now clearly different from what may be seen through untinted glasses. This difference is a clue which may open a new *telic direction* for development.

2) Now say that somewhere, sometime, someone notes that certain kinds of tinting reveal unexpected results. Such is a much more complex technique now used in infrared satellite photos. (For the moment, I shall ignore the fact that part of this process is a combined embodiment and hermeneutic relation.) If the photo is of the peninsula of Baja California, it will remain recognizable in shape. Geography, whatever depth and height representations, etc., remain but vary in a direction different from any ordinary vision. The infrared photo enhances the difference between vegetation and non-vegetation beyond the limits of any isomorphic color photography. This difference corresponds, in the analogue example, to something like a pictograph. It simultaneously leaves certain analogical structures there and begins to modify the representation into a different, non-perceived "representation."

3) Very sophisticated versions of still representative but non-ordinary forms of visual recognition occur in the new heat-sensitive and light-enhanced technologies employed by the military and police. Night scopes which enhance a person's heat radiation still look like a person but with entirely different regions of what stands out and what recedes. In high-altitude observations, "heat shadows" on the ground can indicate an airplane that has recently had its engines running compared to others which have not. Here visual technologies bring into visibility what was not visible, but in a distinctly now perceivable way.

4) If now one takes a much larger step to spectrographic astronomy, one can see the acceleration of this development. The spectrographic picture of a star no longer "resembles" the star at all. There is no point of light, no disk size, no spatial isomorphism at all—merely a band of differently colored rainbow stripes. The naive reader would not know that this was a picture of a star at all—the reader would have to know the language, the alphabet, that has coded the star. The astronomer-hermeneut does know the language and "reads" the visual "ABCs" in such a way that he knows the chemical composition of the star, its internal makeup, rather than its shape or external configuration. We are here in the presence of a more fully hermeneutic relation, the star mediated not only instrumentally but in a transformation such that we must now thematically *read* the result. And only the informed reader can do the reading.

There remains, of course, the *reference* to the star. The spectrograph is *of* Rigel or *of* Polaris, but the individuality of the star is now made present hermeneutically. Here we have a beginning of a special transformation of perception, a transformation which deliberately enhances differences rather than similarities in order to get at what was previously unperceived.

5) Yet even the spectrograph is but a more radical transformation of perception. It, too, can be transformed by a yet more radical *hermeneutic* analogue to the *digital* transformation which lies embedded in the preferred quantitative praxis of science. The "alphabet" of science is, of course, mathematics, a mathematics that separates itself by yet another hermeneutic step from perception embodied.

There are many ways in which this transformation can and does occur, most of them interestingly involving a particular act of *translation* that often goes unnoticed. To keep the example as simple as possible, let us assume *mechanical* or *electronic* "translation." Suppose our spectrograph is read by a machine that yields not a rainbow spectrum but a set of numbers. Here we would arrive at the final hermeneutic accomplishment, the transformation of even the analogue to a digit. But in the process of hermeneuticization, the "transparency" to the object referred to becomes itself enigmatic. Here more explicit and thematic interpretation must occur.

C. ALTERITY RELATIONS

Beyond hermeneutic relations there lie *alterity relations.* The first suggestions of such relations, which I shall characterize as relations *to* or *with* a technology, have already been suggested in different ways from within the embodiment and hermeneutic contexts. Within embodiment relations, were the technology to intrude upon rather than facilitate one's perceptual and bodily extension into the world, the technology's objectness would necessarily have appeared negatively. Within hermeneutic relations, however, there emerged a certain positivity to the objectness of instrumental technologies. The bodily-perceptual focus *upon* the instrumental text is a condition of its own peculiar hermeneutic transparency. But what of a positive or presentential sense of relations with technologies? In what phenomenological senses can a technology be *other?*

The analysis here may seem strange to anyone limited to the habits of objectivist accounts, for in such accounts technologies as objects usually come first rather than last. The problem for a phenomenological account is that objectivist ones are non-relativistic and thus miss or submerge what is distinctive about human-technology relations.

A naive objectivist account would likely begin with some attempt to circumscribe or define technologies by object characteristics. Then, what I have called the technical properties of technologies would become focal. Some combination of physical and material properties would be taken to be definitional. (This is an inherent tendency of the standard nomological positions such as those of Bunge and Hacking). The definition will often serve a secondary purpose by being stipulative: only those technologies that are obviously dependent upon or strongly related to contemporary scientific and industrial productive practices will count.

This is not to deny that objectivist accounts have their own distinctive strengths. For example, many such accounts recognize that technological or "artificial" products are different from the simply found object or the natural object. But the submergence of the human-technology relation remains hidden, since either object may enter into praxis and both will have their material, and thus limited, range of technical usability within the relation. Nor is this to deny that the objectivist accounts of types of technologies, types of organization, or types of designed purposes should be considered. But the focus in this first program remains the phenomenological derivation of the set of human-technology relations.

There is a tactic behind my placing alterity relations last in the order of focal human-technology relations. The tactic is designed, on the one side, to circumvent the tendency succumbed to by Heidegger and his more orthodox followers to see the otherness of technology only in negative terms or through negative derivations. The hammer example, which remains paradigmatic for this approach, is one that derives objectness from breakdown. The broken or missing or malfunctioning technology could be *discarded.* From being an obtrusion it could become *junk.* Its objectness would be clear—but only partly so. Junk is not a focal object of use relations (except in certain limited situations). It is more ordinarily a background phenomenon, that which has been put out of use.

Nor, on the other side, do I wish to fall into a naively objectivist account that would simply concentrate upon the material properties of the technology as an object of knowledge.

Such an account would submerge the relativity of the intentionality analysis, which I wish to preserve here. What is needed is an analysis of the positive or presentential senses in which humans relate to technologies as relations *to* or with technologies, to technology-as-other. It is this sense which is included in the term "alterity."

Philosophically, the term "alterity" is borrowed from Emmanuel Levinas. Although Levinas stands within the traditions of phenomenology and hermeneutics, his distinctive work, *Totality and Infinity,* was "anti-Heideggerian." In that work, the term "alterity" came to mean the radical difference posed to any human by another human, an *other* (and by the ultimately other, God). Extrapolating radically from within the tradition's emphasis upon the non-reducibility of the human to either objectness (in epistemology) or as a means (in ethics), Levinas poses the otherness of humans as a kind of *infinite* difference that is concretely expressed in an ethical, face-to-face encounter.

I shall retain but modify this radical Levinasian sense of human otherness in returning to an analysis of human-technology relations. How and to what extent do technologies become other or, at least, *quasi-other?* At the heart of this question lie a whole series of well-recognized but problematic interpretations of technologies. On the one side lies the familiar problem of anthropomorphism, the personalization of artifacts. This range of anthropomorphism can reach from serious artifact-human analogues to trivial and harmless affections for artifacts.

An instance of the former lies embedded in much AI research. To characterize computer "intelligence" as human-like is to fall into a peculiarly contemporary species of anthropomorphism, however sophisticated. An instance of the latter is to find oneself "fond" of some particular technofact as, for instance, a long-cared-for automobile which one wishes to keep going and which may be characterized by quite deliberate anthropomorphic terms. Similarly, in ancient or non-Western cultures, the role of sacredness attributed to artifacts exemplifies another form of this phenomenon.

The religious object (idol) does not simply "represent" some absent power but is endowed with the sacred. Its aura of sacredness is spatially and temporally present within the range of its efficacy. The tribal devotee will defend, sacrifice to, and care for the sacred artifact. Each of these illustrations contains the seeds of an alterity relation.

A less direct approach to what is distinctive in human-technology alterity relations may perhaps better open the way to a phenomenologically relativistic analysis. My first example comes from a comparison to a technology and to an animal "used" in some practical (although possibly sporting) context: the spirited horse and the spirited sports car.

To ride a spirited horse is to encounter a lively animal *other.* In its pre- or nonhuman context, the horse has a life of its own within the environment that allowed this form of life. Once domesticated, the horse can be "used" as an "instrument" of human praxis—but only to a degree and in a way different from counterpart technologies; in this case, the "spirited" sports car.

There are, of course, analogues which may at first stand out. Both horse and car give the rider/driver a magnified sense of power. The speed and the experience of speed attained in riding/driving are dramatic extensions of my own capacities. Some prominent features of embodiment relations can be found analogously in riding/driving. I experience the trail/road through horse/car and guide/steer the mediating entity under way. But there are equally prominent differences. No matter how well trained, no horse displays the same "obedience" as the car. Take malfunction: in the car, a malfunction "resists" my command—I push the accelerator, and because of a clogged gas line, there is not the response I expected. But the animate resistance of a spirited horse is more than such a mechanical lack of response—the response is more than malfunction, it is *dis*obedience. (Most experienced riders, in fact, prefer spirited horses over the more passive ones, which might more nearly approximate a mechanical obedience.) This life of the other in a horse may be carried much further—it may live without me in the proper environment; it does not need the *deistic* intervention of turning the starter to be "animated." The car will not shy at the rabbit springing up in the path any more than most horses will obey the "command" of the driver to hit the stone wall when he is too drunk to notice. The horse, while approximating some features of a mediated embodiment situation, never fully enters such a relation in the way a technology does. Nor does the car ever attain the sense of animation to be found in horseback riding. Yet the analogy is so deeply embedded in our contemporary consciousness (and perhaps the lack of sufficient experience with horses helps) that we might be tempted to emphasize the similarities rather than the differences.

Anthropomorphism regarding the technology on the one side and the contrast with horseback riding on the other point to a first approximation to the unique type of otherness that relations to technologies hold. Technological otherness is a *quasi-otherness,* stronger than mere objectness but weaker than the otherness found within the animal kingdom or the human one; but the phenomenological derivation must center upon the positive experiential aspects outlining this relation.

In yet another familiar phenomenon, we experience technologies as *toys* from childhood. A widely cross-cultural example is the spinning top. Prior to being put into use, the top may appear as a top-heavy object with a certain symmetry of design (even early tops approximate the more purely functional designs of streamlining, etc.), but once "deistically" animated through either stick motion or a string spring, the now spinning top appears to take on a life of its own. On its tip (or "foot") the top appears to defy its top-heaviness and gravity itself. It traces unpredictable patterns along its pathway. It is an object of *fascination.*

Note that once the top has been set to spinning, what was imparted through an embodiment relation now exceeds it. What makes it fascinating is this property of quasi-animation, the life of its own. Also, of course, once "automatic" in its motion, the top's movements may be entered into a whole series of possible contexts. I might enter a game of warring tops in which mine (suitably marked) represents me. If I-as-top am successful in knocking down the other tops, then this game of hermeneutics has the top winning for me. Similarly, if I take its quasi-autonomous motion to be a hermeneutic predictor, I may enter a divination context in which the path traced or the eventual point of stoppage indicates some fortune. Or, entering the region of scientific instrumentation, I may transform the top into a gyroscope, using its constancy of direction within its now-controlled confines as a better-than-magnetic compass. But in each of these cases, the top may become the focal center of attention as a quasi-other to which I may relate. Nor need the object of fascination carry either an embodiment or hermeneutic referential transparency.

To the ancient and contemporary top, compare briefly the fascination that occurs around video games. In the actual use of video games, of course, the embodiment and hermeneutic relational dimensions are present. The joystick that embodies hand and eye coordination skills extends the player into the displayed field. The field itself displays some hermeneutic context (usually either some "invader" mini-world or some sports analogue), but this context does not refer beyond itself into a worldly reference.

In addition to these dimensions, however, there is the sense of *interacting with* something other than me, the technological *competitor.* In competition there is a kind of dialogue or exchange. It is the quasi-animation, the quasi-otherness of the technology that fascinates and challenges. I must beat the machine or it will beat me.

Although the progression of the analysis here moves from embodiment and hermeneutic relations to alterity ones, the interjection of film or cinema examples is of suggestive interest. Such technologies are transitional between hermeneutic and alterity phenomena. When I first introduced the notion of hermeneutic relations, I employed what could be called a "static" technology: writing. The long and now ancient technologies of writing result in fixed texts (books, manuscripts, etc., all of which, barring decay or destruction, remain stable in themselves). With film, the "text" remains fixed only in the sense that one can repeat, as with a written text, the seeing and hearing of the cinema text. But the mode of presentation is dramatically different. The "characters" are now animate and theatrical, unlike the fixed alphabetical characters of the written text. The dynamic "world" of the cinema-text, while retaining many of the functional features of writing, also now captures the semblance of real-time, action, etc. It remains to be "read" (viewed and heard), but the object-correlate necessarily appears more "life-like" than its analogue—written text. This factor, naively experienced by the current generations of television addicts, is doubtless one aspect in the problems that emerge between television watching habits and the state of reading skills. James Burke has pointed out that "the majority of the people in the advanced industrialized nations spend more time watching television than doing anything else beside work."[3] The same balance of time use also has shown up in surveys regarding students. The hours spent watching television among college and university students, nationally, are equal to or exceed those spent in doing homework or out-of-class preparation.

Film, cinema, or television can, in its hermeneutic dimension, refer in its unique way to a "world." The strong negative response to the Vietnam War was clearly due in part to the virtually unavoidable "presence" of the war in virtually everyone's living room. But films, like readable technologies, are also *presentations,* the focal terminus of a perceptual situation. In that emergent sense, they are more dramatic forms of perceptual immediacy in which the presented display has its own characteristics conveying quasi-alterity. Yet the engagement with the film normally remains short of an engagement with an *other.* Even in the anger that comes through in outrage about civilian atrocities or the pathos experienced in seeing starvation epidemics in Africa, the emotions are not directed to the screen but, indirectly, through it, in more appropriate forms of political or charitable action. To this extent there is retained a hermeneutic reference elsewhere than at the technological instrument. Its quasi-alterity, which is also present, is not fully focal in the case of such media technologies.

A high-technology example of breakdown, however, provides yet another hint at the emergence of alterity phenomena. Word processors have become familiar technologies, often strongly liked by their users (including many philosophers who fondly defend their choices, profess knowledge about the relative abilities of their machines and programs, etc.). Yet in breakdown, this quasi-love relationship reveals its quasi-hate underside as well. Whatever form of "crash" may occur, particularly if some fairly large section of text is involved, it occasions frustration and even rage. Then, too, the programs have their idiosyncrasies, which allow or do not allow certain movements; and another form of human-technology competition may emerge. (Mastery in the highest sense most likely comes from learning to program and thus overwhelm the machine's previous brainpower. "Hacking" becomes the game-like competition in which an entire system is the alterity correlate.) Alterity relations may be noted to emerge in a wide range of computer technologies that, while failing quite strongly to mimic bodily incarnations, nevertheless display a quasi-otherness within the limits of linguistics and, more particularly, of logical behaviors. Ultimately, of course, whatever contest emerges, its sources lie opaquely with other humans as well but also with the transformed technofact, which itself now plays a more obvious role within the overall relational net.

I have suggested that the computer is one of the stronger examples of a technology which may be positioned within alterity relations. But its otherness remains a quasi-otherness, and its genuine usefulness still belongs to the borders of its hermeneutic capacities. Yet in spite of this, the tendency to fantasize its quasi-otherness into an authentic otherness is pervasive. Romanticizations such as the portrayal of the emotive, speaking "Hal" of the movie *2001: A Space Odyssey,* early fears that the "brain power" of computers would soon replace human thinking, fears that political or military decisions will not only be informed by but also made by computers—all are symptoms revolving around the positing of otherness to the technology.

These romanticizations are the alterity counterparts to the previously noted dreams that wish for total embodiment. Were the technofact to be genuinely an other, it would both be and not be a *technology.* But even as quasi-other, the technology falls short of such totalization. It retains its unique role in the human-technology continuum of relations as the medium of transformation, but as a recognizable medium.

The wish-fulfillment desire occasioned by embodiment relations—the desire for a fully transparent technology that would *be* me while at the same time giving me the powers that the use of the technology makes available—here has its counterpart fantasy, and this new fantasy has the same internal contradiction: It both reduces or, here, extrapolates the technology into that which is not a technology (in the first case, the magical transformation is *into me;* in this case, *into the other*), and at the same time, it desires what is not identical with me or the other. The fantasy is for the transformational effects. Both fantasies, in effect, deny technologies playing the roles they do in the human-technology continuum of relations; yet it is only on the condition that there be some detectable differentiation within the relativity that the unique ways in which technologies transform human experience can emerge.

In spite of the temptation to accept the fantasy, what the quasi-otherness of alterity relations does show is that humans may relate positively or presententially *to* technologies. In that respect and to that degree, technologies emerge as focal entities that may receive the multiple attentions humans give the different forms of the other. For this reason, a third formalization may be employed to distinguish this set of relations:

$$I \rightarrow \text{technology-(-world)}$$

I have placed the parentheses thusly to indicate that in alterity relations there may be, but need not be, a relation through the technology to the world (although it might well be expected that the *usefulness* of any technology will necessarily entail just such a referentiality). The world, in this case, may remain context and background, and the technology may emerge as the foreground and focal quasi-other with which I momentarily engage.

This disengagement of the technology from its ordinary-use context is also what allows the technology to fall into the various disengaged engagements which constitute such activities as play, art, or sport.

A first phenomenological itinerary through direct and focal human-technology relations may now be considered complete. I have argued that the three sets of distinguishable relations occupy a continuum. At the one extreme lie those relations that approximate technologies to a quasi-me (embodiment relations). Those technologies that I can so take into my experience that through their semi-transparency they allow the world to be made immediate thus enter into the existential relation which constitutes my self. At the other extreme of the continuum lie alterity relations in which the technology becomes quasi-other, or technology "as" other *to* which I relate. Between lies the relation with technologies that both mediate and yet also fulfill my perceptual and bodily relation with technologies, hermeneutic relations. The variants may be formalized thus:

Human-technology-World Relations
Variant 1, Embodiment Relations
(Human-technology) → World
Variant 2, Hermeneutic Relations
Human → (technology-World)
Variant 3, Alterity Relations
Human → technology-(-World)

Although I have characterized the three types of human-technology relations as belonging to a continuum, there is also a sense in which the elements within each type of relation are differently distributed. There is a *ratio* between the objectness of the technology and its transparency in use. At the extreme height of embodiment, a background presence of the technology may still be detected. Similarly but with a different ratio, once the technology has emerged as a quasi-other, its alterity remains within the domain of human invention through which the world is reached. Within all the types of relations, technology remains artifactual, but it is also its very artifactual formation which allows the transformations affecting the earth and ourselves.

All the relations examined heretofore have also been focal ones. That is, each of the forms of action that occur through these relations have been marked by an implicated self-awareness. The engagements through, with, and to technologies stand within the very core of praxis. Such an emphasis, while necessary, does not exhaust the role of technologies nor the experiences of them. If focal activities are central and foreground, there are also fringe and background phenomena that are no more neutral than those of the foreground. It is for that reason that one final foray in this phenomenology of technics must be undertaken. That foray must be an examination of technologies in the background and at the horizons of human-technology relations.

D. BACKGROUND RELATIONS

With background relations, this phenomenological survey turns from attending to technologies in a foreground to those which remain in the background or become a kind of near-technological environment itself. Of course, there are discarded or no-longer-used technologies, which in an extreme sense occupy a background position in human experience—junk. Of these, some may be recuperated into non-use but focal contexts such as in technology museums or in the transformation into junk art. But the analysis here points to specifically functioning technologies which ordinarily occupy background or field positions.

First, let us attend to certain individual technologies designed to function in the background—automatic and semiautomatic machines, which are so pervasive today—as good candidates for this analysis. In the mundane context of the home, lighting, heating, and cooling systems, and the plethora of semiautomatic appliances are good examples. In each case, there is some necessity for an instant of deistic intrusion to program or set the machinery into motion or to its task. I set the thermostat; then, if the machinery is high-tech, the heating/cooling system will operate independently of ongoing action. It may employ time-temperature changes, external sensors to adjust to changing weather, and other cybernetic operations. (While this may function well in the home situation, I remain amused at the still-primitive state of the art in the academic complex I occupy. It takes about two days for the system to adjust to the sudden fall and spring weather changes, thus making offices which actually have opening windows—a rarity—highly desirable.) Once operating, the technology functions as a barely detectable background presence; for example, in the form of background noise, as when the heating kicks in. But in operation, the technology does not call for focal attention.

Note two things about this human-technology relation: First, the machine activity in the role of background presence is not displaying either what I have termed a transparency or an opacity. The "withdrawal" of this technological function is phenomenologically distinct as a kind of "absence." The technology is, as it were, "to the side." Yet as a present absence, it nevertheless becomes part of the experienced field of the inhabitant, a piece of the immediate environment.

Somewhat higher on the scale of semiautomatic technologies are task-oriented appliances that call for explicit and repeated deistic interventions. The washing machine, dryer, microwave, toaster, etc., all call for repeated programming and then for dealing with the processed product (wash, food, etc.). Yet like the more automated systems, the semiautomatic machine remains in the background while functioning.

In both systems and appliances, however, one also may detect clues to the ways in which background relations texture the immediate environment. In the electric home, there is virtually a constant hum of one sort or the other, which is part of the technological texture. Ordinarily, this "white noise" may go unnoticed, although I am always reassured that it remains part of fringe awareness, as when guests visit my mountain home in Vermont. The inevitable comment is about the silence of the woods. At once, the absence of background hum becomes noticeable.

Technological texturing is, of course, much deeper than the layer of background noise which signals its absent presence. Before turning to further implications, one temptation which could occur through the too-narrow selection of contemporary examples must be avoided. It might be thought that only, or predominantly, the high-technology contemporary world uses and experiences technologies as backgrounds. That is not the case, even with respect to automated or semiautomatic technologies.

The scarecrow is an ancient "automated" device. Its mimicry of a human, with clothes flapping in the breeze, is a specifically designed automatic crow scarer, made to operate in the absence of humans. Similarly, in ancient Japan there were automated deer scarers, made of bamboo tubes, pivoted on a pin and placed so that a waterfall or running stream would slowly fill the tube. When it is full enough, the device would trip and its other end strike a sounding board or drum, the noise of which would frighten away any marauding deer. We have already noted the role automation plays in religious rituals (prayer wheels and worship representations thought to function continuously).

Interpreted technologically, there are even some humorous examples of "automation" to be found in ancient religious praxes. The Hindu prayer windmill "automatically" sends its prayers when the wind blows; and in the ancient Sumerian temples there were idols with large eyes at the altars (the gods), and in front of them were smaller, large-eyed human statues representing worshipers. Here was an ancient version of an "automated" worship. (Its contemporary counterpart would be the joke in which the professor leaves his or her lecture on a tape recorder for the class—which students could also "automatically" hear, by leaving their own cassettes to tape the master recording.)

While we do not often conceptualize such ancient devices in this way, part of the purpose of an existential analysis is precisely to take account of the identity of function and of the "ancientness" of all such existential relations. This is in no way to deny the differences of context or the degree of complexity pertaining to the contemporary, as compared to the ancient, versions of automation.

Another form of background relation is associated with various modalities of the technologies that serve to insulate humans from an external environment. Clothing is a borderline case. Clothing clearly insulates our bodies from temperature, wind, and other external weather phenomena that could become dangerous to life; but clothing experienced is borderline with embodiment relations, for we do feel the external environment through clothing, albeit in a particularly damped-down mode. Clothing is not designed, in most cases, to be "transparent" in the way the previous instrument examples were but rather to have a certain opacity without restricting movement. Yet clothing is part of a fringe awareness in most of our daily activities (I am obviously not addressing fashion aspects of clothing here).

A better example of a background relation is a shelter technology. Although shelters may be found (caves) and thus enter untransformed into human praxis, most are constructed, as are most technological artifacts; but once constructed and however designed to insulate or account for external weather, they become a more field-like background phenomenon. Here again, human cultures display an amazing continuum from minimalist to maximalist strategies with respect to this version of a near-background.

Many traditional cultures, particularly in Southern Hemisphere areas, practice an essentially open shelter technology, perhaps with primarily a roof to keep off rain and sun. Such peoples frequently find distasteful such items as windows and, particularly, glassed windows. They do not wish to be too isolated or insulated from the elements. At the other extreme is the maximalist strategy, which most extremely wishes to totalize shelter technology into a virtual life-support system, autonomous and enclosed. I shall call this a technological cocoon.

A contemporary example of a near-cocoon is the nuclear submarine. Its crew lives inside, and the vessel is designed to remain at sea for prolonged periods, even underwater for long stretches of time. There are sophisticated recycling systems for waste, water, and air. Contact with the outside, obviously important in this case, is primarily through monitoring equally sophisticated hermeneutic devices (sonar, low-frequency radio, etc.). All ordinary duties take place in the cocoon-like interior. A multibillion-dollar projection to a greater degree of cocoonhood is the long-term space station now under debate.

Part of the very purpose of the space station is to experiment with creating a mini-environment, or artificial "earth," which would be totally technologically mediated. Yet contemporary high-tech suburban homes show similar features. Fully automated for temperature and humidity, tight air structures, some with glass that adjusts to glare, all such homes lie on the same trajectory of self-containment. But while these illustrations are uniquely high-technology textured, there remain, as before, differently contexted but similar examples from the past.

Totally enclosed spaces have frequently been associated with ritual and religious praxis. The Kiva of past southwestern native American cultures was dug deep into the ground, windowless and virtually sealed. It was the site for important initiatory and secret societies, which gathered into such ancient cocoons for their own purposes. The enclosure bespeaks different kinds of totalization.

What is common to the entire range of examples pointed to here is the position occupied by such technology, background position, the position of an absent presence as a part of or a total field of immediate technology.

In each of the examples, the background role is a field one, not usually occupying focal attention but nevertheless conditioning the context in which the inhabitant lives. There are, of course, great differences to be detailed in terms of the types of contexts which such background technologies play. Breakdown, again, can play a significant indexical role in pointing out such differences.

The involvement implications of contemporary, high-technology society are very complex and often so interlocked as to fall into major disruption when background technology fails. In 1985 Long Island was swept by Hurricane Gloria with massive destruction of power lines. Most areas went without electricity for at least a week, and in some cases, two. Lighting had to be replaced by older technologies (lanterns, candles, kerosene lamps), supplies for which became short immediately. My own suspicion is that a look at birth statistics at the proper time after this radical change in evening habits will reveal the same glitch which actually did occur during the blackouts of earlier years in New York.

Similarly, with the failure of refrigeration, eating habits had to change temporarily. The example could be expanded quite indefinitely; a mass purchase of large generators by university buyers kept a Minnesota company in full production for several months after, to be prepared the "next time." In contrast, while the same effects on a shorter-term basis were experienced in the grid-wide blackouts of 1965, I was in Vermont at my summer home, which is lighted by kerosene lamps and even refrigerated with a kerosene refrigerator. I was simply unaware of the massive disruption until the Sunday *Times* arrived. Here is a difference between an older, loose-knit and a contemporary, tight-knit system.

Despite their position as field or background relations, technologies here display many of the same transformational characteristics found in the previous explicit focal relations. Different technologies texture environments differently. They exhibit unique forms of non-neutrality through the different ways in which they are interlinked with the human lifeworld. Background technologies, no less than focal ones, transform the gestalts of human experience and, precisely because they are absent presences, may exert more subtle indirect effects upon the way a world is experienced. There are also involvements both with wider circles of connection and amplification/reduction selectivities that may be discovered in the roles of background relations; and finally, the variety of minimalist to maximalist strategies remains as open to this dimension of human-technology relations as each of the others.

NOTES

1. This illustration is my version of a similar one developed by Patrick Heelan in his more totally hermeneuticized notion of perception in *Space Perception and the Philosophy of Science* (Berkeley: University of California Press, 1983), p. 193.

2. Ian Hacking, *Representing and Intervening* (Cambridge: Cambridge University Press, 1983), p. 195. Hacking develops a very excellent and suggestive history of the use of microscopes. His focus, however, is upon the technical properties that were resolved before microscopes could be useful in the sciences. He and Heelan, however, along with Robert Ackermann, have been among the pioneers dealing with perception and instrumentation in instruments. Cf. also my *Technics and Praxis* (Dordrecht: Reidel Publishers, 1979).

3. James Burke, *Connections,* (Boston: Little, Brown, and Co., 1978), p. 5.

11

A Cyborg Manifesto

Donna J. Haraway

AN IRONIC DREAM OF A COMMON LANGUAGE FOR WOMEN IN THE INTEGRATED CIRCUIT

This chapter is an effort to build an ironic political myth faithful to feminism, socialism, and materialism. Perhaps more faithful as blasphemy is faithful, than as reverent worship and identification. Blasphemy has always seemed to require taking things very seriously. I know no better stance to adopt from within the secular-religious, evangelical traditions of United States politics, including the politics of socialist feminism. Blasphemy protects one from the moral majority within, while still insisting on the need for community. Blasphemy is not apostasy. Irony is about contradictions that do not resolve into larger wholes, even dialectically, about the tension of holding incompatible things together because both or all are necessary and true. Irony is about humour and serious play. It is also a rhetorical strategy and a political method, one I would like to see more honoured within socialist-feminism. At the centre of my ironic faith, my blasphemy, is the image of the cyborg.

A cyborg is a cybernetic organism, a hybrid of machine and organism, a creature of social reality as well as a creature of fiction. Social reality is lived social relations, our most important political construction, a world-changing fiction. The international women's movements have constructed 'women's experience', as well as uncovered or discovered this crucial collective object. This experience is a fiction and fact of the most crucial, political kind. Liberation rests on the construction of the consciousness, the imaginative apprehension, of oppression, and so of possibility. The cyborg is a matter of fiction and lived experience that changes what counts as women's experience in the late twentieth century. This is a struggle over life and death, but the boundary between science fiction and social reality is an optical illusion.

Contemporary science fiction is full of cyborgs—creatures simultaneously animal and machine, who populate worlds ambiguously natural and crafted. Modern medicine is also full

From Donna J. Haraway, "A Cyborg Manifesto: Science, Technology, and Socialist-Feminism in the Late Twentieth Century," in *Simians, Cyborgs and Women: The Reinvention of Nature*, New York: Routledge, 1991, pp. 149–181. Reprinted by permission of Routledge, Inc., a part of The Taylor Francis Group.

of cyborgs, of couplings between organism and machine, each conceived as coded devices, in an intimacy and with a power that was not generated in the history of sexuality. Cyborg 'sex' restores some of the lovely replicative baroque of ferns and invertebrates (such nice organic prophylactics against heterosexism). Cyborg replication is uncoupled from organic reproduction. Modern production seems like a dream of cyborg colonization work, a dream that makes the nightmare of Taylorism seem idyllic. And modern war is a cyborg orgy, coded by C³I, command-control-communication-intelligence, an $84 billion item in 1984's US defence budget. I am making an argument for the cyborg as a fiction mapping our social and bodily reality and as an imaginative resource suggesting some very fruitful couplings. Michael Foucault's biopolitics is a flaccid premonition of cyborg politics, a very open field.

By the late twentieth century, our time, a mythic time, we are all chimeras, theorized and fabricated hybrids of machine and organism; in short, we are cyborgs. The cyborg is our ontology; it gives us our politics. The cyborg is a condensed image of both imagination and material reality, the two joined centres structuring any possibility of historical transformation. In the traditions of 'Western' science and politics—the tradition of racist, male-dominant capitalism; the tradition of progress; the tradition of the appropriation of nature as resource for the productions of culture; the tradition of reproduction of the self from the reflections of the other—the relation between organism and machine has been a border war. The stakes in the border war have been the territories of production, reproduction, and imagination. This chapter is an argument for *pleasure* in the confusion of boundaries and for *responsibility* in their construction. It is also an effort to contribute to socialist-feminist culture and theory in a postmodernist, non-naturalist mode and in the utopian tradition of imagining a world without gender, which is perhaps a world without genesis, but maybe also a world without end. The cyborg incarnation is outside salvation history. Nor does it mark time on an oedipal calendar, attempting to heal the terrible cleavages of gender in an oral symbiotic utopia or post-oedipal apocalypse. As Zoe Sofoulis argues in her unpublished manuscript on Jacques Lacan, Melanie Klein, and nuclear culture, *Lacklein,* the most terrible and perhaps the most promising monsters in cyborg worlds are embodied in non-oedipal narratives with a different logic of repression, which we need to understand for our survival.

The cyborg is a creature in a post-gender world; it has no truck with bisexuality, pre-oedipal symbiosis, unalienated labour, or other seductions to organic wholeness through a final appropriation of all the powers of the parts into a higher unity. In a sense, the cyborg has no origin story in the Western sense—a 'final' irony since the cyborg is also the awful apocalyptic *telos* of the 'West's' escalating dominations of abstract individuation, an ultimate self untied at last from all dependency, a man in space. An origin story in the 'Western', humanist sense depends on the myth of original unity, fullness, bliss and terror, represented by the phallic mother from whom all humans must separate, the task of individual development and of history, the twin potent myths inscribed most powerfully for us in psychoanalysis and Marxism. Hilary Klein has argued that both Marxism and psychoanalysis, in their concepts of labour and of individuation and gender formation, depend on the plot of original unity out of which difference must be produced and enlisted in a drama of escalating domination of woman/nature. The cyborg skips the step of original unity, of identification with nature in the Western sense. This is its illegitimate promise that might lead to subversion of its teleology as star wars.

The cyborg is resolutely committed to partiality, irony, intimacy, and perversity. It is oppositional, utopian, and completely without innocence. No longer structured by the polarity of public and private, the cyborg defines a technological polis based partly on a revolution of social relations in the *oikos,* the household. Nature and culture are reworked; the one can no longer be the resource for appropriation or incorporation by the other. The relationships for forming wholes from parts, including those of polarity and hierarchical domination, are at issue in the cyborg world. Unlike the hopes of Frankenstein's monster, the cyborg does not expect its father to save it through a restoration of the garden; that is, through the fabrication of a heterosexual mate, through its completion in a finished whole, a city and cosmos. The cyborg does not dream of community on the model of the organic family, this time without the oedipal project. The cyborg would not recognize the Garden of Eden; it is not made of mud and cannot dream of returning to dust. Perhaps that is why I want to see if cyborgs can subvert the apocalypse of returning to nuclear dust in the manic compulsion to name the Enemy. Cyborgs are not reverent; they do not re-member the cosmos. They are wary of holism, but needy for connection—they seem to have a natural feel for united front politics, but without the vanguard party. The main trouble with cyborgs, of course, is that they are the illegitimate offspring of militarism and patriarchal capitalism, not to mention state socialism. But illegitimate offspring are often exceedingly unfaithful to their origins. Their fathers, after all, are inessential.

I will return to the science fiction of cyborgs at the end of this chapter, but now I want to signal three crucial boundary breakdowns that make the following political-fictional (political-scientific) analysis possible. By the late twentieth century in United States scientific culture, the boundary between human and animal is thoroughly breached. The last beachheads of uniqueness have been polluted if not turned into amusement parks—language, tool use, social behaviour, mental events, nothing really convincingly settles the separation of human and animal. And many people no longer feel the need for such a separation; indeed, many branches of feminist culture affirm the pleasure of connection of human and other living creatures. Movements for animal rights are not irrational denials of human uniqueness; they are a clear-sighted recognition of connection across the discredited breach of nature and culture. Biology and evolutionary theory over the last two centuries have simultaneously produced modern organisms as objects of knowledge and reduced the line between humans and animals to a faint trace re-etched in ideological struggle or professional disputes between life and social science. Within this framework, teaching modern Christian creationism should be fought as a form of child abuse.

Biological-determinist ideology is only one position opened up in scientific culture for arguing the meanings of human animality. There is much room for radical political people to contest the meanings of the breached boundary.[1] The cyborg appears in myth precisely where the boundary between human and animal is transgressed. Far from signaling a walling off of people from other living beings, cyborgs signal disturbingly and pleasurably tight coupling. Bestiality has a new status in this cycle of marriage exchange.

The second leaky distinction is between animal-human (organism) and machine. Pre-cybernetic machines could be haunted; there was always the spectre of the ghost in the machine. This dualism structured the dialogue between materialism and idealism that was settled by a dialectical progeny, called spirit or history, according to taste. But basically machines were not self-moving, self-designing, autonomous. They could not achieve man's dream, only mock it. They were not man, an author to himself, but only a caricature of that masculinist reproductive dream. To think they were otherwise was paranoid. Now we are not so sure. Late twentieth-century machines have made thoroughly ambiguous the difference between natural and artificial, mind and body, self-developing and externally designed, and many other distinctions that used to apply to organisms and machines. Our machines are disturbingly lively, and we ourselves frighteningly inert.

Technological determination is only one ideological space opened up by the reconceptions of machine and organism as coded texts through which we engage in the play of writing and reading the world.[2] 'Textualization' of everything in poststructuralist, postmodernist theory has been damned by Marxists and socialist feminists for its utopian disregard for the lived relations of domination that ground the 'play' of arbitrary reading.[3] It is certainly true that postmodernist strategies, like my cyborg myth, subvert myriad organic wholes (for example, the poem, the primitive culture, the biological organism). In short, the certainty of what counts as nature—a source of insight and promise of innocence—is undermined, probably fatally. The transcendent authorization of interpretation is lost, and with it the ontology grounding 'Western' epistemology. But the alternative is not cynicism or faithlessness, that is, some version of abstract existence, like the accounts of technological determinism destroying 'man' by the 'machine' or 'meaningful political action' by the 'text'. Who cyborgs will be is a radical question; the answers are a matter of survival. Both chimpanzees and artefacts have politics, so why shouldn't we (de Waal, 1982; Winner, 1980)?

The third distinction is a subset of the second: the boundary between physical and non-physical is very imprecise for us. Pop physics books on the consequences of quantum theory and the indeterminacy principle are a kind of popular scientific equivalent to Harlequin romances* as a marker of radical change in American white heterosexuality: they get it wrong, but they are on the right subject. Modern machines are quintessentially microelectronic devices: they are everywhere and they are invisible. Modern machinery is an irreverent upstart god, mocking the Father's ubiquity and spirituality. The silicon chip is a surface for writing; it is etched in molecular scales disturbed only by atomic noise, the ultimate interference for nuclear scores. Writing, power, and technology are old partners in Western stories of the origin of civilization, but miniaturization has changed our experience of mechanism. Miniaturization has turned out to be about power; small is not so much beautiful as pre-eminently dangerous, as in cruise missiles. Contrast the TV sets of the 1950s or the news cameras of the 1970s with the TV wrist bands or hand-sized video cameras now advertised. Our best machines are made of sunshine; they are all light and clean because they are nothing but signals, electromagnetic waves, a section of a spectrum, and these machines are eminently portable, mobile—a matter of immense human pain in Detroit and Singapore. People are nowhere near so fluid, being both material and opaque. Cyborgs are ether, quintessence.

The ubiquity and invisibility of cyborgs is precisely why these sunshine-belt machines are so deadly. They are as hard to see politically as materially. They are about consciousness—or its simulation.[4] They are floating signifiers moving in pickup trucks across Europe, blocked more effectively by the witch-weavings of the displaced and so unnatural Greenham women, who read the cyborg webs of power so very well, than by the militant labour of older masculinist politics, whose natural constituency needs defence jobs. Ultimately the 'hardest' science is about the realm of greatest boundary confusion, the realm of pure number, pure spirit, C^3I, cryptography, and the preservation of potent secrets. The new machines are so clean and light. Their engineers are sun-worshippers mediating a new scientific revolution associated with the night dream of post-industrial society. The diseases evoked by these clean machines are 'no more' than the minuscule coding changes of an antigen in the immune system, 'no more' than the experience of stress. The nimble fingers of 'Oriental' women, the old fascination of little Anglo-Saxon Victorian girls with doll's houses, women's enforced attention to the small take on quite new dimensions in this world. There might be a cyborg Alice taking

---

*The US equivalent of Mills & Boon.

account of these new dimensions. Ironically, it might be the unnatural cyborg women making chips in Asia and spiral dancing in Santa Rita jail* whose constructed unities will guide effective oppositional strategies.

So my cyborg myth is about transgressed boundaries, potent fusions, and dangerous possibilities which progressive people might explore as one part of needed political work. One of my premises is that most American socialists and feminists see deepened dualisms of mind and body, animal and machine, idealism and materialism in the social practices, symbolic formulations, and physical artefacts associated with 'high technology' and scientific culture. From *One-Dimensional Man* (Marcuse, 1964) to *The Death of Nature* (Merchant, 1980), the analytic resources developed by progressives have insisted on the necessary domination of technics and recalled us to an imagined organic body to integrate our resistance. Another of my premises is that the need for unity of people trying to resist world-wide intensification of domination has never been more acute. But a slightly perverse shift of perspective might better enable us to contest for meanings, as well as for other forms of power and pleasure in technologically mediated societies.

From one perspective, a cyborg world is about the final imposition of a grid of control on the planet, about the final abstraction embodied in a Star Wars apocalypse waged in the name of defence, about the final appropriation of women's bodies in a masculinist orgy of war (Sofia, 1984). From another perspective, a cyborg world might be about lived social and bodily realities in which people are not afraid of their joint kinship with animals and machines, not afraid of permanently partial identities and contradictory standpoints. The political struggle is to see from both perspectives at once because each reveals both dominations and possibilities unimaginable from the other vantage point. Single vision produces worse illusions than double vision or many-headed monsters. Cyborg unities are monstrous and illegitimate; in our present political circumstances, we could hardly hope for more potent myths for resistance and recoupling. I like to imagine LAG, the Livermore Action Group, as a kind of cyborg society, dedicated to realistically converting the laboratories that most fiercely embody and spew out the tools of technological apocalypse, and committed to building a political form that actually manages to hold together witches, engineers, elders, perverts, Christians, mothers, and Leninists long enough to disarm the state. Fission Impossible is the name of the affinity group in my town. (Affinity: related not by blood but by choice, the appeal of one chemical nuclear group for another, avidity.)[5]

THE INFORMATICS OF DOMINATION

In this attempt at an epistemological and political position, I would like to sketch a picture of possible unity, a picture indebted to socialist and feminist principles of design. The frame for my sketch is set by the extent and importance of rearrangements in world-wide social relations tied to science and technology. I argue for a politics rooted in claims about fundamental changes in the nature of class, race, and gender in an emerging system of world order analogous in its novelty and scope to that created by industrial capitalism; we are living through a movement from an organic, industrial society to a polymorphous, information system—from all work to all play, a deadly game. Simultaneously material and ideological, the dichotomies

---

*A practice at once both spiritual and political that linked guards and arrested anti-nuclear demonstrators in the Alameda County jail in California in the early 1980s.

may be expressed in the following chart of transitions from the comfortable old hierarchical dominations to the scary new networks I have called the informatics of domination:

Representation	Simulation
Bourgeois novel, realism	Science fiction, postmodernism
Organism	Biotic component
Depth, integrity	Surface, boundary
Heat	Noise
Biology as clinical practice	Biology as inscription
Physiology	Communications engineering
Small group	Subsystem
Perfection	Optimization
Eugenics	Population Control
Decadence, *Magic Mountain*	Obsolescence, *Future Shock*
Hygiene	Stress Management
Microbiology, tuberculosis	Immunology, AIDS
Organic division of labour	Ergonomics/cybernetics of labour
Functional specialization	Modular construction
Reproduction	Replication
Organic sex role specialization	Optimal genetic strategies
Biological determinism	Evolutionary inertia, constraints
Community ecology	Ecosystem
Racial chain of being	Neo-imperialism, United Nations humanism
Scientific management in home/ factory	Global factory / Electronic cottage
Family / Market / Factory	Women in the Integrated Circuit
Family wage	Comparable worth
Public / Private	Cyborg citizenship
Nature / Culture	Fields of difference
Co-operation	Communications enhancement
Freud	Lacan
Sex	Genetic engineering
Labour	Robotics
Mind	Artificial Intelligence
Second World War	Star Wars
White Capitalist Patriarchy	Informatics of Domination

This list suggests several interesting things.[12] First, the objects on the right-hand side cannot be coded as 'natural', a realization that subverts naturalistic coding for the left-hand side as well. We cannot go back ideologically or materially. It's not just that 'god' is dead; so is the 'goddess'. Or both are revivified in the worlds charged with microelectronic and biotechnological politics. In relation to objects like biotic components, one must think not in terms of essential properties, but in terms of design, boundary constraints, rates of flows, systems logics, costs of lowering constraints. Sexual reproduction is one kind of reproductive strategy among many, with costs and benefits as a function of the system environment. Ideologies of sexual reproduction can no longer reasonably call on notions of sex and sex role as organic aspects in natural objects like organisms and families. Such reasoning will be unmasked as irrational, and ironically corporate executives reading *Playboy* and anti-porn radical feminists will make strange bedfellows in jointly unmasking the irrationalism.

Likewise for race, ideologies about human diversity have to be formulated in terms of frequencies of parameters, like blood groups or intelligence scores. It is 'irrational' to invoke concepts like primitive and civilized. For liberals and radicals, the search for integrated social systems gives way to a new practice called 'experimental ethnography' in which an organic object dissipates in attention to the play of writing. At the level of ideology, we see translations of racism and colonialism into languages of development and under-development, rates and constraints of modernization. Any objects or persons can be reasonably thought of in terms of disassembly and reassembly; no 'natural' architectures constrain system design. The financial districts in all the world's cities, as well as the export-processing and free-trade zones, proclaim this elementary fact of 'late capitalism'. The entire universe of objects that can be known scientifically must be formulated as problems in communications engineering (for the managers) or theories of the text (for those who would resist). Both are cyborg semiologies.

One should expect control strategies to concentrate on boundary conditions and interfaces, on rates of flow across boundaries—and not on the integrity of natural objects. 'Integrity' or 'sincerity' of the Western self gives way to decision procedures and expert systems. For example, control strategies applied to women's capacities to give birth to new human beings will be developed in the languages of population control and maximization of goal achievement for individual decision-makers. Control strategies will be formulated in terms of rates, costs of constraints, degrees of freedom. Human beings, like any other component or subsystem, must be localized in a system architecture whose basic modes of operation are probabilistic, statistical. No objects, spaces, or bodies are sacred in themselves; any component can be interfaced with any other if the proper standard, the proper code, can be constructed for processing signals in a common language. Exchange in this world transcends the universal translation effected by capitalist markets that Marx analysed so well. The privileged pathology affecting all kinds of components in this universe is stress—communications breakdown (Hogness, 1983). The cyborg is not subject to Foucault's biopolitics; the cyborg simulates politics, a much more potent field of operations.

This kind of analysis of scientific and cultural objects of knowledge which have appeared historically since the Second World War prepares us to notice some important inadequacies in feminist analysis which has proceeded as if the organic, hierarchical dualisms ordering discourse in 'the West' since Aristotle still ruled. They have been cannibalized, or as Zoe Sofia (Sofoulis) might put it, they have been 'techno-digested'. The dichotomies between mind and body, animal and human, organism and machine, public and private, nature and culture, men and women, primitive and civilized are all in question ideologically. The actual situation of women is their integration/exploitation into a world system of production/reproduction and communication called the informatics of domination. The home, workplace, market, public arena, the body itself—all can be dispersed and interfaced in nearly infinite, polymorphous ways, with large consequences for women and others—consequences that themselves are very different for different people and which make potent oppositional international movements difficult to imagine and essential for survival. One important route for reconstructing socialist-feminist politics is through theory and practice addressed to the social relations of science and technology, including crucially the systems of myth and meanings structuring our imaginations. The cyborg is a kind of disassembled and reassembled, postmodern collective and personal self. This is the self feminists must code.

Communications technologies and biotechnologies are the crucial tools recrafting our bodies. These tools embody and enforce new social relations for women world-wide. Technologies and scientific discourses can be partially understood as formalizations, i.e., as frozen moments, of the fluid social interactions constituting them, but they should also be viewed as instruments for enforcing meanings. The boundary is permeable between tool and myth, instrument and concept, historical systems of social relations and historical anatomies of possible bodies, including objects of knowledge. Indeed, myth and tool mutually constitute each other.

Furthermore, communications sciences and modern biologies are constructed by a common move—*the translation of the world into a problem of coding,* a search for a common language in which all resistance to instrumental control disappears and all heterogeneity can be submitted to disassembly, reassembly, investment, and exchange.

In communications sciences, the translation of the world into a problem in coding can be illustrated by looking at cybernetic (feedback-controlled) systems theories applied to telephone technology, computer design, weapons deployment, or data base construction and maintenance. In each case, solution to the key questions rests on a theory of language and control; the key operation is determining the rates, directions, and probabilities of flow of a quantity called information. The world is subdivided by boundaries differentially permeable to information. Information is just that kind of quantifiable element (unit, basis of unity) which allows universal translation, and so unhindered instrumental power (called effective communication). The biggest threat to such power is interruption of communication. Any system breakdown is a function of stress. The fundamentals of this technology can be condensed into the metaphor C^3I, command-control-communication-intelligence, the military's symbol for its operations theory.

In modern biologies, the translation of the world into a problem in coding can be illustrated by molecular genetics, ecology, sociobiological evolutionary theory, and immunobiology. The organism has been translated into problems of genetic coding and read-out. Biotechnology, a writing technology, informs research broadly.[13] In a sense, organisms have ceased to exist as objects of knowledge, giving way to biotic components, i.e., special kinds of information-processing devices. The analogous moves in ecology could be examined by probing the history and utility of the concept of the ecosystem. Immunobiology and associated medical practices are rich exemplars of the privilege of coding and recognition systems as objects of knowledge, as constructions of bodily reality for us. Biology here is a kind of cryptography. Research is necessarily a kind of intelligence activity. Ironies abound. A stressed system goes awry; its communication processes break down; it fails to recognize the difference between self and other. Human babies with baboon hearts evoke national ethical perplexity— for animal rights activists at least as much as for the guardians of human purity. In the US gay men and intravenous drug users are the 'privileged' victims of an awful immune system disease that marks (inscribes on the body) confusion of boundaries and moral pollution (Treichler, 1987).

But these excursions into communications sciences and biology have been at a rarefied level; there is a mundane, largely economic reality to support my claim that these sciences and technologies indicate fundamental transformations in the structure of the world for us. Communications technologies depend on electronics. Modern states, multinational corporations, military power, welfare state apparatuses, satellite systems, political processes, fabrication of our imaginations, labour-control systems, medical constructions of our bodies, commercial pornography, the international division of labour, and religious evangelism depend intimately upon electronics. Microelectronics is the technical basis of simulacra; that is, of copies without originals.

Microelectronics mediates the translations of labour into robotics and word processing, sex into genetic engineering and reproductive technologies, and mind into artificial intelligence and decision procedures. The new biotechnologies concern more than human reproduction. Biology as a powerful engineering science for redesigning materials and processes has revolutionary implications for industry, perhaps most obvious today in areas of fermentation, agriculture, and energy. Communications sciences and biology are constructions of natural-technical objects of knowledge in which the difference between machine and organism is thoroughly blurred; mind, body, and tool are on very intimate terms. The 'multinational' material organization of the production and reproduction of daily life and the symbolic organization of the production and reproduction of culture and imagination seem equally implicated. The boundary-maintaining images of base and superstructure, public and private, or material and ideal never seemed more feeble.

I have used Rachel Grossman's (1980) image of women in the integrated circuit to name the situation of women in a world so intimately restructured through the social relations of science and technology.[14] I used the odd circumlocution, 'the social relations of science and technology', to indicate that we are not dealing with a technological determinism, but with a historical system depending upon structured relations among people. But the phrase should also indicate that science and technology provide fresh sources of power, that we need fresh sources of analysis and political action (Latour, 1984). Some of the rearrangements of race, sex, and class rooted in high-tech-facilitated social relations can make socialist-feminism more relevant to effective progressive politics.

CYBORGS: A MYTH OF POLITICAL IDENTITY

I want to conclude with a myth about identity and boundaries which might inform late twentieth-century political imaginations. I am indebted in this story to writers like Joanna Russ, Samuel R. Delany, John Varley, James Tiptree, Jr, Octavia Butler, Monique Wittig, and Vonda McIntyre.[15] These are our story-tellers exploring what it means to be embodied in high-tech worlds. They are theorists for cyborgs. Exploring conceptions of bodily boundaries and social order, the anthropologist Mary Douglas (1966, 1970) should be credited with helping us to consciousness about how fundamental body imagery is to world view, and so to political language. French feminists like Luce Irigaray and Monique Wittig, for all their differences, know how to write the body; how to weave eroticism, cosmology, and politics from imagery of embodiment, and especially for Wittig, from imagery of fragmentation and reconstitution of bodies.[16]

American radical feminists like Susan Griffin, Audre Lorde, and Adrienne Rich have profoundly affected our political imaginations—and perhaps restricted too much what we allow as a friendly body and political language.[17] They insist on the organic, opposing it to the technological. But their symbolic systems and the related positions of ecofeminism and feminist paganism, replete with organicisms, can only be understood in Sandoval's terms as oppositional ideologies fitting the late twentieth century. They would simply bewilder anyone not preoccupied with the machines and consciousness of late capitalism. In that sense they are part of the cyborg world. But there are also great riches for feminists in explicitly embracing the possibilities inherent in the breakdown of clean distinctions between organism and machine and similar distinctions structuring the Western self. It is the simultaneity of breakdowns that cracks the matrices of domination and opens geometric possibilities. What might be learned from personal and political 'technological' pollution? I look briefly at two overlapping groups of texts for their insight into the construction of a potentially helpful cyborg myth: constructions of women of colour and monstrous selves in feminist science fiction.

Earlier I suggested that 'women of colour' might be understood as a cyborg identity, a potent subjectivity synthesized from fusions of outsider identities and in the complex political-historical layerings of her 'biomythography', *Zami* (Lorde, 1982; King, 1987a, 1987b). There are material and cultural grids mapping this potential, Audre Lorde (1984) captures the tone in the title of her *Sister Outsider*. In my political myth, Sister Outsider is the offshore woman, whom US workers, female and feminized, are supposed to regard as the enemy preventing their solidarity, threatening their security. Onshore, inside the boundary of the United States, Sister Outsider is a potential amidst the races and ethnic identities of women manipulated for division, competition, and exploitation in the same industries. 'Women of colour' are the preferred labour force for the science-based industries, the real women for whom the world-wide sexual market, labour market, and politics of reproduction kaleidoscope into daily life. Young Korean women hired in the sex industry and in electronics assembly are recruited from high schools, educated for the integrated circuit. Literacy, especially in English, distinguishes the 'cheap' female labour so attractive to the multinationals.

Contrary to orientalist stereotypes of the 'oral primitive', literacy is a special mark of women of colour, acquired by US black women as well as men through a history of risking death to learn and to teach reading and writing. Writing has a special significance for all colonized groups. Writing has been crucial to the Western myth of the distinction between oral and written cultures, primitive and civilized mentalities, and more recently to the erosion of that distinction in 'postmodernist' theories attacking the phallogocentrism of the West, with its worship of the monotheistic, phallic, authoritative, and singular work, the unique and perfect name.[18] Contests for the meanings of writing are a major form of contemporary political struggle. Releasing the play of writing is deadly serious. The poetry and stories of US women of colour are repeatedly about writing, about access to the power to signify; but this time that power must be neither phallic nor innocent. Cyborg writing must not be about the Fall, the imagination of a once-upon-a-time wholeness before language, before writing, before Man. Cyborg writing is about the power to survive, not on the basis of original innocence, but on the basis of seizing the tools to mark the world that marked them as other.

The tools are often stories, retold stories, versions that reverse and displace the hierarchical dualisms of naturalized identities. In retelling origin stories, cyborg authors subvert the central myths of origin of Western culture. We have all been colonized by those origin myths, with their longing for fulfillment in apocalypse. The phallogocentric origin stories most crucial for feminist cyborgs are built into the literal technologies—technologies that write the world, biotechnology and microelectronics—that have recently textualized our bodies as code problems on the grid of C^{3}I. Feminist cyborg stories have the task of recoding communication and intelligence to subvert command and control.

Figuratively and literally, language politics pervade the struggles of women of colour; and stories about language have a special power in the rich contemporary writing by US women of colour. For example, retellings of the story of the indigenous woman Malinche, mother of the mestizo 'bastard' race of the new world, master of languages, and mistress of Cortés, carry special meaning for Chicana constructions of identity. Cherríe Moraga (1983) in *Loving in the War Years* explores the themes of identity when one never possessed the original language, never told the original story, never resided in the harmony of legitimate heterosexuality in the garden of culture, and so cannot base identity on a myth or a fall from innocence and right to natural names, mother's or father's.[19] Moraga's writing, her superb literacy, is presented in her poetry as the same kind of violation as Malinche's mastery of the conqueror's language—a violation, an illegitimate production, that allows survival. Moraga's language is not 'whole'; it is self-consciously spliced, a chimera of English and Spanish, both conqueror's languages. But it is this chimeric monster, without claim to an original language before violation, that crafts the erotic, competent, potent identities of women of colour. Sister Outsider hints at the possibility of world survival not because of her innocence, but because of her ability to live on the boundaries, to write without the founding myth of original wholeness, with its inescapable apocalypse of final return to a deathly oneness that Man has imagined to be the innocent and all-powerful Mother, freed at the End from another spiral of appropriation by her son. Writing marks Moraga's body, affirms it as the body of a woman of colour, against the possibility of passing into the unmarked category of the Anglo father or into the orientalist myth of 'original illiteracy' of a mother that never was. Malinche was mother here, not Eve before eating the forbidden fruit. Writing affirms Sister Outsider, not the Woman-before-the-Fall-into-Writing needed by the phallogocentric Family of Man.

Writing is pre-eminently the technology of cyborgs, etched surfaces of the late twentieth century. Cyborg politics is the struggle for language and the struggle against perfect communication, against the one code that translates all meaning perfectly, the central dogma of phallogocentrism. That is why cyborg politics insist on noise and advocate pollution, rejoicing in the illegitimate fusions of animal and machine. These are the couplings which make Man and Woman so problematic, subverting the structure of desire, the force imagined to generate language and gender, and so subverting the structure and modes of reproduction of 'Western' identity, of nature and culture, of mirror and eye, slave and master, body and mind. 'We' did not originally choose to be cyborgs, but choice grounds a liberal politics and epistemology that imagines the reproduction of individuals before the wider replications of 'texts'.

From the perspective of cyborgs, freed of the need to ground politics in 'our' privileged position of the oppression that incorporates all other dominations, the innocence of the merely violated, the ground of those closer to nature, we can see powerful possibilities. Feminisms and Marxisms have run aground on Western epistemological imperatives to construct a revolutionary subject from the perspective of a hierarchy of oppressions and/or a latent position of moral superiority, innocence, and greater closeness to nature. With no available original dream of a common language or original symbiosis promising protection from hostile 'masculine' separation, but written into the play of a text that has no finally privileged reading or salvation history, to recognize 'oneself' as fully implicated in the world, frees us of the need to root politics in identification, vanguard parties, purity, and mothering. Stripped of identity, the bastard race teaches about the power of the margins and the importance of a mother like Malinche. Women of colour have transformed her from the evil mother of masculinist fear into the originally literate mother who teaches survival.

This is not just literary deconstruction, but liminal transformation. Every story that begins with original innocence and privileges the return to wholeness imagines the drama of life to be individuation, separation, the birth of the self, the tragedy of autonomy, the fall into writing, alienation; that is, war, tempered by imaginary respite in the bosom of the Other. These plots are ruled by a reproductive politics—rebirth without flaw, perfection, abstraction. In this plot women are imagined either better or worse off, but all agree they have less selfhood, weaker individuation, more fusion to the oral, to Mother, less at stake in masculine autonomy. But there is another route to having less at stake in masculine autonomy, a route that does not pass through Woman, Primitive, Zero, the Mirror Stage and its imaginary. It passes through women and other present-tense, illegitimate cyborgs, not of Woman born, who refuse the ideological resources of victimization so as to have a real life. These cyborgs are the people who refuse to disappear on cue, no matter how many times a 'Western' commentator remarks on the sad passing of another primitive, another organic group done in by 'Western' technology, by writing.[20] These real-life cyborgs (for example, the Southeast Asian village women workers in Japanese and US electronics firms described by Aihwa Ong) are actively rewriting the texts of their bodies and societies. Survival is the stakes in this play of readings.

To recapitulate, certain dualisms have been persistent in Western traditions; they have all been systemic to the logics and practices of domination of women, people of colour, nature, workers, animals—in short, domination of all constituted as others, whose task is to mirror the self. Chief among these troubling dualisms are self/other, mind/body, culture/nature, male/female, civilized/primitive, reality/appearance, whole/part, agent/resource, maker/made, active/passive, right/wrong, truth/illusion, total/partial, God/man. The self is the One who is not dominated, who knows that by the service of the other, the other is the one who holds the future, who knows that by the experience of domination, which gives the lie to the autonomy of the self. To be One is to be autonomous, to be powerful, to be God; but to be One is to be an illusion, and so to be involved in a dialectic of apocalypse with the other. Yet to be other is to be multiple, without clear boundary, frayed, insubstantial. One is too few, but two are too many.

High-tech culture challenges these dualisms in intriguing ways. It is not clear who makes and who is made in the relation between human and machine. It is not clear what is mind and what body in machines that resolve into coding practices. In so far as we know ourselves in both formal discourse (for example, biology) and in daily practice (for example, the homework economy in the integrated circuit), we find ourselves to be cyborgs, hybrids, mosaics, chimeras. Biological organisms have become biotic systems, communications devices like others. There is no fundamental, ontological separation in our formal knowledge of machine and organism, of technical and organic. The replicant Rachel in the Ridley Scott film *Blade Runner* stands as the image of a cyborg culture's fear, love, and confusion.

One consequence is that our sense of connection to our tools is heightened. The trance state experienced by many computer users has become a staple of science-fiction film and cultural jokes. Perhaps paraplegics and other severely handicapped people can (and sometimes do) have the most intense experiences of complex hybridization with other communication devices.[21] Anne McCaffrey's pre-feminist *The Ship Who Sang* (1969) explored the consciousness of a cyborg, hybrid of girl's brain and complex machinery, formed after the birth of a severely handicapped child. Gender, sexuality, embodiment, skill: all were reconstituted in the story. Why should our bodies end at the skin, or include at best other beings encapsulated by skin? From the seventeenth century till now, machines could be animated—given ghostly souls to make them speak or move or to account for their orderly development and mental capacities. Or organisms could be mechanized—reduced to body understood as resource of mind. These machine/organism relationships are obsolete, unnecessary. For us, in imagination and in other practice, machines can be prosthetic devices, intimate components, friendly selves. We don't need organic holism to give impermeable wholeness, the total woman and her feminist variants (mutants?). Let me conclude this point by a very partial reading of the logic of the cyborg monsters of my second group of texts, feminist science fiction.

The cyborgs populating feminist science fiction make very problematic the statuses of man or woman, human, artefact, member of a race, individual entity, or body. Katie King clarifies how pleasure in reading these fictions is not largely based on identification. Students facing Joanna Russ for the first time, students who have learned to take modernist writers like James Joyce or Virginia Woolf without flinching, do not know what to make of *The Adventures of Alyx* or *The Female Man*, where characters refuse the reader's search for innocent wholeness while granting the wish for heroic quests, exuberant eroticism, and serious politics. *The Female Man* is the story of four versions of one genotype, all of whom meet, but even taken together do not make a whole, resolve the dilemmas of violent moral action, or remove the growing scandal of gender. The feminist science fiction of Samuel R. Delany, especially *Tales of Nevèrÿon*, mocks stories of origin by redoing the neolithic revolution, replaying the founding moves of Western civilization to subvert their plausibility. James Tiptree, Jr, an author whose fiction was regarded as particularly manly until her 'true' gender was revealed, tells tales of reproduction based on non-mammalian technologies like alternation of generations of male brood pouches and male nurturing. John Varley constructs a supreme cyborg in his arch-feminist exploration of Gaea, a mad goddess-planet-trickster-old woman-technological device on whose surface an extraordinary array of post-cyborg symbioses are spawned. Octavia Butler writes of an African sorceress pitting her powers of transformation against the genetic manipulations of her rival (*Wild Seed*), of time warps that bring a modern US black woman into slavery where her actions in relation to her white master-ancestor determine the possibility of her own birth (*Kindred*), and of the illegitimate insights into identity and community of an adopted cross-species child who came to know the enemy as self (*Survivor*). In *Dawn* (1987), the first installment of a series called *Xenogenesis,* Butler tells the story of Lilith Iyapo, whose personal name recalls Adam's first and repudiated wife and whose family name marks her status as the widow of the son of Nigerian immigrants to the US. A black woman and a mother whose child is dead, Lilith mediates the transformation of humanity through genetic exchange with extra-terrestrial lovers/rescuers/destroyers/genetic engineers, who reform earth's habitats after the nuclear holocaust and coerce surviving humans into intimate fusion with them. It is a novel that interrogates reproductive, linguistic, and nuclear politics in a mythic field structured by late twentieth-century race and gender.

Because it is particularly rich in boundary transgressions, Vonda McIntyre's *Superluminal* can close this truncated catalogue of promising and dangerous monsters who help redefine the pleasures and politics of embodiment and feminist writing. In a fiction where no character is 'simply' human, human status is highly problematic. Orca, a genetically altered diver, can speak with killer whales and survive deep ocean conditions, but she longs to explore space as a pilot, necessitating bionic implants jeopardizing her kinship with the divers and cetaceans. Transformations are effected by virus vectors carrying a new developmental code, by transplant surgery, by implants of microelectronic devices, by analogue doubles, and other means. Laenea becomes a pilot by accepting a heart implant and a host of other alterations allowing survival in transit at speeds exceeding that of light. Radu Dracul survives a virus-caused plague in his outerworld planet to find himself with a time sense that changes the boundaries of spatial perception for the whole species. All the characters explore the limits of language; the dream of communicating experience; and the necessity of limitation, partiality, and intimacy even in this world of protean transformation and connection. *Superluminal* stands also for the defining contradictions of a cyborg world in another sense; it embodies textually the intersection of feminist theory and colonial discourse in the science fiction I have alluded to in this chapter. This is a conjunction with a long history that many 'First World' feminists have tried to repress, including myself in my readings of *Superluminal* before being called to account by Zoe Sofoulis, whose different location in the world system's informatics of domination made her acutely alert to the imperialist moment of all science fiction cultures, including women's science fiction. From an Australian feminist sensitivity, Sofoulis remembered more readily McIntyre's role as writer of the adventures of Captain Kirk and Spock in TV's *Star Trek* series than her rewriting the romance in *Superluminal.*

Monsters have always defined the limits of community in Western imaginations. The Centaurs and Amazons of ancient Greece established the limits of the centred polis of the Greek male human by their disruption of marriage and boundary pollutions of the warrior with animality and woman. Unseparated twins and hermaphrodites were the confused human material in early modern France who grounded discourse on the natural and supernatural, medical and legal, portents and diseases—all crucial to establishing modern identity.[22] The evolutionary and behavioural sciences of monkeys and apes have marked the multiple boundaries of late twentieth-century industrial identities. Cyborg monsters in feminist science fiction define quite different political possibilities and limits from those proposed by the mundane fiction of Man and Woman.

There are several consequences to taking seriously the imagery of cyborgs as other than our enemies. Our bodies, ourselves; bodies are maps of power and identity. Cyborgs are no exception. A cyborg body is not innocent; it was not born in a garden; it does not seek unitary identity and so generate antagonistic dualisms without end (or until the world ends); it takes irony for granted. One is too few, and two is only one possibility. Intense pleasure in skill, machine skill, ceases to be a sin, but an aspect of embodiment. The machine is not an *it* to be animated, worshipped, and dominated. The machine is us, our processes, an aspect of our embodiment. We can be responsible for machines; *they* do not dominate or threaten us. We are responsible for boundaries; we are they. Up till now (once upon a time), female embodiment seemed to be given, organic, necessary; and female embodiment seemed to mean skill in mothering and its metaphoric extensions. Only by being out of place could we take intense pleasure in machines, and then with excuses that this was organic activity after all, appropriate to females. Cyborgs might consider more seriously the partial, fluid, sometimes aspect of sex and sexual embodiment. Gender might not be global identity after all, even if it has profound historical breadth and depth.

The ideologically charged question of what counts as daily activity, as experience, can be approached by exploiting the cyborg image. Feminists have recently claimed that women are given to dailiness, that women more than men somehow sustain daily life, and so have a privileged epistemological position potentially. There is a compelling aspect to this claim, one that makes visible unvalued female activity and names it as the ground of life. But *the* ground of life? What about all the ignorance of women, all the exclusions and failures of knowledge and skill? What about men's access to daily competence, to knowing how to build things, to take them apart, to play? What about other embodiments? Cyborg gender is a local possibility taking a global vengeance. Race, gender, and capital require a cyborg theory of wholes and parts. There is no drive in cyborgs to produce total theory, but there is an intimate experience of boundaries, their construction and deconstruction. There is a myth system waiting to become a political language to ground one way of looking at science and technology and challenging the informatics of domination—in order to act potently.

One last image: organisms and organismic, holistic politics depend on metaphors of rebirth and invariably call on the resources of reproductive sex. I would suggest that cyborgs have more to do with regeneration and are suspicious of the reproductive matrix and of most birthing. For salamanders, regeneration after injury, such as the loss of a limb, involves regrowth of structure and restoration of function with the constant possibility of twinning or other odd topographical productions at the site of former injury. The regrown limb can be monstrous, duplicated, potent. We have all been injured, profoundly. We require regeneration, not rebirth, and the possibilities for our reconstitution include the utopian dream of the hope for a monstrous world without gender.

Cyborg imagery can help express two crucial arguments in this essay: first, the production of universal, totalizing theory is a major mistake that misses most of reality, probably always, but certainly now; and second, taking responsibility for the social relations of science and technology means refusing an anti-science metaphysics, a demonology of technology, and so means embracing the skilful task of reconstructing the boundaries of daily life, in partial connection with others, in communication with all of our parts. It is not just that science and technology are possible means of great human satisfaction, as well as a matrix of complex dominations. Cyborg imagery can suggest a way out of the maze of dualisms in which we have explained our bodies and our tools to ourselves. This is a dream not of a common language, but of a powerful infidel heteroglossia. It is an imagination of a feminist speaking in tongues to strike fear into the circuits of the supersavers of the new right. It means both building and destroying machines, identities, categories, relationships, space stories. Though both are bound in the spiral dance, I would rather be a cyborg than a goddess.

NOTES

1. Useful references to left and/or feminist radical science movements and theory and to biological/biotechnical issues include: Bleier (1984, 1986), Harding (1986), Fausto-Sterling (1985), Gould (1981), Hubbard *et al.* (1982), Keller (1985), Lewontin *et al.* (1984), *Radical Science Journal* (became *Science as Culture* in 1987), 26 Freegrove Road, London N7 9RQ; *Science for the People,* 897 Main St, Cambridge, MA 02139.

2. Starting points for left and/or feminist approaches to technology and politics include: Cowan (1983), Rothschild (1983), Traweek (1988), Young and Levidow (1981, 1985), Weizenbaum (1976), Winner (1977, 1986), Zimmerman (1983), Athanasiou (1987), Cohn (1987a, 1987b), Winograd and Flores (1986), Edwards (1985). *Global Electronics Newsletter,* 867 West Dana St, #204, Mountain View, CA 94041; *Processed World,* 55 Sutter St, San Francisco, CA 94104; ISIS, Women's International Information and Communication Service, PO Box 50 (Cornavin), 1211 Geneva 2, Switzerland, and Via Santa Maria Dell'Anima 30, 00186 Rome, Italy. Fundamental approaches to modern social studies of science that do not continue the liberal mystification that it all started with Thomas Kuhn, include: Knorr-Cetina (1981), Knorr-Cetina and Mulkay (1983), Latour and Woolgar (1979), Young (1979). The 1984 Directory of the Network for the Ethnographic Study of Science, Technology, and Organizations lists a wide range of people and projects crucial to better radical analysis; available from NESSTO, PO Box 11442, Stanford, CA 94305.

3. A provocative, comprehensive argument about the politics and theories of 'postmodernism' is made by Fredric Jameson (1984), who argues that postmodernism is not an option, a style among others, but a cultural dominant requiring radical reinvention of left politics from within; there is no longer any place from without that gives meaning to the comforting fiction of critical distance. Jameson also makes clear why one cannot be for or against postmodernism, an essentially moralist move. My position is that feminists (and others) need continuous cultural reinvention, postmodernist critique, and historical materialism; only a cyborg would have a chance. The old dominations of white capitalist patriarchy seem nostalgically innocent now: they normalized heterogeneity, into man and woman, white and black, for example. 'Advanced capitalism' and postmodernism release heterogeneity without a norm, and we are flattened, without subjectivity, which requires depth, even unfriendly and drowning depths. It is time to write *The Death of the Clinic.* The clinic's methods required bodies and works; we have texts and surfaces. Our dominations don't work by medicalization and normalization any more; they work by networking, communications redesign, stress management. Normalization gives way to automation, utter redundancy. Michel Foucault's *Birth of the Clinic* (1963), *History of Sexuality* (1976),

and *Discipline and Punish* (1975) name a form of power at its moment of implosion. The discourse of biopolitics gives way to technobabble, the language of the spliced substantive; no noun is left whole by the multinationals. These are their names, listed from one issue of *Science*: Tech-Knowledge, Genentech, Allergen, Hybritech, Compupro, Genen-cor, Syntex, Allelix, Agrigenetics Corp., Syntro, Codon, Repligen, MicroAngelo from Scion Corp., Percom Data, Inter Systems, Cyborg Corp., Statcom Corp., Intertec. If we are imprisoned by language, then escape from that prison-house requires language poets, a kind of cultural restriction enzyme to cut the code; cyborg heteroglossia is one form of radical cultural politics. For cyborg poetry, see Perloff (1984); Fraser (1984). For feminist modernist/postmodernist 'cyborg' writing, see HOW(-ever), 871 Corbett Ave, San Francisco, CA 94131.

4. Baudrillard (1983). Jameson (1984, p. 66) points out that Plato's definition of the simulacrum is the copy for which there is no original, i.e., the world of advanced capitalism, of pure exchange. See *Discourse* 9 (Spring/Summer 1987) for a special issue on technology (cybernetics, ecology, and the postmodern imagination).

5. For ethnographic accounts and political evaluations, see Epstein (forthcoming), Sturgeon (1986). Without explicit irony, adopting the spaceship earth/whole earth logo of the planet photographed from space, set off by the slogan 'Love Your Mother', the May 1987 Mothers and Others Day action at the nuclear weapons testing facility in Nevada none the less took account of the tragic contradictions of views of the earth. Demonstrators applied for official permits to be on the land from officers of the Western Shoshone tribe, whose territory was invaded by the US government when it built the nuclear weapons test ground in the 1950s. Arrested for trespassing, the demonstrators argued that the police and weapons facility personnel, without authorization from the proper officials, were the trespassers. One affinity group at the women's action called themselves the Surrogate Others; and in solidarity with the creatures forced to tunnel in the same ground with the bomb, they enacted a cyborgian emergence from the constructed body of a large, non-heterosexual desert worm.

12. This chart was published in 1985. My previous efforts to understand biology as a cybernetic command-control discourse and organisms as 'natural-technical objects of knowledge' were Haraway (1979, 1983, 1984).

13. For progressive analyses and action on the biotechnology debates: *GeneWatch, a Bulletin of the Committee for Responsible Genetics,* 5 Doane St, 4th Floor, Boston, MA 02109; Genetic Screening Study Group (formerly the Sociobiology Study Group of Science for the People), Cambridge, MA; Wright (1982, 1986); Yoxen (1983).

14. Starting references for 'women in the integrated circuit': D'Onofrio-Flores and Pfafflin (1982), Fernandez-Kelly (1983), Fuentes and Ehrenreich (1983), Grossman (1980), Nash and Fernandez-Kelly (1983), Ong (1987), Science Policy Research Unit (1982).

15. King (1984). An abbreviated list of feminist science fiction underlying themes of this essay: Octavia Butler, *Wild Seed, Mind of My Mind, Kindred, Survivor,* Suzy McKee Charnas, *Motherliness*; Samuel R. Delany, the Nevèrÿon series; Anne McCaffery, *The Ship Who Sang, Dinosaur Planet*; Vonda McIntyre, *Superluminal, Dreamsnake*; Joanna Russ, *Adventures of Alyx, The Female Man;* James Tiptree, Jr, *Star Songs of an Old Primate, Up the Walls of the World*; John Varley, *Titan, Wizard, Demon.*

16. French feminisms contribute to cyborg heteroglossia. Burke (1981); Irigaray (1977, 1979); Marks and de Courtivron (1980); *Signs* (Autumn 1981); Wittig (1973); Duchen (1986). For English translation of some currents of francophone feminism see *Feminist Issues: A Journal of Feminist Social and Political Theory,* 1980.

17. But all these poets are very complex, not least in their treatment of lying and erotic, decentred collective and personal identities. Griffin (1978), Lorde (1984), Rich (1978).

18. Derrida (1976, especially part II); Lévi-Strauss (1961, especially 'The Writing Lesson'); Gates (1985); Kahn and Neumaier (1985); Ong (1982); Kramarae and Treichler (1985).

19. The sharp relation of women of colour to writing as theme and politics can be approached through: Program for 'The Black Woman and the Diaspora: Hidden Connections and Extended Acknowledgments', An International Literary Conference, Michigan State University, October 1985; Evans (1984); Christian (1985); Carby (1987); Fisher (1980); *Frontiers* (1980, 1983); Kingston (1977); Lerner (1973); Giddings (1985); Moraga and Anzaldúa (1981); Morgan (1984). Anglophone European and Euro-American women have also crafted special relations to their writing as a potent sign: Gilbert and Gubar (1979), Russ (1983).

20. The convention of ideologically taming militarized high technology by publicizing its applications to speech and motion problems of the disabled/differently abled takes on a special irony in monotheistic, patriarchal, and frequently anti-semitic culture when computer-generated speech allows a boy with no voice to chant the Haftorah at his bar mitzvati. See Sussman (1986). Making the always context-relative social definitions of 'ableness' particularly clear, military high-tech has a way of making human beings disabled by definition, a perverse aspect of much automated battlefield and Star Wars R&D. See Welford (1 July 1986).

21. James Clifford (1985, 1988) argues persuasively for recognition of continuous cultural reinvention, the stubborn non-disappearance of those 'marked' by Western imperializing practices.

22. DuBois (1982), Daston and Park (n.d.), Park and Daston (1981). The noun *monster* shares its root with the verb *to demonstrate*.

12

A Collective of Humans and Nonhumans

Bruno Latour

A detailed case study of sociotechnical networks ought to follow at this juncture, but many such studies have already been written, and most have failed to make their new social theory felt, as the science wars have made painfully clear to all. Despite the heroic efforts of these studies, many of their authors are all too often misunderstood by readers as cataloguing examples of the "social construction" of technology. Readers account for the evidence mustered in them according to the dualist paradigm that the studies themselves frequently undermine. The obstinate devotion to "social construction" as an explanatory device, whether by careless readers or "critical" authors, seems to derive from the difficulty of disentangling the various meanings of the catchword *sociotechnical*. What I want to do, then, is to peel away, one by one, these layers of meaning and attempt a genealogy of their associations.

Moreover, having disputed the dualist paradigm for years, I have come to realize that no one is prepared to abandon an arbitrary but useful dichotomy, such as that between society and technology, if it is not replaced by categories that have at least a semblance of providing the same discriminating power as the one jettisoned. Of course, I will never be able to do the same political job with the pair human-nonhuman as the subject-object dichotomy has accomplished, since it was in fact to free science from politics that I embarked on this strange undertaking. In the meantime we can toss around the phrase "sociotechnical assemblages" forever without moving beyond the dualist paradigm that we wish to leave behind. To move forward I must convince the reader that, pending the resolution of the political kidnapping of science, *there is an alternative to the myth of progress*. At the heart of the science wars lies the powerful accusation that those who undermine the objectivity of science and the efficiency of technology are trying to lead us backward into some primitive, barbaric dark age—that, incredibly, the insights of science studies are somehow "reactionary."

In spite of its long and complex history, the myth of progress is based on a very rudimentary mechanism (Figure 12.1). What gives the thrust to the arrow of time is that modernity at last breaks out of a confusion, made in the past, between what objects really are in themselves and what subjectivity of humans believes them to be, projecting onto them passions, biases, and prejudices. What could be called a front of modernization—like the Western Frontier— thus clearly distinguishes the confused past from the future, which will be more and more radiant, no doubt about that, because it will distinguish even more clearly the efficiency and objectivity of the laws of nature from the values, rights, ethical requirements, subjectivity, and politics of the human realm. With this map in their hands, science warriors have no difficulty situating science studies: "Since they are always insisting that objectivity and subjectivity [the science warriors' terms for nonhumans and humans] are mixed up, science students are leading us in only one possible direction, into the obscure past out of which we must extract ourselves by a movement of radical conversion, the conversion through which a barbarian premodernity becomes a civilised modernity."

In an interesting case of cartographic incommensurability, however, science studies uses an entirely different map (Figure 12.2). The arrow of time is *still there,* it still has a powerful and maybe irresistible thrust, but an entirely different mechanism makes it tick. Instead of clarifying even further the relations between objectivity and subjectivity, time enmeshes, at an ever greater level of intimacy and on an ever greater scale, humans and nonhumans with each other. The feeling of time, the definition of where it leads, of what we should do, of what war we should wage, is entirely different in the two maps, since in the one I use, Figure 12.2, the confusion of humans and nonhumans is not only our past *but our future as well.* If there is one thing of which we may be as certain as we are of death and taxation, it is that we will live tomorrow in imbroglios of science, techniques, and society *even more tightly linked* than those of yesterday—as the mad cow affair has demonstrated so clearly to European beefeaters. The difference between the two maps is total, because what the modernist science warriors see as a horror to be avoided at all costs—the mixing up of objectivity and subjectivity—is for us, on the contrary, the hallmark of a civilized life, except that what time mixes up in the future

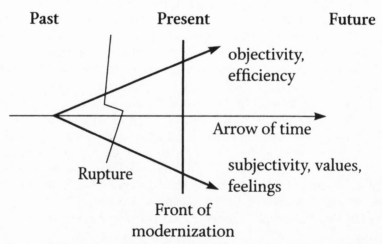

Figure 12.1   What makes the arrow of time thrust forward in the modernist narrative of progress is the certainty that the past will differ from the future because what was confused will become distinct: objectivity and subjectivity will no longer be mixed up. The result of this certainty is a front of modernization that allows one to distinguish slips backward from steps forward.

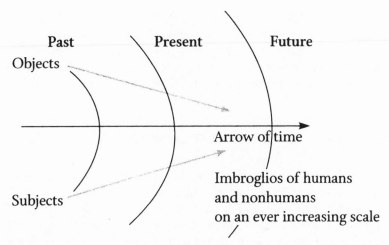

Figure 12.2   In the alternative "servant" narrative there is still an arrow of time, but it is registered very differently from Figure 12.1: the two lines of objects and subjects become more confused in the future than they were in the past, hence the feeling of instability. What is growing instead is the ever expanding scale at which humans and nonhumans are connected together.

even more than in the past *are not objects and subjects at all, but humans and nonhumans,* and that makes a world of difference. Of this difference the science warriors remain blissfully ignorant, convinced that we want to confuse objectivity and subjectivity.

I am now in the usual quandary. I have to offer an alternative picture of the world that can rely on none of the resources of common sense although, in the end, I aim at nothing but common sense. The myth of progress has centuries of institutionalization behind it, and my little pragmatogony is helped by nothing but my miserable diagrams. And yet I have to go on, since the myth of progress is so powerful that it puts any discussion to an end.

Yes, I want to tell another tale. For my present pragmatogony* [mythic origin of technology], I have isolated eleven distinct layers. Of course I do not claim for these definitions, or for their sequence, any plausibility. I simply want to show that the tyranny of the dichotomy between objects and subjects is not inevitable, since it is possible to envision another myth in which it plays no role. If I succeed in opening some space for the imagination, then we are not forever stuck with the implausible myth of progress. If I could even begin to recite this pragmatogony—I use this word to insist on its fanciful character—I would have found an alternative to the myth of progress, that most powerful of all the modernist myths, the one that held my friend under its sway when he asked me, "Do we know more than we used to?" No, we don't know more, if by this expression we mean that every day we extract ourselves further from a confusion between facts, on the one hand, and society, on the other. But yes, we do know a good deal more, if by this we mean that our collectives are tying themselves ever more deeply, more intimately, into imbroglios of humans and nonhumans. Until we have an alternative to the notion of progress, provisional as it may be, science warriors will always be able to attach to science studies the infamous stigma of being "reactionary."

I will build this alternative with the strangest of means. I want to highlight the successive crossovers through which humans and nonhumans have exchanged their properties. Each of those crossovers results in a dramatic change in the scale of the collective, in its composition, and in the degree to which humans and nonhumans are enmeshed. To tell my tale I will open

Pandora's box backward; that is, starting with the most recent types of folding. I will try to map the labyrinth until we find the earliest (mythical) folding. As we will see, contrary to the science warriors' fear, no dangerous regression is involved here, since all of the earlier steps are still with us today. Far from being a horrifying miscegenation between objects and subjects, they are simply the very hybridizations that make us humans and nonhumans.

LEVEL 11: POLITICAL ECOLOGY

Talk of a crossover between techniques and politics does not, in my pragmatogony, indicate belief in the distinction between a material realm and a social one. I am simply unpacking the eleventh layer of what is packed in the definitions of society and technique. The eleventh interpretation of the crossover—the swapping of properties—between humans and nonhumans is the simplest to define because it is the most *literal.* Lawyers, activists, ecologists, business-men, political philosophers, are now seriously talking, in the context of our ecological crisis, of granting to nonhumans some sort of rights and even legal standing. Not so many years ago, contemplating the sky meant thinking of matter, or of nature. These days we look up at a sociopolitical imbroglio, since the depletion of the ozone layer brings together a scientific con-troversy, a political dispute between North and South, and immense strategic changes in indus-try. Political representation of nonhumans seems not only plausible now but necessary, when the notion would have seemed ludicrous or indecent not long ago. We used to deride primitive peoples who imagined that a disorder in society, a pollution, could threaten the natural order. We no longer laugh so heartily, as we abstain from using aerosols for fear the sky may fall on our heads. Like the "primitives," we fear the pollution caused by our negligence—which means of course that neither "they" nor "we" have ever been primitive.

As with all crossovers, all exchanges, this one mixes elements from both sides, the politi-cal with the scientific and technical, and this mixture is not a haphazard rearrangement. Tech-nologies have taught us how to manage vast assemblies of nonhumans; our newest sociotechnical hybrid brings what we have learned to bear on the political system. The new hybrid remains a nonhuman, but not only has it lost its material and objective character, it has acquired properties of citizenship. It has, for instance, the right not to be enslaved. This first layer of meaning—the last in chronological sequence to arrive—is that of political ecology or, to use Michel Serres's term, "the natural contract." *Literally,* not symbolically as before, we have to manage the planet we inhabit, and must now define a politics of things.

LEVEL 10: TECHNOSCIENCE

If I descend to the tenth layer, I see that our current definition of technology is itself due to the crossover between a previous definition of society and a particular version of what a nonhuman can be. To illustrate: some time ago, at the Institut Pasteur, a scientist introduced himself, "Hi, I am the coordinator of yeast chromosome 11." The hybrid whose hand I shook was, all at once, a person (he called himself "I"), a corporate body ("the coordinator"), and a natural phenomenon (the genome, the DNA sequence, of yeast). The dualist paradigm will not allow us to understand this hybrid. Place its social aspect on one side and yeast DNA on the other, and you will bungle not only the speaker's words but also the opportunity to grasp how a genome becomes known to an organization and how an organization is naturalized in a DNA sequence on a hard disk.

We again encounter a crossover here, but it is of a different sort and goes in a different direction, although it could also be called sociotechnical. For the scientist I interviewed there is no question of granting any sort of rights, of citizenship, to yeast. For him yeast is a strictly material entity. Still, the industrial laboratory where he works is a place in which new modes of organization of labor elicit completely new features in nonhumans. Yeast has been put to work for millennia, of course, for instance in the old brewing industry, but now it works for a network of thirty European laboratories where its genome is mapped, humanized, and socialized, as a code, a book, a program of action, compatible with our ways of coding, counting, and reading, retaining none of its material quality, the quality of an outsider. It is absorbed into the collective. Through technoscience—defined, for my purposes here, as a fusion of science, organization, and industry—the forms of coordination learned through "networks of power" (see Level 9) are extended to inarticulate entities. Nonhumans are endowed with speech, however primitive, with intelligence, fore-sight, self-control, and discipline, in a fashion both large-scale and intimate. Socialness is shared with nonhumans in an almost promiscuous way. While in this model, the tenth meaning of sociotechnical (see Figure 12.3), automata have no rights, they are much more than material entities; they are complex organizations.

LEVEL 9: NETWORKS OF POWER

Technoscientific organizations, however, are not purely social, because they themselves recapitulate, in my story, nine prior crossovers of humans and nonhumans. Alfred Chandler and Thomas Hughes have each traced the interpenetration of technical and social factors in what Chandler terms the "global corporation" and Hughes terms "networks of power." Here again the phrase "sociotechnical imbroglio" would be apt, and one could replace the dualist paradigm with the "seamless web" of technical and social factors so beautifully traced by Hughes. But the point of my little genealogy is also to identify, inside the seamless web, properties

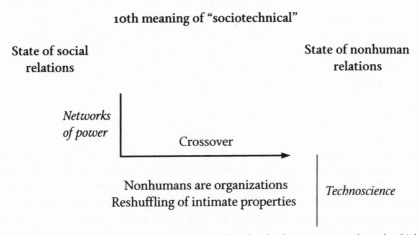

Figure 12.3 Each step in the mythical pragmatogony may be sketched as a crossover through which skills and properties learned in social relations are made relevant for establishing relations within nonhumans. By convention, the next step will be understood as going in the opposite direction.

borrowed from the social world in order to socialize nonhumans and properties borrowed from nonhumans in order to naturalize and expand the social realm. For each layer of meaning, whatever happens happens as if we are learning, in our contacts with one side, ontological properties that are then reimported to the other side, generating new, completely unexpected effects.

The extension of networks of power in the electrical industry, in telecommunications, in transportation, is impossible to imagine without a massive mobilization of material entities. Hughes's book is exemplary for students of technology because it shows how a technical invention (electric lighting) led to the establishment (by Edison) of a corporation of unprecedented scale, its scope directly related to the physical properties of electrical networks. Not that Hughes in any way talks of the infrastructure triggering changes in the superstructure; on the contrary, his networks of power are complete hybrids, though hybrids of a peculiar sort— they lend their nonhuman qualities to what were until then weak, local, and scattered corporate bodies. The management of large masses of electrons, clients, power stations, subsidiaries, meters, and dispatching rooms acquires the formal and universal character of scientific laws.

This ninth layer of meaning resembles the eleventh, since in both cases the crossover goes roughly from nonhumans to corporate bodies. (What can be done with electrons can be done with electors.) But the intimacy of human and nonhuman is less apparent in networks of power than in political ecology. Edison, Bell, and Ford mobilized entities that looked like matter, that seemed nonsocial, whereas political ecology involves the fate of nonhumans already socialized, so closely related to us that they have to be protected by delineation of their legal rights.

LEVEL 8: INDUSTRY

Philosophers and sociologists of techniques tend to imagine that there is no difficulty in defining material entities because they are objective, unproblematically composed of forces, elements, atoms. Only the social, the human realm, is difficult to interpret, we often think, because it is complexly historical and, as they say, "symbolic." But whenever we talk of matter we are really considering, as I am trying to show here, a *package* of former crossovers between social and natural elements, so that what we take to be primitive and pure terms are belated and mixed ones. Already we have seen that matter varies greatly from layer to layer— matter in the layer I have called "political ecology" differs from that in the layers called "technology" and "networks of power." Far from being primitive, immutable, and ahistorical, matter too has a complex genealogy and is handed down to us through a long and convoluted pragmatogony.

The extraordinary feat of what I will call *industry* is to extend to matter a further property that we think of as exclusively social, the capacity to relate to others of one's kind, to conspecifics, so to speak. Nonhumans have this capacity when they are made part of the assembly of actants that we call a machine: an automaton endowed with autonomy of some sort and submitted to regular laws that can be measured with instruments and accounting procedures. From tools held in the hands of human workers, the shift historically was to assemblies of machines, where tools relate to one another, creating a massive array of labor and material relations in factories that Marx described as so many circles of hell. The paradox of this stage of relations between humans and nonhumans is that it has been termed "alienation," dehumanization, as if this were the first time that poor and exploited human weakness was confronted by an all-powerful objective force. However, to relate nonhumans together in an assembly of machines, ruled by laws and accounted for by instruments, is to grant them a sort of social life.

Indeed, the modernist project consists in creating this peculiar hybrid: a fabricated non-human that has nothing of the character of society and politics yet builds the body politic all the more effectively because it seems completely estranged from humanity. This famous shapeless matter, celebrated so fervently throughout the eighteenth and nineteenth centuries, which is there for Man's—but rarely Woman's—ingenuity to mold and fashion, is only one of many ways to socialize nonhumans. They have been socialized to such an extent that they now have the capacity to create an assembly of their own, an automaton, checking and surveying, pushing and triggering other automata, as if with full autonomy. In effect, however, the proper-ties of the "megamachine" (see Level 7) have been extended to nonhumans.

It is only because we have not undertaken an anthropology of our modern world that we can overlook the strange and hybrid quality of matter as it is seized and implemented by indus-try. We take matter as mechanistic, forgetting that mechanism is one half of the modern defini-tion of society. A society of machines? Yes, the eighth meaning of the word sociotechnical, though it seems to designate an unproblematic industry, dominating matter through machin-ery, is the strangest sociotechnical imbroglio yet. Matter is not a given but a recent historical creation.

LEVEL 7: THE MEGAMACHINE

But where does industry come from? It is neither a given nor the sudden discovery by capital-ism of the objective laws of matter. We have to imagine its genealogy through earlier and more primitive meanings of the term sociotechnical. Lewis Mumford has made the intriguing suggestion that the megamachine—the organization of large numbers of humans via chains of command, deliberate planning, and accounting procedures—represents a change of scale that had to be made before wheels and gears could be developed. At some point in history human interactions come to be mediated through a large, stratified, externalized body politic that keeps track, through a range of "intellectual techniques" (writing and counting, basically), of the many nested subprograms for action. When some, though not all, of these subprograms are replaced by nonhumans, machinery and factories are born. The nonhumans, in this view, enter an organization that is already in place and take on a role rehearsed for centuries by obedient human servants enrolled in the imperial megamachine.

In this seventh level, the mass of nonhumans assembled in cities by an internalized ecol-ogy (I will define this expression shortly) has been brought to bear on empire building. Mum-ford's hypothesis is debatable, to say the least, when our context of discussion is the history of technology; but the hypothesis makes excellent sense in the context of my pragmatogony. Before it is possible to delegate action to nonhumans, and possible to relate nonhumans to one another in an automaton, it must first be possible to nest a range of subprograms for action into one another without losing track of them. Management, Mumford would say, precedes the expansion of material techniques. More in keeping with the logic of my story, one might say that *whenever we learn something about the management of humans, we shift that knowledge to nonhumans and endow them with more and more organizational properties.* The even-numbered episodes I have recounted so far follow this pattern: industry shifts to nonhumans the management of people learned in the imperial machine, much as technoscience shifts to nonhumans the large-scale management learned through networks of power. In the odd-numbered levels, the opposite process is at work: *what has been learned from nonhumans is reimported so as to reconfigure people.*

LEVEL 6: INTERNALIZED ECOLOGY

In the context of layer seven, the megamachine seems a pure and even final form, composed entirely of social relations; but, as we reach layer six and examine what underlies the megamachine, we find the most extraordinary extension of social relations to nonhumans: agriculture and the domestication of animals. The intense socialization, reeducation, and reconfiguration of plants and animals—so intense that they change shape, function, and often genetic makeup—is what I mean by the term "internalized ecology." As with our other even-numbered levels, domestication cannot be described as a sudden access to an objective material realm that exists *beyond* the narrow limits of the social. In order to enroll animals, plants, proteins in the emerging collective, one must first endow them with the social characteristics necessary for their integration. This shift of characteristics results in a manmade landscape for society (villages and cities) that completely alters what was until then meant by social and material life. In describing the sixth level we may speak of urban life, empires, and organizations, but not of society and techniques—or of symbolic representation and infrastructure. So profound are the changes entailed at this level that we pass beyond the gates of history and enter more profoundly those of prehistory, of mythology.

LEVEL 5: SOCIETY

What is a society, the starting point of all social explanations, the *a priori* of all social science? If my pragmatogony is even vaguely suggestive, society cannot be part of our final vocabulary, since the term had itself to be made—"socially constructed" as the misleading expression goes. But according to the Durkheimian interpretation, a society is primitive indeed: it precedes individual action, lasts very much longer than any interaction does, dominates our lives; it is that in which we are born, live, and die. It is externalized, reified, more real than ourselves, and hence the origin of all religion and sacred ritual, which for Durkheim are nothing but the return, through figuration and myth, of the transcendent to individual interactions.

And yet society itself is constructed only through such quotidian interactions. However advanced, differentiated, and disciplined society becomes, we still repair the social fabric out of our own, immanent knowledge and methods. Durkheim may be right, but so is Harold Garfinkel. Perhaps the solution, in keeping with the generative principle of my genealogy, is to look for nonhumans. (This explicit principle is: look for nonhumans when the emergence of a social feature is inexplicable; look to the state of social relations when a new and inexplicable type of object enters the collective.) What Durkheim mistook for the effect of a *sui generis* social order is simply the effect of having brought so many techniques to bear on our social relations. It was from techniques, that is, the ability to nest several subprograms, that we learned what it means to subsist and expand, to accept a role and discharge a function. By reimporting this competence into the definition of society, we taught ourselves to reify it, to make society stand independent of fast-moving interactions. We even learned how to delegate to society the task of relegating us to roles and functions. Society exists, in other words, *but is not socially constructed.* Nonhumans proliferate below the bottom line of social theory.

LEVEL 4: TECHNIQUES

By this stage in our speculative genealogy we can no longer speak of humans, of anatomically modern humans, but only of social prehumans. At last we are in a position to define technique, in the sense of a *modus operandi,* with some precision. Techniques, we learn from archaeologists, are articulated subprograms for actions that subsist (in time) and extend (in space). Techniques imply not society (that late-developing hybrid) but a semisocial organization that brings together nonhumans from very different seasons, places, and materials. A bow and arrow, a javelin, a hammer, a net, an article of clothing are composed of parts and pieces that require recombination in sequences of time and space that bear no relation to their original settings. Techniques are what happen to tools and nonhuman actants when they are processed through an organization that extracts, recombines, and socializes them. Even the simplest techniques are sociotechnical; even at this primitive level of meaning, forms of organization are inseparable from technical gestures.

LEVEL 3: SOCIAL COMPLICATION

But what form of organization can explain these recombinations? Recall that at this stage there is no society, no overarching framework, no dispatcher of roles and functions; there are merely interactions among prehumans. Shirley Strum and I call this third layer of meaning *social complication.* Here complex interactions are marked and followed by nonhumans enrolled for a specific purpose. What purpose? Nonhumans stabilize social negotiations. Nonhumans are at once pliable and durable; they can be shaped very quickly but, once shaped, last far longer than the interactions that fabricated them. Social interactions are extremely labile and transitory. More precisely, either they are negotiable but transient or, if they are encoded (for instance) in the genetic makeup, they are extremely durable but difficult to renegotiate. The involvement of nonhumans resolves the contradiction between durability and negotiability. It becomes possible to follow (or "blackbox") interactions, to recombine highly complicated tasks, to nest subprograms into one another. What was impossible for complex social animals to accomplish becomes possible for prehumans—who use tools not to acquire food but to fix, underline, materialize, and keep track of the social realm. Though composed only of interactions, the social realm becomes visible and attains through the enlistment of nonhumans—tools—some measure of durability.

LEVEL 2: THE BASIC TOOL KIT

The tools themselves, wherever they came from, offer the only testimony on behalf of hundreds of thousands of years. Many archaeologists proceed on the assumption that the basic tool kit (as I call it) and techniques are directly related by an evolution of tools into composite tools. But there is no *direct* route from flints to nuclear power plants. Further, there is no direct route, as many social theorists presume there to be, from social complication to society, megamachines, networks. Finally, there is not a set of parallel histories, the history of infrastructure and the history of superstructure, but only one sociotechnical history.

What, then, is a tool? The extension of social skills to nonhumans. Machiavellian monkeys and apes possess little in the way of techniques, but can devise social tools (as Hans Kummer has called them) through complex strategies of manipulating and modifying one another. If you grant the prehumans of my own mythology the same kind of social complexity, you grant as well that they may generate tools by *shifting* that competence to nonhumans, by treating a stone, say, as a social partner, modifying it, then using it to act on a second stone. Prehuman tools, in contrast to the ad hoc implements of other primates, also represent the extension of a skill rehearsed in the realm of social interactions.

LEVEL 1: SOCIAL COMPLEXITY

We have finally reached the level of the Machiavellian primates, the last circumvolution in Daedalus's maze. Here they engage in social interactions to repair a constantly decaying social order. They manipulate one another to survive in groups, with each group of conspecifics in a state of constant mutual interference. We call this state, this level, social complexity. I will leave it to the ample literature of primatology to show that this stage is no more free of contact with tools and techniques than any of the later stages.

AN IMPOSSIBLE BUT NECESSARY RECAPITULATION

I know I should not do it. I more than anyone ought to see that it is madness, not only to peel away the different meanings of sociotechnical, but also to recapitulate all of them in a single diagram, as if we could read off the history of the world at a glance. And yet it is always surprising to see how few alternatives we have to the grandiose scenography of progress. We may tell a lugubrious countertale of decay and decadence as if, at each step in the extension of science and technology, we were stepping down, away from our humanity. This is what Heidegger did, and his account has the somber and powerful appeal of all tales of decadence. We may also abstain from telling any master narrative, under the pretext that things are always local, historical, contingent, complex, multiperspectival, and that it is a crime to hold them all in one pathetically poor scheme. But this ban on master narratives is never very effective, because, in the back of our minds, no matter how firmly we are convinced of the radical multiplicity of existence, something surreptitiously gathers everything into one little bundle which may be even cruder than my diagrams—including the postmodern scenography of multiplicity and perspective. This is why, against the ban on master narratives, I cling to the right to tell a "servant" narrative. My aim is not to be reasonable, respectable, or sensible. It is to fight modernism by finding the hideout in which science has been held since being kidnapped for political purposes I do not share.

If we gather in one table the different layers I have briefly outlined—one of my other excuses is how brief the survey, covering so many millions of years, has been!—we may give some sense to a story in which the further we go the more articulated are the collectives we live in (see Figure 12.4). To be sure, we are not ascending toward a future made of more sub-

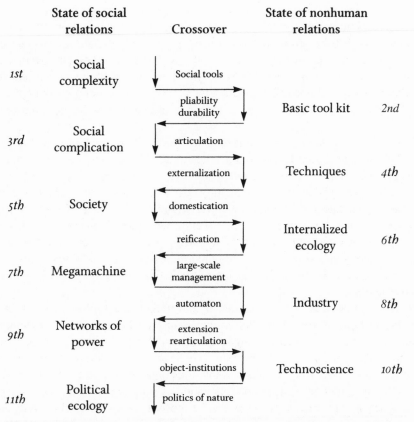

Figure 12.4   If the successive crossovers are summed up, a pattern emerges: relations among humans are made out of a previous set of relations that related nonhumans to one another; these new skills and properties are then reused to pattern new types of relations among nonhumans, and so on; at each (mythical) stage the scale and the entanglement increase. The key feature of this myth, is that, at the final stage, the definitions we can make of humans and nonhumans should recapitulate all the earlier layers of history. The further we go, the less pure are the definitions of humans and nonhumans.

jectivity and more objectivity. But neither are we descending, chased ever further from the Eden of humanity and *poesis*.

Even if the speculative theory I have outlined is entirely false, it shows, at the very least, the possibility of imagining a genealogical alternative to the dualist paradigm. We are not forever trapped in a boring alternation between objects or matter and subjects or symbols. We are not limited to "not only . . . but also" explanations. My little origin myth makes apparent the impossibility of having an artifact that does not incorporate social relations, as well as the impossibility of defining social structures without accounting for the large role played in them by nonhumans.

Second, and more important, the genealogy demonstrates that it is false to claim, as so many do, that once we abandon the dichotomy between society and techniques we are faced with a seamless web of factors in which all is included in all. The properties of humans and nonhumans cannot be swapped haphazardly. Not only is there an order in the exchange of properties, but in each of the eleven layers the meaning of the word "sociotechnical" is clarified if we consider the exchange: that which has been learned from nonhumans and reimported into the social realm, that which has been rehearsed in the social realm and exported back to the nonhumans. Nonhumans too have a history. They are not material objects or constraints. Sociotechnical[1] is different from sociotechnical[6] or[7] or[8] or[11]. By adding superscripts we are able to qualify the meanings of a term that until now has been hopelessly confused. In place of the great vertical dichotomy between society and techniques, there is conceivable (in fact, now, available) a range of horizontal distinctions between very different meanings of the sociotechnical hybrids. It is possible to have our cake and eat it too—to be monists and make distinctions.

All this is not to claim that the old dualism, the previous paradigm, had nothing to say for itself. We do indeed alternate between states of social and states of nonhuman relations, but this is not the same as alternating between humanity and objectivity. The mistake of the dualist paradigm was its definition of humanity. Even the shape of humans, our very body, is composed to a great extent of sociotechnical negotiations and artifacts. To conceive of humanity and technology as polar opposites is, in effect, to wish away humanity: we are sociotechnical animals, and each human interaction is sociotechnical. We are never limited to social ties. We are never faced only with objects. This final diagram relocates humanity right where we belong—in the crossover, the central column, the articulation, the possibility of mediating between mediators.

But my main point is that, in each of the eleven episodes I have traced, an increasingly large number of humans are mixed with an increasingly large number of nonhumans, to the point that, today, the whole planet is engaged in the making of politics, law, and soon, I suspect, morality. The illusion of modernity was to believe that the more we grew, the more separate objectivity and subjectivity would become, thus creating a future radically different from our past. After the paradigm shift in our conception of science and technology, we now know that this will never be the case, indeed that this has never *been* the case. Objectivity and subjectivity are not opposed, they grow together, and they do so irreversibly. At the very least, I hope I have convinced the reader that, if we are to meet our challenge, we will not meet it by considering artifacts as things. They deserve better. They deserve to be housed in our intellectual culture as full-fledged social actors. Do they mediate our actions? No, they are us. The goal of our philosophy, social theory, and morality is to invent political institutions that can absorb this much history, this vast spiraling movement, this labyrinth, this fate.

The nasty problem we now have to deal with is that, unfortunately, we do *not* have a definition of politics that can answer the specifications of this nonmodern history. On the contrary, every single definition we have of politics comes from the modernist settlement and from the polemical definition of science that we have found so wanting. Every one of the weapons used in the science wars, including *the very distinction* between science and politics, has been handed down to the combatants by the side we want to oppose. No wonder we always lose and are accused of politicizing science! It is not only the practice of science and technology that epistemology has rendered opaque, but also that of politics. As we shall soon see, the fear of mob rule, the proverbial scenography of might versus right, is what holds the old settlement together, is what has rendered us modern, is what has kidnapped the practice of science, all for the most implausible political project: that of doing away with politics.

13

Ecological Restoration and the Culture of Nature: A Pragmatic Perspective

Andrew Light

Most environmental philosophers have failed to understand the theoretical and practical importance of ecological restoration. I believe this failure is primarily due to the mistaken impression that ecological restoration is only an attempt to restore nature itself, rather than an effort to restore an important part of the human relationship with nonhuman nature. In investigating this claim, I will first discuss the possibility of transforming environmental philosophy into a more pragmatic discipline, better suited to contributing to the formation of sound environmental policies, including ecological restoration. In particular, I will advocate an alternative philosophical approach to the kind of work on the value of ecological restoration raised by Eric Katz and other philosophers who claim that restored nature can never reproduce the actual value of nature. Here, I will make this contrast more explicit and go on to further argue that Katz's views in particular are not sufficiently sensitive to the values at work in the variety of projects falling within the category of ecological restoration. A more practically oriented philosophical contribution to future discussions of our policies concerning ecological restoration is needed than has been provided by environmental philosophers so far. A richer description of the ethical implications of restoration will identify a large part of its value in the revitalization of the human culture of nature. Before reaching this conclusion, however, I will briefly consider an alternative framework for environmental philosophy as a whole.

ENVIRONMENTAL PHILOSOPHY: WHAT AND FOR WHOM?

Two underlying questions that I believe still confound most environmental philosophers are, What is our discipline actually for? and, consequently, Who is our audience? So far, most work in environmental ethics has been concerned with describing the nonanthropocentric value of nature—that is, the value of nature independent of human concerns and reasons for valuing nature—and determining the duties, obligations, or rights that follow from that description. But one can easily wonder whether such work is directed only toward other environmental philosophers as a contribution to the literature on value theory or whether it has a broader aim. Certainly, given the history of the field—formally beginning in the early 1970s with the work of thinkers as diverse as Arne Naess, Val Plumwood, Holmes Rolston, Peter Singer, and Richard Sylvan, all concerned with how philosophers could make some sort of contribution to the resolution of environmental problems—one would think that the aspirations of environmental philosophy would be greater than simply continuing an intramural discussion about the value of nature.

But if environmental philosophy is more than a discussion among philosphers about natural value, to what broader purposes and audiences should it reach? Taking a cue from the content and expected readership of this book, I pose at least four responses. Environmental philosophy might serve as (1) a guide for environmental activists searching for ethical justifications for their activities in defense of other animals and ecosystems; (2) an applied ethic for resource managers; (3) a general tool for policy makers, helping them to shape more responsible environmental policies; and (4) a beacon for the public at large, attempting to expand their notions of moral obligation beyond the traditional confines of anthropocentric (human-centered) moral concerns.

Environmental philosophy should, of course, aim to serve all of these purposes and groups, although I think that most importantly we should focus our energies on guiding policy makers and the public. My rationale is this: if the original reason for philosophers establishing this field was to make a philosophical contribution to the resolution of environmental problems (consistent with the response by other professionals in the early 1970s around environmental concerns), then the continuation, indeed the urgency, of those problems demands that philosophers do all that they can to actually help change present policies and attitudes involving environmental problems. If we talk only to each other about value theory, we have failed as environmental professionals, but if we can help convince policy makers to form better policies and make the case to the public at large to support these policies for ethical reasons, then we can join other environmental professionals in making more productive contributions to the resolution of environmental problems.

But as it now stands, the current focus in environmental philosophy on describing the nonanthropocentric value of nature often ends up separating environmental philosophy from other forms of environmental inquiry. As a prime example of this disconnection from practical considerations, many environmental philosophers do not think of restoration ecology in a positive light. My friend and colleague Eric Katz comes near the top of this list of philosophers; his chapter, "Another Look at Restoration: Technology and Artificial Nature" in *Restoring Ecology*, is the latest in a series of articles in which he argues that ecological restoration does not result in a restoration of nature and, in fact, may even create a disvalue in nature. Robert Elliot is another influential thinker in this camp, although his views have moderated significantly in recent years. Katz, Elliot, and others maintain that if the goal of environmental philosophy is to describe the nonhuman-centered value of nature and to distinguish nature from human appreciation of it, then presumably nature cannot be the sort of thing that is associated with human creation or manipulation. Thus, if restorations are human creations, so the arguments of the philosophical critics like Katz go, they can never count as the sort of thing that contains natural value.

In this view, restorations are not natural—they are artifacts. To claim that environmental philosophers should be concerned with ecological restoration is therefore to commit a kind of category mistake: it is to ask that they talk about something that is not part of nature. But to label ecological restorations a philosophical category mistake is the best case scenario of their assessment. At worst, restorations in their view represent the tyranny of humans over nature and shouldn't be practiced at all. Katz has put it most emphatically in arguing that "the practice of ecological restoration *can only* represent a misguided faith in the hegemony and infallibility of the human power to control the natural world" (Katz 1996, 222, my emphasis).

I have long disagreed with claims like this one. My early response to such positions was to simply set them aside in my search for broader ethical and political questions useful for a more public discussion of policies concerning ecological restoration (e.g., Light and Higgs 1996). But I now think it is dangerous to ignore the arguments of Katz and Elliot, for at least two reasons. First, the arguments of Katz and Elliot represent the most sustained attempt yet to make a philosophical contribution to the overall literature on restoration and thus ought to be answered by philosophers also interested in restoration. Second, the larger restoration community is increasingly coming to believe that the sorts of questions being addressed by Katz and Elliot are the only kind of contribution that philosophy as a discipline can make to discussions of restoration. And since Katz has explicitly rejected the idea that ecological restoration is an acceptable environmental practice, the restoration community's assumption that environmental ethicists tend to be hostile to the idea of ecological restoration is a fair one. Given this disjunction, there would be no ground left for a philosophical contribution to public policy questions concerning ecological restoration since none of these issues would count as moral or ethical questions.[1]

I believe that philosophers can make constructive contributions to ecological restoration and to environmental issues in general by helping to articulate the normative foundations for environmental policies in ways that are translatable to the public. But making such contributions requires doing environmental philosophy in some different ways. Specifically, it requires a more public philosophy, one focused on making the kinds of arguments that resonate with the moral intuitions that most people carry around with them every day. Such intuitions usually resonate more with human-centered notions of value than with abstract nonanthropocentric conceptions of natural value.

I call the view that makes it plausible for me to make this claim about the importance of appealing to human motivations in valuing nature *environmental pragmatism*. By this I do not mean an application of the traditional writings of the American pragmatists—Dewey, James, and Pierce, for example—to environmental problems. Instead, I simply mean the recognition that a responsible and complete environmental philosophy includes a public component with a clear policy emphasis (see, for example, Light 1996a, 1996b, 1996c). It is certainly appropriate for philosophers to continue their search for a true and foundational nonanthropocentric description of the value of nature. But environmental philosophers would be remiss if they did not also try to make other, perhaps more appealing ethical arguments that may have an audience in an anthropocentric public. Environmental pragmatism in my sense is agnostic concerning the existence of nonanthropocentric natural value. It is simply a methodology permitting environmental philosophers to endorse a pluralism allowing for one kind of philosophical task inside the philosophy community—searching for the "real" value of nature; and another task outside of that community—articulating a value to nature that resonates with the public and therefore has more impact on discussions of projects such as ecological restorations that may be performed by the public.

This approach modifies the philosophical contribution to questions about restoration ecology in a positive way. As mentioned, many philosophers have criticized ecological restoration because it is a human intervention into natural processes. In contrast, I have argued that such projects as the prairie restorations at the University of Wisconsin–Madison Arboretum would be fully supported by a pragmatic environmental philosophy (Light 1996b). Restoration makes sense because on the whole it results in many advantages over mere preservation of ecosystems that have been substantially damaged by humans. More significantly, this pragmatic approach exposes other salient ethical issues involving the practice of ecological restoration beyond the discussion of natural value, such as whether there are moral grounds that justify encouraging public participation in restoration (see Light and Higgs 1996). It is therefore the duty of the pragmatic environmental philosopher to get involved in debates with practitioners about what the value of restoration is in human terms, rather than restricting the discussion to a private debate among philosophers on whether restored nature is really nature. In the rest of this chapter, I will both offer a specific critique of Katz's claims about the value of restoration that does not rely on a pragmatist foundation for environmental philosophy as well as go on to discuss some pragmatic issues that contribute to a fuller philosophical analysis of the practice and ethics of ecological restoration.

ECOLOGICAL RESTORATION: A PRELIMINARY DISTINCTION

Following the project described above, in previous work I have outlined some preliminary distinctions that paint a broader picture of the philosophical terrain up for grabs in restoration than that presented by Katz and Elliot. Specifically, in response to Elliot's early critique of restoration (1995), I have tried to distinguish between two categories of ecological restoration that have differing moral implications.

Elliot begins his seminal article on restoration, "Faking Nature," by identifying a particularly pernicious kind of restoration—restoration that is used to rationalize the destruction of nature. On this claim, any harm done to nature by humans is ultimately repairable through restoration, so the harm should be discounted. Elliot calls this view the "restoration thesis" and states that "the destruction of what has value [in nature] is compensated for by the later creation (re-creation) of something of equal value" (Elliot 1995, 76). Elliot rejects the restoration thesis through an analogy based on the relationship between original and replicated works of art and nature. Just as we would not value a replication of a work of art as much as we would value the original, we would not value a replicated bit of nature as much as we would the original thing, such as some bit of wilderness. Elliot is persuasive that the two sorts of value choices are similar.

In responding to Elliot's (1995) criticisms of the value of restoration, I suggested a distinction implicit in his analysis of restoration to help us think through the value of ecological restoration (Light 1997). The distinction is based on an acknowledgment Elliot makes in his 1995 article (and expands upon in his 1997 book):

> Artificially transforming an utterly barren, ecologically bankrupt landscape into something richer and more subtle may be a good thing. That is a view quite compatible with the belief that replacing a rich natural environment with a rich artificial one is a bad thing. (Elliot 1995, 82)[2]

Following Elliot's lead that some kinds of restoration may be beneficial, I distinguished between two sorts of restorations: (1) malicious restorations, such as the kind described in the restoration thesis; and (2) benevolent restorations, or those undertaken to remedy a past harm done to nature although not offered as a justification for harming nature. Benevolent restorations, unlike malicious restorations, cannot serve as justifications for the conditions that would warrant their engagement.

If this distinction holds, then we can claim that Elliot's original target was not all of restoration, but only a particular kind of restoration, namely, malicious restorations. Although there is mixed evidence to support the claim that Elliot was originally going only after malicious restorations in his first work on the topic, the distinction is nonetheless intuitively plausible. It is certainly not the case, for example, that the sorts of restorations undertaken at the Wisconsin Arboretum or as part of the Chicago Wilderness effort are offered as excuses or rationales for the destruction of nature. In contrast, the restorations involved in mountaintop mining projects in rural West Virginia can certainly be seen as examples of malicious restorations. Mountaintop mining—through which tops of mountains are destroyed and dumped into adjacent valleys—is in part rationalized through a requirement that the damaged streambeds in the adjacent valleys be restored. The presumed ability to restore these streambeds is used as a justification for allowing mountaintop mining, counting this practice as a clear instantiation of Elliot's restoration thesis. The upshot of this malicious-benevolent distinction, however, is that one may be able to grant much of Elliot's claim that restored nature is not original nature while still not denying that there is some kind of positive value to the act of ecological restoration in many cases. Even if benevolent restorations are not restorations of original nature, and hence more akin to art forgeries rather than original works of art, they can still have some kind of positive content.

This positive content for many restorations can be developed more by pushing the art analogy a bit further. If ecological restoration is a material practice like making a piece of art (fake or not), why isn't it more like art restoration rather than art forgery? After all, we know that some parallels can be drawn between restoration projects and mitigation projects. A mitigation often involves the wholesale creation of a new ecosystem designed to look like a bit of nature that may have absolutely no historical continuity with the natural history of the land on which it is placed. For example, in order to meet an environmental standard that demands no net loss of wetlands, some environmental managers will sanction the creation of a wetland to replace a destroyed one on a piece of land where there had been no wetland. Conversely, a restoration must be tied to some claim about the historical continuity of the land on which the restoration is taking place. In some cases, this might simply entail linking original pieces of nature together to restore the integrity of the original ecosystem without creating a new landscape altogether (as in the case of the Wildlands Project to link the great Western parks in the United States and Canada with protected corridors). In that sense, a restoration could be more like repairing a damaged work of art than creating a fake one.[3]

The possibility of having benevolent restorations does much to clear the way for a positive philosophical contribution to questions of restoration. Katz, however, unlike Elliot, denies the positive value of any kind of restoration. For him, all restorations "can only" be malicious because they all represent evidence of human domination and arrogance toward nature. But surprisingly, even though Katz draws on Elliot's work in formulating his own position, he seems to ignore the fact that Elliot's original description of the restoration thesis was primarily directed against particular kinds of restorations. In his earliest and most famous article on restoration, "The Big Lie: Human Restoration of Nature," Katz acknowledged that although Elliot claimed that the restoration thesis mostly was advocated as a way of undermining conservation efforts by big business, he (Katz) was surprised to see environmental thinkers (such as forest biologist Chris Maser) advocating "a position similar to Elliot's 'restoration thesis.'" This position as Katz interprets it is that "restoration of damaged nature is seen not only as a practical option for environmental policy but also as a moral obligation for right-thinking environmentalists" (Katz 1997, 96). But Maser's position is not the restoration thesis as Elliot defines it. Katz never does show that Maser, or any other restoration advocate that he analyzes, actually argues for restoration as a rationale for destruction of nature. As such, Katz never demonstrates that those in the restoration community that he criticizes endorse restorations for malicious reasons. If that is the case, then what is wrong with restoration in Katz's view?

KATZ AGAINST RESTORATION

Just as Elliot's original target of the "restoration thesis" has faded from philosophical memory, Katz's original target has also been somewhat lost in the years since he began writing on this topic in 1992. At first, Katz seemed most concerned with the arguments of fellow environmental ethicists like Paul Taylor and Peter Wenz, who advocated variously "restitutive justice" and a "principle of restitution" as part of our fulfillment of possible human obligations to nature. If we harmed nature, according to Taylor and Wenz, we would have to compensate it. Restoration would be part of a reasonable package of restitution. According to Katz, on these views humans have an "*obligation* and *ability* to repair or reconstruct damaged ecosystems" (Katz 1997, 95, my emphasis). But I think it is crucial here to pay attention to the argument Katz is actually taking on and the objection he proceeds to make.

As Katz describes it, there are actually two separable questions to put to Taylor, Wenz, and other advocates of restoration: (1) do we have an obligation to try to restore damaged nature? and (2) do we have the ability to restore damaged nature? Katz argues quite forcefully that we do not have the ability to restore nature because what we actually create in ecological restorations are humanly produced artifacts and not nature, nonanthropocentrically conceived. Based on this claim, he assumes that the first question—whether we have an obligation to try to restore nature—is moot. Katz's logic is simple: we do not have an obligation to do what we cannot in principle do.

But even if we were to grant Katz the argument that it is impossible to restore nature, we may still have moral obligations to *try* to restore nature. How can this be true? There are a number of reasons, but before fully explicating this position we need to first better understand Katz's arguments.

Katz's chapter in *Restoring Ecology* reviews and expands upon several arguments he has made against restoration over the years. In examining his papers on this topic,[4] I have identified five separable, but often overlapping, arguments he has made against both the idea that we can restore nature and the practice of trying to restore it. I call these arguments KR1–5. They are listed below in order of how they arise in his work, accompanied with an example of supporting evidence from Katz's various papers on restoration.

KR1. The Duplicitous Argument
"I am outraged by the idea that a technologically created 'nature' will be passed off as reality" (Katz 1997, 97).[5]
KR2. The Arrogance (or Hubris) Argument
"The human presumption that we are capable of this technological fix demonstrates (once again) the arrogance with which humanity surveys the natural world" (Katz 1997, 97).

KR3. The Artifact Argument

"The re-created natural environment that is the end result of a restoration project is nothing more than an artifact created for human use" (Katz 1997, 97). [6]

KR4. The Domination Argument

"The attempt to redesign, recreate and restore natural areas and objects is a radical intervention in natural processes. Although there is an obvious spectrum of possible restoration[s] . . . all of these projects involve the manipulation and domination of natural areas. All of these projects involve the creation of artifactual realities, the imposition of anthropocentric interests on the processes and objects of value. Nature is not permitted to be free, to pursue its own independent course of development" (Katz 1997, 105). [7]

KR5. The Replacement Argument

"If a restored environment is an adequate replacement for the previously existing natural environment [which, for Katz, it can never be], then humans can use, degrade, destroy, and replace natural entities and habitats with no moral consequence whatsoever. The value in the original natural entity does not require preservation" (Katz 1997, 113).[8]

I disagree with all of these arguments and have what I hope are thorough answers to all of them elsewhere. Here, I will focus on KR4, the domination argument, which is perhaps the argument that comes up the most throughout all of Katz's restoration papers. It is arguably the case that one can answer all of Katz's arguments by conceding one important premise of all of his claims as long as KR4 can be independently answered. KR4 is also interesting to me because his original articulation of it involved a very slim bit of admission that there is some sort of difference between various kinds of restoration projects. Even though these differences are not ultimately important for Katz, they are still nonetheless acknowledged, and they give me a space in which I can critique his position.

As I said above, I believe that KR1–3 and KR5 can be ignored in rejecting Katz's position as long as we are prepared to concede for now one important premise to all of his arguments. This is Katz's ontological assumption (a claim concerning the nature or essence of a thing) that humans and nature can be meaningfully separated so as to definitively argue that restored nature is an artifact, a part of human culture, rather than a part of nature. As Katz has admitted in an as yet unpublished public forum on his work, he is a nature–culture dualist. This means that for Katz, nature and culture are separate things entirely.[9] If one rejects this overall ontological view, then one may reject most of Katz's objections to restoration. But it is incredibly difficult to disprove another philosopher's ontology, let alone get him or her to concede this point.[10] Thus, even though I disagree with it, I will accept Katz's underlying assumption that restored nature does not reproduce nature.

But even if I grant this point that restored nature is not really nature, KR4 is still false because it is arguably the case that restoration does not "dominate" nature in any coherent sense but instead often helps nature to be "free" of just the sort of domination that Katz is worried about. The reasoning here is straightforward enough. If I can show that restorations are valuable for nature, even if I concede that they do not re-create nature, then the various motivations for restoration will distinguish whether a restoration is duplicitous (KR1) or arrogant (KR2). A benevolent restoration, for example, would not risk KR1 or KR2 because in principle it is not trying to fool anyone nor is it necessarily arrogant. Further, and more simply, conceding Katz's ontological claim about the distinction between nature and culture eliminates the significance of KR3—since we no longer care that what is created may or may not be an artifact—as well as KR5, since we have given up hope that a restoration could ever actually serve as a replacement for "real" nature.

Now, back to the domination argument. KR4 is a claim that could hold even for a view that conceded Katz's nature–culture distinction. The reason, following Katz, would be that even a failed attempt to duplicate natural value—or create something akin to nature while conceding that in principle "real" nature can never be restored by humans—could still count as an instance of "domination," as Katz has described it. An *attempt* at restoration, according to Katz's logic, would still prohibit nature from ever being able to pursue its own development. The reason is that for Katz, restoration is always a substitute for whatever would have occurred at a particular site without human interference. The idea is that even if humans can produce a valuable landscape of some sort on a denuded acreage, this act of production is still an instance of domination over the alternative of a natural evolution of this same acreage, even if a signifi-cant natural change would take ten times as long as the human-induced change and would be arguably less valuable for the species making use of it. Still, one can muster several arguments against KR4 (I will provide four) and still play largely within Katz's biggest and most conten-tious assumption about the ontological status of restored nature. After going through these arguments, we will see that these claims can lead to a new philosophical context for the evalua-tion of restoration, which I believe in the end also undermines the other KR arguments.

(1) We can imagine cases in which nature cannot pursue its own interests (however one wishes to understand this sense of nature having interests) because of something we have done to it. For example, many instances of restoration are limited to bioactivation of soil that has become contaminated by one form or another of hazardous industrial waste. If restoration nec-essarily prohibits nature from being "free," as KR4 maintains, then how do we reconcile the relative freedom that bioactivation makes possible with this claim? Restoration need not deter-mine exactly what grows in a certain place, but may in fact simply be the act of allowing nature to again pursue its own interests rather than shackling it to perpetual human-induced trauma. In many cases of restoration, this point can be driven home further when we see how anthropogenically damaged land (or soil) can be uniquely put at risk of invasion by anthropo-genically introduced exotic plants. South African ice plant, an exotic in southern California that destroys the soil it is introduced into, is highly opportunistic and can easily spread onto degraded land, thus ensuring that native plants will not be able to reestablish themselves. I highlight here this contentious native-exotic distinction because I suspect that given Katz's strong nature–culture distinction he would necessarily have to prefer a landscape of native plants over a landscape of exotics where the existence of the exotics is a result of an act of human (cultural) interference in nature. Allowing nature to pursue its own interests, given prior anthropogenic interference, thus involves at least as strong a claim to protect it from further anthropogenic risk through restoration practices as the case Katz makes for leaving it alone.

(2) Even if we do agree with Katz that restorations only produce artifacts, can't it still be the case that the harm we cause nature still requires us to engage in what Katz would term "attempted restorations"? It simply does not follow from the premise that something is more natural when it is relatively free of human interference that we must therefore always avoid interfering with nature (this is actually a point that Katz finally recognizes in a later paper, "Imperialism and Environmentalism"). It is a classic premise of holism in environmental eth-ics (the theory that obligations to the nonhuman natural world are to whole ecosystems and not to individual entities, a view that Katz endorses) that some interference is warranted when we are the cause of an imbalance in nature. For example, hunting white-tailed deer is thought to be permissible under holism since humans have caused that species' population explosion.

If such interventions are permissible to help "rectify the balance of nature," then why are there not comparable cases with the use of restoration as an aid to the "original," "real" nature? We can even imagine that such cases would be less controversial than holist defenses of hunting.

There are good cases in which restoration, even if it results in the production of an artifact, does not lead to the domination described by Katz. Imagine the case where the restoration project is one that will restore a corridor between two wilderness preserves. If there is positive natural value in the two preserves that is threatened because wildlife is not allowed to move freely between them, then restoration projects that would restore a corridor (by removing roads, for example) would actually not only be morally permissible but also possibly ethically required depending on one's views of the value of the nature in the preserves. This is not restoration as a "second best" to preservation or a distraction away from preservation; it is restoration as an integral and critical part of the maintenance of natural value. So, even if we agree with Katz that humans cannot really restore nature, it does not follow that they ought not to engage in restoration projects that actually repair the damage caused by past domination rather than furthering that domination.

Given objections like the two discussed so far, it is important to try to get a better handle on exactly what sort of damage is caused by domination in the sense described by Katz. It turns out that the worst damage to nature for Katz is domination that prevents the "self-realization" of nature:

> The fundamental error is thus domination, the denial of freedom and autonomy. Anthropocentrism, the major concern of most environmental philosophers, is only one species of the more basic attack on the preeminent value of self-realization. From within the perspective of anthropocentrism, humanity believes it is justified in dominating and molding the nonhuman world to its own human purposes. (Katz 1997, 105)

Thus, the problem with restoration is that it restricts natural self-realization in order to force nature onto a path that we would find more appealing.

(3) With this clarification, we can then further object to Katz that his sense of restoration confuses restoration with mitigation. The force of the charge of domination is that we mold nature to fit our "own human purposes." But most restorationists would counter that it is non-anthropocentric nature that sets the goals for restoration, not humans. Although there is indeed some subjectivity in determining what should be restored at a particular site (which period do we restore to?) and uncertainty in how we should do it (limitations in scientific and technical expertise), we cannot restore a landscape just any way we wish and still have a good restoration in scientific terms. If Katz objects that when we restore a denuded bit of land we are at least making something that fits our need of having more attractive "natural" surroundings—an argument that Katz often makes—we can reply that because of the constraints on restoration, as opposed to mitigation, the fact that we find a restored landscape appealing is only contingently true. It is often the case that what we must restore to is not the preferred landscape of most people. The Chicago Wilderness project is a good example of this: many local residents see restoration activities as destroying the aesthetically pleasing forests that now exist in order to restore the prairie and oak savanna ecosystems that existed prior to European settlement. But philosophically, because a restored landscape can never necessarily be tied only to our own desires (since our desires are not historically and scientifically determined in the same way as the parameters of a restoration), then those desires cannot actually be the direct cause of any restriction on the self-realization of nature.

(4) Finally, we must wonder about this value of self-realization. Setting aside the inherent philosophical problems with understanding what this claim to self-realization means in the case of nature, one has to wonder how we could know what natural self-realization would be in any particular case and why we would totally divorce a human role in helping to make it happen if we could discern it. In an analogous case involving two humans, we do not say that a human right to (or value of) self-realization is abrogated when a criminal who harms someone is forced to pay restitution. Even if the restitution is forced against the will of the victim, and even if the compensation in principle can never make up for the harm done, we would not say that somehow the victim's self-realization has been restricted by the act of restitution by the criminal. Again, there seems to be no clear argument here for why the moral obligation to try to restore has been diminished by Katz's arguments that we do not have the ability to really restore nature or pass off an artifact as nature.

RESTORING ENVIRONMENTAL PHILOSOPHY

If I am justified in setting aside the rest of Katz's arguments (KR1–3 and KR5) by accepting his claim that humans really cannot restore "real" nature, then what sort of conclusions could we draw about the role of philosophy in sorting out the normative issues involved in restoration? As it turns out, Katz gives us an insight in figuring out the next step.

After explaining the harm we do to nature in the domination we visit upon it through acts of restoration, Katz briefly assesses the harm that we do to *ourselves* through such actions:

> But a policy of domination transcends the anthropocentric subversion of natural processes. A policy of domination subverts both nature and human existence; it denies both the cultural and natural realization of individual good, human and nonhuman. Liberation from all forms of domination is thus the chief goal of any ethical or political system. (Katz 1997, 105)

Although not very clearly explained by Katz, this intuition represents a crucial point for proceeding further. In addition to connecting environmental philosophy to larger projects of social liberation, Katz here opens the door to a consideration of the consequences of restoration on humans and human communities. As such, Katz allows an implicit assertion that there is a value involved in restoration that must be evaluated in addition to the value of the objects that are produced by restoration.

But the problem with drawing this conclusion is that this passage is also perhaps the most cryptic in all of Katz's work on restoration. What does Katz mean by this claim? How exactly does restoration deny the realization of an individual human, or cultural, good? This claim can only be made understandable by assuming that some kind of cultural value connected to nature is risked through the act of domination, or otherwise causing harm to nature. But what is this value?

I think the value Katz is alluding to here, although he never explores it seriously, describes the value of that part of human culture that is connected to external, nonhuman nature. This is not simply a suggestion that we humans are part of nature; it also points out that we have a *relationship with* nature that exists on moral as well as physical terrain in such a way that our actions toward nature can reciprocally harm us. If this is the view implicit in this claim, then it is still consistent with much of the rest of Katz's larger views about the value of nature. We have a relationship with nature even if we are separable from it. I will accept this basic tenet of Katz's argument: we do exist in some kind of moral relationship with nature. And without fully explicating the content of that relationship, it seems that Katz is right in assuming that somehow the way in which we act toward nature morally implicates us in a particular way. In the same sense, when we morally mistreat another human, we not only harm them but also harm ourselves (by diminishing our character, by implicating ourselves in evil, or however you want to put it). Katz is suggesting that our relationship with nature has a determinant effect on our moral character. Or, perhaps more accurately, this is a suggestion necessary for Katz's comment to make sense, even though he never expresses this view himself.

Now if this assumption is correct, and if there is anything to the arguments I have put forward so far that there can be some kind of positive value to our interaction with nature, then doing right by nature will have the same reciprocal effect of morally implicating us in a positive value as occurs when we do right by other persons. Perhaps Katz would agree. Where Katz would disagree is with the suggestion I would want to add to this: that there is some part of many kinds of restorations (if not most kinds) that contains positive value. Aside from the other suggestions I have already made concerning the possible positive content of restoration, one can also consider that the relationship with nature that is implied in Katz's view has a moral content in itself that is not reducible to the value of fulfilling this relationship's concomitant obligations. The relationship between humans and nature imbues restoration with a positive value even if it cannot replicate natural value in its products. But understanding this point will require some explanation.

Consider that if I have a reciprocal relationship with another human (in which I do right by them and they do right by me), then, to generalize Katz's account, there is a moral content to both of our actions that implicates each of us as persons. Each of us is a better person morally because of the way we interact with each other in the relationship. But the relationship itself, or rather just the fact of the existence of the relationship, also has a moral content of its own (or what we could call a *normative content,* meaning that the relationship can be assessed as being in a better or worse state) that is independent of the fulfillment of any obligations. If this point of the possible separation between the value of a relationship and the value of the fulfillment of obligations does not follow intuitively, imagine the case where two people act according to duty toward each other without building a relationship of substantive normative content between them. Consider the following example. I have a brother with whom I am not terribly close. Although I always act according to duty to him—I never knowingly do harm to him and I even extend special family obligations to him—I do not have a substantive relationship with him that in itself has a normative content. Thus, if I do not speak to him for a year, nothing is lost because there is no relationship there to maintain or that requires maintenance for normative reasons. But if my brother needed a kidney transplant, I would give him my kidney unhesitatingly out of a sense of obligation—something I would not feel obliged to do for non–family members—even though I still do not feel intimately comfortable around him in the same way I do with my closest friends. Our relationship as persons—that sense of intimate affection and care for another person that I have experienced with other people—has no positive value for me (it isn't necessarily a disvalue, only a sense of indifference and a lack of closeness). So, I can have interaction with another person, even interaction that involves substantial components of obligation and duty (and, in Katz's terms, I will never put myself in a position to dominate that other person) but still not have a relationship with that person that involves any kind of positive value or that has normative standards of maintenance.

I do not think that I have any obligation to have a relationship in this sense with my brother. I, in fact, do not, even though my mother would like it if I did. But if I did have a relationship with my brother in this sense, then it would have a value above and beyond the moral interaction that I have with him now (the obligations that I have to him that can be iterated) that aids in a determination of our moral character.[11] If we had a relationship with normative content, there would be a positive or negative value that could be assessed if I lost touch with my brother or ceased to care about his welfare. (I could very well claim that it would be better for me to have such a relationship with him, but this would require an additional argument.)

Consider further that if I wanted to rectify or create anew a substantive normative relationship with my brother, like the relationship I have with several close friends, how would I do it? One thing I could do would be to engage in activities with him—the same sorts of activities (let us call them *material interactions*) that I do with my friends now. I might work with him to put up a fence or help him plant his garden. I might begin to talk over my personal and professional problems with him. I might go on a long journey with him that demanded some kind of mutual reliance such as whitewater rafting or visiting a foreign city where neither of us spoke the native language. In short, although there are, of course, no guarantees, I could begin to have some kind of material relationship with him as a prelude to having some kind of substantive normative relationship with him. Many factors might limit the success of such a project: for one thing, the distance between the two of us—he lives in our hometown of Atlanta and I live in New York. So, if I was really serious about this project of building a relationship between us that had value independent of the value of the fulfillment of our mutual obligations to each other that already exist, I'd have to come up with ways to bridge these interfering factors. Importantly, though, I couldn't form a substantive normative relationship with him merely by respecting his right of self-realization and autonomy as a person; I would have to somehow become actively involved with him.

Now, when we compare the case of the estranged brother to that of nature, many parallels arise. We know that we can fulfill obligations to nature in terms of respecting its autonomy and self-realization as a subject (in Katz's terms) without ever forming a substantive normative relationship with it. Assuming also that there is a kind of relationship with nature possible according to Katz's scheme (for this is in part what we harm when we dominate nature), it is fair to say that a relationship consisting of positive normative value with nature is compatible with Katz's overall view of the human–nature relationship. Because he says so little about what our positive relationship to nature could be, he is in no position to restrict it a priori. We also know that, as in the case of the estranged brother, we need some kind of material bridge to create a relationship with nature in order to see that relationship come about.

How do we build that bridge? Suggesting ways to overcome the gap between humans and nature (without necessarily disvaluing it) seems in part to be the restored role of environmental philosophy in questions of ecological restoration. Certainly, as in the case of my brother, distance is a problem. Numerous environmental professionals have emphasized the importance of being in nature in order to care for nature. Also, acts of preservation are important for there to be nature to have a relationship with. But what about restoration? Can restoration help engender such a positive normative relationship with nature? It seems clear to me that it can. When we engage in acts of benevolent restoration, we are *bound by* nature in the same sense that we are obligated to respect what it once was attempting to realize before we interfered with it. In Katz's terms, we are attempting to respect it as an autonomous subject. But we are also *bound to* nature in the act of restoring. In addition to the substantial personal and social benefits that accrue to people who engage in benevolent forms of restoration,[12] we can also say that restoration restores the human connection to nature by restoring that part of culture that has historically contained a connection to nature. This kind of relationship goes well beyond mere reciprocity; it involves the creation of a value in relationship with nature beyond obligation. Although it would take further argument to prove, I believe that this kind of relationship is a necessary condition for encouraging people to protect natural systems and landscapes around them rather than trade them off for short-term monetary gains from development. If I am in a normative relationship with the land around me (whether it is "real" nature or not), I am less likely to allow it to be harmed further. Specifying the parameters of restoration that help to achieve this moral relationship with nature will be the task of a more pragmatic environmental philosophy. As mentioned at the outset of this chapter, environmental pragmatism allows for and encourages the development of human-centered notions of the value of nature. Pragmatists are not restricted to identifying obligations to nature in the existence of nonanthropocentric conceptions of value but may embrace an expression of environmental values in human terms. More adequately developing the idea of restoration in terms of the human–nature relationship is thus appropriately under the pragmatist's purview. More importantly, however, the value articulated here exists between anthropocentrism and nonanthropocentrism, fully relying on the capacities of both sides of the human–nature relationship.[13]

We can even look to Katz for help in completing this pragmatic task. We don't want restorations that try to pass themselves off as the real thing when they are really "fakes" (KR1) or are pursued through arrogance (KR2); nor are we interested in those that are offered as justifications for replacing or destroying nature (KR5). We would not want our comparable human relationships to exhibit those properties either. But even given the legacy of inhuman treatment of each other, we know that it is possible to restore human relationships that do not resemble KR1, KR2, or KR5. There is, however, one possible worry to attend to in KR3, the artifact argument. Although earlier I said that the importance of KR3 is diminished by granting Katz's nature–culture distinction, there is a way that it can still cause us problems in grounding attempts at restoration in the positive value of strengthening the human–nature relationship.

Katz may object to my relationship argument that if we allow his claim that what has been restored is not really nature then we are not restoring a cultural relationship with nature but, in a sense, only extending the artifactual material culture of humans. At best, all we can have with restoration is a relationship with artifacts, not nature. Maybe he will allow that we improve relations with each other through cooperative acts of restoration, but this is not the same as a restoration of a relationship with nature itself.

But it should be clear by now that Katz would be mistaken to make such an objection for several reasons stemming in part from my earlier remarks.

(1) Even if we admit that restored nature is an artifact and not real nature, restored nature can also serve as a way for real nature to free itself from the shackles we have previously placed upon it. Restoration can allow nature to engage in its own autonomous restitution. Of the different sorts of restoration projects that I have sketched above, many amount to aids to nature rather than creations of new nature.

(2) Even if restoration is the production of an artifact, these artifacts do bear a striking resemblance to the real thing. This is not to say that restorations can be good enough to fool us (KR1). Rather, it is simply to point out that an opportunity to interact with the flora and fauna of the sort most common in benevolent restorations will increase the bonds of care that people will have with nonrestored nature. If a denuded and abandoned lot in the middle of an inner-city ghetto is restored by local residents who have never been outside of their city, then it will help them better appreciate the fragility and complexity of the natural processes of

nature itself should they encounter them. The fact that restorationists are engaged in a techno-logical process does not necessarily mean that their practices do not serve the broader purpose of restoring a relationship with nature. Just as beginning some form of mediated communica-tion with my brother (such as e-mail or regular phone calls) does not restore a fully healthy communicative relationship with him that could be found through face-to-face conversation, it still helps me get used to the idea of some form of immediate and substantive communication.

And, finally, (3) if Katz persists in his worry that the act of restoration reifies domination by reaffirming our power over nature through the creation of artifacts, we can say that exactly the opposite is likely the case (at least in the case of benevolent restorations) when the goal is restoring the culture of nature, if not nature itself. Restorationists get firsthand (rather than anecdotal and textbook) exposure to the actual consequences of human domination of nature. A better understanding of the problems of bioactivating soil, for example, gives us a better idea of the complexity of the harm we have caused to natural processes. In a much healthier way than Katz seems willing to admit, knowing about that harm can empower us to know more precisely why we should object to the kinds of activities that can cause that harm to nature in the first place. As a parallel human case, imagine a carrier of a deadly and contagious disease (that she cannot die from) who ignores warnings about how to take precautions against spreading the disease to other people. If that person passes on her deadly disease to other people, would it not in the end benefit her to have the opportunity to volunteer to work in a hospital ward full of people dying from this particular disease? If the disease was incurable, she could never restore health to its victims (either out of reciprocity or a desire to form helpful normative relationships with others), but she might learn through her experience in the hospital ward to respect the importance of not risking giving this disease to others. Restoration simi-larly teaches us the actual consequences of our actions rather than allowing us to ignore them by restricting our interaction with nature to those parts we have not yet damaged.[14]

CONCLUSION

In a followup essay to "The Big Lie" called "The Call of the Wild," which used the figure of the *wildness* in the white-tailed deer population at Katz's summer home on Fire Island to help distinguish nature from culture, Katz embraced a kind of reciprocal relationship with nature. The wild white-tailed deer, which Katz admits in the essay are now quite tame, are described as

> members of [Katz's] moral and natural community. The deer and I are partners in the continuous struggle for the preservation of autonomy, freedom, and integrity. This shared partnership creates obligations on the part of humanity for the preservation and protection of the natural world. (Katz 1997, 117)

Surely we would respond that this relationship also creates obligations of benevolent restora-tion as well. If the deer were threatened with harm without a needed restoration of a breeding ground, for example, would Katz not be obliged to do it? And, in doing this restoration, would he not help to generate positive value in his relationship with those deer?

It seems clear that benevolent restorations of this sort are valuable because they help us restore our relationship with nature, by restoring what could be termed our "culture of nature." This is true even if Katz is correct that restored nature has the ontological property of an arti-fact. Restoration is an obligation exercised in the interests of forming a positive community with nature and thus is well within the boundaries of a positive, pragmatic environmental philosophy. Just as artifacts can serve valuable relationship goals by creating material bridges to other subjects, artifactual landscapes can help restore the culture of nature. Further defining the normative ground of benevolent restorations should be the contribution that philosophy can make to the public consideration and practice of ecological restoration. It is a contribution directed at a larger audience, beyond the professional philosophy community, and aimed toward the practical end of helping to resolve environmental problems.[15]

NOTES

1. If we accept Katz's position, a philosophical inquiry into restoration would actually be an investigation of some other kinds of questions than those legitimately posed by environmental philosophers. Since Katz argues that restored nature is only an artifact, philosophers of technology would presumably still be doing philosophy when they were involved in an investigation of ecological restoration. This possibility of trying to define out certain practices from environmental ethics is no red herring. In a public forum discussing his work at the Central Division meeting of the American Philosophical Association in Chicago in 1998, Katz stated publicly that agriculture was not the proper purview of environmental ethics. Philosophers working on questions of ethics and agriculture could be doing agricultural ethics but not environmental ethics.

2. Elliot strengthens the more charitable view of restoration in his 1997 book.

3. From the early aesthetic theory of Mark Sagoff (before he ever turned to environmental questions), one can also pull the following distinction to help further deepen the discussion of different kinds of benevolent restorations: (1) integral restorations—restorations that "put new pieces in the place of original fragments that have been lost"; and (2) purist restorations—restorations that "limit [themselves] to clearing works of art and to reattaching original pieces that may have fallen" (Sagoff 1978, 457). As it turns out, one can argue that integral restorations are aesthetically (and possibly ethically) worrisome since they seem to create hybrid works of art (created by both the artist and the restorationist). But this is not too much of a problem for the analogy with ecological restoration since many of these restorations amount to something more akin to purist restorations—for example, cleaning land by bioactivating soil. Perhaps more common would be a subclass of purist restoration that we might call *rehabilitative* restoration. Examples of such projects would include cleaning out exotic plants that were introduced at some time into a site and allowing the native plants to reestablish themselves. Such activity is akin to the work of a purist art restorationist who corrects the work of a restorationist who had come before her. If a restorationist, for example, were to remove an eighteenth-century integral addition to a sixteenth-century painting, then we would assume that this rehabilitative act was consistent with a purist restoration. I provide a much more thorough discussion of the import of this distinction for ecological restoration in Light (1997).

4. Katz has four main papers on restoration: "The Big Lie: Human Restoration of Nature" (1992), "The Call of the Wild: The Struggle against Domination and the Technological Fix of Nature" (1992), "Artifacts and Functions: A Note on the Value of Nature" (1993), and "Imperialism and Environmentalism" (1993). All of these papers are collected in Katz (1997), and it is these versions that I have drawn on for this chapter.

5. Originally in Katz, "The Big Lie" (as are KR2–KR4). KR 1 is restated later in "The Call of the Wild": "What makes value in the artifactually restored natural environment questionable is its ostensible claim to be the original" (Katz 1997, 114).

6. KR3 is most thoroughly elaborated later in Katz, "Artifacts and Functions."

7. The domination argument is repeated in Katz, "The Call of the Wild" (1997, 115) with the addition of an imported quote from Eugene Hargrove: domination "reduces [nature's] ability to be creative." The argument is also repeated in Katz, "Artifacts and Functions," and further specified in Katz, "Imperialism and Environmentalism." As far as I can tell, though, the argument for domination is not really expanded on in this last paper, except that imperialism is deemed wrong because it makes nature into an artifact (KR3).

8. Originally in Katz, "The Call of the Wild," and repeated in Katz, "Imperialism and Environmentalism" (1997, 139).

9. The forum here is the same as the one referenced in note 1: a public forum discussing Katz's work at the Central Division meeting of the American Philosophical Association in Chicago in 1998.

10. The absence of any perceptible progress in Katz's views following his debate with Donald Scherer is a case in point. Scherer spends too much time, I think, trying to push a critique of Katz's ontology and metaphysics. The resulting debate appears intractable. See Scherer (1995) and Katz (1996).

11. On a broader scale, just as there can be a town full of decent, law-abiding citizens, those citizens may not constitute a moral community in any significant sense.

12. Herbert W. Schroeder, "Psychological Benefits of Volunteering in Stewardship Programs," in *Restoring Ecology*. Robert E. Grese, Rachel Kaplan. Robert L. Ryan, and Jane Buxton, "Lessons for Restoration in the Traditions of Stewardship: Sustainable Land Management in Northern New Mexico," in *Restoring Ecology*.

13. It is also the case that restoration will be only one out of a large collection of practices available for adaptive management. Indeed, there could even be cases where something akin to mitigation (albeit a benevolent kind) would be justified rather than restoration if a claim to sustaining some form of natural value warranted it. In a project to clean up an abandoned mine site, for example, we can imagine a case where restoring the site to a landscape that was there before would not be the best choice and that instead some other sustainable landscape that would help to preserve an endangered species now in the area would be more appropriate. But overall, environmentalists must accept human interaction with nature as an acceptable practice to begin the ethical assessment of any case of environmental management. I am indebted to Anne Chapman for pressing me to clarify this point.

14. Katz can legitimately respond here that there seems to be no unique reason why people couldn't get these kinds of experiences that generate a closer relationship with nature out of some other kinds of activities. Why couldn't we just use this sort of argument to encourage more acts of preservation, or to simply take more walks though nature? Such an objection would, however, miss a crucial point. Even if it can be proved that we can get these kinds of positive experiences with nature in forms other than acts of restoration (and I see no reason why we couldn't), this does not diminish the case being built here: that restoration does not necessarily result in the domination of nature. The goal of my argument here is not to show that restoration provides a unique value compared with other environmental practices, but only to reject the claim that there is no kind of positive value that restoration can contribute to nature in some sense. So, an objection by Katz of this sort would miss the target of our substantive disagreement. Additionally, one could also argue that (1) restoration does, in fact, produce some unique values in our relationship with nature (see Andrew Light, "Negotiating Nature: Making Restoration Happen in an Urban Park Context," in *Restoring Ecology* for such a case in relation to the potential democratic values in restoration); and that (2) even if not unique in itself, restoration helps to improve other sorts of unique values in nature. A case for (2) could be made, for example, in Allen Carlson's work on the importance of scientific understanding for appreciating the aesthetic value of nature (Carlson 1995). Arguably, our experiences as restorationists give us some of the kinds of understandings of natural processes required for aesthetic appreciation according to Carlson's account. Importantly, this understanding is a transitive property: it gives us an ability to aesthetically appreciate not only the nature we are trying to restore, but also the nature we are not trying to restore. Restoration thus could provide a unique avenue into the aesthetic appreciation of all of nature, restored or not. The main point, however, should not be lost: restoration is an important component in a mosaic of efforts to revive the culture of nature. Without any reason to believe that it has other disastrous effects, restoration seems warranted within a prescribed context even if it is not a cure-all.

15. This chapter is based on a presentation originally given at a plenary session (with Eric Katz and William Jordan) of the International Symposium on Society and Resource Management, University of Missouri, Columbia, May 1998. Subsequent versions were presented as the keynote address of the Eastern Pennsylvania Philosophy Association annual meeting, Bloomsburg University, November 1998; and at Georgia State University, SUNY Binghamton, and Lancaster University (UK). I have benefited much from the discussions at all of these occasions and especially from the helpful comments provided by Cari Dzuris, Cheryl Foster, Warwick Fox, Paul Gobster, Leslie Heywood, Bruce Hull, Bryan Norton, George Rainbolt, and Christopher Wellman.

REFERENCES

Carlson, A. 1995. "Nature, Aesthetic Appreciation, and Knowledge." *Journal of Aesthetics and Art Criticism* 53: 393–400.

Elliot, R. 1995. "Faking Nature." In *Environmental Ethics,* edited by R. Elliot, 76–88. Oxford: Oxford University Press.

———. 1997. *Faking Nature.* London: Routledge.

Katz, E. 1996. "The Problem of Ecological Restoration." *Environmental Ethics* 18: 222–224.

———. 1997. *Nature as Subject: Human Obligation and Natural Community.* Lanham, MD: Rowman & Littlefield Publishers.

Light, A. 1996a. "Environmental Pragmatism as Philosophy or Metaphilosophy." In *Environmental Pragmatism,* edited by A. Light and E. Katz, 325–338. London: Routledge.

———. 1996b. "Compatibilism in Political Ecology." In *Environmental Pragmatism,* edited by A. Light and E. Katz, 161–184. London: Routledge.

———. 1996c. "Callicott and Naess on Pluralism." *Inquiry* 39: 273–94.

———. 1997. "Restoration and Reproduction." Unpublished manuscript presented at the Symposium on Ethics and Environmental Change: Recognizing the Autonomy of Nature, St. John's, Newfoundland, Canada, June 4–5.

Light, A., and E. Higgs. 1996. "The Politics of Ecological Restoration." *Environmental Ethics* 18: 227–247.

Sagoff, M. 1978. "On Restoring and Reproducing Art." *The Journal of Philosophy* 75: 453–470.

Scherer, D. 1995. "Evolution, Human Living, and the Practice of Ecological Restoration." *Environmental Ethics* 17: 359–379.

14

Democratic Rationalization: Technology, Power, and Freedom

Andrew Feenberg

THE LIMITS OF DEMOCRATIC THEORY

Technology is one of the major sources of public power in modern societies. So far as decisions affecting our daily lives are concerned, political democracy is largely overshadowed by the enormous power wielded by the masters of technical systems: corporate and military leaders, and professional associations of groups such as physicians and engineers. They have far more to do with control over patterns of urban growth, the design of dwellings and transportation systems, the selection of innovations, and our experience as employees, patients, and consumers than all the governmental institutions of our society put together.

Marx saw this situation coming in the middle of the nineteenth century. He argued that traditional democratic theory erred in treating the economy as an extrapolitical domain ruled by natural laws such as the law of supply and demand. He claimed that we will remain disenfranchised and alienated so long as we have no say in industrial decision making. Democracy must be extended from the political domain into the world of work. This is the underlying demand behind the idea of socialism.

Modern societies have been challenged by this demand for over a century. Democratic political theory offers no persuasive reason of principle to reject it. Indeed, many democratic theorists endorse it.[1] What is more, in a number of countries, socialist parliamentary victories or revolutions have brought parties to power dedicated to achieving it. Yet today we do not appear to be much closer to democratizing industrialism than in Marx's time.

This state of affairs is usually explained in one of the following two ways.

On the one hand, the common sense view argues that modern technology is incompatible with workplace democracy. Democratic theory cannot reasonably press for reforms that would destroy the economic foundations of society. For evidence, consider the Soviet case: although they were socialists, the communists did not democratize industry, and the current democrati-

A revised version of Andrew Feenberg, "Subversive Rationality: Technology, Power, and Democracy," in *Inquiry* 35/3–4 (1992): 301–22. Copyright © 1992 Taylor & Francis. Reprinted by permission of Taylor & France, Oslo: Norway. (www.tandf.no/inquiry)

zation of Soviet society extends only to the factory gate. At least regarding the ex–Soviet Union, everyone can agree on the need for authoritarian industrial management.

On the other hand, a minority of radical theorists claims that technology is not responsible for the concentration of industrial power. That is a political matter, due to the victory of capitalist and communist elites in struggles with the underlying population. No doubt modern technology lends itself to authoritarian administration, but in a different social context it could just as well be operated democratically.

In what follows, I will argue for a qualified version of this second position, somewhat different from both the usual Marxist and democratic formulations. The qualification concerns the role of technology, which I see as *neither* determining nor neutral. I will argue that modern forms of hegemony are based on the technical mediation of a variety of social activities, whether it be production or medicine, education or the military, and that, consequently, the democratization of our society requires radical technical as well as political change.

This is a controversial position. The common sense view of technology limits democracy to the state. By contrast, I believe that unless democracy can be extended beyond its traditional bounds into the technically mediated domains of social life, its use value will continue to decline, participation will wither, and the institutions we identify with a free society will gradually disappear.

Let me turn now to the background to my argument. I will begin by presenting an overview of various theories that claim that insofar as modern societies depend on technology, they require authoritarian hierarchy. These theories presuppose a form of technological determinism that is refuted by historical and sociological arguments I will briefly summarize. I will then present a sketch of a nondeterministic theory of modern society I call *critical theory of technology*. This alternative approach emphasizes contextual aspects of technology ignored by the dominant view. I will argue that technology is not just the rational control of nature; both its development and impact are also intrinsically social. I will then show that this view undermines the customary reliance on efficiency as a criterion of technological development. That conclusion, in turn, opens broad possibilities of change foreclosed by the usual understanding of technology.

DYSTOPIAN MODERNITY

Max Weber's famous theory of rationalization is the original argument against industrial democracy. The title of this chapter implies a provocative reversal of Weber's conclusions. He defined rationalization as the increasing role of calculation and control in social life, a trend leading to what he called the "iron cage" of bureaucracy.[2] "Democratic" rationalization is thus a contradiction in terms.

Once traditionalist struggle against rationalization has been defeated, further resistance in a Weberian universe can only reaffirm irrational life forces against routine and drab predictability. This is not a democratic program but a romantic antidystopian one, the sort of thing that is already foreshadowed in Dostoyevsky's *Notes from Underground* and various back-to-nature ideologies.

My title is meant to reject the dichotomy between rational hierarchy and irrational protest implicit in Weber's position. If authoritarian social hierarchy is truly a contingent dimension of technical progress, as I believe, and not a technical necessity, then there must be an alternative way of rationalizing society that democratizes rather than centralizes control. We need not go underground or native to preserve threatened values such as freedom and individuality.

But the most powerful critiques of modern technological society follow directly in

Weber's footsteps in rejecting this possibility. I am thinking of Heidegger's formulation of "the question of technology" and Ellul's theory of "the technical phenomenon."[3] According to these theories, we have become little more than objects of technique, incorporated into the mechanism we have created. As Marshall McLuhan once put it, technology has reduced us to the "sex organs of machines." The only hope is a vaguely evoked spiritual renewal that is too abstract to inform a new technical practice.

These are interesting theories, important for their contribution to opening a space of reflection on modern technology. I will return to Heidegger's argument in the conclusion to this chapter. But first, to advance my own argument, I will concentrate on the principal flaw of *dystopianism,* the identification of technology in general with the specific technologies that have developed in the last century in the West. These are technologies of conquest that pretend to an unprecedented autonomy; their social sources and impacts are hidden. I will argue that this type of technology is a particular feature of our society and not a universal dimension of "modernity" as such.

TECHNOLOGICAL DETERMINISM

Determinism rests on the assumption that technologies have an autonomous functional logic that can be explained without reference to society. *Technology* is presumably social only through the purpose it serves, and purposes are in the mind of the beholder. Technology would thus resemble science and mathematics by its intrinsic independence of the social world.

Yet unlike science and mathematics, technology has immediate and powerful social impacts. It would seem that society's fate is at least partially dependent on a nonsocial factor that influences it without suffering a reciprocal influence. This is what is meant by *technological determinism.* Such a deterministic view of technology is commonplace in business and government, where it is often assumed that progress is an exogenous force influencing society rather than an expression of changes in culture and values.

The dystopian visions of modernity I have been describing are also deterministic. If we want to affirm the democratic potentialities of modern industrialism, we will therefore have to challenge their deterministic premises. These I will call the *thesis of unilinear progress* and the *thesis of determination by the base.* Here is a brief summary of these two positions.

(1) Technical progress appears to follow a unilinear course, a fixed track, from less to more advanced configurations. Although this conclusion seems obvious from a backward glance at the development of any familiar technical object, in fact it is based on two claims of unequal plausibility: first, that technical progress proceeds from lower to higher levels of development; and second, that that development follows a single sequence of necessary stages. As we will see, the first claim is independent of the second and is not necessarily deterministic.

(2) Technological determinism also affirms that social institutions must adapt to the "imperatives" of the technological base. This view, which no doubt has its source in a certain reading of Marx, is now part of the common sense of the social sciences.[4] Below, I will discuss one of its implications in detail: the supposed "trade-off" between prosperity and environmental values.

These two theses of technological determinism present decontextualized, self-generating technology as the unique foundation of modern society. Determinism thus implies that our technology and its corresponding institutional structures are universal, indeed, planetary in scope. There may be many forms of tribal society, many feudalisms, and even many forms of early capitalism, but there is only one modernity and it is exemplified in our society for good or ill. Developing societies should take note: as Marx once said, calling the attention of his backward German compatriots to British advances: *De te fabula narratur*—of you the tale is told.[5]

CONSTRUCTIVISM

The implications of determinism appear so obvious that it is surprising to discover that neither of its two theses can withstand close scrutiny. Yet contemporary sociology of technology undermines the first thesis of unilinear progress, whereas historical precedents are unkind to the second thesis of determination by the base.

Recent constructivist sociology of technology grows out of new social studies of science. These studies challenge our tendency to exempt scientific theories from the sort of sociological examination to which we submit nonscientific beliefs. They affirm the *principle of symmetry,* according to which all contending beliefs are subject to the same type of social explanation regardless of their truth or falsity.[6] A similar approach to technology rejects the usual assumption that technologies succeed on purely functional grounds.

Constructivism argues that theories and technologies are underdetermined by scientific and technical criteria. Concretely, this means two things: first, there is generally a surplus of workable solutions to any given problem, and social actors make the final choice among a batch of technically viable options; and second, the problem–definition often changes in the course of solution. The latter point is the more conclusive but also more difficult of the two.

Two sociologists of technology, Pinch and Bijker, illustrate it with the early history of the bicycle.[7] The object we take to be a self-evident "black box" actually started out as two very different devices: a sportsman's racer and a utilitarian transportation vehicle. The high front wheel of the sportsman's bike was necessary at the time to attain high speeds, but it also caused instability. Equal-sized wheels made for a safer but less exciting ride. These two designs met different needs and were in fact different technologies with many shared elements. Pinch and Bijker call this original ambiguity (of the object designated as a "bicycle") "interpretative flexibility."

Eventually the "safety" design won out, and it benefited from all the later advances that occurred in the field. In retrospect, it seems as though the high wheelers were a clumsy and less efficient stage in a progressive development leading through the old "safety" bicycle to current designs. In fact, the high wheeler and the safety shared the field for years, and neither was a stage in the other's development. The high wheeler represents a possible alternative path of bicycle development that addressed different problems at the origin.

Determinism is a species of Whig history that makes it seem as though the end of the story was inevitable from the very beginning by projecting the abstract technical logic of the finished object back into the past as a cause of development. That approach confuses our understanding of the past and stifles the imagination of a different future. Constructivism can open up that future, although its practitioners have hesitated so far to engage the larger social issues implied in their method.[8]

INDETERMINISM

If the thesis of unilinear progress falls, the collapse of the notion of determination by the technological base cannot be far behind. Yet it is still frequently invoked in contemporary political debates.

I shall return to these debates later in this chapter. For now, let us consider the remarkable anticipation of current attitudes in the struggle over the length of the workday and over child labor in mid-nineteenth-century England. The debate on the Factory Bill of 1844 was entirely structured around the deterministic opposition of technological imperatives and ideology. Lord Ashley, the chief advocate of regulation, protested in the name of familial ideology that

> the tendency of the various improvements in machinery is to supersede the employment of adult males, and substitute in its place, the labour of children and females. What will be the effect on future generations, if their tender frames be subjected, without limitation or control, to such destructive agencies.[9]

He went on to deplore the decline of the family consequent upon the employment of women, which "disturbs the order of nature" and deprives children of proper upbringing. "It matters not whether it be prince or peasant, all that is best, all that is lasting in the character of a man, he has learnt at his mother's knees." Lord Ashley was outraged to find that

> females not only perform the labour, but occupy the places of men; they are forming various clubs and associations, and gradually acquiring all those privileges which are held to be the proper portion of the male sex. . . . They meet together to drink, sing, and smoke; they use, it is stated, the lowest, most brutal, and most disgusting language imaginable.

Proposals to abolish child labor met with consternation on the part of factory owners, who regarded the little (child) worker as an "imperative" of the technologies created to employ him. They denounced the "inefficiency" of using full-grown workers to accomplish tasks done as well or better by children, and they predicted all the usual catastrophic economic consequences—increased poverty, unemployment, loss of international competitiveness—from the substitution of more costly adult labor. Their eloquent representative, Sir J. Graham, therefore urged caution:

> We have arrived at a state of society when without commerce and manufactures this great community cannot be maintained. Let us, as far as we can, mitigate the evils arising out of this highly artificial state of society; but let us take care to adopt no step that may be fatal to commerce and manufactures.

He further explained that a reduction in the workday for women and children would conflict with the depreciation cycle of machinery and lead to lower wages and trade problems. He concluded that "in the close race of competition which our manufacturers are now running with foreign competitors . . . such a step would be fatal." Regulation, he and his fellows maintained in words that echo still, is based on a "false principle of humanity, which in the end is certain to defeat itself." One might almost believe that Ludd had risen again in the person of Lord Ashley: the issue is not really the length of the workday, "but it is in principle an argument to get rid of the whole system of factory labour." Similar protestations are heard today on behalf of industries threatened with what they call environmental "Luddism."

Yet what actually happened once the regulators succeeded in imposing limitations on the workday and expelling children from the factory? Did the violated imperatives of technology come back to haunt them? Not at all. Regulation led to an intensification of factory labor that was incompatible with the earlier conditions in any case. Children ceased to be workers and were redefined socially as learners and consumers. Consequently, they entered the labor market with higher levels of skill and discipline that were soon presupposed by technological design. As a result, no one is nostalgic for a return to the good old days when inflation was held down by child labor. That is simply not an option (at least not in the developed capitalist world).

This example shows the tremendous flexibility of the technical system. It is not rigidly constraining but on the contrary can adapt to a variety of social demands. This conclusion should not be surprising given the responsiveness of technology to social redefinition discussed previously. It means that technology is just another dependent social variable, albeit an increasingly important one, and not the key to the riddle of history.

Determinism, I have argued, is characterized by the principles of unilinear progress and determination by the base; if determinism is wrong, then technology research must be guided by the following two contrary principles. In the first place, technological development is not unilinear but branches in many directions, and it could reach generally higher levels along more than one different track. And, secondly, technological development is not determining for society but is overdetermined by both technical and social factors.

The political significance of this position should also be clear by now. In a society where determinism stands guard on the frontiers of democracy, indeterminism cannot but be political. If technology has many unexplored potentialities, no technological imperatives dictate the current social hierarchy. Rather, technology is a scene of social struggle, a *parliament of things* on which civilizational alternatives contend.

INTERPRETING TECHNOLOGY

In the next sections of this chapter, I would like to present several major themes of a nondeterminist approach to technology. The picture sketched so far implies a significant change in our definition of technology. It can no longer be considered as a collection of devices, nor, more generally, as the sum of rational means. These are tendentious definitions that make technology seem more functional and less social than in fact it is.

As a social object, technology ought to be subject to interpretation like any other cultural artifact, but it is generally excluded from humanistic study. We are assured that its essence lies in a technically explainable function rather than a hermeneutically interpretable meaning. At most, humanistic methods might illuminate extrinsic aspects of technology, such as packaging and advertising, or popular reactions to controversial innovations such as nuclear power or surrogate motherhood. Technological determinism draws its force from this attitude. If one ignores most of the connections between technology and society, it is no wonder that technology then appears to be self-generating.

Technical objects have two hermeneutic dimensions that I call their *social meaning* and their *cultural horizon*.[10] The role of social meaning is clear in the case of the bicycle introduced above. We have seen that the construction of the bicycle was controlled in the first instance by a contest of interpretations: was it to be a sportsman's toy or a means of transportation? Design features such as wheel size also served to signify it as one or another type of object.[11]

It might be objected that this is merely an initial disagreement over goals with no herme-neutic significance. Once the object is stabilized, the engineer has the last word on its nature, and the humanist interpreter is out of luck. This is the view of most engineers and managers; they readily grasp the concept of *goal* but they have no place for *meaning*.

In fact, the dichotomy of goal and meaning is a product of functionalist professional culture, which is itself rooted in the structure of the modern economy. The concept of a goal strips technology bare of social contexts, focusing engineers and managers on just what they need to know to do their job.

A fuller picture is conveyed, however, by studying the social role of the technical object and the lifestyles it makes possible. That picture places the abstract notion of a goal in its concrete social context. It makes technology's contextual causes and consequences visible rather than obscuring them behind an impoverished functionalism.

The functionalist point of view yields a decontextualized temporal cross-section in the life of the object. As we have seen, determinism claims implausibly to be able to get from one such momentary configuration of the object to the next on purely technical terms. But in the real world, all sorts of unpredictable attitudes crystallize around technical objects and influ-ence later design changes. The engineer may think these are extrinsic to the device he or she is working on, but they are its very substance as a historically evolving phenomenon.

These facts are recognized to a certain extent in the technical fields themselves, espe-cially in computers. Here we have a contemporary version of the dilemma of the bicycle dis-cussed above. Progress of a generalized sort in speed, power, and memory goes on apace while corporate planners struggle with the question of what it is all for. Technical development does not point definitively toward any particular path. Instead, it opens branches, and the final deter-mination of the "right" branch is not within the competence of engineering because it is sim-ply not inscribed in the nature of the technology.

I have studied a particularly clear example of the complexity of the relation between the technical function and meaning of the computer in the case of French videotext.[12] Called *Tele-tel*, this system was designed to bring France into the Information Age by giving telephone subscribers access to databases. Fearing that consumers would reject anything resembling office equipment, the telephone company attempted to redefine the computer's social image; it was no longer to appear as a calculating device for professionals but was to become an informational network for all.

The telephone company designed a new type of terminal, the *Minitel*, to look and feel like an adjunct to the domestic telephone. The telephonic disguise suggested to some users that they ought to be able to talk to each other on the network. Soon the *Minitel* underwent a further redefinition at the hands of these users, many of whom employed it primarily for anon-ymous online chatting with other users in the search for amusement, companionship, and sex.

Thus, the design of the *Minitel* invited communications applications that the company's engineers had not intended when they set about improving the flow of information in French society. Those applications, in turn, connoted the *Minitel* as a means of personal encounter, the very opposite of the rationalistic project for which it was originally created. The "cold" computer became a "hot" new medium.

At issue in the transformation is not only the computer's narrowly conceived technical function but also the very nature of the advanced society it makes possible. Does networking open the doors to the Information Age in which, as rational consumers hungry for data, we pursue strategies of optimization? Or is it a postmodern technology that emerges from the breakdown of institutional and sentimental stability, reflecting, in Lyotard's words, the "atomisation of society into flexible networks of language games?"[13] In this case, technology is not merely the servant of some predefined social purpose; it is an environment within which a way of life is elaborated.

In sum, differences in the way social groups interpret and use technical objects are not merely extrinsic but also make a difference in the nature of the objects themselves. *What* the object *is* for the groups that ultimately decide its fate determines what it *becomes* as it is redesigned and improved over time. If this is true, then we can only understand technological development by studying the sociopolitical situation of the various groups involved in it.

TECHNOLOGICAL HEGEMONY

In addition to the sort of assumptions about individual technical objects that we have been discussing so far, that situation also includes broader assumptions about social values. This is where the study of the cultural horizon of technology comes in. This second hermeneutic dimension of technology is the basis of modern forms of social hegemony; it is particularly relevant to our original question concerning the inevitability of hierarchy in technological society.

As I will use the term, *hegemony* is a form of domination so deeply rooted in social life that it seems natural to those it dominates. One might also define it as that aspect of the distribution of social power that has the force of culture behind it.

The term *horizon* refers to culturally general assumptions that form the unquestioned background to every aspect of life.[14] Some of these support the prevailing hegemony. For example, in feudal societies, the *chain of being* established hierarchy in the fabric of God's universe and protected the caste relations of the society from challenge. Under this horizon, peasants revolted in the name of the king, the only imaginable source of power. Rationalization is our modern horizon, and technological design is the key to its effectiveness as the basis of modern hegemonies.

Technological development is constrained by cultural norms originating in economics, ideology, religion, and tradition. We discussed earlier how assumptions about the age composition of the labor force entered into the design of nineteenth-century production technology. Such assumptions seem so natural and obvious that they often lie below the threshold of conscious awareness.

This is the point of Herbert Marcuse's important critique of Weber.[15] Marcuse shows that the concept of rationalization confounds the control of labor by management with control of nature by technology. The search for control of nature is generic, but management only arises against a specific social background, the capitalist wage system. Workers have no immediate interest in output in this system, unlike earlier forms of farm and craft labor, since their wage is not essentially linked to the income of the firm. Control of human beings becomes all-important in this context.

Through mechanization, some of the control functions are eventually transferred from human overseers and parcelized work practices to machines. Machine design is thus socially relative in a way that Weber never recognized, and the "technological rationality" it embodies is not universal but particular to capitalism. In fact, it is the horizon of all the existing industrial societies, communist as well as capitalist, insofar as they are managed from above. (In a later section, I discuss a generalized application of this approach in terms of what I call the *technical code*.)

If Marcuse is right, it ought to be possible to trace the impress of class relations in the very design of production technology as has indeed been shown by such Marxist students of the labor process as Harry Braverman and David Noble.[16] The assembly line offers a particularly clear instance because it achieves traditional management goals, such as deskilling and pacing work, through technical design. Its technologically enforced labor discipline increases productivity and profits by increasing control. However, the assembly line only appears as technical progress in a specific social context. It would not be perceived as an advance in an economy based on workers' cooperatives in which labor discipline was more self-imposed than imposed from above. In such a society, a different technological rationality would dictate different ways of increasing productivity.[17]

This example shows that technological rationality is not merely a belief, an ideology, but is also effectively incorporated into the structure of machines. Machine design mirrors back the social factors operative in the prevailing rationality. The fact that the argument for the social relativity of modern technology originated in a Marxist context has obscured its most radical implications. We are not dealing here with a mere critique of the property system, but have extended the force of that critique down into the technical "base." This approach goes well beyond the old economic distinction between capitalism and socialism, market and plan. Instead, one arrives at a very different distinction between societies in which power rests on the technical mediation of social activities and those that democratize technical control and, correspondingly, technological design.

DOUBLE ASPECT THEORY

The argument to this point might be summarized as a claim that social meaning and functional rationality are inextricably intertwined dimensions of technology. They are not ontologically distinct, for example, with meaning in the observer's mind and rationality in the technology proper. Rather they are *double aspects* of the same underlying technical object, each aspect revealed by a specific contextualization.

Functional rationality, like scientific-technical rationality in general, isolates objects from their original context in order to incorporate them into theoretical or functional systems. The institutions that support this procedure, such as laboratories and research centers, themselves form a special context with their own practices and links to various social agencies and powers. The notion of "pure" rationality arises when the work of decontextualization is not itself grasped as a social activity reflecting social interests.

Technologies are selected by these interests from among many possible configurations. Guiding the selection process are social codes established by the cultural and political struggles that define the horizon under which the technology will fall. Once introduced, technology offers a material validation of the cultural horizon to which it has been preformed. I call this the *bias* of technology: apparently neutral, functional rationality is enlisted in support of a hegemony. The more technology society employs, the more significant is this support.

As Foucault argues in his theory of "power/knowledge," modern forms of oppression are not so much based on false ideologies as on the specific technical "truths" that form the basis of the dominant hegemony and that reproduce it.[18] So long as the contingency of the choice of "truth" remains hidden, the deterministic image of a technically justified social order is projected.

The legitimating effectiveness of technology depends on unconsciousness of the cultural-political horizon under which it was designed. A recontextualizing critique of technology can uncover that horizon, demystify the illusion of technical necessity, and expose the relativity of the prevailing technical choices.

THE SOCIAL RELATIVITY OF EFFICIENCY

These issues appear with particular force in the environmental movement today. Many environmentalists argue for technical changes that would protect nature and in the process improve human life as well. Such changes would enhance efficiency in broad terms by reducing harmful and costly side effects of technology. However, this program is very difficult to impose in a capitalist society. There is a tendency to deflect criticism from technological processes to products and people, from a priori prevention to a posteriori cleanup. These preferred strategies are generally costly and reduce efficiency under the horizon of the given technology. This situation has political consequences.

Restoring the environment after it has been damaged is a form of collective consumption, financed by taxes or higher prices. These approaches dominate public awareness. This is why environmentalism is generally perceived as a cost involving trade-offs, and not as a rationalization increasing overall efficiency. But in a modern society obsessed by economic well-being, that perception is damning. Economists and businesspeople are fond of explaining the price we must pay in inflation and unemployment for worshipping at Nature's shrine instead of Mammon's. Poverty awaits those who will not adjust their social and political expectations to technology.

This trade-off model has environmentalists grasping at straws for a strategy. Some hold out the pious hope that people will turn from economic to spiritual values in the face of the mounting problems of industrial society. Others expect enlightened dictators to impose technological reform even if a greedy populace shirks its duty. It is difficult to decide which of these solutions is more improbable, but both are incompatible with basic democratic values.[19]

The trade-off model confronts us with dilemmas—environmentally sound technology versus prosperity, workers' satisfaction and control versus productivity, and so on—when what we need are syntheses. Unless the problems of modern industrialism can be solved in ways that both enhance public welfare and win public support, there is little reason to hope that they will ever be solved. But how can technological reform be reconciled with prosperity when it places a variety of new limits on the economy?

The child labor case shows how apparent dilemmas arise on the boundaries of cultural change, specifically, where the social definition of major technologies is in transition. In such situations, social groups excluded from the original design network articulate their unrepresented interests politically. New values the outsiders believe would enhance their welfare appear as mere ideology to insiders who are adequately represented by the existing designs.

This is a difference of perspective, not of nature. Yet the illusion of essential conflict is renewed whenever major social changes affect technology. At first, satisfying the demands of new groups after the fact has visible costs and, if it is done clumsily, will indeed reduce efficiency until better designs are found. But usually better designs can be found and what appeared to be an insuperable barrier to growth dissolves in the face of technological change.

This situation indicates the essential difference between economic exchange and technique. Exchange is all about trade-offs: more of A means less of B. But the aim of technical advance is precisely to avoid such dilemmas by elegant designs that optimize several variables at once. A single cleverly conceived mechanism may correspond to many different social demands, one structure to many functions.[20] Design is not a zero-sum economic game but an ambivalent cultural process that serves a multiplicity of values and social groups without necessarily sacrificing efficiency.

THE TECHNICAL CODE

That these conflicts over social control of technology are not new can be seen from the interesting case of the "bursting boilers."[21] Steamboat boilers were the first technology regulated in the United States. In the early nineteenth century, the steamboat was a major form of transportation similar to the automobile or airlines today. Steamboats were necessary in a big country without paved roads and lots of rivers and canals. But steamboats frequently blew up when the boilers weakened with age or were pushed too hard. After several particularly murderous accidents in 1816, the city of Philadelphia consulted with experts on how to design safer boilers, the first time an American governmental institution interested itself in the problem. In 1837, at the request of Congress, the Franklin Institute issued a detailed report and recommendations based on rigorous study of boiler construction. Congress was tempted to impose a safe boiler code on the industry, but boilermakers and steamboat owners resisted and government hesitated to interfere with private property.

It took from that first inquiry in 1816 to 1852 for Congress to pass effective laws regulating the construction of boilers. In that time, 5,000 people were killed in accidents on steamboats. Is this many casualties or few? Consumers evidently were not too alarmed to continue traveling by riverboat in ever increasing numbers. Understandably, the ship owners interpreted this as a vote of confidence and protested the excessive cost of safer designs. Yet politicians also won votes demanding safety.

The accident rate fell dramatically once technical changes such as thicker walls and safety valves were mandated. Legislation would hardly have been necessary to achieve this outcome had it been technically determined. But, in fact, boiler design was relative to a social judgment about safety. That judgment could have been made on strictly market grounds, as the shippers wished, or politically, with differing technical results. In either case, those results *constitute* a proper boiler. What a boiler *is* was thus defined through a long process of political struggle culminating finally in uniform codes issued by the American Society of Mechanical Engineers.

This example shows just how technology adapts to social change. What I call the *technical code* of the object mediates the process. That code responds to the cultural horizon of the society at the level of technical design. Quite down-to-earth technical parameters such as the choice and processing of materials are *socially* specified by the code. The illusion of technical necessity arises from the fact that the code is thus literally "cast in iron," at least in the case of boilers.[22]

Conservative antiregulatory social philosophies are based on this illusion. They forget that the design process always already incorporates standards of safety and environmental compatibility; similarly, all technologies support some basic level of user or worker initiative. A properly made technical object simply *must* meet these standards to be recognized as such. We do not treat conformity as an expensive add-on, but regard it as an intrinsic production cost. Raising the standards means altering the definition of the object, not paying a price for an alternative good or ideological value as the trade-off model holds.

But what of the much discussed cost–benefit ratio of design changes such as those mandated by environmental or other similar legislation? These calculations have some application to transitional situations, before technological advances responding to new values fundamentally alter the terms of the problem. But, all too often, the results depend on economists' very rough estimates of the monetary value of such things as a day of trout fishing or an asthma attack. If made without prejudice, these estimates may well help to prioritize policy alternatives. But one cannot legitimately generalize from such policy applications to a universal theory of the costs of regulation.

Such fetishism of efficiency ignores our ordinary understanding of the concept that alone is relevant to social decision making. In that everyday sense, efficiency concerns the narrow range of values that economic actors routinely affect by their decisions. Unproblematic aspects of technology are not included. In theory, one can decompose any technical object and account for each of its elements in terms of the goals it meets, whether it be safety, speed, reliability, and the like, but in practice no one is interested in opening the "black box" to see what is inside.

For example, once the boiler code is established, such things as the thickness of a wall or the design of a safety valve appear as essential to the object. The cost of these features is not broken out as the specific "price" of safety and compared unfavorably with a more efficient but less secure version of the technology. Violating the code in order to lower costs is a crime, not a trade-off. And since all further progress takes place on the basis of the new safety standard, soon no one looks back to the good old days of cheaper, insecure designs.

Design standards are only controversial while they are in flux. Resolved conflicts over technology are quickly forgotten. Their outcomes, a welter of taken-for-granted technical and legal standards, are embodied in a stable code and form the background against which economic actors manipulate the unstable portions of the environment in the pursuit of efficiency. The code is not varied in real world economic calculations but treated as a fixed input.

Anticipating the stabilization of a new code, one can often ignore contemporary arguments that will soon be silenced by the emergence of a new horizon of efficiency calculations. This is what happened with boiler design and child labor; presumably, the current debates on environmentalism will have a similar history, and we will someday mock those who object to cleaner air as a "false principle of humanity" that violates technological imperatives.

Noneconomic values intersect the economy in the technical code. The examples we are dealing with illustrate this point clearly. The legal standards that regulate workers' economic activity have a significant impact on every aspect of their lives. In the child labor case, regulation helped to widen educational opportunities with consequences that are not primarily economic in character. In the riverboat case, Americans gradually chose high levels of security, and boiler design came to reflect that choice. Ultimately, this was no trade-off of one good for another, but a noneconomic decision about the value of human life and the responsibilities of government.

Technology is thus not merely a means to an end; technical design standards define major portions of the social environment, such as urban and built spaces, workplaces, medical activities and expectations, life patterns, and so on. The economic significance of technical change often pales beside its wider human implications in framing a way of life. In such cases, regulation defines the cultural framework *of* the economy; it is not an act *in* the economy.

HEIDEGGER'S "ESSENCE" OF TECHNOLOGY

The theory sketched here suggests the possibility of a general reform of technology. But dystopian critics object that the mere fact of pursuing efficiency or technical effectiveness already does inadmissible violence to human beings and nature. Universal functionalization destroys the integrity of all that is. As Heidegger argues, an "objectless" world of mere resources replaces a world of "things" treated with respect for their own sake as the gathering places of our manifold engagements with "being."[23]

This critique gains force from the actual perils with which modern technology threatens the world today. But my suspicions are aroused by Heidegger's famous contrast between a dam on the Rhine and a Greek chalice. It would be difficult to find a more tendentious comparison. No doubt, modern technology is immensely more destructive than any other. And Heidegger is right to argue that means are not truly neutral, and that their substantive content affects society independent of the goals they serve. But I have argued here that this content is not *essentially* destructive; rather, it is a matter of design and social insertion.

However, Heidegger rejects any merely social diagnosis of the ills of technological societies and claims that the source of their problems dates back at least to Plato, that modern societies merely realize a *telos* immanent in Western metaphysics from the beginning. His originality consists in pointing out that the ambition to control being is itself a way of being and hence subordinate at some deeper level to an ontological dispensation beyond human control. But the overall effect of his critique is to condemn human agency, at least in modern times, and to confuse essential differences between types of technological development.

Heidegger distinguishes between the *ontological* problem of technology, which can only be addressed by achieving what he calls "a free relation" to technology, and the merely *ontic* solutions proposed by reformers who wish to change technology itself. This distinction may have seemed more interesting in years gone by than it does today. In effect, Heidegger is asking for nothing more than a change in attitude toward the selfsame technical world. But that is an idealistic solution in the bad sense, and one that a generation of environmental action would seem decisively to refute.

Confronted with this argument, Heidegger's defenders usually point out that his critique of technology is not merely concerned with human attitudes but also with the way being reveals itself. Roughly translated out of Heidegger's language, this means that the modern world has a technological form in something like the sense in which, for example, the medieval world had a religious form. *Form* is no mere question of attitude but takes on a material life of its own: power plants are the gothic cathedrals of our time. But this interpretation of Heidegger's thought raises the expectation that he will offer criteria for a reform of technology. For example, his analysis of the tendency of modern technology to accumulate and store up nature's powers suggests the superiority of another technology that would not challenge nature in Promethean fashion.

Unfortunately, Heidegger's argument is developed at such a high level of abstraction he literally cannot discriminate between electricity and atom bombs, agricultural techniques and the Holocaust. In a 1949 lecture, he asserted: "Agriculture is now the mechanized food industry, in essence the same as the manufacturing of corpses in gas chambers and extermination camps, the same as the blockade and starvation of nations, the same as the production of hydrogen bombs."[24] All are merely different expressions of the identical enframing that we are called to transcend through the recovery of a deeper relation to being. And since Heidegger rejects technical regression while leaving no room for a better technological future, it is difficult to see in what that relation would consist beyond a mere change of attitude.

HISTORY OR METAPHYSICS

Heidegger is perfectly aware that technical activity was not "metaphysical" in his sense until recently. He must therefore sharply distinguish modern technology from all earlier forms of technique, obscuring the many real connections and continuities. I would argue, on the contrary, that what is new about modern technology can only be understood against the background of the traditional technical world from which it developed. Furthermore, the saving potential of modern technology can only be realized by recapturing certain traditional features of technique. Perhaps this is why theories that treat modern technology as a unique phenomenon lead to such pessimistic conclusions.

Modern technology differs from earlier technical practices through significant shifts in emphasis rather than generically. There is nothing unprecedented in its chief features, such as the reduction of objects to raw materials, the use of precise measurement and plans, and the technical control of some human beings by others, large scales of operation. It is the centrality of these features that is new, and of course the consequences of that are truly without precedent.

What does a broader historical picture of technology show? The privileged dimensions of modern technology appear in a larger context that includes many currently subordinated features that were defining for it in former times. For example, until the generalization of Taylorism, technical life was essentially about the choice of a vocation. Technology was associated with a way of life, with specific forms of personal development, virtues, and so on. Only the success of capitalist deskilling finally reduced these human dimensions of technique to marginal phenomena.

Similarly, modern management has replaced the traditional collegiality of the guilds with new forms of technical control. Just as vocational investment in work continues in certain exceptional settings, so collegiality survives in a few professional or cooperative workplaces. Numerous historical studies show that these older forms are not so much incompatible with the *essence* of technology as with capitalist economics. Given a different social context and a different path of technical development, it might be possible to recover these traditional technical values and organizational forms in new ways in a future evolution of modern technological society.

Technology is an elaborate complex of related activities that crystallizes around tool making and using in every society. Matters such as the transmission of techniques or the management of its natural consequences are not extrinsic to technology per se but are dimensions of it. When, in modern societies, it becomes advantageous to minimize these aspects of technology, that too is a way of accommodating it to a certain social demand, not the revelation of its preexisting essence. In so far as it makes sense to talk about an essence of technology at all, it must embrace the whole field revealed by historical study, and not only a few traits ethnocentrically privileged by our society.

There is an interesting text in which Heidegger shows us a jug "gathering" the contexts in which it was created and functions. This image could be applied to technology as well, and in fact there is one brief passage in which Heidegger so interprets a highway bridge. Indeed, there is no reason why modern technology cannot also gather its multiple contexts, albeit with less romantic pathos than jugs and chalices. This is in fact one way of interpreting contemporary demands for such things as environmentally sound technology, applications of medical technology that respect human freedom and dignity, urban designs that create humane living spaces, production methods that protect workers' health and offer scope for their intelligence, and so on. What are these demands if not a call to reconstruct modern technology so that it gathers a wider range of contexts to itself rather than reducing its natural, human, and social environment to mere resources?

Heidegger would not take these alternatives very seriously because he reifies modern technology as something separate from society, as an inherently contextless force aiming at pure power. If this is the essence of technology, reform would be merely extrinsic. But at this point, Heidegger's position converges with the very Prometheanism he rejects. Both depend on the narrow definition of technology that, at least since Bacon and Descartes, has emphasized its destiny to control the world to the exclusion of its equally essential contextual embeddedness. I believe that this definition reflects the capitalist environment in which modern technology first developed.

The exemplary modern master of technology is the entrepreneur, single-mindedly focused on production and profit. The enterprise is a radically decontextualized platform for action, without the traditional responsibilities for persons and places that went with technical power in the past. It is the autonomy of the enterprise that makes it possible to distinguish so sharply between intended and unintended consequences, between goals and contextual effects, and to ignore the latter.

The narrow focus of modern technology meets the needs of a particular hegemony; it is not a metaphysical condition. Under that hegemony, technological design is unusually decontextualized and destructive. It is that hegemony that is called to account, not technology per se, when we point out that today technical means form an increasingly threatening life environment. It is that hegemony, as it has embodied itself in technology, that must be challenged in the struggle for technological reform.

DEMOCRATIC RATIONALIZATION

For generations, faith in progress was supported by two widely held beliefs: that technical necessity dictates the path of development, and that the pursuit of efficiency provides a basis for identifying that path. I have argued here that both these beliefs are false and that, furthermore, they are ideologies employed to justify restrictions on opportunities to participate in the institutions of industrial society. I conclude that we can achieve a new type of technological society that can support a broader range of values. Democracy is one of the chief values a redesigned industrialism could better serve.

What does it mean to democratize technology? The problem is not primarily one of legal rights but of initiative and participation. Legal forms may eventually routinize claims that are asserted informally at first, but the forms will remain hollow unless they emerge from the experience and needs of individuals resisting a specifically technological hegemony.

That resistance takes many forms, from union struggles over health and safety in nuclear power plants to community struggles over toxic waste disposal to political demands for regulation of reproductive technologies. These movements alert us to the need to take technological externalities into account and demand design changes responsive to the enlarged context revealed in that accounting.

Such technological controversies have become an inescapable feature of contemporary political life, laying out the parameters for official "technology assessment."[25] They prefigure the creation of a new public sphere embracing the technical background of social life, and a new style of rationalization that internalizes unaccounted costs born by "nature," in other words, something or somebody exploitable in the pursuit of profit. Here, respect for nature is not antagonistic to technology but enhances efficiency in broad terms.

As these controversies become commonplace, surprising new forms of resistance and new types of demands emerge alongside them. Networking has given rise to one among many such innovative public reactions to technology. Individuals who are incorporated into new types of technical networks have learned to resist through the net itself in order to influence the powers that control it. This is not a contest for wealth or administrative power, but a struggle to subvert the technical practices, procedures, and designs structuring everyday life.

The example of the *Minitel* can serve as a model of this new approach. In France, the computer was politicized as soon as the government attempted to introduce a highly rationalistic information system to the general public. Users "hacked" the network in which they were inserted and altered its functioning, introducing human communication on a vast scale where only the centralized distribution of information had been planned.

It is instructive to compare this case to the movements of AIDS patients.[26] Just as a rationalistic conception of the computer tends to occlude its communicative potentialities, so in medicine, caring functions have become mere side effects of treatment, which is itself understood in exclusively technical terms. Patients become objects of this technique, more or less "compliant" to management by physicians. The incorporation of thousands of incurably ill AIDS patients into this system destabilized it and exposed it to new challenges.

The key issue was access to experimental treatment. In effect, clinical research is one way in which a highly technologized medical system can care for those it cannot yet cure. But until quite recently, access to medical experiments has been severely restricted by paternalistic concern for patients' welfare. AIDS patients were able to open up access because the networks of contagion in which they were caught were paralleled by social networks that were already mobilized around gay rights at the time the disease was first diagnosed.

Instead of participating in medicine individually as objects of a technical practice, they challenged it collectively and politically. They "hacked" the medical system and turned it to new purposes. Their struggle represents a counter tendency to the technocratic organization of medicine, an attempt at a recovery of its symbolic dimension and caring functions.

As in the case of the *Minitel,* it is not obvious how to evaluate this challenge in terms of the customary concept of politics. Nor do these subtle struggles against the growth of silence in technological societies appear significant from the standpoint of the reactionary ideologies that contend noisily with capitalist modernism today. Yet the demand for communication these movements represent is so fundamental that it can serve as a touchstone for the adequacy of our concept of politics to the technological age.

These resistances, like the environmental movement, challenge the horizon of rationality under which technology is currently designed. Rationalization in our society responds to a particular definition of technology as a means to the goal of profit and power. A broader understanding of technology suggests a very different notion of rationalization based on responsibility for the human and natural contexts of technical action. I call this *democratic rationalization* because it requires technological advances that can only be made in opposition to the dominant hegemony. It represents an alternative to both the ongoing celebration of technocracy triumphant and the gloomy Heideggerian counterclaim that "only a God can save us" from techno-cultural disaster.[27]

Is democratic rationalization in this sense socialist? There is certainly room for discussion of the connection between this new technological agenda and the old idea of socialism. I believe there is significant continuity. In socialist theory, workers' lives and dignity stood for the larger contexts that modern technology ignores. The destruction of their minds and bodies on the workplace was viewed as a contingent consequence of capitalist technical design. The implication that socialist societies might design a very different technology under a different cultural horizon was perhaps given only lip service, but at least it was formulated as a goal.

We can make a similar argument today over a wider range of contexts in a broader variety of institutional settings with considerably more urgency. I am inclined to call such a position socialist and to hope that, in time, it can replace the image of socialism projected by the failed communist experiment.

More important than this terminological question is the substantive point I have been trying to make. Why has democracy not been extended to technically mediated domains of social life despite a century of struggles? Is it because technology excludes democracy, or because it has been used to suppress it? The weight of the argument supports the second conclusion. Technology can support more than one type of technological civilization, and may someday be incorporated into a more democratic society than ours.

NOTES

This chapter expands a presentation of my book *Critical Theory of Technology* (New York: Oxford University Press, 1991), delivered at the American Philosophical Association, December 28, 1991, and first published in an earlier version in *Inquiry* 35, nos. 3–4 (1992): 301–22.

1. See, for example, Joshua Cohen and Joel Rogers, *On Democracy: Toward a Transformation of American Society* (Harmondsworth, UK: Penguin, 1983); and Frank Cunningham, *Democratic Theory and Socialism* (Cambridge: Cambridge University Press, 1987).

2. Max Weber, *The Protestant Ethic and the Spirit of Capitalism,* trans. T. Parsons (New York: Scribners, 1958), 181–82.

3. Martin Heidegger, *The Question Concerning Technology,* trans. W. Lovitt (New York: Harper & Row, 1977); and Jacques Ellul, *The Technological Society,* trans. J. Wilkinson (New York: Vintage, 1964).

4. Richard W. Miller, *Analyzing Marx: Morality, Power and History* (Princeton, N.J.: Princeton University Press, 1984), 188–95.

5. Karl Marx, *Capital* (New York: Modern Library, 1906), 13.

6. See, for example, David Bloor, *Knowledge and Social Imagery* (Chicago: University of Chicago Press, 1991), 175–79. For a general presentation of constructivism, see Bruno Latour, *Science in Action* (Cambridge, Mass.: Harvard University Press, 1987).

7. Trevor Pinch and Wiebe Bijker, "The Social Construction of Facts and Artefacts: or How the Sociology of Science and the Sociology of Technology Might Benefit Each Other," *Social Studies of Science*, no. 14 (1984).

8. See Langdon Winner's blistering critique of the characteristic limitations of the position, entitled, "Upon Opening the Black Box and Finding It Empty: Social Constructivism and the Philosophy of Technology," in *The Technology of Discovery and the Discovery of Technology: Proceedings of the Sixth International Conference of the Society for Philosophy and Technology* (Blacksburg, Va.: The Society for Philosophy and Technology, 1991).

9. *Hansard's Debates, Third Series: Parliamentary Debates 1830–1891* 73 (February 22–April 22, 1844). The quoted passages are found between 1088 and 1123.

10. A useful starting point for the development of a hermeneutics of technology is offered by Paul Ricoeur in "The Model of the Text: Meaningful Action Considered as a Text," in P. Rabinow and W. Sullivan, eds., *Interpretive Social Science: A Reader* (Berkeley: University of California Press, 1979).

11. Michel de Certeau used the phrase "rhetorics of technology" to refer to the representations and practices that contextualize technologies and assign them a social meaning. De Certeau chose the term "rhetoric" because that meaning is not simply present at hand but communicates a content that can be articulated by studying the connotations that technology evokes. See the special issue of *Traverse*, no. 26 (October 1982), entitled *Les Rhetoriques de la Technologie,* and, in that issue, especially, Marc Guillaume's article, "Telespectres": 22–23.

12. See chapter 7, "From Information to Communication: The French Experience with Videotext," in Andrew Feenberg, *Alternative Modernity* (Berkeley: University of California Press, 1995).

13. Jean-François Lyotard, *La Condition Postmoderne* (Paris: Editions de Minuit, 1979), 34.

14. For an approach to social theory based on this notion (called, however, *doxa,* by the author), see Pierre Bourdieu, *Outline of a Theory of Practice,* trans. R. Nice (Cambridge: Cambridge University Press, 1977), 164–70.

15. Herbert Marcuse, "Industrialization and Capitalism in the Work of Max Weber," in *Negations,* trans. J. Shapiro (Boston: Beacon, 1968).

16. Harry Braverman, *Labor and Monopoly Capital* (New York: Monthly Review, 1974); and David Noble, *Forces of Production* (New York: Oxford University Press, 1984).

17. Bernard Gendron and Nancy Holstrom, "Marx, Machinery and Alienation," *Research in Philosophy and Technology* 2 (1979).

18. Foucault's most persuasive presentation of this view is *Surveiller et Punir* (Paris: Gallimard, 1975).

19. See, for example, Robert Heilbroner, *An Inquiry into the Human Prospect* (New York: W. W. Norton, 1975). For a review of these issues in some of their earliest formulations, see Andrew Feenberg, "Beyond the Politics of Survival," *Theory and Society*, no. 7 (1979).

20. This aspect of technology, called *concretization,* is explained in Gilbert Simondon, *Du Mode d'Existence des Objets Techniques* (Paris: Aubier, 1958), ch. 1.

21. John G. Burke, "Bursting Boilers and the Federal Power," in M. Kranzberg and W. Davenport, eds., *Technology and Culture* (New York: New American Library, 1972).

22. The technical code expresses the "standpoint" of the dominant social groups at the level of design and engineering. It is thus relative to a social position without, for that matter, being a mere ideology or psychological disposition. As I will argue in the last section of this chapter, struggle for sociotechnical change can emerge from the subordinated standpoints of those dominated within technological systems. For more on the concept of standpoint epistemology, see Sandra Harding, *Whose Science? Whose Knowledge?* (Ithaca, N.Y.: Cornell University Press, 1991).

23. The texts by Heidegger discussed here are, in order, "The Question Concerning Technology," "The Thing," and "Building Dwelling Thinking," all in *Poetry, Language, Thought,* trans. A. Hofstadter (New York: Harper & Row, 1971).

24. Quoted in T. Rockmore, *On Heidegger's Nazism and Philosophy* (Berkeley: University of California Press, 1992), 241.

25. Alberto Cambrosio and Camille Limoges, "Controversies as Governing Processes in Technology Assessment," in *Technology Analysis & Strategic Management* 3, no. 4 (1991).

26. For more on the problem of AIDS in this context, see Andrew Feenberg, "On Being a Human Subject: Interest and Obligation in the Experimental Treatment of Incurable Disease," *The Philosophical Forum* 23, no. 3 (spring 1992).

27. "Only a God Can Save Us Now," Martin Heidegger interviewed in *Der Spiegel*, translated by D. Schendler, *Graduate Philosophy Journal* 6, no. 1 (winter 1977).

Part III

TECHNOLOGY AND ETHICS

All technologies raise implicit ethical questions. Anything humans make and do is subject to ethical evaluation about appropriate uses, acceptable consequences, and right or wrong actions and intentions. The crucial ethical issue with technology is not whether we *can* make or use something but whether we *should* make or use something. Most often, traditional ethical theories provide an adequate framework for assessing ethical problems associated with technology. We can usually resolve questions easily in terms of either a utilitarian framework of weighing consequences with the aim of maximizing happiness, a deontological framework of rights and responsibilities, or a virtue ethics framework emphasizing good character development and good community membership. In such cases, it is only a matter of applying our traditional moral principles to the situations created by technologies. But often, new technological innovations test the limits of traditional moral principles. The situations created by novel technologies can raise moral questions that are so different that we need to develop new standards and new moral rules. Technologies now allow us to create and extend life in an unprecedented fashion. They equip us to gather, store, and manipulate information in ways that affect our privacy, freedom, and property rights. These technologies also radically alter the life prospects for entire species including *Homo sapiens*. Arguably, innovations in technology require similar innovations in our moral reasoning. The readings in this section examine the ways that technology transforms the very way we think about ethical questions.

In "Technology and Responsibility: Reflections on the New Tasks of Ethics," Hans Jonas argues that modern technology has changed the very nature of human action. Because ethics is concerned with action, Jonas goes on, it needs to change as well. Traditional ethical theories presume four characteristics of human action, none of which are relevant any longer: (1) action on nonhuman things—the whole realm of *techne*—is ethically neutral; (2) ethics is anthropocentric, concerned only with human relations; (3) human nature remains unchanged by *techne;* and (4) ethics is concerned with a limited time span and immediate circumstances. For traditional ethical theories, the domain of moral conduct includes only our contemporaries and extends only to a finite temporal and spatial horizon. The moral knowledge requisite for responsible action is limited in a corresponding fashion; we are not accountable for what we cannot know—in other words, a distant future and distant horizon. Yet, as Jonas notes, modern technology changes everything. Our new powers to affect the entire planet open up a new dimension of responsibility that was previously inconceivable. Accordingly, ethics must grow to encompass this new dimension.

Each of the four characteristics of human action no longer holds. *Techne* in the form of modern technology has made activity in nature ethically significant. Consequently, we need to move beyond anthropocentrism and respect nature as an end-in-itself. Human nature changes as we technologically transform ourselves and our environment; therefore politics has to change to include a longer, broader horizon of responsibility. Finally, our moral responsibilities now include obligations to consider and respect future generations, the environment, and the entire planet to ensure a world fit for habitation.

Unfortunately, Jonas notes, our technical capacities have outstripped our moral knowledge, leaving us responsible for technologies with unforeseeable consequences. In response to these challenges, Jonas proposes a revision of Kant's categorical imperative. This revision establishes an obligation to respect the continuation of humanity into the indefinite future. He concludes the article by applying this new moral imperative to technologies that transform human nature itself—for example, to technologies that prolong life and postpone death, behavior control through chemicals and implants, and genetic control over life. The task for the new ethics Jonas proposes is to balance our technological powers, limited knowledge, and broadened scope of responsibility.

In "Technology, Demography, and the Anachronism of Traditional Rights," Robert E. McGinn describes a problematic technology-related influence on society: the interplay of technology, individual rights, and increasing numbers. As increasing numbers of people exercise their right to own or use the optimal technologies available, the result is likely to diminish the quality of life for everyone. There are three components to the problematic pattern McGinn identifies. First is "technological maximality" (TM)—technology or technology-related phenomena embodying one or more aspects that are the greatest scale or highest degree previously attained. TM can be manifested in devices and systems or in aspects of their production, use, size, scope, and other "maximal" properties. Second is "traditional rights"—natural, universal, and inviolable entitlements owed to individuals. Third is the increasing number of rights-bearing citizens engaged in technologically maximalist practices. McGinn argues that the combination of these three factors often puts the quality of life of an entire society at risk. Examples of the "troubling triad" include the right to life-prolonging medical technologies (problematic related effect: drains resources), motorized vehicles (problematic related effect: pollutes the environment), and private property ownership (problematic related effect: development threatens the biosphere).

The basic conundrum is that we cannot reduce the number of people and thus diminish the burden on the environment, and limits to individual rights to TM are usually seen as illegal if not immoral. McGinn argues that we need a new *contextualized* theory of human rights based on needs that are vital to all human life. This new moral theory of rights is not absolute; it occasionally can be restricted depending on technological and demographic circumstances, as well as based on the effect on societal quality of life. A contextualized theory of rights attempts to balance the rights and needs of individuals as well as the welfare and happiness of society with the problematic realities of technological development and maximality in advanced industrialized nations. McGinn identifies six grounds for revising the absolute character of individual rights by making them somewhat context-dependant. He then tests these grounds in cases involving urban planning and medicine to show how a balance between rights and TM can be achieved.

In "The Constitution in Cyberspace," Laurence H. Tribe inquires whether the changed physical and temporal context of cyberspace requires changes in the Constitution itself. Are we in danger of losing core values like freedom, privacy, and equality because the notions of space and time in cyberspace are very different from what they were when the Constitution was written? Or can the core values of the Constitution be applied to new technologies that the Framers never could have even imagined? Tribe states his belief that the Constitution continues to be relevant regardless of how new technologies change our lives. He examines five basic assumptions underlying constitutional interpretation to show how they can be adapted to suit our current technological landscape.

The first axiom of the Constitution is that it regulates actions by the government, not actions undertaken by individuals and groups. New technologies do not change the function of the Constitution to limit the powers of the government and to protect private groups. The second axiom is that a person's mind, body, and property belong to that person, not to the public as a whole. The Constitution, however, only regulates some commercial activity. Most questions about new technologies—copying software, patent protection, and other issues of cyber-property—are political, not constitutional. The third axiom is that the government should remain neutral as to the value or content of information regardless of its physical or virtual status. The fourth axiom is that the Constitution is founded on normative principles that are not affected by developments in science and technology. Morality is concerned only with what *should be,* not with *what is.* The fifth axiom is that the Constitution's norms must be invariant despite technological transformations. At its core, the Constitution protects people, not places or things. Tribe concludes that the Constitution "must be read through technologically transparent lenses." New technologies may raise new moral, legal, and political challenges, but they do not change the core values of the Constitution.

In "Technological Ethics in a Different Voice," Diane P. Michelfelder considers Borgmann's proposed reform of technology from the perspective of care ethics. Like Borgmann, Michelfelder believes that technology can fulfill its promise of freedom and the good life but only if we use technology to live a life of engagement. Also like Borgmann, she believes that ordinary, everyday life (of work, play, family, and friendships) has moral significance. But unlike Borgmann, Michelfelder turns to feminist care ethics to make sense of our everyday, interpersonal relationships and, thus, technology. *Care ethics* emphasizes precisely what traditional, Enlightenment moral philosophy, with its emphasis on abstract, universalist, impartial rational procedures, overlooks: face-to-face encounters, personal life, compassion and love, and attention to unique situations. Care ethics, she claims, also is a better approach for evaluating the moral aspects of everyday life that Enlightenment morality in principle excludes, like the act of mothering, the maintenance of friendships, or the value of sharing a home-cooked meal together.

Michelfelder then uses the framework of care ethics to analyze our moral relationships with technologies to see if Borgmann's distinction between devices (that disengage us) and focal things (that engage us) is valid. She finds his distinction to be problematic and suggests that rather than classify technologies as either devices or things, we examine the ways in which people—in particular women—actually experience material objects. She agrees with the goal of technological reform but questions the usefulness of his distinction if we can use devices to relate more fully to one another and to the world, hence as focal things. Some devices, like telephones, may foster rather than threaten our engagement with each other. When devices are used in a context that builds relationships, they help deliver the promise of technology to make our lives better. Ultimately what is important, Michelfelder argues, is not whether an object is a device or focal thing, or if it is designed in a more democratic process, but the role a technology plays in making our everyday lives more meaningful.

15

Technology and Responsibility

Hans Jonas

All previous ethics—whether in the form of issuing direct enjoinders to do and not to do certain things, or in the form of defining principles for such enjoinders, or in the form of establishing the ground of obligation for obeying such principles—had these interconnected tacit premises in common: that the human condition, determined by the nature of man and the nature of things, was given once for all; that the human good on that basis was readily determinable; and that the range of human action and therefore responsibility was narrowly circumscribed. It will be the burden of my argument to show that these premises no longer hold, and to reflect on the meaning of this fact for our moral condition. More specifically, it will be my contention that with certain developments of our powers the *nature of human action* has changed, and since ethics is concerned with action, it should follow that the changed nature of human action calls for a change in ethics as well: this not merely in the sense that new objects of action have added to the case material on which received rules of conduct are to be applied, but in the more radical sense that the qualitatively novel nature of certain of our actions has opened up a whole new dimension of ethical relevance for which there is no precedent in the standards and canons of traditional ethics.

I

The novel powers I have in mind are, of course, those of modern *technology*. My first point, accordingly, is to ask how this technology affects the nature of our acting, in what ways it makes acting under its dominion *different* from what it has been through the ages. Since throughout those ages man was never without technology, the question involves the human difference of *modern* from previous technology. Let us start with an ancient voice on man's

powers and deed which in an archetypal sense itself strikes, as it were, a technological note—
the famous Chorus from Sophocles' *Antigone.*

> Many the wonders but nothing more wondrous than man.
> This thing crosses the sea in the winter's storm, making his path through the roaring waves. And
>   she, the greatest of gods, the Earth—deathless she is, and unwearied—he wears her away as
>   the ploughs go up and down from year to year and his mules turn up the soil.
> The tribes of the lighthearted birds he ensnares, and the races of all the wild beasts and the salty
>   brood of the sea, with the twisted mesh of his nets, he leads captive, this clever man.
> He controls with craft the beasts of the open air, who roam the hills. The horse with his shaggy
>   mane he holds and harnesses, yoked about the neck, and the strong bull of the mountain.
> Speech and thought like the wind and the feelings that make the town, he has taught himself,
>   and shelter against the cold, refuge from rain. Ever resourceful is he. He faces no future help-
>   less. Only against death shall he call for aid in vain. But from baffling maladies has he con-
>   trived escape.
> Clever beyond all dreams the inventive craft that he has which may drive him one time or another
>   to well or ill.
> When he honors the laws of the land the gods' sworn right high indeed in his city; but stateless
>   the man who dares to do what is shameful.

This awestruck homage to man's powers tells of his violent and violating irruption into the
cosmic order, the self-assertive invasion of nature's various domains by his restless cleverness;
but also of his building—through the self-taught powers of speech and thought and social
sentiment—the home for his very humanity, the artifact of the city. The raping of nature and
the civilizing of himself go hand in hand. Both are in defiance of the elements, the one by
venturing into them and overpowering their creatures, the other by securing an enclave against
them in the shelter of the city and its laws. Man is the maker of his life *qua* human, bending
circumstances to his will and needs, and except against death he is never helpless.

Yet there is a subdued and even anxious quality about this appraisal of the marvel that is
man, and nobody can mistake it for immodest bragging. With all his boundless resourceful-
ness, man is still small by the measure of the elements: precisely this makes his sallies into
them so daring and allows those elements to tolerate his forwardness. Making free with the
denizens of land and sea and air, he yet leaves the encompassing nature of those elements
unchanged, and their generative powers undiminished. Them he cannot harm by carving out
his little dominion from theirs. They last, while his schemes have their short lived way. Much
as he harries Earth, the greatest of gods, year after year with his plough—she is ageless and
unwearied; her enduring patience he must and can trust, and must conform. And just as ageless
is the sea. With all his netting of the salty brood, the spawning ocean is inexhaustible, nor is
it hurt by the plying of ships, nor sullied by what is jettisoned into its deeps. And no matter
how many illnesses he contrives to cure, mortality does not bow to cunning.

All this holds because man's inroads into nature, as seen by himself, were essentially
superficial, and powerless to upset its appointed balance. Nor is there a hint, in the *Antigone*
chorus or anywhere else, that this is only a beginning and that greater things of artifice and
power are yet to come—that man is embarked on an endless course of conquest. He had gone
thus far in reducing necessity, had learned by his wits to wrest that much from it for the
humanity of his life, and there he could stop. The room he had thus made was filled by the
city of men—meant to enclose and not to expand—and thereby a new balance was struck
within the larger balance of the whole. All the well or ill to which man's inventive craft may
drive him one time or another is inside the human enclave and does not touch the nature of things.

The immunity of the whole, untroubled in its depth by the importunities of man, that is, the essential immutability of Nature as the cosmic order, was indeed the backdrop to all of mortal man's enterprises, between the abiding and the changing: the abiding was Nature, the changing his own works. The greatest of these works was the city, and on it he could offer some measure of abidingness by the laws he made for it and undertook to honor. But no long-range certainty pertained to this contrived abidingness. As a precarious artifact, it can lapse or go astray. Not even within its artificial space, with all the freedom it gives to man's determination of self, can the arbitrary ever supersede the basic terms of his being. The very inconstancy of human fortunes assures the constancy of the human condition. Chance and luck and folly, the great equalizers in human affairs, act like an entropy of sorts and make all definite designs in the long run revert to the perennial norm. Cities rise and fall, rules come and go, families prosper and decline; no change is there to stay, and in the end, with all the temporary deflections balancing each other out, the state of man is as it always was. So here too, in his very own artifact, man's control is small and his abiding nature prevails.

Still, in this citadel of his own making, clearly set off from the rest of things and entrusted to him, was the whole and sole domain of man's responsible action. Nature was not an object of human responsibility—she taking care of herself and, with some coaxing and worrying, also of man: not ethics, only cleverness applied to her. But in the city, where men deal with men, cleverness must be wedded to morality, for this is the soul of its being. In this intra-human frame dwells all traditional ethics and matches the nature of action delimited by this frame.

II

Let us extract from the preceding those characteristics of human action which are relevant for a comparison with the state of things today.

1. All dealing with the non-human world, i.e., the whole realm of *techne* (with the exception of medicine), was ethically neutral—in respect to both the object and the subject of such action: in respect to the object, because it impinged but little on the self-sustaining nature of things and thus raised no question of permanent injury to the integrity of its object, the natural order as a whole; and in respect to the agent subject it was ethically neutral because *techne* as an activity conceived itself as a determinate tribute to necessity and not as an indefinite, self-validating advance to mankind's major goal, claiming in its pursuit man's ultimate effort and concern. The real vocation of man lay elsewhere. In brief, action on non-human things did not constitute a sphere of authentic ethical significance.
2. Ethical significance belonged to the direct dealing of man with man, including the dealing with himself: all traditional ethics is *anthropocentric.*
3. For action in this domain, the entity "man" and his basic condition was considered constant in essence and not itself an object of reshaping *techne.*
4. The good and evil about which action had to care lay close to the act, either in the praxis itself or in its immediate reach, and were not a matter for remote planning. This proximity of ends pertained to time as well as space. The effective range of action was small, the time-span of foresight, goal-setting and accountability was short, control of circumstances limited. Proper conduct had its immediate criteria and

almost immediate consummation. The long run of consequences beyond was left to change, fate or providence. Ethics accordingly was of the here and now, of occasions as they arise between men, of the recurrent, typical situations of private and public life. The good man was he who met these contingencies with virtue and wisdom, cultivating these powers in himself, and for the rest resigning himself to the unknown.

All enjoinders and maxims of traditional ethics, materially different as they may be, show this confinement to the immediate setting of the action. "Love thy neighbor as thyself"; "Do unto others as you would wish them to do unto you"; "Instruct your child in the way of truth"; "Strive for excellence by developing and actualizing the best potentialities of your being *qua* man"; "Subordinate your individual good to the common good"; "Never treat your fellow man as a means only but always *also* as an end in himself"—and so on. Note that in all those maxims the agent and the "other" of his action are sharers of a common present. It is those alive now and in some commerce with me that have a claim on my conduct as it affects them by deed or omission. The ethical universe is composed of contemporaries, and its horizon to the future is confined by the foreseeable span of their lives. Similarly confined is its horizon of place, within which the agent and the other meet as neighbor, friend or foe, as superior and subordinate, weaker and stronger, and in all the other roles in which humans interact with one another. To this proximate range of action all morality was geared.

III

It follows that the *knowledge* that is required—besides the moral will—to assure the morality of action, fitted these limited terms: it was not the knowledge of the scientist or the expert, but knowledge of a kind readily available to all men of good will. Kant went so far as to say that "human reason can, in matters of morality, be easily brought to a high degree of accuracy and completeness even in the most ordinary intelligence";[1] that "there is no need of science or philosophy for knowing what man has to do in order to be honest and good, and indeed to be wise and virtuous. . . . [Ordinary intelligence] can have as good hope of hitting the mark as any philosopher can promise himself";[2] and again: "I need no elaborate acuteness to find out what I have to do so that my willing be morally good. Inexperienced regarding the course of the world, unable to anticipate all the contingencies that happen in it," I can yet know how to out in accordance with the moral law.[3]

Not every thinker in ethics, it is true, went so far in discounting the cognitive side of moral action. But even when it received much greater emphasis, as in Aristotle, where the discernment of the situation and what is fitting for it makes considerable demands on experience and judgment, such knowledge has nothing to do with the science of things. It implies, of course, a general conception of the human good as such, a conception predicated on the presumed invariables of man's nature and condition, which may or may not find expression in a theory of its own. But its translation into practice requires a knowledge of the here and now, and this is entirely no-theoretical. This "knowledge" proper to virtue (of the "where, when, to whom, and how") stays with the immediate issue, in whose defined context the action *as the agent's own* takes its course and within which it terminates. The good or bad of the action is wholly decided within that short-term context. Its moral quality shines forth from it, visible to its witnesses. No one was held responsible for the unintended later effects of his well-intentioned, well-considered, and well-performed act. The short arm of human power did not call for a long arm of predictive knowledge; the shortness of the one is as little culpable as that of the other. Precisely because the human good, known in its generality, is the same for all time, its relation or violation takes place at each time, and its complete locus is always the present.

IV

All this has decisively changed. Modern technology has introduced actions, objects, and consequences of such novel scale that the framework of former ethics can no longer contain them. The *Antigone* chorus on the *deinotes,* the wondrous power, of man would have to read differently now; and its admonition to the individual to honor the laws of the land would no longer be enough. To be sure, the old prescriptions of the "neighbor" ethics—of justice, charity, honesty, and so on—still hold in their intimate immediacy of the nearest, day by day sphere of human interaction. But this sphere is overshadowed by a growing realm of collective action where doer, deed, and effect are no longer the same as they were in the proximate sphere, and which by the enormity of its powers forces upon ethics a new dimension of responsibility never dreamt of before.

Take, for instance, as the first major change in the inherited picture, the critical *vulnerability* of nature to man's technological intervention—unsuspected before it began to show itself in damage already done. This discovery, whose shock led to the concept and nascent science of ecology, alters the very concept of ourselves as a causal agency in the larger scheme of things. It brings to light, through the effects, that the nature of human action has *de facto* changed, and that an object of an entirely new order—no less than the whole biosphere of the planet—has been added to what we must be responsible for because of our power over it. And of what surpassing importance an object, dwarfing all previous objects of active man! Nature as a human responsibility is surely a *novum* to be pondered in ethical theory. What kind of obligation is operative in it? Is it more than a utilitarian concern? Is it just prudence that bids us not to kill the goose that lays the golden eggs, or saw off the branch on which we sit? But the "we" that here sits and may fall into the abyss is all future mankind, and the survival of the species is more than a prudential duty of its present members. Insofar as it is the fate of *man,* as affected by the condition of nature, which makes us care about the preservation of nature, such care admittedly still retains the anthropocentric focus of all classical ethics. Even so, the difference is great. The containment of nearness and contemporaneity is gone, swept away by the spatial spread and time-span of the cause-effect trains which technological practice sets afoot, even when undertaken for proximate ends. Their irreversibility conjoined to their aggregate magnitude injects another novel factor into the moral equation. To this take their cumulative character: their effects add themselves to one another, and the situation for later acting and being becomes increasingly different from what it was for the initial agent. The cumulative self-propagation of the technological change of the world thus constantly overtakes the conditions of its contributing acts and moves through none but unprecedented situations, for which the lessons of experience are powerless. And not even content with changing its beginning to the point of unrecognizability, the cumulation as such may consume the basis of the whole series, the very condition of itself. All this would have to be co-intended in the will of the single action if this is to be a morally responsible one. Ignorance no longer provides it with an alibi.

Knowledge, under these circumstances, becomes a prime duty beyond anything claimed for it heretofore, and the knowledge must be commensurate with the causal scale of our action. The fact that it cannot really be thus commensurate, i.e., that the predictive knowledge falls behind the technical knowledge which nourishes our power to act, itself assumes ethical importance. Recognition of ignorance becomes the obverse of the duty to know and thus part of the ethics which must govern the ever more necessary self-policing of our out-sized might. No previous ethics had to consider the global condition of human life and the far-off future, even existence, of the race. Their now being an issue demands, in brief, a new concept of duties and rights, for which previous ethics and metaphysics provide not even the principles, let alone a ready doctrine.

And what if the new kind of human action would mean that more than the interest of man alone is to be considered—that our duty extends further and the anthropocentric confinement of former ethics no longer holds? It is at least not senseless anymore to ask whether the condition of extra-human nature, the biosphere as a whole and in its parts, now subject to our power, has become a human trust and has something of a moral claim on us not only for our ulterior sake but for its own and in its own right. If this were the case it would require quite some rethinking in basic principles of ethics. It would mean to seek not only the human good, but also the good of things extra-human, that is, to extend the recognition of "ends in themselves" beyond the sphere of man and make the human good include the care for them. For such a role of stewardship no previous ethics has prepared us—and the dominant, scientific view of *Nature* even less. Indeed, the latter emphatically denies us all conceptual means to think of Nature as something to be honored, having reduced it to the indifference of necessity and accident, and divested it of any dignity of ends. But still, a silent plea for sparing its integrity seems to issue from the threatened plenitude of the living world. Should we heed this plea, should we grant its claim as sanctioned by the nature of things, or dismiss it as a mere sentiment on our part, which we may indulge as far as we wish and can afford to do? If the former, it would (if taken seriously in its theoretical implications) push the necessary rethinking beyond the doctrine of action, i.e., ethics, into the doctrine of being, i.e., metaphysics, in which all ethics must ultimately be grounded. On this speculative subject I will here say no more than that we should keep ourselves open to the thought that natural science may not tell the whole story about Nature.

V

Returning to strictly intra-human considerations, there is another ethical aspect to the growth of *techne* as a pursuit beyond the pragmatically limited terms of former times. Then, so we found, *techne* was a measured tribute to necessity, not the road to mankind's chosen goal—a means with a finite measure of adequacy to well-defined proximate ends. Now, *techne* in the form of modern technology has turned into an infinite forward-thrust of the race, its most significant enterprise, in whose permanent, self-transcending advance to ever greater things the vocation of man tends to be seen, and whose success of maximal control over things and himself appears as the consummation of his destiny. Thus the triumph of *homo faber* over his external object means also his triumph in the internal constitution of *homo sapiens,* of whom he used to be a subsidiary part. In other words, technology, apart from its objective works, assumes ethical significance by the central place it now occupies in human purpose. Its cumulative creation, the expanding artificial environment, continuously reinforces the particular powers in man that created it, by compelling their unceasing inventive employment in its management and further advance, and by rewarding them with additional success—which only adds to the relentless claim. This positive feedback of functional necessity and reward—in whose dynamics pride of achievement must not be forgotten—assures the growing ascendancy of one side of man's nature over all the others, and inevitably at their expense. If nothing succeeds like success, nothing also entraps like success. Outshining in prestige and starving in resources whatever else belongs to the fullness of man, the expansion of his power is accompanied by a contraction of his self-conception and being. In the image he entertains of himself—the potent self-formula which determines his actual being as much as it reflects it—man now is evermore the maker of what he has made and the doer of what he can do, and most of all the preparer of what he will be able to do next. But not you or I: it is the aggregate, not the individual doer or deed that matters here; and the indefinite future, rather than the contemporary context of the action, constitutes the relevant horizon of responsibility. This requires imperatives of a new sort. If the realm of making has invaded the space of essential action, then morality must invade the realm of making, from which it had formerly stayed aloof, and must do so in the form of public policy. With issues of such inclusiveness and such lengths of anticipation public policy has never had to deal before. In fact, the changed nature of human action changes the very nature of politics.

For the boundary between "city" and "nature" has been obliterated: the city of men, once an enclave in the non-human world, spreads over the whole of terrestrial nature and usurps its place. The difference between the artificial and the natural has vanished, the natural is swallowed up in the sphere of the artificial, and at the same time the total artifact, the works of man working on and through himself, generates a "nature" of its own, i.e., a necessity with which human freedom has to cope in an entirely new sense. Once it could be said *Fiat justitia, pereat mundus,* "Let justice be done, and may the world perish"—where "world," of course, meant the renewable enclave in the imperishable whole. Not even rhetorically can the like be said anymore when the perishing of the whole through the doings of man—be they just or unjust—has become a real possibility. Issues never legislated on come into the purview of the laws which the total city must give itself so that there will be a world for the generations of man to come.

That there *ought* to be through all future time such a world fit for human habitation, and that it ought in all future time to be inhabited by a mankind worthy of the human name, will be readily affirmed as a general axiom or a persuasive desirability of speculative imagination (as persuasive and undemonstrable as the proposition that there being a world at all is "better" than there being none): but as a *moral* proposition, namely, a practical *obligation* toward the posterity of a distant future, and a principle of decision in present action, it is quite different from the imperatives of the previous ethics of contemporaneity; and it has entered the moral scene only with our novel powers and range of prescience.

The *presence of man in the world* had been a first and unquestionable given, from which all idea of obligation in human conduct started out. Now it has itself become an *object* of obligation—the obligation namely to ensure the very premise of all obligation, i.e., the *foot-hold* for a moral universe in the physical world—the existence of mere *candidates* for a moral order. The difference this makes for ethics may be illustrated in one example.

VI

Kant's categorical imperative said: "Act so that you *can* will that the maxim of our action be made the principle of a universal law." The "can" here invoked is that of reason and its consistency with itself: *Given* the existence of a community of human agents (acting rational beings), the action must be such that it can without self-contradiction be imagined as a general practice of that community. Mark that the basic reflection of morals here is not itself a moral but a logical one: The "I *can* will" or "I *cannot* will" expresses logical compatibility or incompatibility, not moral approbation or revulsion. But there is no self-contradiction in the thought that humanity would once come to an end, therefore also none in the thought that the happiness of present and proximate generations would be bought with the unhappiness or even non-existence of later ones—as little as, after all, in the inverse thought that the existence or happiness of later generations would be bought with the unhappiness or even partial extinction of present ones. The sacrifice of the future for the present is *logically* no more open to attack than the sacrifice of the present for the future. The difference is only that in the one case the series goes on, and in the other it does not. But that it *ought to go on,* regardless of the distribution of happiness or unhappiness, even with a persistent preponderance of unhappiness over happiness, nay, even of immorality over morality[4]—this cannot be derived from the rule of self-consistency *within* the series, long or short as it happens to be: it is a commandment of a very different kind, lying outside and "prior" to the series as a whole, and its ultimate grounding can only be metaphysical.

An imperative responding to the new type of human action and addressed to the new type of agency that operates it might run thus: "Act so that the effects of your action are compatible with the permanence of genuine human life"; or expressed negatively: "Act so that the effects of your action are not destructive of the future possibility of such life"; or simply: "Do not compromise the conditions for an indefinite continuation of humanity on earth"; or most generally: "In your present choices, include the future wholeness of Man among the objects of your will."

It is immediately obvious that no rational contradiction is involved in the violation of this kind of imperative. I *can* will the present good with sacrifice of the future good. It is also evident that the new imperative addresses itself to public policy rather than private conduct, which is not in the causal dimension to which that imperative applies. Kant's categorical imperative was addressed to the individual, and its criterion was instantaneous. It enjoined each of us to consider what would happen *if* the *maxim* of my present action were made, or at this moment already were, the principle of a universal legislation; the self-consistency or inconsistency of such a *hypothetical* universalization is made the test for my *private* choice. But it was no part of the reasoning that there is any probability of my private choice *in fact* becoming universal law, or that it might contribute to its becoming that. The universalization is a thought-experiment by the private agent not to test the immanent morality of his action. Indeed, real consequences are not considered at all, and the principle is one not of objective responsibility but of the subjective quality of my self-determination. The new imperative invokes a different consistency: not that of the act with itself, but that of its eventual *effects* with the continuance of human agency in times to come. And the "universalization" it contemplates is by no means hypothetical—i.e., a purely logical transference from the individual "me" to an imaginary, causally unrelated "all" ("*if* everybody acted like that"); on the contrary, the actions subject to the new imperative—actions of the collective whole—have their universal reference in their actual scope of efficacy: they "totalize" themselves in the progress of their momentum and thus are bound to terminate in shaping the universal dispensation of things. This adds a *time* horizon to the moral calculus which is entirely absent from the instantaneous logical operation of the Kantian imperative: whereas the latter extrapolates into an ever-present order of abstract compatibility, our imperative extrapolates into a predictable real *future* as the open-ended dimension of our responsibility.

VII

Similar comparisons could be made with all the other historical forms of the ethics of contemporaneity and immediacy. The new order of human action requires a commensurate ethics of foresight and responsibility, which is as new as are the issues with which it has to deal. We have seen that these are the issues posed by the works of *homo faber* in the age of technology. But among those novel works we haven't mentioned yet the potentially most ominous class. We have considered *techne* only as applied to the non-human realm. But man himself has been added to the objects of technology. *Homo faber* is turning upon himself and gets ready to make over the maker of all the rest. This consummation of his power, which may well portend the overpowering of man, this final imposition of art on nature, calls upon the utter resources of ethical thought, which never before has been faced with elective alternatives to what were considered the definite terms of the human condition.

a. Take, for instance, the most basic of these "givens," man's mortality. Who ever before had to make up his mind on its desirable and *eligible* measure? There was nothing to choose about the upper limit, the "three score years and ten, or by reason of strength fourscore." Its inexorable rule was the subject of lament, submission, or vain (not to say foolish) wish-dreams about possible exceptions—strangely enough, almost never of affirmation. The intellectual imagination of a George Bernard Shaw and a Jonathan Swift speculated on the privilege of not having to die, or the curse of not being able to die. (Swift with the latter was the more perspicacious of the two.) Myth and legend toyed with such themes against the acknowledged background of the unalterable, which made the earnest man rather pray "teach us to number our days that we may get a heart of wisdom" (Psalm 90). Nothing of this was in the realm of doing, and effective decision. The question was only how to relate to the stubborn fact.

But lately, the dark cloud of inevitability seems to lift. A practical hope is held out by certain advances in cell biology to prolong, perhaps indefinitely extend the span of life by counteracting biochemical processes of aging. Death no longer appears as a necessity belonging to the nature of life, but as an avoidable, at least in principle tractable and long-delayable, organic malfunction. A perennial yearning of mortal man seems to come nearer fulfillment, and for the first time we have in earnest to ask the question "How desirable is this? How desirable for the individual, and how for the species?" These questions involve the very meaning of our finitude, the attitude toward death, and the general biological significance of the balance of death and procreation. Even prior to such ultimate questions are the more pragmatic ones of who should be eligible for the boon: persons of particular quality and merit? of social eminence? those that can pay for it? everybody? The last would seem the only just course. But it would have to be paid for at the opposite end, at the source. For clearly, on a population-wide scale, the price of extended age must be a proportional slowing of replacement, i.e., a diminished access of new life. The result would be a decreasing proportion of youth in an increasingly aged population. How good or bad would that be for the general condition of man? Would the species gain or lose? And how *right* would it be to preempt the place of youth? Having to die is bound up with having been born: mortality is but the other side of the perennial spring of "a natality" (to use Hannah Arendt's term). This had always been ordained; now its meaning has to be pondered in the sphere of decision.

To take the extreme (not that it will ever be obtained): if we abolish death, we must abolish procreation as well, for the latter is life's answer to the former, and so we would have a world of old age with no youth, and of known individuals with no surprises of such that had never been before. But this perhaps is precisely the wisdom in the harsh dispensation of our mortality: that it grants us the eternally renewed promise of the freshness, immediacy and eagerness of youth, together with the supply of otherness as such. There is no substitute for this in the greater accumulation of prolonged experience: it can never recapture the unique privilege of seeing the world for the first time and with new eyes, never relive the wonder which, according to Plato, is the beginning of philosophy, never the curiosity of the child, which rarely enough lives on as thirst for knowledge in the adult, until it wanes there too. This ever renewed beginning, which is only to be had at the price of ever repeated ending, may well be mankind's hope, its safeguard against lapsing into boredom and routine, its chance of retaining the spontaneity of life. Also, the role of the *memento mori* in the individual's life must be considered, and what its attenuation to indefiniteness may do to it. Perhaps a non-negotiable limit to our expected time is necessary for each of us as the incentive to number our days and make them count.

So it could be that what by intent is a philanthropic gift of science to man, the partial granting of his oldest wish—to escape the curse of mortality—turns out to be to the detriment of man. I am not indulging in prediction and, in spite of my noticeable bias, not even in valuation. My point is that already the promised gift raises questions that had never to be asked before in terms of practical choice, and that no principle of former ethics, which took the human constants for granted, is competent to deal with them. And yet they must be dealt with ethically and by principle and not merely by the pressure of interest.

b. It is similar with all the other, quasi-utopian powers about to be made available by the advances of biomedical science as they are translated into technology. Of these, *behavior control* is much nearer to practical readiness than the still hypothetical prospect I have just been discussing, and the ethical questions it raises are less profound but have a more direct bearing on the moral conception of man. Here again, the new kind of intervention exceeds the old ethical categories. They have not equipped us to rule, for example, on mental control by chemical means or by direct electrical action of the brain via implanted electrodes—undertaken, let us assume, for defensible and even laudable ends. The mixture of beneficial and dangerous potentials is obvious, but the lines are not easy to draw. Relief of mental patients from distressing and disabling symptoms seems unequivocally beneficial. But from the relief of the *patient,* a goal entirely in the tradition of the medical art, there is an easy passage to the relief of *society* from the inconvenience of difficult individual behavior among its members: that is, the passage from medical to social application; and this opens up an indefinite field with grave potentials. The troublesome problems of rule and unruliness in modern mass society make the extension of such control methods to non-medical categories extremely tempting for social management. Numerous questions of human rights and dignity arise. The difficult question of preemption care versus enabling care insists on concrete answers. Shall we induce learning attitudes in school children by the mass administration of drugs, circumventing the appeal to autonomous motivation? Shall we overcome aggression by electronic pacification of brain areas? Shall we generate sensations of happiness or pleasure or at least contentment through independent stimulation (or tranquilizing) of the appropriate centers—independent, that is, of the objects of happiness, pleasure, or content and their attainment in personal living and achieving? Candidacies could be multiplied. Business firms might become interested in some of these techniques for performance-increase among their employees.

Regardless of the question of compulsion or consent, and regardless also of the question of undesirable side-effects, each time we thus bypass the human way of dealing with human problems, short-circuiting it by an impersonal mechanism, we have taken away something from the dignity of personal self hood and advanced a further step on the road from responsible subjects to programmed behavior systems. Social functionalism, important as it is, is only one side of the question. Decisive is the question of what kind of individuals the society is composed of to make its existence valuable as a whole. Somewhere along the line of increasing social manageability at the price of individual autonomy, the question of the worthwhileness of the human enterprise must pose itself. Answering it involves the image of man we entertain. We must think its anew in light of the things we can do to it now and could never do before.

c. This holds even more with respect to the last object of a technology applied on man himself—the genetic control of future men. This is too wide a subject for cursory treatment. Here I merely point to this most ambitious dream of *homo faber,* summed up in the phrase that man will take his own evolution in hand, with the aim of not just preserving the integrity of the species but of modifying it by improvements of his own design. Whether we have the right to do it, whether we are qualified for that creative role, is the most serious question that can be posed to man finding himself suddenly in possession of such failed powers. Who will be the image-makers, by what standards, and on the basis of what knowledge? Also, the question of the moral right to experiment on future human beings must be asked. These and similar questions, which demand an answer before we embark on a journey into the unknown, show most vividly how far our powers to act are pushing us beyond the terms of all former ethics.

VIII

The ethically relevant common feature in all the examples adduced is what I like to call the inherently "utopian" drift of our actions under the conditions of modern technology, whether it works on non-human or on human nature, and whether the "utopia" at the end of the road be planned or unplanned. By the kind and size of its snowballing effects, technological power propels us into goals of a type that was formerly the preserve of Utopias. To put it differently, technological power has turned what used and ought to be tentative, perhaps enlightening, plays of speculative reason into competing blueprints for projects, and in choosing between them we have to choose between extremes of remote effects. The one thing we can really know of them is their extremism as such—that they concern the total condition of nature on our globe and the very kind of creatures that shall, or shall not, populate it. In consequence of the inevitably "utopian" scale of modern technology, the salutary gap between everyday and ultimate issues, between occasions is closing. Living now constantly in the shadow of unwanted, built-in, automatic utopianism, we are constantly confronted with issues whose positive choice requires supreme wisdom—an impossible, and in particular for contemporary man, who denies the very existence of its object: viz., objective value and truth. We need wisdom most when we believe in it least.

If the new nature of our acting then calls for a new ethics of long-range responsibility, coextensive with the range of our power, it calls in the name of that very responsibility also for a new kind of humility—a humility not like former humility, i.e., owing to the littleness, but owing to the excessive magnitude of our power, which is the excess of our power to act over our power to foresee and our power to evaluate and to judge. In the face of the quasi-eschatological potentials of our technological processes, ignorance of the ultimate implications becomes itself a reason for responsible restraint—as the second best to the possession of wisdom itself.

One other aspect of the required new ethics of responsibility for and to a distant future is worth mentioning: the insufficiency of representative government to meet the new demands on its normal principles and by its normal mechanics. For according to these, only *present* interests make themselves heard and felt and enforce their condition. It is to them that public agencies are accountable, and this is the way in which concretely the respecting of rights comes about (as distinct from their abstract acknowledgement). But the *future* is not represented, it is not a force that can throw its weight into the scales. The non-existent has no lobby, and the unborn are powerless. Thus accountability to them has no political reality behind it yet in present decision-making, and when they can make their complaint, then we, the culprits, will not longer be there.

This raises to an ultimate pitch the old question of the power of the wise, or the force of ideas not allied to self-interest, in the body politic. What *force* shall represent the future in the present? However, before *this* question can become earnest in practical terms, the new ethics must find its theory, on which dos and don'ts can be based. That is: before the question of what *force,* comes the question of what *insight* or value-knowledge shall represent the future in the present.

IX

And here is where I get stuck, and where we all get stuck. For the very same movement which put us in possession of the powers that have now to be regulated by norms—the movement of modern knowledge called science—has by a necessary complementarity eroded the foundations from which norms could be derived; it has destroyed the very idea of norm as such. Not, fortunately, the feeling for norm and even for particular norms. But this feeling, becomes uncertain of itself when contradicted by alleged knowledge or at least denied all sanction by it. Anyway and always does it have a difficult enough time against the loud clamors of greed and fear. Now it must in addition blush before the frown of superior knowledge, as unfounded and incapable of foundation. First, Nature has been "neutralized" with respect to value, then man himself. Now we shiver in the nakedness of a nihilism in which near-omnipotence is paired with near-emptiness, greatest capacity with knowing least what for. With the apocalyptic pregnancy of our actions, that very knowledge which we lack has become more urgently needed than at any other stage in the adventure of mankind. Alas, urgency is no promise of success. On the contrary, it must be avowed that to seek for wisdom today requires a good measure of unwisdom. The very nature of the age which cries out for an ethical theory makes it suspiciously look like a fool's errand. Yet we have no choice in the matter but to try.

It is a question whether without restoring the category of the sacred, the category most thoroughly destroyed by the scientific enlightenment, we can have an ethics able to cope with the extreme powers which we possess today and constantly increase and are almost compelled to use. Regarding those consequences imminent enough still to hit ourselves, fear can do the job—so often the best substitute for genuine virtue or wisdom. But this means fails us towards the more distant prospects, which here matter the most, especially as the beginnings seem mostly innocent in their smallness. Only awe of the sacred with its unqualified veto is independent to fit computations of mundane fear and the solace of uncertainty about distant consequences. But religion as a soul-determining force is no longer there to be summoned to the aid of ethics. The latter must stand on its worldly feet—that is, on reason and its fitness for philosophy. And while of faith it can be said that it either is there or is not, of ethics it holds that it must be there.

It must be there because men act, and ethics is for the reordering of actions and for regulating the power to act. It must be there all the more, then, the greater the powers of acting that are to be regulated; and with their size, the ordering principle must also fit their kind. Thus, novel powers to act require novel ethical rules and perhaps even a new ethics.

"Thou shalt not kill" was enunciated because man has the power to kill and often the occasion and even inclination for it—in short, because killing is actually done. It is only under the *pressure* of real habits of action, and generally of the fact that always action already takes place, without *this* having to be commanded first, that ethics as the ruling of such acting under the standard of the good or the permitted enters the stage. Such a *pressure* emanates from the novel technological powers of man, whose exercise is given with their existence. *If* they really are as novel in kind as here contended, and if by the kind of their potential consequences they really have abolished the moral neutrality which the technical commerce with matter hitherto enjoyed—then their pressure bids to seek for new prescriptions in ethics which are competent to assume their guidance, but which first of all can hold their own theoretically against that very pressure. To the demonstration of those premises this chapter was devoted. If they are accepted, then we who make thinking our business have a task to last us for our time. We must do it in time, for since we act anyway, we shall have some ethic or other in any case, and without a supreme effort to determine the right one, we may be left with a wrong one by default.

NOTES

1. Immanuel Kant, *Groundwork of the Metaphysics of Morals,* preface.
2. *Op. cit.,* chapter I.
3. *Ibid.* (I have followed H. J. Paton's translation with some changes.)
4. On this last point, the biblical God changed his mind to an all-encompassing "yes" after the Flood.

16

Technology, Demography, and the Anachronism of Traditional Rights

Robert E. McGinn

INTRODUCTION

Critics of the influence of technology on society debit the unhappy outcomes they decry to different causal accounts. Some target *specific characteristics or purposes* of technologies that they hold are inherently objectionable. For example, certain critics believe biotechnologies such as human *in vitro* fertilisation and the genetic engineering of transgenic animals to be morally wrong, regardless of who controls or uses them, and attribute what they see as the negative social consequences of these innovations to their defining characteristics or informing purposes. Others, eschewing the technological determinism implicit in such a viewpoint, find fault with the *social contexts* of technological developments and hold these contexts—more precisely, those who shape and control them—responsible for such unhappy social outcomes as result. For example, some critics have blamed the tragic medical consequences of silicone gel breast implants on a profit-driven rush to market these devices and on lax government regulation. Still other critics focus on *users* of technologies, pointing to problematic aspects of the use and operation of the technics and technical systems at their disposal. For example, some attributed the fatal crash of a DC-10 in Chicago in 1979 and the rash of reports in the mid 1980s of spontaneous acceleration of Audi automobiles upon braking to the alleged carelessness of maintenance workers and consumers.

Questions of the validity and relative value of these theory-laden approaches aside, this chapter identifies and analyses an important source of problematic technology-related influence on society of a quite different nature. Neither wholly technical, nor wholly social, nor wholly individual in nature, the source discussed below combines technical, social, and individual elements.

The source in question is a recurrent pattern of sociotechnical practice characteristic of

Originally published as "Technology, Demography, and the Anachronism of Traditional Rights," *Journal of Applied Philosophy,* Vol. 11, No. 1, 1994. Copyright © 1994 Society for Applied Philosophy, Blackwell Publishers, 108 Cowley Road, Oxford, OX4 1JF, UK and 3 Cambridge Center, Cambridge, MA 02142, USA.

contemporary Western societies. The pattern poses a challenge to professionals in fields as diverse as medicine, city planning, environmental management, and engineering. While not intrinsically problematic—indeed, the pattern sometimes yields beneficial consequences—the pattern is *potentially* problematic. Its manifestations frequently dilute or jeopardise the quality of life in societies in which they unfold. Unless appropriate changes are forthcoming, the pattern's effects promise to be even more destructive in the future. In what follows, I shall describe and clarify the general pattern, explore its sources of strength, elaborate a conceptual/theoretical change that will be necessary to bring the pattern under control and mitigate its negative effects, and survey some conflicts over recent efforts to do just that in two social arenas: urban planning and medicine.

THE PATTERN: NATURE AND MEANING

The pattern in question involves the interplay of technology, rights, and numbers. It may be characterised thus:

> 'technological maximality,' unfolding under the auspices of 'traditional rights' supposedly held and exercised by a large and increasing number of parties, is apt to dilute or diminish contemporary societal quality of life.

Let us begin by defining the three key expressions in this formulation.

First, in speaking of an item of technology or a technology-related phenomenon as exhibiting 'technological maximality'(TM), I mean the *quality of embodying in one or more of its aspects or dimensions the greatest scale or highest degree previously attained or currently possible in that aspect or dimension.*

Thus understood, TM can be manifested in various forms. Some hinge on the characteristics of technological products and systems, while others have to do with aspects of their production, diffusion, use, or operation. Making material artifacts (technics) and sociotechnical systems of hitherto unequalled or unsurpassed scale or performance might be viewed as paradigmatic forms of TM. However, the TM concept is also intended to encompass maximalist phenomena having to do with processes as well as products. Examples of technological maximality of process include producing or diffusing as many units as possible of a technic in a given time interval or domain, and using a technic or system as intensively or extensively as possible in a given domain or situation. It is important to recognise that technological maximality can obtain even where no large-scale or super-powerful technics or technical systems are involved. Technological maximality can be reflected in *how* humans interact with and use their technics and systems as much as in technic and system characteristics proper. TM, one might say, has adverbial as well as substantive modes. In sum, technology can be maximalist in one or more of the following nine senses:

- product size or scale
- product performance (power, speed, efficiency, scope, etc.)
- speed of production of a technic or system
- volume of production of a technic or system
- speed of diffusion of a technic or system
- domain of diffusion of a technic or system

- intensity of use or operation of a technic or system
- domain of use or operation of a technic or system
- duration of use or operation of a technic or system

Secondly, 'traditional rights' are entitlements of individuals as traditionally conceived in modern Western societies. For example, in the traditional Western conception individual rights have been viewed as timelessly valid and morally inviolable. Traditional individual rights often interpreted in this absolutist way include the right to life as well as liberty, property, and pro-creative rights.

Thirdly, the 'large and increasing number of parties' factor refers to the presence in most kinds of context in contemporary Western societies of many, indeed also a growing number of, parties—usually individual humans—each of whom supposedly holds rights of the above sort and may exercise them in, among other ways, technologically maximalist behaviour.

Before proceeding, I want to stress that this paper is neither a critique of 'technological maximality' per se nor a celebration of E. F. Schumacher's 'small is beautiful' idea. For, like the above triadic pattern, TM (or, for that matter, technological minimality) per se is not inherently morally objectionable or problematic. Specimens of technological maximality such as the then unprecedentedly large medieval Gothic cathedrals and the mammoth Saturn V rockets of the kind that took Apollo XI toward the moon suffice to refute any such claim. Rather, it is the *conjunction* of the three above-mentioned factors in repeated patterns of sociotechnical practice—*large and increasing numbers of parties engaged in technologically maximalist practices as something that each party supposedly has a morally inviolable right to do*—that is apt to put societal quality of life at risk. With this in mind, in what follows we shall refer to the combustible mixture of these three interrelated factors as 'the troubling triad.'

The triadic pattern is surprisingly widespread. Consider the following examples:

1. the intensive, often protracted use of life-prolonging technologies or technological procedures in thousands of cases of terminally ill or irreversibly comatose patients, or in the case of those needing an organ transplant or other life-sustaining treatment, such uses supposedly being called for by the inviolable right to life;
2. the proliferation of mopeds, all-terrain, snowmobile, and other kinds of versatile transport vehicles in special or fragile environmental areas, such use supposedly being sanctioned by rider mobility rights; and
3. the erection of growing numbers of high-rise buildings in city centres, as supposedly permitted by owner or developer property rights.

As suggested by these examples, our pattern of sociotechnical practice unfolds in diverse spheres of human activity. Problematic phenomena exemplifying the pattern in other arenas include the infestation of American national parks by tens of thousands of small tourist aircraft overflights per year; the depletion of ocean fishing areas through the use of hundreds of enormous, mechanically operated, nylon monofilament nets; and the decimation of old-growth forests in the northwestern U.S. through the use of myriad potent chain saws. The untoward effects exacted by the unfolding of our triadic pattern include steep financial and psychological tolls, the depletion and degradation of environmental resources, and the dilution and disappearance of urban amenities. In short, the costs of the ongoing operation of the triadic pattern are substantial and increasing.

To this point, the pattern identified above makes reference to a number of individual agents, each of whom engages in or is involved with a specimen of technologically maximal behaviour, e.g., having life-prolonging technologies applied intensively to herself or himself, using a technic 'extensively'—meaning either 'in a spatially widespread manner' or 'frequently'—in a fragile, limited, or distinctive domain; or erecting a megastructure.

However, as characterised above, our pattern obscures the fact that TM can be present in *aggregative* as well as non-aggregative situations. Each of a large number of individuals, acting under the auspices of a right construed in traditional fashion, can engage in behaviour that while not technologically maximal in itself becomes so when aggregated over all relevant agents. Of course, aggregating over a number of cases each of which is *already* technologically maximal compounds the maximality in question, and probably also its effects on society. We may say, therefore, that there is *individual* TM (where the individual behaviour in question is technologically maximal) and *aggregative* TM, the latter having two subspecies: *simple-aggregative,* where the individual behaviour is *not* technologically maximal, and *compound-aggregative,* where the individual behaviour is *already* technologically maximal.

One reason why simple-aggregative TM is troubling is that individual agents may have putative rights to engage in specimens of non-technologically-maximal behaviour that, taken individually, seem innocuous or of negligible import. However, contemporary environments or contexts do not automatically become larger or more robust in proportion to technic performance improvements, increasing costs of contemporary technics and systems, or the increasing number of those with access to or affected by these items. Hence, the aggregation of individually permissible behaviour over all relevant agents with access to technics can result in substantial harm to societal quality of life. One can therefore speak of 'public harms of aggregation.' For example, the failure of each of a large number of people to recycle their garbage is technological behaviour that when aggregated provides an instance of problematic technological maximality of use. Aggregating the individually innocuous effects of a large number of people driving motor vehicles that emit pollutants yields the same story: individually innocuous behaviour can, when aggregated over a large group, yield a significant, noxious outcome.

THE PATTERN: SOURCES OF STRENGTH

How is the power of the pattern under discussion to be accounted for? Put differently, why does the troubling triad come under so little critical scrutiny when it has such untoward effects on individual and societal quality of life? In the case of simple-aggregative TM, the reason is that the effects of the behaviour of the individual agent are negligibly problematic. It is difficult to induce a person to restrict her or his behaviour when it is not perceivably linkable to the doing of significant harm to some recognised protectable individual or societal interest.

More generally, the strength of the pattern derives from the effects of factors of various sorts that lend impetus to its constituent elements. Let us examine each element in the pattern in turn.

Technological Maximality

The modern drive to achieve increases in efficiency and economies of scale and thereby reap enhanced profits is unquestionably an important factor that fuels various modes of technological maximality. One thinks in this connection of maximalist technics such as the Boeing 747 and the Alaska pipeline as well as the diffusion speed and scope modes of TM for personal technics like the VCR and CD player.

However, economic considerations do not tell the entire causal story. Cultural phenomena also play an important role and help explain the low level of resistance to our pattern. Technological maximality is encouraged by the 'technological fix' mentality deeply entrenched in Western countries. Should anything go awry as a result of some technological maximalist practice, one can always, it is assumed, concoct a technological fix to remedy or at least patch up the situation in time. Moreover, there is much individual and group prestige to be garnered in modern Western societies by producing, possessing, or using the biggest, fastest, or more potent technic or technological project; more generally, by being, technologically speaking, 'the-firstest-with-the-mostest.' Further, influential sectors of Western opinion gauge societal progress and even a society's level of civilisation by the degree to which it attains and practises certain forms of technological maximality. Small-scale, appropriate technology may be fine for developing countries but resorting to it would be seen as culturally retrogressive for a technologically 'advanced' society such as the U.S.

Technological maximality is often associated with construction projects. In 1985, American developer Donald Trump announced what proved to be abortive plans for a 150-storey, 1,800-foot-tall Television City on the West Side of Manhattan, a megastructure that he revealingly called 'the world's greatest building.'[1] The demise in contemporary Western society of shared qualitative standards for making comparative value judgments has created a vacuum often filled by primarily quantitative standards of value. This, in turn, has fuelled technological maximality as a route to invidious distinction. If a building is quantitatively 'the greatest,' it must surely be qualitatively 'the best,' a convenient confusion of quality and quantity.

TM in the sense of virtually unrestricted technic use throughout special environments is greatly encouraged by modern Western cultural attitudes toward nature. Unlike in many traditional societies, land and space are typically perceived as homogeneous in character. No domains of land or space are sacred areas, hence possibly off limits to certain technological activity. On the contrary, in the contemporary U.S., nature is often regarded more as a playground for technology-intensive human activity. Dune buggy riders were incensed when environmentalists sued to force the National Park Service to ban off-road vehicles from the fragile dunes at Cape Cod National Seashore. The leader of the Massachusetts Beach Buggy Association lamented that 'it seems like every year they come up with more ways to deprive people of recreational activities,'[2] a comment that comes close to suggesting that rider rights have been violated.

The U.S. has no monopoly on TM. For example, France has a long tradition of technological maximality. The country's fascination with 'grands travaux,' large-scale technical projects conceived by politicians, public engineers, or civil servants, is several centuries old.[3] Encompassing classic projects such as Napoleon's Arc de Triomphe, Hausmann's transformation of central Paris, and Eiffel's Tower, the maximalist trend has also been manifested in the nation-wide SNCF electric railroad system and the Anglo-French Concorde supersonic transport airplane. More recent specimens suggesting that TM is alive and well in France include audacious undertakings such as the Channel Tunnel, ever more potent nuclear power stations, the T.G.V. *(très grand vitesse)* train, and the Mitterand government's plan for building the world's largest library, dubbed by critics the 'T.G.B.' *(très grand bibliothèque).*[4] Such projects are not pursued solely or primarily for economic motives but for reasons of national prestige and grandeur, certification of governmental power and competence, as symbols of cultural superiority, and as monuments to individual politicians.

Another cultural factor that fosters certain modes of TM is the relatively democratic consumer culture established in the U.S. and other Western countries in the twentieth century. For the American people, innovative technics should not be reserved for the competitive advantage and enjoyment of the privileged few. Rather, based on experience with technics such as the automobile, the phone, and the television, it is believed and expected that such items should and will become available to the great mass of the American people. This expectation, cultivated by corporate advertising in order to ensure sufficient demand for what industry has the capacity to produce, in turn greatly facilitates technological maximality of production and diffusion.

Traditional Rights

Many modern Western societies are founded on belief in what were once called the 'rights of man,' a term that succeeded the earlier phrase 'natural rights.' Building on Locke's thought about natural rights, the Bill of Rights enacted by the British Parliament in 1689 provided for rights to, among other things, life, liberty, and property. The U.S. Declaration of Independence of 1776 declared that 'all men . . . are endowed by their Creator with certain inalienable rights; that among these are life, liberty, and the pursuit of happiness.' The French 'Declaration des droits de I'homme et du citoyen' of 1789 asserts that 'the purpose of all political association is the conservation of the natural and inalienable rights of man: these rights are liberty, property, security and resistance to oppression.' In the 1940s Eleanor Roosevelt promoted use of the current expression 'human rights' when she determined through her work in the United Nations 'that the rights of men were not understood in some parts of the world to include the rights of women.'[5] Although later articles of the 1948 U.N. Universal Declaration of Human Rights make reference to novel 'economic and social rights' that are more clearly reflections of a particular stage of societal development, the document's Preamble refers to inalienable human rights and its early articles are couched in the language of 'the old natural rights tradition.'[6] Thus, in the dominant modern Western conception, individual rights are immutable, morally inviolable, and, for many, God-given.

What has this development to do with technological maximality? Things are declared as rights in a society under particular historical circumstances. When the declaration that something is a fundamental right in a society is supported by that society's dominant political-economic forces, it is safe to assume that recognition of and respect for that right is congruent with and adaptive in relation to prevailing social conditions. However, given the millennial history of the perceived close relationship between morality and religion, to get citizens of a society to take a declared 'right of man' seriously, it has often seemed prudent to represent rights thus designated as having some kind of transcendental seal of approval: e.g., God's blessing, correspondence with the alleged inherent fabric of the universe, or reference to them in some putatively sacred document. The right in question is thereby imbued with an immutable character, as if, although originating in specific historical circumstances, the right was nevertheless timelessly valid. Such a conception of rights can support even technologically maximal exercises of particular rights of this sort.

However, the specific sets of social-historical circumstances that gave birth to such rights eventually changed, whereas, on the whole, the perceived nature of the rights in question has not. Continuing to affirm the same things as categorical rights can become dysfunctional under new, downstream social-historical conditions; in particular, when the technics and systems to which citizens and society have access have changed radically. Endowing traditional rights with a quasi-sacred status to elicit respect for them has made it more difficult to delimit or retire them as rights further down the historical road, e.g., in the present context of rampant technological maximality. In essence, the cultural strategy used to legitimate traditional rights has bestowed on them considerable intellectual inertia, something which has proved difficult to alter even though technological and demographic changes have radically transformed the context in which those rights are exercised and take effect.

A recent example of how continued affirmation of traditional rights in the context of unprecedented technological maximality can impede or disrupt societal functioning is that of the automated telephone dialer. These devices can systematically call and leave prerecorded messages at every number in a telephone exchange, including listed and unlisted numbers, cellular telephones, pagers, corporate switchboards, and unattended answering machines. By one estimate, at least 20,000 such machines are likely to be at work each day dialing some 20 million numbers around the United States. As a consequence of the potency and number of autodialers, significant communications breakdowns have already occurred.[7]

When Oregon legislators banned the commercial use of autodialers, two small-business owners who used the devices in telemarketing brought suit to invalidate the legislation on the grounds that, among other things, it violated their right to free speech. One issue here is whether U.S. society should leave its traditional robust right to free speech intact when threats to or violations of other important protectable interests, e.g., privacy, emergency preparedness, and efficient organisational operation, result from exercise of this right in revolutionary technological contexts such as those created by the intensive commercial use of autodialers and fax machines for 'junk calls' and 'junk mail.' Significantly, the American Civil Liberties Union, which supported the plaintiffs in the Oregon case, argued for preserving the traditional free speech right unabridged.

In November 1991 Congress passed the Telephone Consumer Protection Act that banned the use of autodialers for calling homes, except for emergency notification or if a party had explicitly agreed to receive such calls. However, the decision to ban turned on the annoying personal experiences of Congressional representatives and their constituents with unsolicited sales calls, not on any principled confrontation with the tension between traditional rights, technological maximality, and increasing numbers.[8] Not surprisingly, in 1993 a U.S. District Court Judge blocked enforcement of the law, ruling that it violated the constitutional right to free speech.[9]

Increasing Numbers

The positive attitude in the U.S. toward an increasing national population was adaptive in the early years of the Republic when more people were needed to settle the country and fuel economic growth. Today, even while evidencing concern over rapid population increases in less developed countries, the U.S. retains strongly pronatalist tax policies and evidences residues of the long-standing belief that when it comes to population 'more is better.' The 'land of unlimited opportunity' myth, belief that America has an unlimited capacity to absorb population increases without undermining its quality of life, and conviction that intergenerational fairness requires that just as America opened its doors widely to earlier generations of impoverished or persecuted peoples so also should it continue to do so today; these and other beliefs militate against taking the difficult steps that might decrease or further slow the rate of increase of the American population, hence of the number of rights claimants.

THE PATTERN AS SELF-REINFORCING

Not only do powerful cultural factors foster each of the three elements of the pattern, it is also self-reinforcing. For example, reproductive freedom, derived from the right of freedom or liberty, is sacrosanct in contemporary Western societies. This belief aids and abets the increasing numbers factor, something that in turn fuels technological maximality (e.g., in technic and system size and production and diffusion rates) to support the resultant growing population. Put differently, the increasing numbers factor intensifies the interaction of the elements of the troubling triad. Under such circumstances the latter can undergo a kind of chain reaction: increases in any of its elements tend to evoke increases in one or both of the other two, and so forth. The rights to life, liberty, property and the pursuit of happiness have traditionally been construed as 'negative rights,' i.e., as entitlements *not to be done to* in certain ways: not to be physically attacked, constrained, deprived of one's property, etc. But, in the context of new technologies, some such rights have also taken on a positive facet: entitlement of the individual *to be done to* in certain ways, e.g., to be provided with access to various kinds of life-sustaining medical technologies and to be provided with certain kinds of information in possession of another party. A positive-faceted right to life encourages further technological maximality in both development and use, something that in turn increases the number of rights holders.

TOWARD A CONTEXTUALISED THEORY OF HUMAN RIGHTS

A society that generates an ever more potent technological arsenal and, in the name of democratic consumerism, makes its elements available in ever larger numbers to a growing citizenry whose members believe they have inviolable rights to make, access, and use those items in individually or aggregatively technologically maximalist ways, risks and may even invite progressive impairment of its quality of life. Substantial changes will be necessary if this scenario is to be avoided, especially in the U.S.

What changes might help avoid this outcome?

Decrease, stabilise, or at least substantially cut the rate of increase in the number of rights holders.

To think that any such possibility could be achieved in the foreseeable future is utopian at this juncture in Western cultural history. In spite of projections about the environmental consequences of a doubled or trebled world population, no politician of standing has raised the question of population limitation as a desirable goal for the U.S. or any other Western society. For this possibility to be realisable, it would seem that certain traditional rights, viz., those relating to reproductive behaviour and mobility, would have to be significantly reined in, a most unlikely prospect.

Put a tighter leash on individual technologically maximal behaviour.

As with the previous possibility, this option too would seem to require abridging certain traditional exercise rights or changing the underlying, quasi-categorical traditional conception of individual rights to a more conditional one. Alternatively, if one could demonstrate that untrammeled operation of the pattern is producing effects that undermine various intangible individual or societal interests, this might furnish a reason for leash-tightening. However, for various reasons, such demonstrations, even if feasible, are rarely socially persuasive.[10]

This situation suggests that one thing that may be crucial to avoiding the above scenario is elaboration and diffusion of *a new theory of moral rights*. While detailed elaboration and defence of such a theory is not feasible here, an acceptable theory should at least include accounts of the basis, function, status, and grounds for limitation of individual rights. Brief remarks on these components follow.

Basis

Western intellectual development has reached a stage in which individual moral rights can be given a more empirical, naturalistic basis. It should be acknowledged that the epistemological plausibility of rights talk need not, indeed should not, depend upon untestable beliefs in the existence and largesse of a deity interested in protecting the vital interests of individual human beings by endowing them with inalienable rights. Human rights can be plausibly anchored in basic human needs, i.e., universal features of human 'wiring' that must be satisfied to an adequate degree if the individual is to survive or thrive.[11] The notion then would be that something qualifies as an individual human right if and only if its protection is vital to the fulfilment of one or more underlying basic human needs. This *bottom-up* approach has the virtue of making discourse about moral rights more empirically grounded than traditional top-down theological or metaphysical approaches.

Function

In the new theory I propose, moral rights have a mundane though important function: to serve as conceptual spotlights that focus attention on aspects of human life that are essential to individual survival or thrival. The reason why such searchlights are needed is that such aspects of human life are ever at risk of being neglected because of political or social inequalities, socially conditioned preoccupation with ephemera, or the tendency of human agents to overlook or discount the interests of parties outside of their respective immediate geographical and temporal circles.

Status

Joel Feinberg has distinguished three degrees of absoluteness for individual moral rights:[12]

1. a right can be absolute in the sense of 'bounded exceptionlessness,' i.e., binding without exception in a finite, bounded domain, as with, e.g., the right to freedom of speech;
2. a right can be absolute in the (higher) sense of an 'ideal directive,' i.e., always deserving of respectful, favourable consideration, even when, after all things have been considered, it is concluded that the right must regrettably be overridden, as with, e.g., the right to privacy, and
3. a right can be absolute in the (still higher) sense of 'unbounded exceptionlessness' and 'non-conflictability,' i.e., binding without exception in an unbounded domain and not intrinsically susceptible to conflict with itself or another right, in the way that, e.g., the right to free speech is conflictable, as exemplified in the hectoring of a speaker. The right not to be subjected to gratuitous torture is a plausible candidate for a right that is absolute in this third sense.

In the theory we propose, individual moral rights will not be absolute in the third, highest degree, only in the first or second degree, depending on technological and demographic circumstances and on the effects on societal quality of life of aggregated maximalist exercise of the right in question.

Grounds for Decreasing the Absoluteness of Individual Rights

There are at least six kinds of circumstantial grounds that may justify restriction or limitation of an individual moral right because of the bearing of its technologically maximal exercise on societal quality of life:

1. if the very existence of society is called into question by the exercise of a putative right, e.g., exercise of the right to self-defence by the acquisition of the capability of making and using weapons or other technologies of mass destruction;
2. if continued effective social functioning is threatened by the exercise of a right, e.g., the disruption of telecommunication by the operation of automatic phone dialers operated under the auspices of the right of free speech;
3. if some natural resource vital to society is threatened through the exercise of a right, e.g., the reduction of fishing areas or forests to non-sustainable conditions by technologically maximal harvesting practices;
4. if a seriously debilitating financial cost is imposed on society by the widespread or frequent exercise of a right, as with mushrooming public health care payments for private kidney dialysis treatment in the name of the right to life;
5. if some phenomenon of significant aesthetic, cultural, historical, or spiritual value to a people is jeopardised by the exercise of a right, e.g., the destruction of a recognised architectural landmark by affixing its façade to a newly built, incongruous, mega-structure under the aegis of a private property right; and
6. if some highly valued social amenity would be seriously damaged or eliminated through the exercise of a right. For example, between 1981 and 1989 convivial public space at the Federal Plaza in downtown Manhattan was effectively eliminated by the installation of an enormous sculpture (Richard Serra's 120 ft long by 14 ft high 'Tilted Arc'). The artist unsuccessfully sued the government attempting to halt removal of the work as a violation of his First Amendment right of free speech, while many of his supporters cited the right to free artistic expression.[13]

In the case of simple-aggregative TM, the only option to acquiescence is to demonstrate the significant harm done to a protectable societal interest by the aggregated act and attempt to effect an ethical revaluation of putatively harmless individual behaviour; in other words, to lower the threshold of individual wrongdoing to reflect the manifest wrong effected by aggregation. With such a revaluation, the individual would have no right to act as he or she once did because of the newly declared immorality of the individual act. This process may be underway vis-à-vis the individual's disposal of home refuse without separation for recyling.

In sum, we need a *contextualised theory of human rights.* An acceptable theory of rights in contemporary technological society must be able to take on board the implications of their exercise in a context in which a rapidly changing, potent technological arsenal is diffused throughout a populous, materialistic, democratic society. Use of such a technological arsenal by a large and growing number of rights holders has considerable potential for diluting or diminishing societal quality of life. Indeed, insistence on untrammeled, entitled use of potent or pervasive technics by a large number of individuals can be self-defeating, e.g., by yielding a state of social affairs incompatible with other social goals whose realisation the group also highly values.

At a deeper level, what is called into question here is the viability of modern Western individualism. Can, say, contemporary U.S. society afford to continue to promote technology-based individualism in the context of the diffusion and use of multiple potent technics by a large and ever growing population? Or is the traditional concept of individualism itself in need of revision or retirement? The ideology of individualism in all areas of life may have been a viable one in the early modern era, one with a less potent and diverse technological arsenal and a less populous society. But can contemporary Western societies have their ideological cake and eat it too? Can individualism continue to be celebrated and promoted even as a greater and greater number of citizens have access to powerful technics and systems that they, however technologically unsocialised, believe themselves entitled to use in maximalist ways?

RECENT STRUGGLES TO ADAPT INDIVIDUAL RIGHTS TO TECHNOLOGICAL MAXIMALITY AND INCREASING NUMBERS

In recent years, struggles to adapt individual rights to the realities of technological maximality in populous democratic societies have been waged incessantly on several professional fronts. Let us briefly discuss some pertinent developments in two such fields: urban planning and medicine.

Urban Planning

Two urban planning concerns involving our pattern, over which there were protracted struggles in the 1980s, are building construction and the unrestricted movement of cars. In 1986, the city of San Francisco, California, became the first large city in U.S. history to impose significant limits on the proliferation of downtown high-rise buildings. After several unsuccessful previous efforts, a citizen initiative was finally approved that established a building height limit and a cap on the amount of new high-rise floor space that can be added to the downtown area each year. The majority of San Francisco voters came to believe that the aggregate effects of the continued exercise of essentially unrestricted individual property rights by land owners and developers in technologically maximal ways—entitled erection of numerous highrises—was undermining the quality of city life. In 1990, voters of Seattle, Washington, reached the same conclusion and approved a similar citizen initiative.

As for cars, the 1980s saw the adoption in a few Western countries of substantial limits on their use in cities. For example, to combat air pollution and enhance the quality of urban social life, citizens of Milan and Florence voted overwhelmingly in the mid 1980s to impose limits on the use of cars. In Milan, they are prohibited from entering the *centro storico* between 7:30 a.m. and 6:30 p.m., while in Florence much of the *centro storico* has been turned into a pedestrians-only zone. In California, the cities of Berkeley and Palo Alto installed barriers to prevent drivers from traversing residential streets in the course of cross-town travel. Revealingly, in a debate in the California State Senate over legislation authorising Berkeley to keep its barriers, one senator argued that 'We should be *entitled* to use all roadways . . . Certain individuals think they're too good to have other people drive down their streets' (emphasis added).[14] The phenomenon combated by the road barriers is a clear instance of aggregative TM of use unfolding under the auspices of traditional mobility rights exercised by large numbers of car-drivers. The senator's mind reading notwithstanding, it would seem that citizens, perceiving this pattern as jeopardising the safety of children and diluting the neighbourhood's residential character (read: quality of social life), prevailed on authorities to diminish the long-established domain of driver mobility rights.

Efforts to restrict individual property and mobility rights in urban settings in light of the quality-of-life consequences of their aggregated, technologically maximalist exercise have initiated a high-stakes struggle that promises to grow in importance and be vigorously contested for the foreseeable future.

Medicine

An important issue in the area of medicine that involves our pattern is the ongoing tension between the right to life and the widespread intensive use of life-prolongation technology. Following World War II, the change in the locus of dying from the home to the technology-intensive hospital enabled the full arsenal of modern medical technology to be mobilised in service of the right to life. However, the quality of the prolonged life was often so abysmal that efforts to pull back from application of technologically maximal life-extending medical care eventually surfaced.

The Karen Ann Quinlin case (1975–1985) was a landmark in the United States. The Quinlins asked their comatose daughter's doctor to disconnect her respirator. He refused, as did the New Jersey Court of Appeals.[15] The latter argued, significantly for our purposes, that 'the right to life and the preservation of it are "interests of the highest order."' In other words, in the Appeals Court's view, respecting the traditional individual right to life was held to require ongoing provision of technologically maximal medical care. The New Jersey Supreme Court eventually found for the Quinlins, not by revoking this idea but by finding that a patient's privacy interest grows in proportion to the invasiveness of the medical care to which the patient is subjected, and that that interest can be exercised in a proxy vein by the patient's parents.[16]

The equally celebrated Nancy Cruzan case (1983–1990) was essentially an extension of Quinlin, except that the technological means of life extension that Nancy's parents sought to terminate were her food and hydration tubes. Many who opposed the Cruzans believed that removal of these tubes was tantamount to killing their comatose daughter, i.e., to violating her right to life. In their view, respect for Nancy's right to life required continued application of these technological means without limitation of time, regardless of the quality of life being sustained. The Missouri Supreme Court concluded that the state's interest in the preservation of life is 'unqualified,' i.e., that the right to life is inviolable. The Court held that in the absence of 'clear and convincing evidence' that a patient would not want to be kept alive by machines in the state into which he or she had fallen, i.e., would not wish to exercise her or his right to life under such circumstances, the perceived absoluteness of the right to life drove continued application of the life-prolonging technology.[17]

The 1990 case of Helga Wanglie, seemingly commonplace at the outset, took on revolutionary potential. Hospitalised after fracturing her hip, Mrs Wanglie suffered a respiratory attack that cut off oxygen to her brain. By the time she could be resuscitated, the patient had incurred severe brain damage and lapsed into a vegetative state believed irreversible by hospital doctors. Despite this prognosis and after extensive consultation with the doctors, Mrs Wanglie's family refused to authorise disconnection of the respirator that prolonged her life, asserting that the patient 'want[ed] everything done.' According to Mr Wanglie, 'she told me, "Only He who gave life has the right to take life."'[18]

Unprecedentedly, believing that further medical care was inappropriate, the hospital brought suit in court to obtain authorisation to disconnect the patient's respirator against her family's wishes. Predictably, this suit was unsuccessful, but Mrs Wanglie died shortly thereafter.[19, 20] Had the suit succeeded, it would have marked a significant departure from traditional thinking and practice concerning the right to life. Care would have been terminated not at the behest of patient or guardians, something increasingly familiar in recent years, but rather as the result of a conclusion by a care-providing institution that further treatment was 'futile.' Projected quality of patient life would have taken precedence over the patient's inviolable right to life as asserted by guardians and the absoluteness of the right to life would have been diminished. Consensus that further treatment, however intensive or extensive, offered no reasonable chance of restoring cognitive functioning would have been established as a sufficient condition for mandatory cessation of care.

There are thousands of adults and children in the U.S. and other Western societies whose lives of grim quality are sustained by technological maximality in the name of the right to life, understood by many as categorically binding.[21] The financial and psychological tolls exacted by this specimen of compound-aggregative TM are enormous and will continue to grow until the right to life—its nature and limits—is adapted to the individual and aggregative implications of the technologies used on its behalf.

The troubling triadic pattern should be of concern to many kinds of professional practitioners, not just public officials. Professionals such as urban designers, environmental managers, engineers, and physicians are increasingly confronted in their respective practices with problematic consequences of the continued operation of the troubling triad. Each such individual must decide whether to conduct her or his professional practice—processing building permits, managing natural resource use, designing technics and sociotechnical systems, and treating patients—on the basis of traditional individualistic conceptions of rights unmodified by contemporary technological capabilities and demographic realities, or to alter the concepts and constraints informing her or his practice to reflect extant forms of technological maximality. The fundamental reason why the triadic pattern should be of concern to practising professionals is that failure to combat it is tantamount to acquiescing in the increasingly serious individual and societal harms apt to result from its predictable repeated manifestations. Professionals have an important role to play in raising societal consciousness about the costs of continuing to rely on anachronistic concepts of individual rights in contemporary technological societies. To date, doctors have made some progress in this effort but other professional groups have not even begun to rise to the challenge.

CONCLUSION

In the coming years U.S. citizens and other Westerners will face some critical choices. If we persist in gratifying our seemingly insatiable appetite for technological maximality, carried out under the auspices of anachronistic conceptions of rights claimed by ever increasing numbers of people, we shall pay an increasingly steep price in the form of a diminishing societal quality of life. Consciousness-raising, through education and responsible activism, though maddeningly slow, seems the most viable route to developing the societal ability to make discriminating choices about technological practices and their aggregated effects. However accomplished, developing that ability is essential if we are to secure a future of quality for our children and theirs. Taming the troubling triadic pattern would be an excellent place to begin this quest. The technodemographic anachronism of selected traditional rights should be recognised and a new, naturalistic, non-absolutist theory of human rights should be elaborated, one that stands in dynamic relationship to evolving technological capabilities and demographic trends. Whether or not such a new theory of rights emerges, becomes embodied in law, and alters the contours of professional practice in the next few decades will be critically important to society in the twenty-first century and beyond.

NOTES

1. *New York Times,* 19 November, 1985, p. 1.

2. *Newsweek,* 25 July, 1983, p. 22.

3. See for example: Cecil O. Smith Jr (1990) The Longest Run: Public Engineers and Planning in France, *American Historical Review,* 95, No. 3, pp. 657–692.

4. *New York Times,* Section II, 22 December, 1991, p. 36.

5. Maurice Cranston (1983) Are There Any Human Rights?, *Daedalus,* 112, No. 4, p. 1.

6. Maurice Cranston (1973) *What Are Human Rights?* (New York, Taplinger), pp. 53–54.

7. *New York Times,* 30 October, 1991, p. A1.

8. *New York Times,* 28 November, 1991, pp. D1 and D3.

9. *New York Times,* 23 May, 1993, I, p. 26.

10. Robert E. McGinn (1979) In Defense of Intangibles: the Responsibility-Feasibility Dilemma in Modern Technological Innovation, *Science, Technology, and Human Values,* No. 29, pp. 4–10.

11. See for example: David Braybrooke (1968) Let Needs Diminish That Preferences May Prosper, in *Studies in Moral Philosophy* (Oxford, Blackwell), pp. 86–107, and the same author's (1987) *Meeting Needs* (Princeton, Princeton University Press) for careful analysis of the concept of basic human needs. For discussion of the testability of claims that something is a bona fide basic human need, see Amatai Etzioni (1968), Basic Human Needs, Alienation, and Inauthenticity, *American Sociological Review,* 33, pp. 870–885. On the relationship between human rights and human needs, see also C. B. Macpherson, quoted in D. D. Raphael (1967), *Political Theory and the Rights of Man* (London, Macmillan), p. 14.

12. Joel Feinberg (1973), *Social Philosophy* (Englewood Cliffs, NJ: Prentice-Hall), pp. 85–88.

13. See for example: J. Hitt, (ed.) The Storm in the Plaza, *Harper's Magazine,* July 1985, pp. 27–33.

14. *San Francisco Chronicle,* 2 July, 1983, p. 6.

15. *In re Quinlin,* 137 N.J. super 227 (1975).

16. *In re Quinlin,* 70 N.J. 10,335 A. 2d 647 (1976).

17. After the U.S. Supreme Court decision upholding the Missouri Supreme Court was handed down, three of the patient's friends provided new evidence of her expressed wish to be spared existence in a technologically sustained vegetative state. This led, in a lower court rehearing, to a judgment permitting parental exercise of the patient's recognised privacy interest through the withdrawal of her food and hydration tubes. Nancy Cruzan expired twelve days after this decision was announced. See *New York Times,* 15 December, 1990, A1 and A9, and 27 December, 1990, A1 and A13.

18. *New York Times,* 10 January, 1991, A16.

19. Ibid., 2 July, 1991, A12.

20. Ibid., 6 July, 1991, I, 8.

21. U.S. Congress, Office of Technology Assessment (1987) *Technology-Dependent Children: Hospital v. Home Care Sustaining Technologies and the Elderly* (Washington, D.C., U.S. Government Printing Office).

17

The Constitution in Cyberspace: Law and Liberty beyond the Electronic Frontier

Laurence H. Tribe

My topic is how to "map" the text and structure of our Constitution onto the texture and topology of *cyberspace.* That's the term coined by cyberpunk novelist William Gibson, which many now use to describe the "place"—a place without physical walls or even physical dimensions—where ordinary telephone conversations "happen," where voice-mail and e-mail messages are stored and sent back and forth, and where computer-generated graphics are transmitted and transformed, all in the form of interactions, some real-time and some delayed, among countless users, and between users and the computer itself.

Some use the cyberspace concept to designate fantasy worlds or *virtual realities* of the sort Gibson described in his novel *Neuromancer,* in which people can essentially turn their minds into computer peripherals capable of perceiving and exploring the data matrix. The whole idea of virtual reality, of course, strikes a slightly odd note. As one of Lily Tomlin's most memorable characters once asked, "What's reality, anyway, but a collective hunch?" Work in this field tends to be done largely by people who share the famous observation that reality is overrated!

However that may be, cyberspace connotes to some users the sorts of technologies that people in Silicon Valley (like Jaron Lanier at VPL Research, for instance) work on when they try to develop "virtual racquetball" for the disabled, computer-aided design systems that allow architects to walk through "virtual buildings" and remodel them *before* they are built, "virtual conferencing" for business meetings, or maybe someday even "virtual day care centers" for latchkey children. The user snaps on a pair of goggles hooked up to a high-powered computer terminal, puts on a special set of gloves (and perhaps other gear) wired into the same computer system, and, looking a little bit like Darth Vader, pretty much steps into a computer-driven, drug-free, three-dimensional, interactive, infinitely expandable hallucination complete with sight, sound, and touch—allowing the user literally to move through, and experience, information.

From Laurence H. Tribe, "The Constitution in Cyberspace," *The Humanist* (Sept./Oct., 1991) 51.5, pp. 15–21.

I'm using the term cyberspace much more broadly, as many have lately. I'm using it to encompass the full array of computer-mediated audio and/or video interactions that are already widely dispersed in modern societies—from things as ubiquitous as the ordinary telephone to things that are still coming online like computer bulletin boards and networks like Prodigy or the WELL (Whole Earth 'Lectronic Link), based in San Francisco. My topic, broadly put, is the implications of that rapidly expanding array for our constitutional order. It is a constitutional order that tends to carve up the social, legal, and political universe along lines of *physical place* or *temporal proximity.* The critical thing to note is that these very lines, in cyberspace, either get bent out of shape or fade out altogether. The question, then, becomes: when the lines along which our Constitution is drawn warp or vanish, what happens to the Constitution itself?

SETTING THE STAGE

To set the stage with a perhaps unfamiliar example, consider a decision handed down in 1990, *Maryland v. Craig,* in which the U.S. Supreme Court upheld the power of a state to put an alleged child abuser on trial with the defendant's accuser testifying not in the defendant's presence but by one-way, closed-circuit television. The Sixth Amendment, which of course antedated television by a century and a half, says: "In all criminal prosecutions, the accused shall enjoy the right . . . to be confronted with the witnesses against him." Justice O'Connor wrote for a bare majority of five justices that the state's procedures nonetheless struck a fair balance between costs to the accused and benefits to the victim and to society as a whole. Justice Scalia, joined by the three "liberals" then on the Court (Justices Brennan, Marshall, and Stevens), dissented from that cost–benefit approach to interpreting the Sixth Amendment. He wrote:

> The Court has convincingly proved that the Maryland procedure serves a valid interest, and gives the defendant virtually everything the Confrontation Clause guarantees (everything, that is, except confrontation). I am persuaded, therefore, that the Maryland procedure is virtually constitutional. Since it is not, however, actually constitutional I [dissent].

Could it be that the high-tech, closed-circuit TV context, almost as familiar to the Court's youngest justice as to his even younger law clerks, might've had some bearing on Justice Scalia's sly invocation of "virtual" constitutional reality? Even if Justice Scalia wasn't making a pun on virtual reality, and I suspect he wasn't, his dissenting opinion about the Confrontation Clause requires *us* to "confront" the recurring puzzle of how constitutional provisions written two centuries ago should be construed and applied in ever-changing circumstances.

Should contemporary society's technology-driven cost–benefit fixation be allowed to water down the old-fashioned value of direct confrontation that the Constitution seemingly enshrined as basic? I would hope not. In that respect, I find myself in complete agreement with Justice Scalia.

But new technological possibilities for seeing your accuser clearly without having your accuser see you at all—possibilities for sparing the accuser any discomfort in ways that the accuser couldn't be spared before one-way mirrors or closed-circuit TVs were developed— *should* lead us at least to ask ourselves whether *two*-way confrontation, in which your accuser is supposed to be made uncomfortable, and thus less likely to lie, really *is* the core value of the Confrontation Clause. If so, virtual confrontation should be held constitutionally insufficient. If not—if the core value served by the Confrontation Clause is just the ability to *watch* your accuser say that you did it—then "virtual" confrontation should suffice. New technologies should lead us to look more closely at just what *values* the Constitution seeks to preserve. New technologies should *not* lead us to react reflexively *either* way—either by assuming that technologies the Framers didn't know about make their concerns and values obsolete, or by assuming that those new technologies couldn't possibly provide new ways out of old dilemmas and therefore should be ignored altogether.

The one-way mirror yields a fitting metaphor for the task we confront. As the Supreme Court said in a different context several years ago, "The mirror image presented [here] requires us to step through an analytical looking glass to resolve it" (*NCAA v. Tarkanian,* 109 S. Ct. at 462). The world in which the Sixth Amendment's Confrontation Clause was written and ratified was a world in which "being confronted with" your accuser *necessarily* meant a simultaneous physical confrontation so that your accuser had to *perceive* you being accused by him. Closed-circuit television and one-way mirrors changed all that by *decoupling* those two dimensions of confrontation, marking a shift in the conditions of information transfer that is in many ways typical of cyberspace.

What does that sort of shift mean for constitutional analysis? A common way to react is to treat the pattern as it existed *prior* to the new technology (the pattern in which doing "A" necessarily *included* doing "B") as essentially arbitrary or accidental. Taking this approach, once the technological change makes it possible to do "A" *without* "B"—to see your accuser without having him or her see you, or to read someone's mail without her knowing it, to switch examples—one concludes that the "old" Constitution's inclusion of "B" is irrelevant; one concludes that it is enough for the government to guarantee "A" alone. Sometimes that will be the case; but it's vital to understand that, sometimes, it won't be.

A characteristic feature of modernity is the subordination of purpose to accident—an acute appreciation of just how contingent and coincidental the connections we are taught to make often are. We understand, as moderns, that many of the ways we carve up and organize the world reflect what our social history and cultural heritage, and perhaps our neurological wiring, bring to the world, and not some irreducible "way things are." A wonderful example comes from a 1966 essay by Jorge Luis Borges, "Other Inquisitions." There, the essayist describes the following taxonomy of the animal kingdom, which he purports to trace to an ancient Chinese encyclopedia entitled *The Celestial Emporium of Benevolent Knowledge:*

On those remote pages it is written that animals are divided into:

(a) those belonging to the Emperor
(b) those that are embalmed
(c) those that are trained
(d) suckling pigs
(e) mermaids
(f) fabulous ones
(g) stray dogs
(h) those that are included in this classification
(i) those that tremble as if they were mad
(j) innumerable ones
(k) those drawn with a very fine camel's hair brush
(l) others

(m) those that have just broken a water pitcher
(n) those that, from a great distance, resemble flies

Contemporary writers from Michel Foucault, in *The Archaeology of Knowledge,* through George Lakoff, in *Women, Fire, and Dangerous Things,* use Borges's Chinese encyclopedia to illustrate a range of different propositions, but the *core* proposition is the supposed arbitrariness—the political character, in a sense—of all culturally imposed categories.

At one level, that proposition expresses a profound truth and may encourage humility by combating cultural imperialism. At another level, though, the proposition tells a dangerous lie: it suggests that we have descended into the nihilism that so obsessed Nietzsche and other thinkers—a world where *everything* is relative, all lines are up for grabs, and all principles and connections are just matters of purely subjective preference or, worse still, arbitrary convention. Whether we believe that killing animals for food is wrong, for example, becomes a question indistinguishable from whether we happen to enjoy eating beans, rice, and tofu.

This is a particularly pernicious notion in a era when we pass more and more of our lives in cyberspace, a place where, almost by definition, our most familiar landmarks are rearranged or disappear altogether—because there is a pervasive tendency, even (and perhaps especially) among the most enlightened, to forget that the human values and ideals to which we commit ourselves may indeed be universal and need not depend on how our particular cultures, or our latest technologies, carve up the universe we inhabit. It was my very wise colleague from Yale, the late Art Leff, who once observed that, even in a world without an agreed-upon God, we can still agree—even if we can't "prove" mathematically—that "napalming babies is wrong."

The Constitution's core values, I'm convinced, need not be transmogrified, or metamorphosed into oblivion, in the dim recesses of cyberspace. But to say that they *need* not be lost there is hardly to predict that they *will* not be. On the contrary, the danger is clear and present that they *will* be.

The "event horizon" against which this transformation might occur is already plainly visible:

Electronic trespassers like Kevin Mitnick don't stop with cracking pay phones, but break into NORAD—the North American Defense Command computer in Colorado Springs—not in a *WarGames* movie, but in real life.

Less challenging to national security but more ubiquitously threatening, computer crackers download everyman's credit history from institutions like TRW, start charging phone calls (and more) to everyman's number, set loose "worm" programs that shut down thousands of linked computers, and spread "computer viruses" through everyman's work or home PC.

It is not only the government that feels threatened by "computer crime"; both the owners and the users of private information services, computer bulletin boards, gateways, and networks feel equally vulnerable to this new breed of invisible trespasser. The response from the many who sense danger has been swift, and often brutal, as a few examples illustrate.

In March 1990, U.S. Secret Service agents staged a surprise raid on Steve Jackson Games, a small games manufacturer in Austin, Texas, and seized all paper and electronic drafts of its newest fantasy role-playing game, *GURPS Cyberpunk,* calling the game a "handbook for computer crime."

By the spring of 1990, up to one quarter of the U.S. Treasury Department's investigators had become involved in a project of eavesdropping on computer bulletin boards, apparently tracking notorious hackers like "Acid Phreak" and "Phiber Optik" through what one journalist dubbed "the dark canyons of cyberspace."

In May 1990, in the now famous (or infamous) "Operation Sun Devil," more than 150 secret service agents teamed up with state and local law enforcement agencies, and with security personnel from AT&T, American Express, U.S. Sprint, and a number of the regional Bell telephone companies, armed themselves with over two dozen search warrants and more than a few guns, and seized 42 computers and 23,000 floppy discs in 14 cities from New York to Texas. Their target: a looseknit group of people in their teens and twenties, dubbed the "Legion of Doom."

I am not describing an Indiana Jones movie. I'm talking about America in the 1990s.

THE PROBLEM

The Constitution's architecture can too easily come to seem quaintly irrelevant, or at least impossible to take very seriously, in the world as reconstituted by the microchip. I propose today to canvass five axioms of our constitutional law—five basic assumptions that I believe shape the way American constitutional scholars and judges view legal issues—and to examine how they can adapt to the cyberspace age. My conclusion (and I will try not to give away too much of the punch line here) is that the Framers of our Constitution were very wise indeed. They bequeathed us a framework for all seasons, a truly astonishing document whose principles are suitable for all times and all technological landscapes.

Axiom 1: There Is a Vital Difference between Government and Private Action

The first axiom I will discuss is the proposition that the Constitution, with the sole exception of the Thirteenth Amendment prohibiting slavery, regulates action by the *government* rather than the conduct of *private* individuals and groups. In an article I wrote in the *Harvard Law Review* in November 1989, "The Curvature of Constitutional Space," I discussed the Constitution's metaphor-morphosis from a Newtonian to an Einsteinian and Heisenbergian paradigm. It was common, early in our history, to see the Constitution as "Newtonian in design with its carefully counterpoised forces and counterforces, its [geographical and institutional] checks and balances" (103 at 3).

Indeed, in many ways contemporary constitutional law is still trapped within and stunted by that paradigm. But today at least, some postmodern constitutionalists tend to think and talk in the language of relativity, quantum mechanics, and chaos theory. This may quite naturally suggest to some observers that the Constitution's basic strategy of decentralizing and diffusing power by constraining and fragmenting governmental authority in particular has been rendered obsolete.

The institutional separation of powers among the three federal branches of government, the geographical division of authority between the federal government and the fifty state governments, the recognition of national boundaries, and, above all, the sharp distinction between the public and private spheres become easy to deride as relics of a simpler, precomputer age. Thus Eli Noam, in the First Ithiel de Sola Pool Memorial Lecture, delivered in October 1990 at MIT, notes that computer networks and network associations acquire quasi-governmental powers as they necessarily take on such tasks as mediating their members' conflicting interests, establishing cost shares, creating their own rules of admission and access and expulsion, even establishing their own de facto taxing mechanisms. In Professor Noam's words, "Networks become political entities," global nets that respect no state or local boundaries. Restrictions on the use of information in one country (to protect privacy, for example) tend to lead to export of that information to other countries, where it can be analyzed and then used on a selective basis in the country attempting to restrict it. *Data havens* reminiscent of the role played by the Swiss in banking may emerge, with few restrictions on the storage and manipulation of information.

A tempting conclusion is that, to protect the free speech and other rights of *users* in such private networks, judges must treat these networks not as associations that have rights of their own *against* the government but as virtual "governments" in themselves—as entities against which individual rights must be defended in the Constitution's name. Such a conclusion would be misleadingly simplistic. There are circumstances, of course, when nongovernmental bodies like privately owned "company towns" or even huge shopping malls should be subjected to legislative and administrative controls by democratically accountable entities, or even to judicial controls as though they were arms of the state—but that may be as true (or as false) of multinational corporations or foundations, or transnational religious organizations, or even small-town communities, as it is of computer-mediated networks. It's a fallacy to suppose that just because a computer bulletin board or network or gateway is something *like* a shopping mall, government has as much constitutional duty—or even authority—to guarantee open public access to such a network as it has to guarantee open public access to a privately owned shopping center like the one involved in the U.S. Supreme Court's famous *PruneYard Shopping Center* decision of 1980, arising from San Jose, California.

The rules of law, both statutory and judge-made, through which each state *allocates* private powers and responsibilities themselves represent characteristic forms of government action. That's why a state's rules for imposing liability on private publishers, or for deciding which private contracts to enforce and which ones to invalidate, are all subject to scrutiny for their consistency with the federal Constitution. But as a general proposition, it is only what *governments* do, either through such rules or through the actions of public officials, that the U.S. Constitution constrains. And nothing about any new technology suddenly erases the Constitution's enduring value of restraining *government* above all else, and of protecting all private groups, large and small, from government.

It's true that certain technologies may become socially indispensable—so that equal or at least minimal access to basic computer power, for example, might be as significant a constitutional goal as equal or at least minimal access to the franchise, or to dispute resolution through the judicial system, or to elementary and secondary education. But all this means (or should mean) is that the Constitution's constraints on government must at times take the form of imposing *affirmative duties* to assure access rather than merely enforcing *negative prohibitions* against designated sorts of invasion or intrusion.

Today, for example, the government is under an affirmative obligation to open up criminal trials to the press and the public, at least where there has not been a particularized finding that such openness would disrupt the proceedings. The government is also under an affirmative obligation to provide free legal assistance for indigent criminal defendants, to assure speedy trials, to underwrite the cost of counting ballots at election time, and to desegregate previously segregated school systems. But these occasional affirmative obligations don't, or shouldn't, mean that the Constitution's axiomatic division between the realm of public power and the realm of private life should be jettisoned.

Nor would the "indispensability" of information technologies provide a license for government to impose strict content, access, pricing, and other types of regulation. *Books* are indispensable to most of us, for example—but it doesn't follow that government should therefore be able to regulate the content of what goes onto the shelves of *bookstores*. The right of a private bookstore owner to decide which books to stock and which to discard, which books to display openly and which to store in limited access areas, should remain inviolate. And note, incidentally, that this needn't make the bookstore owner a *publisher* who is liable for the words printed in the books on her shelves. It's a common fallacy to imagine that the moment a computer gateway or bulletin board begins to exercise powers of selection to control who may be online, it must automatically assume the responsibilities of a newscaster, a broadcaster, or an author. For computer gateways and bulletin boards are really the "bookstores" of cyberspace; most of them organize and present information in a computer format, rather than generating more information content of their own.

Axiom 2: The Constitutional Boundaries of Private Property and Personality Depend on Variables Deeper Than Social Utility and Technological Feasibility

The second constitutional axiom, one closely related to the private-public distinction of the first axiom, is that a person's mind, body, and property belong *to that person* and not to the public as a whole. Some believe that cyberspace challenges that axiom because its entire premise lies in the existence of computers tied to electronic transmission networks that process digital information. Because such information can be easily replicated in series of 1s and 0s, anything that anyone has come up with in virtual reality can be infinitely reproduced. I can log on to a computer library, copy a "virtual book" to my computer disk, and send a copy to your computer without creating a gap on anyone's bookshelf. The same is true of valuable computer programs, costing hundreds of dollars, creating serious piracy problems. This feature leads some, like Richard Stallman of the Free Software Foundation, to argue that in cyberspace everything should be free—that information can't be owned. Others, of course, argue that copyright and patent protections of various kinds are needed in order for there to be incentives to create *cyberspace property* in the first place.

Needless to say, there are lively debates about what the optimal incentive package should be as a matter of legislative and social policy. But the only *constitutional* issue, at bottom, isn't the utilitarian or instrumental selection of an optimal policy. Social judgments about what ought to be subject to individual appropriation, in the sense used by John Locke and Robert Nozick, and what ought to remain in the open public domain, are first and foremost *political* decisions.

To be sure, there are some constitutional constraints on these political decisions. The Constitution does not permit anything and everything to be made into a private commodity. Votes, for example, theoretically cannot be bought and sold. Whether the Constitution itself should be read (or amended) so as to permit all basic medical care, shelter, nutrition, legal assistance, and, indeed, computerized information services to be treated as mere commodities, available only to the highest bidder, are all terribly hard questions—as the Eastern Europeans discovered in the early 1990s as they drafted their own constitutions. But these are not questions that should ever be confused with issues of what is technologically possible, about what is realistically enforceable, or about what is socially desirable.

Similarly, the Constitution does not permit anything and everything to be *socialized* and made into a public good available to whoever needs or "deserves" it most. I would hope, for example, that the government could not use its powers of eminent domain to "take" live body parts like eyes or kidneys or brain tissue for those who need transplants and would be expected to lead particularly productive lives. In any event, I feel certain that whatever constitutional right each of us has to inhabit his or her own body and to hold onto his or her own thoughts and creations should not depend solely on cost–benefit calculations, or on the availability of technological methods for painlessly effecting transfers or for creating good artificial substitutes.

Axiom 3: Government May Not Control Information Content

A third constitutional axiom, like the first two, reflects a deep respect for the integrity of each individual and a healthy skepticism toward government. The axiom is that, although information and ideas have real effects in the social world, it's not up to government to pick and choose for us in terms of the *content* of that information or the *value* of those ideas.

This notion is sometimes mistakenly reduced to the naïve child's ditty that "sticks and stones may break my bones, but words can never hurt me." Anybody who's ever been called something awful by children in a schoolyard knows better than to believe any such thing. The real basis for First Amendment values isn't the false premise that information and ideas have no real impact, but the belief that information and ideas are too *important* to entrust to any government censor or overseer.

If we keep that in mind, and *only* if we keep that in mind, will we be able to see through the tempting argument that, in the Information Age, free speech is a luxury we can no longer afford. That argument becomes especially tempting in the context of cyberspace, where sequences of 0s and 1s may become virtual life forms. Computer "viruses" roam the information nets, attaching themselves to various programs and screwing up computer facilities. Creation of a computer virus involves writing a program; the program then replicates itself and mutates. The electronic code involved is very much like DNA. If information content is *speech,* and if the First Amendment is to apply in cyberspace, then mustn't these viruses be speech—and mustn't their writing and dissemination be constitutionally protected? To avoid that nightmarish outcome, mustn't we say that the First Amendment is *inapplicable* to cyberspace?

The answer is no. Speech is protected, but deliberately yelling "Boo!" at a cardiac patient may still be prosecuted as murder. Free speech is a constitutional right, but handing a bank teller a holdup note that says, "Your money or your life," may still be punished as robbery. Stealing someone's diary may be punished as theft—even if you intend to publish it in book form. And the Supreme Court, over the past fifteen years, has gradually brought advertising within the ambit of protected expression without preventing the government from protecting consumers from deceptive advertising. The lesson, in short, is that constitutional principles are subtle enough to bend to such concerns. They needn't be broken or tossed out.

Axiom 4: The Constitution Is Founded on Normative Conceptions of Humanity That Advances in Science and Technology Cannot "Disprove"

A fourth constitutional axiom is that the human spirit is something beyond a physical information processor. That axiom, which regards human thought processes as not fully reducible to the operations of a computer program, however complex, must not be confused with the silly view that, because computer operations involve nothing more than the manipulation of "on" and "off" states of myriad microchips, it somehow follows that government control or outright seizure of computers and computer programs threatens no First Amendment rights because human thought processes are not directly involved. To say that would be like saying that government confiscation of a newspaper's printing press and tomorrow morning's copy has nothing to do with speech but involves only a taking of metal, paper, and ink. Particularly if the seizure or the regulation is triggered by the content of the information being processed or transmitted, the First Amendment is of course fully involved. Yet this recognition that information processing by computer entails something far beyond the mere sequencing of mechanical or chemical steps still leaves a potential gap between what computers can do internally and in communication with one another—and what goes on within and between human minds. It is that gap to which this fourth axiom is addressed; the very existence of any such gap is, as I'm sure you know, a matter of considerable controversy.

What if people like the mathematician and physicist Roger Penrose, author of *The Emperor's New Mind,* are wrong about human minds? In that provocative recent book, Penrose disagrees with those artificial intelligence, or AI, gurus who insist that it's only a matter of time until human thought and feeling can be perfectly simulated or even replicated by a series of purely physical operations—that it's all just neurons firing and neurotransmitters flowing, all subject to perfect modeling in suitable computer systems. Would an adherent of that AI orthodoxy, someone whom Penrose fails to persuade, have to reject as irrelevant for cyberspace those constitutional protections that rest on the anti-AI premise that minds are *not* reducible to really fancy computers?

Consider, for example, the Fifth Amendment, which provides that "no person shall be . . . compelled in any criminal case to be a witness against himself." The Supreme Court has long held that suspects may be required, despite this protection, to provide evidence that is not "testimonial" in nature—blood samples, for instance, or even exemplars of one's handwriting or voice. In 1990, in a case called *Pennsylvania v. Muniz,* the Supreme Court held that answers to even simple questions like "When was your sixth birthday?" are testimonial because such a question, however straightforward, nevertheless calls for the product of mental activity and therefore uses the suspect's mind against him or her. But what if science could eventually describe thinking as a process no more complex than, say, riding a bike or digesting a meal? Might the progress of neurobiology and computer science eventually overthrow the premises of the *Muniz* decision?

I would hope not. For the Constitution's premises, properly understood, are *normative* rather than descriptive. The philosopher David Hume was right in teaching that no "ought" can ever be logically derived from an "is." If we should ever abandon the Constitution's protection for the distinctively and universally human, it won't be because robotics or genetic engineering or computer science has led us to deeper truths, but rather because they have seduced us into more profound confusions. Science and technology open options, create possibilities, suggest incompatibilities, and generate threats. They do not alter what is "right" or what is "wrong." The fact that those notions are elusive and subject to endless debate need not make them totally contingent on contemporary technology.

Axiom 5: Constitutional Principles Should Not Vary with Accidents of Technology

In a sense, that's the fifth and final constitutional axiom I would urge upon this gathering: that the Constitution's norms, at their deepest level, must be invariant under merely *technological* transformations. Our constitutional law evolves through judicial interpretation, case by case, in a process of reasoning by analogy from precedent. At its best, that process is ideally suited to seeing beneath the surface and extracting deeper principles from prior decisions. At its worst, though, the same process can get bogged down in superficial aspects of preexisting examples, fixating upon unessential features while overlooking underlying principles and values.

When the Supreme Court in 1928 first confronted wiretapping and held in *Olmstead v. United States* that such wiretapping involved no "search" or "seizure" within the meaning of the Fourth Amendment's prohibition of "unreasonable searches and seizures," the majority of the Court reasoned that the Fourth Amendment "itself shows that the search is to be of material things—the person, the house, his papers or his effects," and said that "there was no searching" when a suspect's phone was tapped because the Constitution's language "cannot be extended and expanded to include telephone wires reaching to the whole world from the defendant's house or office." After all, said the Court, the intervening wires "are not part of his house or office any more than are the highways along which they are stretched." Even to a law student in the 1960s, as you might imagine, that "reasoning" seemed amazingly artificial. Yet the *Olmstead* doctrine still survived.

It would be illuminating at this point to compare the Supreme Court's initial reaction to new technology in *Olmstead* with its initial reaction to new technology in *Maryland v. Craig,* the 1990 closed-circuit television case with which we began this discussion. In *Craig,* a majority of the justices assumed that, when the eighteenth-century Framers of the Confrontation Clause included a guarantee of two-way *physical* confrontation, they did so solely because it had not yet become technologically feasible for the accused to look his or her accuser in the eye without having the accuser simultaneously watch the accused. Given that this technological obstacle has been removed, the majority assumed, one-way confrontation is now sufficient. It is enough that the accused not be subject to criminal conviction on the basis of statements made outside his presence.

In *Olmstead,* a majority of the justices assumed that, when the eighteenth-century authors of the Fourth Amendment used language that sounded "physical" in guaranteeing against invasions of a person's dwelling or possessions, they did so not solely because physical invasions were at that time the only serious threats to personal privacy, but for the separate and distinct reason that *intangible* invasions simply would not threaten any relevant dimension of Fourth Amendment privacy.

In a sense, *Olmstead* mindlessly read a new technology *out* of the Constitution, while *Craig* absentmindedly read a new technology *into* the Constitution. But both decisions— *Olmstead* and *Craig*—had the structural effect of withholding the protections of the Bill of Rights from threats made possible by new information technologies. *Olmstead* did so by implausibly reading the Constitution's text as though it represented a deliberate decision not to extend protection to threats that eighteenth-century thinkers simply had not foreseen. *Craig* did so by somewhat more plausibly—but still unthinkingly—treating the Constitution's seemingly explicit coupling of two analytically distinct protections as reflecting a failure of technological foresight and imagination, rather than a deliberate value choice.

The *Craig* majority's approach appears to have been driven in part by an understandable sense of how a new information technology could directly protect a particularly sympathetic group, abused children, from a traumatic trial experience. The *Olmstead* majority's approach probably reflected both an exaggerated estimate of how difficult it would be to obtain wiretapping warrants even where fully justified, and an insufficient sense of how a new information technology could directly threaten all of us. Although both *Craig* and *Olmstead* reveal an inadequate consciousness about how new technologies interact with old values, *Craig* at least seems defensible even if misguided, and *Olmstead* seems just plain wrong.

Around twenty-three years ago, as a then-recent law school graduate serving as law clerk to Supreme Court Justice Potter Stewart, I found myself working on a case involving the government's electronic surveillance of a suspected criminal—in the form of a tiny device attached to the outside of a public telephone booth. Because the invasion of the suspect's privacy was accomplished without physical trespass into a "constitutionally protected area," the federal government argued, relying on *Olmstead,* that there had been no "search" or "seizure," and therefore that the Fourth Amendment "right of the people to be secure in their persons, houses, papers, and effects, against unreasonable searches and seizures," simply did not apply.

At first, there were only four votes to overrule *Olmstead* and to hold the Fourth Amendment applicable to wiretapping and electronic eavesdropping. I'm proud to say that, as a 26-year-old kid, I had at least a little bit to do with changing that number from four to seven—and with the argument, formally adopted by a seven-justice majority in December 1967, that the Fourth Amendment "protects people, not places" (389 U.S. at 351). In that decision, *Katz v. United States,* the Supreme Court finally repudiated *Olmstead* and the many decisions that had relied upon it and reasoned that, given the role of electronic telecommunications in modern life, the First Amendment purposes of protecting *free speech* as well as the Fourth Amendment purposes of protecting privacy require treating as a "search" any invasion of a person's confidential telephone communications, with or without physical trespass.

Sadly, nine years later, in *Smith v. Maryland* (1976), the Supreme Court retreated from the *Katz* principle by holding that no search occurs and therefore no warrant is needed when police, with the assistance of the telephone company, make use of a "pen register," a mechanical device placed on someone's phone line that records all numbers dialed from the phone and the times of dialing. The Supreme Court, over the dissents of Justices Stewart, Brennan, and Marshall, found no legitimate expectation of privacy in the numbers dialed, reasoning that the digits one dials are routinely recorded by the phone company for billing purposes. As Justice Stewart, the author of Katz, aptly pointed out,

> That observation no more than describes the basic nature of telephone calls. . . . It is simply not enough to say, after Katz, that there is no legitimate expectation of privacy in the numbers dialed because the caller assumes the risk that the telephone company will expose them to the police. (442 U.S. at 746–747)

Today, the logic of *Smith* is being used to say that people have no expectation of privacy when they use their cordless telephones since they know or should know that radio waves can be easily monitored!

It is easy to be pessimistic about the way in which the Supreme Court has reacted to technological change. In many respects, *Smith* is unfortunately more typical than *Katz* of the way the Court has behaved. For example, when movies were invented, and for several decades thereafter, the Court held that movie exhibitions were not entitled to First Amendment protection. When community access cable TV was born, the Court hindered municipal attempts to provide it at low cost by holding that rules requiring landlords to install small cable boxes on their apartment buildings amounted to a compensable taking of property. And in *Red Lion v. FCC,* decided in 1969 but still not repudiated today, the Court ratified government control of

TV and radio broadcast content with the dubious logic that the scarcity of the electromagnetic spectrum justified not merely government policies to auction off, randomly allocate, or otherwise ration the spectrum according to neutral rules, but also much more intrusive and content-based government regulation in the form of the so-called fairness doctrine.

Although the Supreme Court and the lower federal courts have taken a somewhat more enlightened approach in dealing with cable television, these decisions for the most part reveal a curious judicial blindness, as if the Constitution had to be reinvented with the birth of each new technology. Judges interpreting a late-eighteenth-century Bill of Rights tend to forget that, unless its *terms* are read in an evolving and dynamic way, its *values* will lose even the *static* protection they once enjoyed. Ironically, *fidelity* to original values requires *flexibility* of textual interpretation. It was Judge Robert Bork, not famous for his flexibility, who once urged this enlightened view upon then-Judge (now Justice) Scalia, when the two of them sat as colleagues on the U.S. Court of Appeals for the D.C. Circuit.

Judicial error in this field tends to take the form of saying that, by using modern technology ranging from the telephone to the television to computers, we "assume the risk." But that typically begs the question. Justice Harlan, in a dissent penned three decades ago, wrote: "Since it is the task of the law to form and project, as well as mirror and reflect, we should not . . . merely recite . . . risks without examining the *desirability* of saddling them upon society." (*United States v. White*, 1971 401 U.S. at 786). And, I would add, we should not merely recite risks without examining how imposing those risks comports with the Constitution's fundamental values of freedom, privacy, and equality.

Failing to examine just that issue is the basic error I believe federal courts and Congress have made: in regulating radio and TV broadcasting without adequate sensitivity to First Amendment values; in supposing that the selection and editing of video programs by cable operators might be less than a form of expression; in excluding telephone companies from cable and other information markets; in assuming that the processing of 0s and 1s by computers as they exchange data with one another is something less than speech; and in generally treating information processed electronically as though it were somehow less entitled to protection for that reason.

The lesson to be learned is that these choices and these mistakes are not dictated by the Constitution. They are decisions for us to make in interpreting that majestic charter, and in implementing the principles that the Constitution establishes.

CONCLUSION

If my own life as a lawyer and legal scholar could leave just one legacy, I'd like it to be the recognition that the Constitution *as a whole* "protects people, not places." If that is to come about, the Constitution as a whole must be read through a technologically transparent lens. That is, we must embrace, as a rule of construction or interpretation, a principle one might call the *cyberspace corollary*. It would make a suitable twenty-seventh amendment to the Constitution, one befitting the 200th anniversary of the Bill of Rights. Whether adopted all at once as a constitutional amendment, or accepted gradually as a principle of interpretation that I believe should obtain even without any formal change in the Constitution's language, the corollary I would propose would do for *technology* in the twenty-first century what I believe the Constitution's Ninth Amendment, adopted in 1791, was meant to do for *text*.

The Ninth Amendment says: "The enumeration in the Constitution, of certain rights, shall not be construed to deny or disparage others retained by the people." That amendment provides added support for the long-debated, but now largely accepted, "right of privacy" that the Supreme Court recognized in such decisions as the famous birth control case of 1965, *Griswold v. Connecticut.* The Ninth Amendment's simple message is that the *text* used by the Constitution's authors and ratifiers does not exhaust the values our Constitution recognizes. Perhaps a twenty-seventh amendment could convey a parallel and equally simple message: the *technologies* familiar to the Constitution's authors and ratifiers similarly do not exhaust the *threats* against which the Constitution's core values must be protected.

The most recent amendment, the twenty-sixth, adopted in 1971, extended the vote to 18-year-olds. It would be fitting, in a world where youth has been enfranchised, for a twenty-seventh amendment to spell a kind of "childhood's end" for constitutional law. The twenty-seventh amendment, to be proposed for at least serious debate in 1991, would read simply:

> This Constitution's protections for the freedoms of speech, press, petition, and assembly, and its protections against unreasonable searches and seizures and the deprivation of life, liberty, or property without due process of law, shall be construed as fully applicable without regard to the technological method or medium through which information content is generated, stored, altered, transmitted, or controlled.

NOTE

Copyright 1991 Laurence H. Tribe, Tyler Professor of Constitutional Law, Harvard Law School.

18

Technological Ethics in a Different Voice

Diane P. Michelfelder

The rapid growth of modern forms of technology has brought both a threat and a promise for liberal democratic society. As we grapple to understand the implications of new techniques for extending a woman's reproductive life or the spreading underground landscape of fiber-optic communication networks or any of the other developments of contemporary technology, we see how these changes conceivably threaten the existence of a number of primary goods traditionally associated with democratic society, including social freedom, individual autonomy, and personal privacy. At the same time, we recognize that similar hopes and promises have traditionally been associated with both technology and democracy. Like democratic society itself, technology holds forth the promise of creating expanded opportunities and a greater realm of individual freedom and fulfillment. This situation poses a key question for the contemporary philosophy of technology. How can technology be reformed to pose more promise than threat for democratic life? How can technological society be compatible with democratic values?

One approach to this question is to suggest that the public needs to be more involved with technology not merely as thoughtful consumers but as active participants in its design. We can find an example of this approach in the work of Andrew Feenberg. As he argues, most notably in his recent book, *Alternative Modernity: The Technical Turn in Philosophy and Social Theory*, the advantage of technical politics, of greater public participation in the design of technological objects and technologically mediated services such as health care, is to open up this process to the consideration of a wider sphere of values than if the design process were to be left up to bureaucrats and professionals, whose main concern is with preserving efficiency. Democratic values such as personal autonomy and individual agency are part of this wider sphere. For Feenberg, the route to technological reform and the preservation of democracy thus runs directly through the intervention of nonprofessionals in the early stages of the development of technology (Feenberg 1995).

By contrast, the route taken by Albert Borgmann starts at a much later point. His insightful explorations into the nature of the technological device—that "conjunction of machinery and commodity" (Borgmann 1992b, 296)—do not take us into a discussion of how public participation in the design process might result in a device more reflective of democratic virtues. Borgmann's interest in technology starts at the point where it has already been designed, developed, and ready for our consumption. Any reform of technology, from his viewpoint, must first pass through a serious examination of the moral status of material culture. But why must it start here, rather than earlier, as Feenberg suggests? In particular, why must it start here for the sake of preserving democratic values?

In taking up these questions in the first part of this paper, I will form a basis for turning in the following section to look at Borgmann's work within the larger context of contemporary moral theory. With this context in mind, in the third part of this paper I will take a critical look from the perspective of feminist ethics at Borgmann's distinction between the thing and the device, a distinction on which his understanding of the moral status of material culture rests. Even if from this perspective this distinction turns out to be questionable, it does not undermine, as I will suggest in the final part of this paper, the wisdom of Borgmann's starting point in his evaluation of technological culture.

PUBLIC PARTICIPATION AND TECHNOLOGICAL REFORM

One of the developments that Andrew Feenberg singles out in *Alternative Modernity* to back up his claim that public involvement in technological change can further democratic culture is the rise of the French videotext system known as Teletel (Feenberg 1995, 144–66). As originally proposed, the Teletel project had all the characteristics of a technocracy-enhancing device. It was developed within the bureaucratic structure of the French government-controlled telephone company to advance that government's desire to increase France's reputation as a leader in emerging technology. It imposed on the public something in which it was not interested: convenient access from home terminals (Minitels) to government-controlled information services. However, as Feenberg points out, the government plan for Teletel was foiled when the public (thanks to the initial assistance of computer hackers) discovered the potential of the Minitels as a means of communication. As a result of these interventions, Feenberg reports, general public use of the Minitels for sending messages eventually escalated to the point where it brought government use of the system to a halt by causing it to crash. For Feenberg, this story offers evidence that the truth of social constructivism is best seen in the history of the computer.

Let us imagine it does offer this evidence. What support, though, does this story offer regarding the claim that public participation in technical design can further democratic culture? In Feenberg's mind, there is no doubt that the Teletel story reflects the growth of liberal democratic values. The effect generated by the possibility of sending anonymous messages to others over computers is, according to Feenberg, a positive one, one that "enhances the sense of personal freedom and individualism by reducing the 'existential' engagement of the self in its communications" (Feenberg 1995, 159). He also finds that in the ease of contact and connection building fostered by computer-mediated communication, any individual or group of individuals who is a part of building these connections becomes more empowered (Feenberg 1995, 160).

But as society is strengthened in this way, in other words, as more and more opportunities open up for electronic interaction among individuals, do these opportunities lead to a more meaningful social engagement and exercise of individual freedom? As Borgmann writes in *Technology and the Character of Contemporary Society* (or *TCCL)*: "The capacity for significance is where human freedom should be located and grounded" (Borgmann 1984, 102). Human interaction without significance leads to disengagement; human freedom without significance leads to banality of agency. If computer-mediated communications take one where Feenberg believes they do (and there is little about the more recent development of Internet-based communication to raise doubts about this), toward a point where personal life increasingly becomes a matter of "staging . . . personal performances" (Feenberg 1995, 160), then one wonders what effect this has on other values important for democratic culture: values such as self-respect, dignity, community, and personal responsibility.

The Teletel system, of course, is just one example of technological development, but it provides an illustration through which Borgmann's concern with the limits of public participation in the design process as a means of furthering the democratic development of technological society can be understood. Despite the philosophical foundations of liberal democracy in the idea that the state should promote equality by refraining from supporting any particular idea of the human good, in practice, he writes, "liberal democracy is enacted as technology. It does not leave the question of the good life open but answers it along technological lines" (Borgmann 1984, 92). The example we have been talking about illustrates this claim. Value-neutral on its surface with respect to the good life, Feenberg depicts the Teletel system as encouraging a play of self-representation and identity that develops at an ever-intensifying pace while simultaneously blurring the distinction between private and public life. The value of this displacement, though, in making life more meaningful, is questionable.

To put it in another way, for technology to be designed so that it offers greater opportunities for more and more people, what it offers has to be put in the form of a commodity. But the more these opportunities are put in the form of commodities, the more banal they threaten to become. This is why, in Borgmann's view, technical politics cannot lead to technical reform.

For there truly to be a reform of technological society, Borgmann maintains, it is not enough only to think about preserving democratic values. One also needs to consider how to make these values meaningful contributors to the good life without overly determining what the good life is. "The good life," he writes, "is one of engagement, and engagement is variously realized by various people" (Borgmann 1984, 214). While a technical politics can influence the design of objects so that they reflect democratic values, it cannot guarantee that these values will be more meaningfully experienced. While a technical politics can lead to more individual freedom, it does not necessarily lead to an enriched sense of freedom. For an object to lead to an enriched sense of freedom, it needs, according to Borgmann, to promote unity over dispersement, and tradition over instantaneity. Values such as these naturally belong to objects, or can be acquired by them, but cannot be designed into them.

To take some of Borgmann's favorite examples, a musical instrument such as a violin can reflect the history of its use in the texture of its wood (Borgmann 1992b, 294); with its seasonal variations, a wilderness area speaks of the natural belonging together of time and space (Borgmann 1984, 191). We need to bring more things like these into our lives, and use technology to enhance our direct experience of them (as in wearing the right kinds of boots for a hike in the woods), for technology to deliver on its promise of bringing about a better life. As Borgmann writes toward the end of *TCCL,* "So counterbalanced, technology can fulfill the promise of a new kind of freedom and richness" (Borgmann 1984, 248).

Thus for Borgmann the most critical moral choices that one faces regarding material culture are "material decisions" (Borgmann 1992a, 112): decisions regarding whether to purchase or adopt a technical device or to become more engaged with things. These decisions, like the decisions to participate in the process of design of an artifact, tend to be inconspicuous. The second type of decision, as Wiebe E. Bijker, Thomas P. Hughes, and Trevor Pinch have shown (1987), fades from public memory over time. The end result of design turns into a "black box" and takes on the appearance of having been created solely by technical experts. The moral decisions Borgmann describes are just as inconspicuous because of the nature of the context in which they are discussed and made. This context is called domestic life. "Technology," he observes, "has step by step stripped the household of substance and dignity" (Borgmann 1984, 125). Just as Borgmann recalls our attention to the things of everyday life, he also makes us remember the importance of the household as a locus for everyday moral decision making. Thus Borgmann's reflections on how technology might be reformed can also be seen as an attempt to restore the philosophical significance of ordinary life.

BORGMANN AND THE RENEWAL OF PHILOSOPHICAL INTEREST IN ORDINARY LIFE

In this attempt, Borgmann does not stand alone. Over the course of the past two decades or so in North America, everyday life has been making a philosophical comeback. Five years after the publication of Borgmann's *TCCL* appeared Charles Taylor's *Sources of the Self,* a fascinating and ambitious account of the history of the making of modern identity. Heard throughout this book is the phrase "the affirmation of everyday life," a life characterized in Taylor's understanding by our nonpolitical relations with others in the context of the material world. As he sees it, affirming this life is one of the key features in the formation of our perception of who we are (Taylor 1989, 13). Against the horizons of our lives of work and play, friendship and family, we raise moral concerns that go beyond the questions of duties and obligations familiar to philosophers. What sorts of lives have the character of good lives, lives that are meaningful and worth living? What does one need to do to live a life that would be good in this sense? What can give my life a sense of purpose? In raising these questions, we affirm ordinary life. This affirmation is so deeply woven into the fabric of our culture that its very pervasiveness, Taylor maintains, serves to shield it from philosophical sight (Taylor 1989, 498).

Other signs point as well to a resurgence of philosophical interest in the moral dimensions of ordinary life. Take, for example, two fairly recent approaches to moral philosophy. In one of these approaches, philosophers such as Lawrence Blum, Christina Hoff Summers, John Hartwig, and John Deigh have been giving consideration to the particular ethical problems triggered by interpersonal relationships, those relationships among persons who know each other as friends or as family members or who are otherwise intimately connected. As George Graham and Hugh LaFollette note in their book *Person to Person,* these relationships are ones that almost all of us spend a tremendous amount of time and energy trying to create and sustain (Graham and LaFollette 1989, 1). Such activity engenders a significant amount of ethical confusion. Creating new relationships often means making difficult decisions about breaking off relationships in which one is already engaged. Maintaining interpersonal relationships often means making difficult decisions about what the demands of love and friendship entail. In accepting the challenge to sort through some of this confusion in a philosophically meaningful way, those involved with the ethics of interpersonal relationships willingly pay attention to ordinary life. In the process, they worry about the appropriateness of importing the standard moral point of view and standard moral psychology used for our dealings with others in larger social contexts—the Kantian viewpoint of impartiality and the distrust of emotions as factors in moral decision making—into the smaller and more intimate settings of families and friendships.

Another, related conversation about ethics includes thinkers such as Virginia Held, Nel Noddings, Joan Tronto, Rita Manning, Marilyn Friedman, and others whose work has been influenced by Carol Gilligan's research into the development of moral reasoning among women. I will call the enterprise in which these theorists are engaged feminist ethics, since I believe that description would be agreeable to those whom I have just mentioned, all of whom take the analysis of women's moral experiences and perspectives to be the starting point from which to rethink ethical theory.[1] Like interpersonal ethics, feminist ethics (particularly the ethics of care) places particular value on our relationships with those with whom we come into face-to-face contact in the context of familial and friendly relations. Its key insight lies in the idea that the experience of looking out for those immediately around one, an experience traditionally associated with women, is morally significant, and needs to be taken into account by anyone interested in developing a moral theory that would be a satisfactory and useful guide to the moral dilemmas facing us in all areas of life. Thus this approach to ethics also willingly accepts the challenge of paying philosophical attention to ordinary life. This challenge is summed up nicely by Virginia Held: "Instead of importing into the household principles derived from the marketplace, perhaps we should export to the wider society the relations suitable for mothering persons and children" (Held 1987, 122).

On the surface, these three paths of ethical inquiry—Borgmann's ethics of modern technology, the ethics of interpersonal relationships, and feminist ethics—are occupied with different ethical questions. But they are united, it seems to me, in at least two ways. First, they are joined by their mutual contesting of the values upon which Kantian moral theory in particular and the Enlightenment in general are based. Wherever the modernist project of submitting public institutions and affairs to one's personal scrutiny went forward, certain privileges were enforced: that of reason over emotion, the "naked self" over the self in relation to others, impartiality over partiality, the public realm over the private sphere, culture over nature, procedural over substantive reasoning, and mind over body. In addition to the critique of Kantian ethics already mentioned by philosophers writing within a framework of an ethics of interpersonal relationships, feminist ethics has argued that these privileges led to the construction of moral theories insensitive to the ways in which women represent their own moral experience. Joining his voice to these critiques, Borgmann has written (while simultaneously praising the work of Carol Gilligan), "Universalism neglects . . . ways of empathy and care and is harsh toward the human subtleties and frailties that do not convert into the universal currency. . . . The major liability of moral universalism is its dominance; the consequence of dominance is an oppressive impoverishment of moral life" (Borgmann 1992a, 54–55).

A second feature uniting these relatively new forms of moral inquiry is a more positive one. Each attempts to limit further increases in the "impoverishment of moral life" by calling attention to the *moral* aspects of typical features of ordinary life that have traditionally been overlooked or even denied. The act of mothering (for Virginia Held), the maintenance of friendships (for Lawrence Blum) and the loving preparation of a home-cooked meal (for Borgmann) have all been defended, against the dominant belief to the contrary, as morally significant events.[2]

Despite the similarities and common concerns of these three approaches to moral philosophy, however, little engagement exists among them. Between feminist ethics and the ethics of interpersonal relationships, some engagement can be found: for instance, the "other-centered" model of friendship discussed in the latter is of interest to care ethicists as part of an alternative to Kantian ethics. However, both of these modes of ethical inquiry have shown little interest in the ethical dimensions of material culture. Nel Noddings, for example, believes that while caring can be a moral phenomenon when it is directed toward one's own self and that of others, it loses its moral dimension when it is directed toward things. In her book *Caring,* she defends the absence of discussion of our relations to things in her work: "as we pass into the realm of things and ideas, we move entirely beyond the ethical. . . . My main reason for setting things aside is that we behave ethically only through them and not toward them" (Noddings 1984, 161–62).

And yet in ordinary life ethical issues of technology, gender, and interpersonal relationships overlap in numerous ways. One wonders as a responsible parent whether it is an act of caring to buy one's son a Mighty Morphin Power Ranger. If I wish to watch a television program that my spouse cannot tolerate, should I go into another room to watch it or should I see what else is on television so that we could watch a program together? Is a married person committing adultery if he or she has an affair with a stranger in cyberspace? Seeing these interconnections, one wonders what might be the result were the probing, insightful questioning initiated by Borgmann into the moral significance of our material culture widened to include the other voices mentioned here. What would we learn, for instance, if Borgmann's technological ethics were explored from the perspective of feminist ethics?

In the context of this paper I can do no more than start to answer this question. With this in mind, I would like to look at one of the central claims of *TCCL:* the claim that the objects of material culture fall into the category of either things or devices.

FEMINISM AND THE DEVICE PARADIGM

As Borgmann describes them, things are machines that, in a manner of speaking, announce their own narratives and as a result are generous in the effects they can produce. For example, we can see the heat of the wood burning in the fireplace being produced in front of our eyes—the heat announces its own story, its own history, in which its relation to the world is revealed. In turn, fireplaces give us a place to focus our attention, to regroup and reconnect with one another as we watch the logs burn. In this regard, Borgmann speaks compellingly not only of the fireplace but also of wine: "Technological wine no longer bespeaks the particular weather of the year in which it grew since technology is at pains to provide assured, i.e. uniform, quality. It no longer speaks of a particular place since it is a blend of raw materials from different places" (Borgmann 1984, 49).

Devices, on the other hand, hide their narratives by means of their machinery and as a result produce only the commodity they were intended to produce. When I key the characters of the words I want to write into my portable computer they appear virtually simultaneously on the screen in front of me. I cannot see the connection between the one event and the other, and the computer does not demand that I know how it works in order for it to function. The commodity we call "processed words" is the result. While things lead to "multi-sided experiences," devices produce "one-sided experiences."[3] Fireplaces provide warmth, the possibility of conviviality, and a closer tie to the natural world; a central heating system simply provides warmth.

What thoughts might a philosopher working within the framework of feminist ethics have about this distinction? To begin with, I think she would be somewhat uneasy with the process of thinking used to make decisions about whether a particular object would be classified as a thing or a device. In this process, Borgmann abstracts from the particular context of the object's actual use and focuses his attention directly on the object itself. The view that some wine is "technological," as the example described above shows, is based on the derivation of the wine, the implication being that putting such degraded wine on the table would lead to a "one-sided experience" and further thwart, albeit in a small way, technology's capability to contribute meaningfully to the good life. In a feminist analysis of the moral significance of material culture, a different methodology would prevail. The analysis of material objects would develop under the assumption that understanding people's actual experiences of these objects, and in particular understanding the actual experiences of women who use them, would be an important source of information in deciding what direction a technological reform of society should take.

The attempt to make sense of women's experience of one specific technological innovation is the subject of communication professor Lana Rakow's book *Gender on the Line: Women, the Telephone, and Community Life* (1992). As its title suggests, this is a study of the telephone practices of the women residents of a particular community, a small midwestern town she called, to protect its identity, Prospect.

Two features of Rakow's study are of interest with regard to our topic. One relates to the discrepancy between popular perceptions of women's use of the telephone, and the use revealed in her investigation. She was well aware at the beginning of her study of the popular perception, not just in Prospect but widespread throughout American culture, of women's use of the telephone. In the popular perception, characterized by expressions such as "Women just like to talk on the phone" and "Women are on the phone all the time," telephone conversations among women appear as "productivity sinks," as ways of wasting time. Understandably from this perception the telephone could appear as a device used for the sake of idle chatter that creates distraction from the demands of work and everyday life. This is how Borgmann sees it:

> The telephone network, of course, is an early version of hyperintelligent communication, and we know in what ways the telephone has led to disconnectedness. It has extinguished the seemingly austere communication via letters. Yet this austerity was wealth in disguise. To write a letter one needed to sit down, collect one's thoughts and world, and commit them laboriously to paper. Such labor was a guide to concentration and responsibility. (Borgmann 1992a, 105)

Rakow's study, however, did not support the popular perception. She found that the "womentalk" engaged in by her subjects was neither chatter nor gossip. Rather, it was a means to the end of producing, affirming, and reinforcing the familial and community connections that played a very large role in defining these women's lives. Such "phone work," very often consisting of exchanges of stories, was the stuff of which relations were made: "Women's talk holds together the fabric of the community, building and maintaining relationships and accomplishing important community relations" (Rakow 1992, 34).

Let me suggest some further support for this view from my own experience. While I was growing up, I frequently witnessed this type of phone work on Sunday afternoons as my mother would make and receive calls from other women to discuss "what had gone on at church." Although these women had just seen each other at church several hours before, their phone calls played exactly the role that Rakow discovered they played in Prospect. At the time, they were not allowed to hold any positions of authority within the organizational structure of this particular church. The meaning of these phone calls would be missed by calling them idle talk; at least in part, these phone visits served to strengthen and reinforce their identity within the gendered community to which these women belonged.

Another interesting feature of Rakow's study was its discovery of how women used the telephone to convey care:

> Telephoning functions as a form of care-giving. Frequency and duration of calls . . . demonstrate a need for caring or to express care (or a lack of it). Caring here has the dual implication of caring *about* and caring *for*—that is, involving both affection and service. . . . While this [care-giving role] has been little recognized or valued, the caring work of women over the telephone has been even less noted. (Rakow 1992, 57)

As one of the places where the moral status of the care-giving role of women has been most clearly recognized and valued, feminist ethics is, of course, an exception to this last point. Rakow's recognition of the telephone as a means to demonstrate one's caring for speaks directly to Nel Noddings's understanding of why giving care can be considered a moral activity (Noddings 1984). In caring one not only puts another's needs ahead of one's own, but, in reflecting on how to take care of those needs, one sees oneself as being related to, rather than detached from, the self of the other. In commenting that not only checking on the welfare of another woman or phoning her on her birthday but "listening to others who need to talk is also a form of care" (Rakow 1992, 57), Rakow singles out a kind of caring that well reflects Noddings's description. More often one needs to listen to others who call one than one needs to call others; and taking care of the needs of those who call often involves simply staying on the phone while the other talks. As Rakow correctly points out, this makes this particular practice of telephone caring a form of work. Those who criticize the ethics of care for taking up too much of one's time with meeting the needs of individual others might also be critical of Rakow's subjects who reported that

> they spend time listening on the phone when they do not have the time or interest for it. . . . One elderly woman . . . put a bird feeder outside the window by her telephone so she can watch the birds when she has to spend time with these phone calls. "I don't visit; I just listen to others," she said. (Rakow 1992, 57)

As these features of telephone conversations came to light in the interviews she conducted with the women of Prospect, Rakow began to see the telephone as "a gendered, not a neutral, technology" (Rakow 1992, 33). As a piece of gendered technology, the telephone arguably appears more like a thing than like a device, allowing for, in Borgmann's phrase, the "focal practice" of caring to take place. Looking at the telephone from this perspective raises doubts about Borgmann's assessment of the telephone. Has the telephone in fact become a substitute for the thing of the letter, contributing to our widespread feelings of disconnectedness and to our distraction? Rakow's fieldwork provides support for the idea that phone work, much like letter writing, can be "a guide to concentration and responsibility." By giving care over the phone, the development of both these virtues is supported. Thus, on Borgmann's own terms—"The focal significance of a mental activity should be judged, I believe, by the force and extent with which it gathers and illuminates the tangible world and our appropriation of it" (Borgmann 1984, 217)—it is difficult to see how using the telephone as a means of conveying care could not count as a focal concern.

Along with the question of whether a particular item of our material culture is or is not a device, looking at the device paradigm from a feminist point of view gives rise to at least two other issues. One is connected to an assumption on which this paradigm rests: that the moral significance of an object is directly related to whether or not that object is a substitute for the real thing. This issue is also connected to the idea that because technological objects are always substitutes for the real thing, the introduction of new technology tends to be a step forward in the impoverishment of ordinary life.

Certainly technological objects are always substitutes for *something or another.* A washing machine is a substitute for a washing board, dryers are substitutes for the line out back, krab [*sic*] is often found these days on salad bars, and so forth. In some cases, the older object gradually fades from view, as happened with the typewriter, which (but only as of fairly recently) is no longer being produced. In other cases, however, the things substituted for is not entirely replaced, but continues to coexist alongside the substitute. In these cases, it is harder to see how the technological object is a substitute *for the real thing,* and thus harder to see how the introduction of the new object threatens our sense of engagement with the world. While it is true that telephones substitute for letter writing, as Borgmann observes, the practice of letter writing goes on, even to the point of becoming intertwined with the use of the telephone. Again, from Rakow:

> The calls these women make and the letters they send literally call families into existence and maintain them as a connected group. A woman who talks daily to her two nearby sisters demonstrated the role women play in keeping track of the well-being of family members and changes in their lives. She said, "If we get a letter from any of them (the rest of the family) we always call and read each other the letters." (Rakow 1992, 64)

Perhaps, though, the largest question prompted by Rakow's study has to do with whether Borgmann's distinction itself between things and devices can hold up under close consideration of the experiences and practices of different individuals. There are many devices that can be and are used as the women in this study used the telephone. Stereos, for example, can be a means for someone to share with someone else particular cuts on a record or songs from a CD to which she or he attaches a great deal of personal significance. In this way, stereos can serve as equipment that aid the development of mutual understanding and relatedness, rather than only being mechanisms for disengagement. The same goes for the use of the computer as a communicative device. Empirical investigations into the gendered use of computer-mediated communications suggest that while women do not necessarily use this environment like the telephone, as a means of promoting care, they do not "flame" (send electronic messages critical of another individual) nearly as much as do men, and they are critical of men who do engage in such activity.[4]

In particular, from a feminist perspective one might well wonder whether, in Borgmann's language, the use of those "conjunctions of machinery and commodity" inevitably hamper one's efforts at relating more to others and to the world. Borgmann argues that because devices hide their origins and their connections to the world, they cannot foster our own bodily and social engagement with the world. But as I have tried to show here, this is arguably not the case. Whether or not a material object hides or reveals "its own story" does not seem to have a direct bearing on that object's capacity to bind others together in a narrative web. For instance, older women participating in Rakow's study generally agreed that telephones improved in their ability to serve as a means of social support and caregiving once their machinery became more hidden: when private lines took the place of party lines and the use of an operator was not necessary to place a local call. To generalize, the machinery that clouds the story of a device does not appear to prevent that device from playing a role in relationship building.

DEVICES AND THE PROMISE OF TECHNOLOGY

While a child growing up in New Jersey, I looked forward on Friday evenings in the summer to eating supper with my aunt and uncle. I would run across the yard separating my parents' house from theirs to take my place at a chair placed at the corner of the kitchen table. The best part of the meal, I knew, would always be the same, and that was why I looked forward to these evenings. While drinking lemonade from the multicolored aluminum glasses so popular during the 1950s, we would eat Mrs. Paul's fish sticks topped with tartar sauce. With their dubious nutritional as well as aesthetic value, fish sticks are to fresh fish as, in a contrast described eloquently by Borgmann, Kool Whip is to fresh cream (Borgmann 1987, 239–42). One doesn't know the seas in which the fish that make up fish sticks swim. Nearly anyone can prepare them in a matter of minutes. Still, despite these considerations, these meals were marked by family sociability and kindness, and were not hurried affairs.

I recall these meals now with the following point in mind. One might be tempted by the course of the discussion here to say that the objects of material culture should not be divided along the lines proposed in *TCCL* but divided in another manner. From the perspective of feminist ethics, one might suggest that one needs to divide up contemporary material culture between relational things, things open up the possibility of caring relations to others, and non-relational things: things that open up the possibility of experience but not the possibility of relation. Telephones, on this way of looking at things, would count as relational things. Virtual reality machines, such as the running simulator Borgmann imagines in *CPD*, or golf simulators that allow one to move from the green of the seventeenth hole at Saint Andrews to the tee of the eighteenth hole at Pebble Beach, would be nonrelational things. One can enjoy the experiences a virtual golf course makes possible, but one cannot in turn, for example, act in a caring manner toward the natural environment it so vividly represents. But the drawback of this distinction seems similar to the drawback of the distinction between things and devices: the possibility of using a thing in a relational and thus potentially caring manner seems to depend more on the individual using that thing and less on the thing itself. Depending on who is playing it, a match of virtual golf has the potential of strengthening, rather than undoing, narrative connections between oneself, others and the world.

But if our discussion does not lead in this direction, where does it lead? Let me suggest that although it does not lead one to reject the device paradigm outright, it does lead one to recognize that while any device does use machinery to produce a commodity, the meaning of one's experience associated with this device does not necessarily have to be diminished. And if one can use technology (such as the telephone) to carry out focal practices (such as caregiving), then we might have cause to believe that there are other ways to recoup the promise of technology than Borgmann sees. As mentioned earlier, his hope is that we will give technology more of a supporting role in our lives than it has at present (Borgmann 1984, 247), a role he interprets as meaning that it should support the focal practices centered around focal things.

But if devices can themselves support focal practices, then the ways in which technology can assume a supporting role in our lives are enhanced.

But if the idea that devices can support focal practices is in one way a challenge to the device paradigm, in another way it gives additional weight to the notion that there are limits to reforming technology through the process of democratic design. When they are used in a context involving narrative and tradition, devices can help build engagement and further reinforce the cohesiveness of civil society. Robert Putnam has pointed out the importance of trust and other forms of "social capital" necessary for citizens to interact with each other in a cooperative manner. As social capital erodes, democracy itself, he argues, is threatened (Putnam 1995, 67). While this paper has suggested that devices can under some conditions further the development of social capital, it is difficult to see how they can be deliberately designed to do so. In thinking about how to reform technology from a democratic perspective, we need to remember the role of features of ordinary life such as narrative and tradition in making our experience of democratic values more meaningful. Borgmann's reminder to us of this role is, it seems to me, one of the reasons why *TCCL* will continue to have a significant impact in shaping the field of the philosophy of technology.

NOTES

1. I am not using "feminist ethics" in a technical sense, but as a way of referring to the philosophical approach to ethics that starts from a serious examination of the moral experience of women. For philosophers such as Alison Jaggar, the term feminist ethics primarily means an ethics that recognizes the patriarchal domination of women and the need for women to overcome this system of male domination. Thus she and others might disagree that the ethics of care, as I take it here, is an enterprise of feminist ethics.

2. For example, Virginia Held has written: "[Feminist moral inquiry] pays attention to the neglected experience of women and to such a woefully neglected though enormous area of human moral experience as that of mothering. . . . That this whole vast region of human experience can have been dismissed as 'natural' and thus as irrelevant to morality is extraordinary" (Held 1995, 160).

3. The term "multi-sided experiences" is used by Mihaly Csikszentmihalyi and Eugene Rochbert-Halton in their work *The Meaning of Things,* discussed in Borgmann 1992b.

4. See, for example, Susan Herring, "Gender Differences in Computer-Mediated Communication: Bringing Familiar Baggage to the New Frontier" (unpublished paper).

Part IV

TECHNOLOGY AND POLITICS

Technology plays a vital role in the organization of social life. It shapes the very conditions of our lives. It affects the way we live, the way we work, the way we get our information, the nature of our healthcare, and countless other basic features of social life. Decisions about the design, development, and administration of technologies often have lasting, even irreversible, effects on a society. Technology may bring opportunity and advantages to some members of society but risks and disadvantages to other members. It may give some members of society power and control that other members do not enjoy. Or it may expose some members to hazards and risks from which others are free. Technology policy is always tied up with political problems regarding distributive justice and social equality. The articles in this section address this political dimension of technology. The common thread running through the selections here is the conviction that no firm distinction can be made between technological and political concerns. Both determine the distribution of advantages and disadvantages, benefits and burdens of our shared social life. Like the ethical dimension, the political dimension is intrinsic to technological practice. Technology is neither ethically nor politically neutral. Technology shapes our lives as citizens, and thus is bound up with questions of freedom, democracy, social justice, and our vision of the good life.

In "Do Artifacts Have Politics?" Langdon Winner explores the idea that technical things have political qualities built into them and that these qualities embody specific forms of power and authority. Winner examines two ways in which artifacts have political qualities. First are instances in which a technological device or system is used to settle a social or political problem in a community. In these cases, the technology is far from neutral but designed to produce results that structure social relations, reinforce vested interests, and engineer human relations. Examples include highways with low overpasses designed to prevent bus traffic (thus excluding poor people from certain areas), riot-proof campuses designed to prevent large crowds from amassing, and curb cuts designed to accommodate people in wheelchairs. In each case, the "technological deck was stacked" in advance to favor certain social and political interests. Many technological devices and systems function like laws that build order into our world. They constitute a form of life.

The second manner in which artifacts have political qualities is instances in which a technology requires or is at least compatible with particular political relationships. Inherently, political technologies are far more rigid and inflexible; they unavoidably shape and pattern human relationships into political forms of life—for example, as centralized or decentralized, egalitarian or inegalitarian, or repressive or liberating. The strong version of this claim holds that the technology itself requires specific social conditions in order to operate. The weaker version holds that a given technology is merely compatible with particular social and political relationships. Examples include technological systems of power, communication, transportation, and production, all of which are large-scale, centralized, hierarchical organizations administered by highly skilled managers. The nature of the systems themselves opposes alternate social-political organizations that might be more democratic, egalitarian, and decentralized. Winner invites us to weigh the moral and political claims of practical necessity and efficiency against the moral and political demands of democratic self-management, freedom, justice, and community. If the politics internal to technological management cannot be separated so neatly from public policy, why should we sacrifice our political rights for the sake of technical efficiency? If the technologies themselves have intractable properties, then it doesn't matter what kind of social system they function in—capitalist, socialist, anarchist—because their internal political qualities will remain unchanged. Winner argues that we need to tend to the flexible use of technologies in relation to broader social-political contexts and their inflexible inherent qualities.

In "Strong Democracy and Technology," Richard E. Sclove argues that citizens have the right to participate in decisions about a society's basic organization, structure, and evolution. On this model of "strong democracy," citizens should be included in decisions about more than just legal and political matters; they should be included in decisions about technological design and practice, as well. Something as important as the sociotechnical organization of society should be established by democratic procedures in a way that satisfies the common interests of citizens. Sclove's argument is strictly moral. If a decision is legitimate, it must have the informed free consent of those affected by it. In the United States, decisions about the design, management, and uses of technological systems are made by elected officials and market forces, often influenced by small groups of technically skilled people, who we can only hope have our best interests in mind. The stakes of having such important decisions about our lives made by other people are nothing less than our rights and liberties. If we have little or no say in decisions that shape and pattern our collective fate, Sclove finds, our autonomy is unacceptably compromised.

The implication for public policy is to create mechanisms that enable people to participate in technological design and management, as well as to contest or reject a technology whenever we determine that our rights, liberties, opportunities, and collective well-being are threatened. Such decisions should be made in a democratic process that would include representatives from grassroots organizations, public interest groups, academic scientists from the social and natural sciences, and community organizations. Yet Sclove takes this proposal even further and argues that strong democracy requires democratic background conditions that foster among citizens a sense of civic responsibility and readiness to participate fully in decision-making processes. These background conditions include *democratic politics* to ensure full representation and participation at both the local and national levels, *democratic community* to nurture the bonds of mutual understanding and respect among citizens, and *democratic work* to develop our talents and capacities to participate responsibly—in a way that enhances the common needs and interests of everyone (rather than just private accumulation). Strong democracy implies not only a democratic politics of technology but also the democratic institutions, communities, and workplaces that would reflect, support, and enable such a politics of technology.

Tony Smith develops the idea of democratizing technology in more detail in "Socialism and the Democratic Planning of Technical Change," excerpted from his book *Technology and Capital in the Age of Lean Production.* Like Sclove, Smith asserts that economic, political, and technical institutions and practices should be accountable, democratic, and egalitarian. However, for Smith, the system of production, distribution, and administration that best upholds the values of political liberalism (e.g., justice, equality, and autonomy) is not capitalism but democratic socialism. In this reading, the current form of capitalism we live in, which Smith calls "lean production" and which underlies the so-called new economy or globalization, is examined. Lean production is characterized by a "just in time" system of production and distribution of commodities (i.e., less raw material and more partially assembled parts), elimination of non–"value adding" positions in production (i.e., management and quality control), mass customization to meet consumer desire, the creation of jobs for knowledge workers, and cooperation among and within firms. The claim of proponents of the new economy is that lean production will greatly improve capital/labor relations, capital/consumer relations, the relation among units of capital—and eventually capitalism itself. Smith then considers the kinds of institutional changes that would have to be made for the promises of the new economy to be realized. The answer is public ownership of large enterprises, workplace democracy, and social accountability for all investment decisions—in other words, a transformation from a capitalist market society to a democratic form of socialism.

Democratic planning of technical change plays a crucial role in socialist democracy. Yet since World War II in the United States, there already exists a technology policy based on democratic central planning, through which two thirds of all research and development has been funded by the government. Smith suggests a national technology board be developed to allocate resources for technological research and development, determined in a democratic process, and then that it finally be set in motion at the local level in worker-run enterprises for production and distribution. (These enterprises would be funded either internally or by investments from community banks.) Smith then shows how democratic planning of technical change would not be plagued by the same problems of bureaucratic central planning. He claims that it would foster rather than hamper innovation, create incentives for technical change, and improve the quality of goods by making planners accountable to consumers. This model of democratic socialism would not only be morally preferable and a feasible alternative, but also would not suffer the shortcomings of either bureaucratic state socialism or free market capitalism. This model would be neither as inefficient and undemocratic as a centralized planned economy nor as unstable and inegalitarian as lean production.

In "The Insurgent Architect at Work," excerpted from *Spaces of Hope,* David Harvey asks us to imagine that we are architects striving to change our world. The insurgent architect must consider several "spaciotemporal scales" or "theaters of thought and action" that occur simultaneously in different places shaping and patterning our world and our collective fate. These theaters include fields of action, meaning, geography, and technology, among other factors that shape our lives individually, socially, and politically. Yet, despite the vast complexity of the world, there are many ways of designing new and better forms of life. Harvey describes seven theaters of insurgent action in which we can transform the world through a process of "dialectical utopianism"—in other words, creative action grounded in historical-geographic realities. *Dialectics* is a philosophical method that attempts to trace out the inner law and internal connections that make a continuous whole out of disparate processes of change and development. It is geared toward uncovering the relationships among things, in particular their internal (i.e., necessary) connections and historical geneses. Dialectics lets us find order in diversity without reducing the complexity of the world to a simplistic, casual model.

The first of the seven interrelated spaciotemporal scales described by Harvey is the *theater of private personal life*, where changes in the material organization of the world intimately affect us as individuals. To say that "the personal is political" means that personal transformation is always tied up with social and political issues, such as class interests, public opinion, acts of government, and architectural practices that may foster or diminish private life. In the next two theaters, the focus shifts from the political person to the "socioecological" relations, communities, and collectives that make us who we are. This broader scope of action includes potentially the transformation of such things as systems of property rights, living and working conditions, transportation systems, and urban landscapes with the aim of connecting communities on ever wider spaciotemporal scales.

The next two theaters of action include acts that are geared toward balancing the universal requirements of social justice with the needs and interests of particular communities and places. In order to address site-specific and group-specific problems, Harvey argues that it is crucial for mediating institutions and built environments to mediate the dialectic between particularity and universality. The sixth theater of action involves practices that aim at translating the political aspirations across the wide range of socioecological and political-economic conditions. Insurgent architects must mediate between theory and action, universal and particular, constraints and possibilities. Finally, the seventh theater encompasses actions that touch upon the moment of universality itself—in other words, that which bridges judgment and decision, the personal and political, and natural and social orders. The universal aspect of dialectical utopianism is modeled after the theory of human rights from the United Nations Declaration of Human Rights. The insurgent architect, Harvey writes, must respect human rights in an ongoing process of transforming socioecological changes across all of the other theaters of action. Harvey's work bridges architecture and philosophy of technology by showing that social action is a complicated network of humans, material stuff, ideas, and values. The design of a more just society is inseparable from a critique of social and economic practices, environmental degradation, and architectural processes that may either enable or hinder our ability to live well together.

19

Do Artifacts Have Politics?

Langdon Winner

In controversies about technology and society, there is no idea more provocative than the notion that technical things have political qualities. At issue is the claim that the machines, structures, and systems of modern material culture can be accurately judged not only for their contributions to efficiency and productivity and their positive and negative environmental side effects, but also for the ways in which they can embody specific forms of power and authority. Since ideas of this kind are a persistent and troubling presence in discussions about the meaning of technology, they deserve explicit attention.

It is no surprise to learn that technical systems of various kinds are deeply interwoven in the conditions of modern politics. The physical arrangements of industrial production, warfare, communications, and the like have fundamentally changed the exercise of power and the experience of citizenship. But to go beyond this obvious fact and to argue that certain technologies *in themselves* have political properties seems, at first glance, completely mistaken. We all know that people have politics; things do not. To discover either virtues or evils in aggregates of steel, plastic, transistors, integrated circuits, chemicals, and the like seems just plain wrong, a way of mystifying human artifice and of avoiding the true sources, the human sources of freedom and oppression, justice and injustice. Blaming the hardware appears even more foolish than blaming the victims when it comes to judging conditions of public life.

Hence, the stern advice commonly given those who flirt with the notion that technical artifacts have political qualities: What matters is not technology itself, but the social or economic system in which it is embedded. This maxim, which in a number of variations is the central premise of a theory that can be called the social determination of technology, has an obvious wisdom. It serves as a needed corrective to those who focus uncritically upon such things as "the computer and its social impacts" but who fail to look behind technical devices to see the social circumstances of their development, deployment, and use. This view provides an antidote to naïve technological determinism—the idea that technology develops as the sole result of an internal dynamic and then, unmediated by any other influence, molds society to fit

"Do Artifacts Have Politics?" reprinted by permission of *Dædalus*, Journal of the American Academy of Arts and Sciences, from the issue entitled, "Modern Technology: Problem or Opportunity?" 109 (1) (Winter 1980).

its patterns. Those who have not recognized the ways in which technologies are shaped by social and economic forces have not gotten very far.

But the corrective has its own shortcomings; taken literally, it suggests that technical *things* do not matter at all. Once one has done the detective work necessary to reveal the social origins—power holders behind a particular instance of technological change—one will have explained everything of importance. This conclusion offers comfort to social scientists. It validates what they had always suspected, namely, that there is nothing distinctive about the study of technology in the first place. Hence, they can return to their standard models of social power—those of interest-group politics, bureaucratic politics, Marxist models of class struggle, and the like—and have everything they need. The social determination of technology is, in this view, essentially no different from the social determination of, say, welfare policy or taxation.

There are, however, good reasons to believe that technology is politically significant in its own right, good reasons why the standard models of social science only go so far in accounting for what is most interesting and troublesome about the subject. Much of modern social and political thought contains recurring statements of what can be called a theory of technological politics, an odd mongrel of notions often crossbred with orthodox liberal, conservative, and socialist philosophies.[1] The theory of technological politics draws attention to the momentum of large-scale sociotechnical systems, to the response of modern societies to certain technological imperatives, and to the ways human ends are powerfully transformed as they are adapted to technical means. This perspective offers a novel framework of interpretation and explanation for some of the more puzzling patterns that have taken shape in and around the growth of modern material culture. Its starting point is a decision to take technical artifacts seriously. Rather than insist that we immediately reduce everything to the interplay of social forces, the theory of technological politics suggests that we pay attention to the characteristics of technical objects and the meaning of those characteristics. A necessary complement to, rather than a replacement for, theories of the social determination of technology, this approach identifies certain technologies as political phenomena in their own right. It points us back, to borrow Edmund Husserl's philosophical injunction, *to the things themselves.*

In what follows I will outline and illustrate two ways in which artifacts can contain political properties. First are instances in which the invention, design, or arrangement of a specific technical device or system becomes a way of settling an issue in the affairs of a particular community. Seen in the proper light, examples of this kind are fairly straightforward and easily understood. Second are cases of what can be called "inherently political technologies," man-made systems that appear to require or to be strongly compatible with particular kinds of political relationships. Arguments about cases of this kind are much more troublesome and closer to the heart of the matter. By the term "politics" I mean arrangements of power and authority in human associations as well as the activities that take place within those arrangements. For my purposes here, the term "technology" is understood to mean all of modern practical artifice, but to avoid confusion I prefer to speak of "technologies" plural, smaller or larger pieces or systems of hardware of a specific kind.[2] My intention is not to settle any of the issues here once and for all, but to indicate their general dimensions and significance.

TECHNICAL ARRANGEMENTS AND SOCIAL ORDER

Anyone who has traveled the highways of America and has gotten used to the normal height of overpasses may well find something a little odd about some of the bridges over the parkways on Long Island, New York. Many of the overpasses are extraordinarily low, having as little as nine feet of clearance at the curb. Even those who happened to notice this structural peculiarity would not be inclined to attach any special meaning to it. In our accustomed way of looking at things such as roads and bridges, we see the details of form as innocuous and seldom give them a second thought.

It turns out, however, that some two hundred or so low-hanging overpasses on Long Island are there for a reason. They were deliberately designed and built that way by someone who wanted to achieve a particular social effect. Robert Moses, the master builder of roads, parks, bridges, and other public works of the 1920s to the 1970s in New York, built his overpasses according to specifications that would discourage the presence of buses on his parkways. According to evidence provided by Moses' biographer, Robert A. Caro, the reasons reflect Moses' social class bias and racial prejudice. Automobile-owning whites of "upper" and "comfortable middle" classes, as he called them, would be free to use the parkways for recreation and commuting. Poor people and blacks, who normally used public transit, were kept off the roads because the twelve-foot-tall buses could not handle the overpasses. One consequence was to limit access of racial minorities and low-income groups to Jones Beach, Moses' widely acclaimed public park. Moses made doubly sure of this result by vetoing a proposed extension of the Long Island Railroad to Jones Beach.

Robert Moses' life is a fascinating story in recent U.S. political history. His dealings with mayors, governors, and presidents; his careful manipulation of legislatures, banks, labor unions, the press, and public opinion could be studied by political scientists for years. But the most important and enduring results of his work are his technologies, the vast engineering projects that give New York much of its present form. For generations after Moses' death and the alliances he forged have fallen apart, his public works, especially the highways and bridges he built to favor the use of the automobile over the development of mass transit, will continue to shape that city. Many of his monumental structures of concrete and steel embody a systematic social inequality, a way of engineering relationships among people that, after a time, became just another part of the landscape. As New York planner Lee Koppleman told Caro about the low bridges on Wantagh Parkway, "The old son of a gun had made sure that buses would *never* be able to use his goddamned parkways."[3]

Histories of architecture, city planning, and public works contain many examples of physical arrangements with explicit or implicit political purposes. One can point to Baron Haussmann's broad Parisian thoroughfares, engineered at Louis Napoleon's direction to prevent any recurrence of street fighting of the kind that took place during the revolution of 1848. Or one can visit any number of grotesque concrete buildings and huge plazas constructed on university campuses in the United States during the late 1960s and early 1970s to defuse student demonstrations. Studies of industrial machines and instruments also turn up interesting political stories, including some that violate our normal expectations about why technological innovations are made in the first place. If we suppose that new technologies are introduced to achieve increased efficiency, the history of technology shows that we will sometimes be disappointed. Technological change expresses a panoply of human motives, not the least of which is the desire of some to have dominion over others even though it may require an occasional sacrifice of cost savings and some violation of the normal standard of trying to get more from less.

One poignant illustration can be found in the history of nineteenth-century industrial mechanization. At Cyrus McCormick's reaper manufacturing plant in Chicago in the middle 1880s, pneumatic molding machines, a new and largely untested innovation, were added to the foundry at an estimated cost of $500,000. The standard economic interpretation would lead us to expect that this step was taken to modernize the plant and achieve the kind of efficiencies that mechanization brings. But historian Robert Ozanne has put the development in a broader context. At the time, Cyrus McCormick II was engaged in a battle with the National Union of Iron Molders. He saw the addition of the new machines as a way to "weed out the bad element among the men," namely, the skilled workers who had organized the union local in Chicago.[4] The new machines, manned by unskilled laborers, actually produced inferior castings at a higher cost than the earlier process. After three years of use the machines were, in fact, abandoned, but by that time they had served their purpose—the destruction of the union. Thus, the story of these technical developments at the McCormick factory cannot be adequately understood outside the record of workers' attempts to organize, police repression of the labor movement in Chicago during that period, and the events surrounding the bombing at Haymarket Square. Technological history and U.S. political history were at that moment deeply intertwined.

In the examples of Moses' low bridges and McCormick's molding machines, one sees the importance of technical arrangements that precede the *use* of the things in question. It is obvious that technologies can be used in ways that enhance the power, authority, and privilege of some over others, for example, the use of television to sell a candidate. In our accustomed way of thinking technologies are seen as neutral tools that can be used well or poorly, for good, evil, or something in between. But we usually do not stop to inquire whether a given device might have been designed and built in such a way that it produces a set of consequences logically and temporally *prior to any of its professed uses*. Robert Moses' bridges, after all, were used to carry automobiles from one point to another; McCormick's machines were used to make metal castings; both technologies, however, encompassed purposes far beyond their immediate use. If our moral and political language for evaluating technology includes only categories having to do with tools and uses, if it does not include attention to the meaning of the designs and arrangements of our artifacts, then we will be blinded to much that is intellectually and practically crucial.

Because the point is most easily understood in the light of particular intentions embodied in physical form, I have so far offered illustrations that seem almost conspiratorial. But to recognize the political dimensions in the shapes of technology does not require that we look for conscious conspiracies or malicious intentions. The organized movement of handicapped people in the United States during the 1970s pointed out the countless ways in which machines, instruments, and structures of common use—buses, buildings, sidewalks, plumbing fixtures, and so forth—made it impossible for many handicapped persons to move freely about, a condition that systematically excluded them from public life. It is safe to say that designs unsuited for the handicapped arose more from long-standing neglect than from anyone's active intention. But once the issue was brought to public attention, it became evident that justice required a remedy. A whole range of artifacts have been redesigned and rebuilt to accommodate this minority.

Indeed, many of the most important examples of technologies that have political consequences are those that transcend the simple categories "intended" and "unintended" altogether. These are instances in which the very process of technical development is so thoroughly biased in a particular direction that it regularly produces results heralded as wonderful breakthroughs by some social interests and crushing setbacks by others. In such cases it is neither correct nor insightful to say, "Someone intended to do somebody else harm."

Rather one must say that the technological deck has been stacked in advance to favor certain social interests and that some people were bound to receive a better hand than others.

The mechanical tomato harvester, a remarkable device perfected by researchers at the University of California from the late 1940s to the present offers an illustrative tale. The machine is able to harvest tomatoes in a single pass through a row, cutting the plants from the ground, shaking the fruit loose, and (in the newest models) sorting the tomatoes electronically into large plastic gondolas that hold up to twenty-five tons of produce headed for canning factories. To accommodate the rough motion of these harvesters in the field, agricultural researchers have bred new varieties of tomatoes that are hardier, sturdier, and less tasty than those previously grown. The harvesters replace the system of handpicking in which crews of farm workers would pass through the fields three or four times, putting ripe tomatoes in lug boxes and saving immature fruit for later harvest.[5] Studies in California indicate that the use of the machine reduces costs by approximately five to seven dollars per ton as compared to hand harvesting.[6] But the benefits are by no means equally divided in the agricultural economy. In fact, the machine in the garden has in this instance been the occasion for a thorough reshaping of social relationships involved in tomato production in rural California.

By virtue of their very size and cost of more than $50,000 each, the machines are compatible only with a highly concentrated form of tomato growing. With the introduction of this new method of harvesting, the number of tomato growers declined from approximately 4,000 in the early 1960s to about 600 in 1973, and yet there was a substantial increase in tons of tomatoes produced. By the late 1970s an estimated 32,000 jobs in the tomato industry had been eliminated as a direct consequence of mechanization.[7] Thus, a jump in productivity to the benefit of very large growers has occurred at the sacrifice of other rural agricultural communities.

The University of California's research on and development of agricultural machines such as the tomato harvester eventually became the subject of a lawsuit filed by attorneys for California Rural Legal Assistance, an organization representing a group of farm workers and other interested parties. The suit charged that university officials are spending tax monies on projects that benefit a handful of private interests to the detriment of farm workers, small farmers, consumers, and rural California generally and asks for a court injunction to stop the practice. The university denied these charges, arguing that to accept them "would require elimination of all research with any potential practical application."[8]

As far as I know, no one argued that the development of the tomato harvester was the result of a plot. Two students of the controversy, William Friedland and Amy Barton, specifically exonerate the original developers of the machine and the hard tomato from any desire to facilitate economic concentration in that industry.[9] What we see here instead is an ongoing social process in which scientific knowledge, technological invention, and corporate profit reinforce each other in deeply entrenched patterns, patterns that bear the unmistakable stamp of political and economic power. Over many decades agricultural research and development in U.S. land-grant colleges and universities has tended to favor the interests of large agribusiness concerns.[10] It is in the face of such subtly ingrained patterns that opponents of innovations such as the tomato harvester are made to seem "antitechnology" or "antiprogress." For the harvester is not merely the symbol of a social order that rewards some while punishing others; it is in a true sense an embodiment of that order.

Within a given category of technological change there are, roughly speaking, two kinds of choices that can affect the relative distribution of power, authority, and privilege in a community. Often the crucial decision is a simple "yes or no" choice—are we going to develop and adopt the thing or not? In recent years many local, national, and international disputes about technology have centered on "yes or no" judgments about such things as food additives, pesticides, the building of highways, nuclear reactors, dam projects, and proposed high-tech weapons. The fundamental choice about an antiballistic missile or supersonic transport is whether or not the thing is going to join society as a piece of its operating equipment. Reasons given for and against are frequently as important as those concerning the adoption of an important new law.

A second range of choices, equally critical in many instances, has to do with specific features in the design or arrangement of a technical system after the decision to go ahead with it has already been made. Even after a utility company wins permission to build a large electric power line, important controversies can remain with respect to the placement of its route and the design of its towers; even after an organization has decided to institute a system of computers, controversies can still arise with regard to the kinds of components, programs, modes of access, and other specific features the system will include. Once the mechanical tomato harvester had been developed in its basic form, a design alteration of critical social significance— the addition of electronic sorters, for example—changed the character of the machine's effects upon the balance of wealth and power in California agriculture. Some of the most interesting research on technology and politics at present focuses upon the attempt to demonstrate in a detailed, concrete fashion how seemingly innocuous design features in mass transit systems, water projects, industrial machinery, and other technologies actually mask social choices of profound significance. Historian David Noble has studied two kinds of automated machine tool systems that have different implications for the relative power of management and labor in the industries that might employ them. He has shown that although the basic electronic and mechanical components of the record/playback and numerical control systems are similar, the choice of one design over another has crucial consequences for social struggles on the shop floor. To see the matter solely in terms of cost cutting, efficiency, or the modernization of equipment is to miss a decisive element in the story.[11]

From such examples I would offer some general conclusions. These correspond to the interpretation of technologies as "forms of life" presented earlier, filling in the explicitly political dimensions of that point of view.

The things we call "technologies" are ways of building order in our world. Many technical devices and systems important in everyday life contain possibilities for many different ways of ordering human activity. Consciously or unconsciously, deliberately or inadvertently, societies choose structures for technologies that influence how people are going to work, communicate, travel, consume, and so forth over a very long time. In the processes by which structuring decisions are made, different people are situated differently and possess unequal degrees of power as well as unequal levels of awareness. By far the greatest latitude of choice exists the very first time a particular instrument, system, or technique is introduced. Because choices tend to become strongly fixed in material equipment, economic investment, and social habit, the original flexibility vanishes for all practical purposes once the initial commitments are made. In that sense technological innovations are similar to legislative acts or political foundings that establish a framework for public order that will endure over many generations. For that reason the same careful attention one would give to the rules, roles, and relationships of politics must also be given to such things as the building of highways, the creation of television networks, and the tailoring of seemingly insignificant features on new machines. The issues that divide or unite people in society are settled not only in the institutions and practices of politics proper, but also, and less obviously, in tangible arrangements of steel and concrete, wires and semiconductors, nuts and bolts.

INHERENTLY POLITICAL TECHNOLOGIES

None of the arguments and examples considered thus far addresses a stronger, more troubling claim often made in writings about technology and society—the belief that some technologies are by their very nature political in a specific way. According to this view, the adoption of a given technical system unavoidably brings with it conditions for human relationships that have a distinctive political cast—for example, centralized or decentralized, egalitarian or inegalitarian, repressive or liberating. This is ultimately what is at stake in assertions such as those of Lewis Mumford that two traditions of technology, one authoritarian, the other democratic, exist side by side in Western history. In all the cases cited above the technologies are relatively flexible in design and arrangement and variable in their effects. Although one can recognize a particular result produced in a particular setting, one can also easily imagine how a roughly similar device or system might have been built or situated with very much different political consequences. The idea we must now examine and evaluate is that certain kinds of technology do not allow such flexibility, and that to choose them is to choose unalterably a particular form of political life.

A remarkably forceful statement of one version of this argument appears in Friedrich Engels's little essay "On Authority," written in 1872. Answering anarchists who believed that authority is an evil that ought to be abolished altogether, Engels launches into a panegyric for authoritarianism, maintaining, among other things, that strong authority is a necessary condition in modern industry. To advance his case in the strongest possible way, he asks his readers to imagine that the revolution has already occurred. "Supposing a social revolution dethroned the capitalists, who now exercise their authority over the production and circulation of wealth. Supposing, to adopt entirely the point of view of the anti-authoritarians, that the land and the instruments of labour had become the collective property of the workers who use them. Will authority have disappeared or will it have only changed its form?"[12]

His answer draws upon lessons from three sociotechnical systems of his day, cotton-spinning mills, railways, and ships at sea. He observes that on its way to becoming finished thread, cotton moves through a number of different operations at different locations in the factory. The workers perform a wide variety of tasks, from running the steam engine to carrying the products from one room to another. Because these tasks must be coordinated and because the timing of the work is "fixed by the authority of the steam," laborers must learn to accept a rigid discipline. They must, according to Engels, work at regular hours and agree to subordinate their individual wills to the persons in charge of factory operations. If they fail to do so, they risk the horrifying possibility that production will come to a grinding halt. Engels pulls no punches. "The automatic machinery of a big factory," he writes, "is much more despotic than the small capitalists who employ workers ever have been."[13]

Similar lessons are adduced in Engels's analysis of the necessary operating conditions for railways and ships at sea. Both require the subordination of workers to an "imperious authority" that sees to it that things run according to plan. Engels finds that far from being an idiosyncrasy of capitalist social organization, relationships of authority and subordination arise "independently of all social organization, [and] are imposed upon us together with the material conditions under which we produce and make products circulate." Again, he intends this to be stern advice to the anarchists who, according to Engels, thought it possible simply to eradicate subordination and superordination at a single stroke. All such schemes are nonsense. The roots of unavoidable authoritarianism are, he argues, deeply implanted in the human involvement with science and technology. "If man, by dint of his knowledge and inventive genius, has subdued the forces of nature, the latter avenge themselves upon him by subjecting him, insofar as he employs them, to a veritable despotism independent of all social organization."[14]

Attempts to justify strong authority on the basis of supposedly necessary conditions of technical practice have an ancient history. A pivotal theme in the *Republic* is Plato's quest to borrow the authority of *technē* and employ it by analogy to buttress his argument in favor of authority in the state. Among the illustrations he chooses, like Engels, is that of a ship on the high seas. Because large sailing vessels by their very nature need to be steered with a firm hand, sailors must yield to their captain's commands; no reasonable person believes that ships can be run democratically. Plato goes on to suggest that governing a state is rather like being captain of a ship or like practicing medicine as a physician. Much the same conditions that require central rule and decisive action in organized technical activity also create this need in government.

In Engels's argument, and arguments like it, the justification for authority is no longer made by Plato's classic analogy, but rather directly with reference to technology itself. If the basic case is as compelling as Engels believed it to be, one would expect that as a society adopted increasingly complicated technical systems as its material basis, the prospects for authoritarian ways of life would be greatly enhanced. Central control by knowledgeable people acting at the top of a rigid social hierarchy would seem increasingly prudent. In this respect his stand in "On Authority" appears to be at variance with Karl Marx's position in Volume I of *Capital*. Marx tries to show that increasing mechanization will render obsolete the hierarchical division of labor and the relationships of subordination that, in his view, were necessary during the early stages of modern manufacturing. "Modern Industry," he writes, "sweeps away by technical means the manufacturing division of labor, under which each man is bound hand and foot for life to a single detail operation. At the same time, the capitalistic form of that industry reproduces this same division of labour in a still more monstrous shape; in the factory proper, by converting the workman into a living appendage of the machine."[15] In Marx's view the conditions that will eventually dissolve the capitalist division of labor and facilitate proletarian revolution are conditions latent in industrial technology itself. The differences between Marx's position in *Capital* and Engels's in his essay raise an important question for socialism: What, after all, does modern technology make possible or necessary in political life? The theoretical tension we see here mirrors many troubles in the practice of freedom and authority that had muddied the tracks of socialist revolution.

Arguments to the effect that technologies are in some sense inherently political have been advanced in a wide variety of contexts, far too many to summarize here. My reading of such notions, however, reveals there are two basic ways of stating the case. One version claims that the adoption of a given technical system actually requires the creation and maintenance of a particular set of social conditions as the operating environment of that system. Engels's position is of this kind. A similar view is offered by a contemporary writer who holds that "if you accept nuclear power plants, you also accept a techno-scientific-industrial-military elite. Without these people in charge, you could not have nuclear power."[16] In this conception some kinds of technology require their social environments to be structured in a particular way in much the same sense that an automobile requires wheels in order to move. The thing could not exist as an effective operating entity unless certain social as well as material conditions were met. The meaning of "required" here is that of practical (rather than logical) necessity. Thus, Plato thought it a practical necessity that a ship at sea have one captain and an unquestionably obedient crew.

A second, somewhat weaker, version of the argument holds that a given kind of technology is strongly compatible with, but does not strictly require, social and political relationships of a particular stripe. Many advocates of solar energy have argued that technologies of that variety are more compatible with a democratic, egalitarian society than energy systems based on coal, oil, and nuclear power; at the same time they do not maintain that anything about solar energy requires democracy. Their case is, briefly, that solar energy is decentralizing in both a technical and political sense: technically speaking, it is vastly more reasonable to build solar systems in a disaggregated, widely distributed manner than in large-scale centralized plants; politically speaking, solar energy accommodates the attempts of individuals and local communities to manage their affairs effectively because they are dealing with systems that are more accessible, comprehensible, and controllable than huge centralized sources. In this view solar energy is desirable not only for its economic and environmental benefits, but also for the salutary institutions it is likely to permit in other areas of public life.[17]

Within both versions of the argument there is a further distinction to be made between conditions that are internal to the workings of a given technical system and those that are external to it. Engels's thesis concerns internal social relations said to be required within cotton factories and railways, for example; what such relationships mean for the condition of society at large is, for him, a separate question. In contrast, the solar advocate's belief that solar technologies are compatible with democracy pertains to the way they complement aspects of society removed from the organization of those technologies as such.

There are, then, several different directions that arguments of this kind can follow. Are the social conditions predicated said to be required by, or strongly compatible with, the workings of a given technical system? Are those conditions internal to that system or external to it (or both)? Although writings that address such questions are often unclear about what is being asserted, arguments in this general category are an important part of modern political discourse. They enter into many attempts to explain how changes in social life take place in the wake of technological innovation. More important, they are often used to buttress attempts to justify or criticize proposed courses of action involving new technology. By offering distinctly political reasons for or against the adoption of a particular technology, arguments of this kind stand apart from more commonly employed, more easily quantifiable claims about economic costs and benefits, environmental impacts, and possible risks to public health and safety that technical systems may involve. The issue here does not concern how many jobs will be created, how much income generated, how many pollutants added, or how many cancers produced. Rather, the issue has to do with ways in which choices about technology have important consequences for the form and quality of human associations.

If we examine social patterns that characterize the environments of technical systems, we find certain devices and systems almost invariably linked to specific ways of organizing power and authority. The important question is: Does this state of affairs derive from an unavoidable social response to intractable properties in the things themselves, or is it instead a pattern imposed independently by a governing body, ruling class, or some other social or cultural institution to further its own purposes?

Taking the most obvious example, the atom bomb is an inherently political artifact. As long as it exists at all, its lethal properties demand that it be controlled by a centralized, rigidly hierarchical chain of command closed to all influences that might make its workings unpredictable. The internal social system of the bomb must be authoritarian; there is no other way. The state of affairs stands as a practical necessity independent of any larger political system in which the bomb is embedded, independent of the type of regime or character of its rulers. Indeed, democratic states must try to find ways to ensure that the social structures and mentality that characterize the management of nuclear weapons do not "spin off" or "spill over" into the polity as a whole.

The bomb is, of course, a special case. The reasons very rigid relationships of authority are necessary in its immediate presence should be clear to anyone. If, however, we look for other instances in which particular varieties of technology are widely perceived to need the maintenance of a special pattern of power and authority, modern technical history contains a wealth of examples.

Alfred D. Chandler in *The Visible Hand,* a monumental study of modern business enterprise, presents impressive documentation to defend the hypothesis that the construction and day-to-day operation of many systems of production, transportation, and communication in the nineteenth and twentieth centuries require the development of particular social form—a large-scale centralized, hierarchical organization administered by highly skilled managers. Typical of Chandler's reasoning is his analysis of the growth of the railroads.[18]

> Technology made possible fast, all-weather transportation; but safe, regular, reliable movement of goods and passengers, as well as the continuing maintenance and repair of locomotives, rolling stock, and track, roadbed, stations, roundhouses, and other equipment, required the creation of a sizable administrative organization. It meant the employment of a set of managers to supervise these functional activities over an extensive geographical area; and the appointment of an administrative command of middle and top executives to monitor, evaluate, and coordinate the work of managers responsible for the day-to-day operations.

Throughout his book Chandler points to ways in which technologies used in the production and distribution of electricity, chemicals, and a wide range of industrial goods "demanded" or "required" this form of human association. "Hence, the operational requirements of railroads demanded the creation of the first administrative hierarchies in American business."[19]

Were there other conceivable ways of organizing these aggregates of people and apparatus? Chandler shows that a previously dominant social form, the small traditional family firm, simply could not handle the task in most cases. Although he does not speculate further, it is clear that he believes there is, to be realistic, very little latitude in the forms of power and authority appropriate within modern sociotechnical systems. The properties of many modern technologies—oil pipelines and refineries, for example—are such that overwhelmingly impressive economies of scale and speed are possible. If such systems are to work effectively, efficiently, quickly, and safely, certain requirements of internal social organization have to be fulfilled; the material possibilities that modern technologies make available could not be exploited otherwise. Chandler acknowledges that as one compares sociotechnical institutions of different nations, one sees "ways in which cultural attitudes, values, ideologies, political systems, and social structure affect these imperatives."[20] But the weight of argument and empirical evidence in *The Visible Hand* suggests that any significant departure from the basic pattern would be, at best, highly unlikely.

It may be that other conceivable arrangements of power and authority, for example, those of decentralized, democratic worker self-management, could prove capable of administering factories, refineries, communications systems, and railroads as well as or better than the organizations Chandler describes. Evidence from automobile assembly teams in Sweden and worker-managed plants in Yugoslavia and other countries is often presented to salvage these possibilities. Unable to settle controversies over this matter here, I merely point to what I consider to be their bone of contention. The available evidence tends to show that many large, sophisticated technological systems are in fact highly compatible with centralized, hierarchical managerial control. The interesting question, however, has to do with whether or not this pattern is in any sense a requirement of such systems, a question that is not solely empirical. The matter ultimately rests on our judgments about what steps, if any, are practically necessary in the workings of particular kinds of technology and what, if anything, such measures require of the structure of human associations. Was Plato right in saying that a ship at sea needs steering by a decisive hand and that this could only be accomplished by a single captain and an obedient crew? Is Chandler correct in saying that the properties of large-scale systems require centralized, hierarchical managerial control?

To answer such questions, we would have to examine in some detail the moral claims of practical necessity (including those advocated in the doctrines of economics) and weigh them against moral claims of other sorts, for example, the notion that it is good for sailors to participate in the command of a ship or that workers have a right to be involved in making and administering decisions in a factory. It is characteristic of societies based on large, complex technological systems, however, that moral reasons other than those of practical necessity appear increasingly obsolete, "idealistic," and irrelevant. Whatever claims one may wish to make on behalf of liberty, justice, or equality can be immediately neutralized when confronted with arguments to the effect, "Fine, but that's no way to run a railroad" (or steel mill, or airline, or communication system, and so on). Here we encounter an important quality in modern political discourse and in the way people commonly think about what measures are justified in response to the possibilities technologies make available. In many instances, to say that some technologies are inherently political is to say that certain widely accepted reasons of practical necessity—especially the need to maintain crucial technological systems as smoothly working entities—have tended to eclipse other sorts of moral and political reasoning.

One attempt to salvage the autonomy of politics from the bind of practical necessity involves the notion that conditions of human association found in the internal workings of technological systems can easily be kept separate from the polity as a whole. Americans have long rested content in the belief that arrangements of power and authority inside industrial corporations, public utilities, and the like have little bearing on public institutions, practices, and ideas at large. That "democracy stops at the factory gates" was taken as a fact of life that had nothing to do with the practice of political freedom. But can the internal politics of technology and the politics of the whole community be so easily separated? A recent study of business leaders in the United States, contemporary exemplars of Chandler's "visible hand of management," found them remarkably impatient with such democratic scruples as "one man, one vote." If democracy doesn't work for the firm, the most critical institution in all of society, American executives ask, how well can it be expected to work for the government of a nation—particularly when that government attempts to interfere with the achievements of the firm? The authors of the report observe that patterns of authority that work effectively in the corporation become for businessmen "the desirable model against which to compare political and economic relationships in the rest of society."[21] While such findings are far from conclusive, they do reflect a sentiment increasingly common in the land: what dilemmas such as the energy crisis require is not a redistribution of wealth or broader public participation but, rather, stronger, centralized public and private management.

An especially vivid case in which the operational requirements of a technical system might influence the quality of public life is the debates about the risks of nuclear power. As the supply of uranium for nuclear reactors runs out, a proposed alternative fuel is the plutonium generated as a by-product in reactor cores. Well-known objections to plutonium recycling focus on its unacceptable economic costs, its risks of environmental contamination, and its dangers in regard to the international proliferation of nuclear weapons. Beyond these concerns, however, stands another less widely appreciated set of hazards—those that involve the sacrifice of civil liberties. The widespread use of plutonium as a fuel increases the chance that this toxic substance might be stolen by terrorists, organized crime, or other persons. This raises the prospect, and not a trivial one, that extraordinary measures would have to be taken to safeguard plutonium from theft and to recover it should the substance be stolen. Workers in the nuclear industry as well as ordinary citizens outside could well become subject to background security checks, covert surveillance, wiretapping, informers, and even emergency measures under martial law—all justified by the need to safeguard plutonium.

Russell W. Ayres's study of the legal ramifications of plutonium recycling concludes: "With the passage of time and the increase in the quantity of plutonium in existence will come pressure to eliminate the traditional checks the courts and legislatures place on the activities of the executive and to develop a powerful central authority better able to enforce strict safeguards." He avers that "once a quantity of plutonium had been stolen, the case for literally turning the country upside down to get it back would be overwhelming." Ayres anticipates and worries about the kinds of thinking that, I have argued, characterize inherently political technologies. It is still true that in a world in which human beings make and maintain artificial systems nothing is "required" in an absolute sense. Nevertheless, once a course of action is under way, once artifacts such as nuclear power plants have been built and put in operation, the kinds of reasoning that justify the adaptation of social life to technical requirements pop up as spontaneously as flowers in the spring. In Ayres's words, "Once recycling begins and the risks of plutonium theft become real rather than hypothetical, the case for governmental infringement of protected rights will seem compelling."[22] After a certain point, those who cannot accept the hard requirements and imperatives will be dismissed as dreamers and fools.

*   *   *

The two varieties of interpretation I have outlined indicate how artifacts can have political qualities. In the first instance we noticed ways in which specific features in the design or arrangement of a device or system could provide a convenient means of establishing patterns of power and authority in a given setting. Technologies of this kind have a range of flexibility in the dimensions of their material form. It is precisely because they are flexible that their consequences for society must be understood with reference to the social actors able to influence which designs and arrangements are chosen. In the second instance we examined ways in which the intractable properties of certain kinds of technology are strongly, perhaps unavoidably, linked to particular institutionalized patterns of power and authority. Here the initial choice about whether or not to adopt something is decisive in regard to its consequences. There are no alternative physical designs or arrangements that would make a significant difference; there are, furthermore, no genuine possibilities for creative intervention by different social systems—capitalist or socialist—that could change the intractability of the entity or significantly alter the quality of its political effects.

To know which variety of interpretation is applicable in a given case is often what is at stake in disputes, some of them passionate ones, about the meaning of technology for how we live. I have argued a "both/and" position here, for it seems to me that both kinds of understanding are applicable in different circumstances. Indeed, it can happen that within a particular complex of technology—a system of communication or transportation, for example—some aspects may be flexible in their possibilities for society, while other aspects may be (for better or worse) completely intractable. The two varieties of interpretation I have examined here can overlap and intersect at many points.

These are, of course, issues on which people can disagree. Thus, some proponents of energy from renewable resources now believe they have at last discovered a set of intrinsically democratic, egalitarian, communitarian technologies. In my best estimation, however, the social consequences of building renewable energy systems will surely depend on the specific configurations of both hardware and the social institutions created to bring that energy to us. It may be that we will find ways to turn this silk purse into a sow's ear. By comparison, advocates of the further development of nuclear power seem to believe that they are working on a rather flexible technology whose adverse social effects can be fixed by changing the design parameters of reactors and nuclear waste disposal systems. For reasons indicated above, I believe them to be dead wrong in that faith. Yes, we may be able to manage some of the "risks" to public health and safety that nuclear power brings. But as society adapts to the more dangerous and apparently indelible features of nuclear power, what will be the long-range toll in human freedom?

My belief that we ought to attend more closely to technical objects themselves is not to say that we can ignore the contexts in which those objects are situated. A ship at sea may well require, as Plato and Engels insisted, a single captain and obedient crew. But a ship out of service, parked at the dock, needs only a caretaker. To understand which technologies and which contexts are important to us, and why, is an enterprise that must involve both the study of specific technical systems and their history as well as a thorough grasp of the concepts and controversies of political theory. In our times people are often willing to make drastic changes in the way they live to accommodate technological innovation while at the same time resisting similar kinds of changes justified on political grounds. If for no other reason than that, it is important for us to achieve a clearer view of these matters than has been our habit so far.

NOTES

1. Langdon Winner, *Autonomous Technology: Technics-Out-of-Control as a Theme in Political Thought* (Cambridge: MIT Press, 1977).

2. The meaning of "technology" I employ in this essay does not encompass some of the broader definitions of that concept found in contemporary literature, for example, the notion of "technique" in the writings of Jacques Ellul. My purposes here are more limited. For a discussion of the difficulties that arise in attempts to define "technology," see *Autonomous Technology,* 8–12.

3. Robert A. Caro, *The Power Broker: Robert Moses and the Fall of New York* (New York: Random House, 1974), 318, 481, 514, 546, 951–958, 952.

4. Robert Ozanne, *A Century of Labor-Management Relations at McCormick and International Harvester* (Madison: University of Wisconsin Press, 1967), 20.

5. The early history of the tomato harvester is told in Wayne D. Rasmussen, "Advances in American Agriculture: The Mechanical Tomato Harvester as a Case Study," *Technology and Culture* 9:531–543, 1968.

6. Andrew Schmitz and David Seckler, "Mechanized Agriculture and Social Welfare: The Case of the Tomato Harvester," *American Journal of Agricultural Economics* 52:569–577, 1970.

7. William H. Friedland and Amy Barton, "Tomato Technology," *Society* 13:6, September/October 1976. See also William H. Friedland, *Social Sleepwalkers: Scientific and Technological Research in California Agri-*

culture, University of California, Davis, Department of Applied Behavioral Sciences, Research Monograph No. 13, 1974.

8. *University of California Clip Sheet* 54:36, May 1, 1979.

9. "Tomato Technology."

10. A history and critical analysis of agricultural research in the land-grant colleges is given in James Hightower, *Hard Tomatoes, Hard Times* (Cambridge: Schenkman, 1978).

11. David F. Noble, *Forces of Production: A Social History of Machine Tool Automation* (New York: Alfred A. Knopf, 1984).

12. Friedrich Engels, "On Authority," in *The Marx-Engels Reader,* ed. 2, Robert Tucker (ed.) (New York: W. W. Norton, 1978), 731.

13. Ibid.

14. Ibid., 732, 731.

15. Karl Marx, *Capital,* vol. 1, ed. 3, translated by Samuel Moore and Edward Aveling (New York: Modern Library, 1906), 530.

16. Jerry Mander, *Four Arguments for the Elimination of Television* (New York: William Morrow, 1978), 44.

17. See, for example, Robert Argue, Barbara Emanuel, and Stephen Graham, *The Sun Builders: A People's Guide to Solar, Wind and Wood Energy in Canada* (Toronto: Renewable Energy in Canada, 1978). "We think decentralization is an implicit component of renewable energy; this implies the decentralization of energy systems, communities and of power. Renewable energy doesn't require mammoth generation sources of disruptive transmission corridors. Our cities and towns, which have been dependent on centralized energy supplies, may be able to achieve some degree of autonomy, thereby controlling and administering their own energy needs." (16)

18. Alfred D. Chandler, Jr., *The Visible Hand: The Managerial Revolution in American Business* (Cambridge: Belknap, 1977), 244.

19. Ibid.

20. Ibid., 500.

21. Leonard Silk and David Vogel, *Ethics and Profits: The Crisis of Confidence in American Business* (New York: Simon and Schuster, 1976), 191.

22. Russell W. Ayres, "Policing Plutonium: The Civil Liberties Fallout," *Harvard Civil Rights—Civil Liberties Law Review* 10 (1975):443, 413–414, 374.

20

Strong Democracy and Technology

Richard E. Sclove

> In West Central Minnesota, local farmers have been opposing an electrical trans-
> mission line for over four years . . . The public relations man for the utility said . . . ,
> "You should be proud to have the biggest powerline in the world in your country,"
> but the farmers felt differently.
>> To people who love and care for the land, a transmission line of this size is
> a desecration. People who once felt they lived in a democratic society feel they have
> been betrayed and no longer control their own lives.
>
> —*Minnesota farmer and protester Alice Tripp*[1]

How does the key insight that technologies represent a species of social structure bear on the relationship between technology and democracy? The answer depends partly on one's concept of democracy. One common view is that, as a matter of justice, people should be able to influence the basic social circumstances of their lives. This view implies organizing society along relatively egalitarian and participatory lines, a vision that Benjamin Barber has labeled "strong democracy."[2]

Historic examples approaching this ideal include New England town meetings, the confederation of self-governing Swiss villages and cantons, and the English and American tradition of trial by a jury of peers. Strong democracy is apparent also in the methods or aspirations of various social movements such as the late-19th-century American Farmers Alliance, the 1960s U.S. civil rights movement, and the 1980s Polish Solidarity movement. In each of these cases ordinary people claimed the rights and responsibilities of active citizenship concerning basic social issues.

The strong democratic tradition contrasts with more passive or inegalitarian models of democracy that in practice tend to prevail today, so-called thin democracy.[3] Here the focus shifts from a core concern with substantive political equality and with citizens' active engagement in political discourse, or in seeking their common good, to a preoccupation with representative institutions, periodic elections, and competition among conflicting private interests, elites, and power blocs. Within thin democracies power is less evenly distributed; citizens can vote for representatives but ordinarily have little direct influence on important public decisions.

The contest—both in theory and in practice—between the strong and thin democratic traditions is long-standing and unlikely to be resolved soon. Rather than stopping now to compare and contrast the two, I propose initially to suspend judgment and simply posit a specific, strong democratic model of how societies ought to be organized.

TECHNOLOGY AND DEMOCRACY

The strong democratic ideal envisions extensive opportunities for citizens to participate in important decisions that affect them. A decision qualifies as important particularly insofar as it bears on a society's basic organization or structure. The commitment to egalitarian participation does not preclude continued reliance on some representative institutions, but these should be designed to support and incorporate, rather than to replace, participatory processes.

Complementing this procedural standard of strong democracy is a substantive standard: in their political involvements citizens ought, whatever else they do, to grant precedence to respecting any important concerns or interests common to everyone. Above all, they should perpetuate their society's basic character as a strong democracy. Apart from this one substantive moral obligation, citizens are free to attend as they wish to their diverse and perhaps conflicting personal concerns.

This model of democracy, even in schematic form, is sufficient for deriving a prescriptive theory of democracy and technology: *If citizens ought to be empowered to participate in determining their society's basic structure, and technologies are an important species of social structure, it follows that technological design and practice should be democratized.* Strong democracy's complementary procedural and substantive components entail, furthermore, that technological democratization incorporate two corresponding elements. Procedurally, people from all walks of life require expanded opportunities to shape their evolving technological order. And substantively, the resulting technologies should be compatible with citizens' common interests and affinities—to whatever extent such exist—and particularly with their fundamental interest in strong democracy itself.

Democratic Evaluation, Choice, and Governance

The preceding argument suggests that processes of technological development that are today guided by market forces, economic self-interest, distant bureaucracies, or international rivalry should be subordinated to democratic prerogatives. Only in this way can technologies begin actively to support, rather than to coerce or constrict, people's chosen ways of life. For example, residents of many American cities have grown resigned to daily traffic jams, sprawling shopping malls, the stress associated with combining careers with parenthood, and the television as babysitter. This pattern of sociotechnological organization is largely haphazard.[4]

At other times, an existing technological order, or its process of transformation, reflects the direct intentions of powerful organizations or elites. For instance, this chapter's epigraph alludes to an electric utility consortium that proceeded, despite adamant local opposition, to construct a huge transmission line across prime Minnesota farmland. That outcome was not haphazard or unplanned, but neither did it reflect democratic preferences.

Technological evolution can thus encompass social processes ranging from the haphazard to the bitterly contested or blatantly coercive. None of these processes is strongly democratic. This is not to say that every particular technology must suddenly be subjected to formal political review. Each time one is moved to buy a fork or to sell a pencil sharpener, one should not have to defend the decision before a citizens' tribunal or a congressional committee. Not all technologies exert an equal structural influence. However, consider a modern society's treatment of another genus of social structure: various kinds of law. The rules that parents create for their children are subject to relatively little social oversight. But rule making by federal agencies is governed by extensive formal procedure, and even more stringent procedure is required to amend a national constitution. Why should the treatment of technology be so different?

Whether a technology requires political scrutiny and, if so, where and how exhaustively, should correspond roughly to the degree to which it promises, fundamentally or enduringly, to affect social life. This implies the need for a graduated set of democratic procedures for reviewing existing technological arrangements, monitoring emerging ones, and ensuring that the technological order is compatible with informed democratic wishes.

Within such a system, citizens or polities that believe that a set of technologies may embody significant structural potency ought always to have the opportunity to make that case in an appropriate political forum. Beyond this, there should be a system of ongoing democratic oversight of the entire technological order, scanning for the unanticipated emergence of undemocratic technological consequences or dynamics, and prepared when necessary to intervene remedially in the interest of democratic norms.

This does not mean, however, that everyone has to participate in each technological decision that becomes politicized. Logistical nightmares aside, there is more to life than politics. But in contrast to the present state of affairs, there should be abundant opportunity for widespread and effective participation. Ideally, each citizen would at least occasionally exercise that opportunity, particularly on technological matters significant to him or her.

For example, in the early 1970s the cry rang out that there was natural gas beneath the frigid and remote northwest corner of Canada. Eager to deliver the fuel to urban markets, energy companies began planning to build a high-pressure, chilled pipeline across thousands of miles of wilderness, the traditional home of the Inuit (Eskimos) and various Indian tribes. At that point, a Canadian government ministry, anticipating significant environmental and social repercussions, initiated a public inquiry under the supervision of a respected Supreme Court justice, Thomas R. Berger.

The MacKenzie Valley Pipeline Inquiry (also called the Berger Inquiry) began with preliminary hearings open to participation by any Canadian who felt remotely affected by the pipeline proposal. Responding to what they heard, Berger and his staff then developed a novel format to encourage a thorough, open, and accessible inquiry process. One component involved formal, quasi-judicial hearings comprising conventional expert testimony with cross-examination. But Berger also initiated a series of informal "community hearings." Travelling 17,000 miles to 35 remote villages, towns, and settlements, the Berger Inquiry took testimony from nearly 1,000 native witnesses. The familiarity of a local setting and the company of family and neighbors encouraged witness spontaneity and frankness. One native commented: "It's the first time anybody bothered asking us how we felt."[5]

Disadvantaged groups received funding to support travel and other needs related to competent participation. The Canadian Broadcasting Company carried daily radio summaries of both the community and the formal hearings, in English as well as in six native languages. Thus each community was aware of evidence and concerns that had previously been expressed. Moreover, by interspersing formal hearings with travel to concurrent community hearings, Berger made clear his intention to weigh respectfully the testimony of both Ph.D. scientists and teenaged subsistence fishermen.

Berger's final report quoted generously from the full range of witnesses and became a national bestseller. Based on testimony concerning environmental, socioeconomic, cultural, and other issues, the judge recommended a 10-year delay in any decision to build a pipeline through the MacKenzie Valley, as well as a host of more specific steps (including a major new wilderness park and a whale sanctuary). Within months the original pipeline proposal was rejected, and the Canadian Parliament instead approved an alternate route paralleling the existing Alaska Highway.[6]

Some might fault the MacKenzie Valley Inquiry for depending so much on the democratic sensibilities and good faith of one man—Judge Berger—rather than empowering the affected native groups to play a role in formulating the inquiry's conclusions. Nevertheless, the process was vastly more open and egalitarian than is the norm in industrial societies. It contrasts sharply with the steps forced on those Minnesota farmers, mentioned earlier, who were loathe to see a transmission line strung across their fields:

> The farmers have tried to use every legitimate legal and political channel to make known to the utility company, the government and the public their determination to save the land and to maintain safety in their workplaces. The farmers and their urban supporters have been met with indifference and arrogance by both the utility and the government. Turned away at the state capitol, they have taken their case to the courts again and again, only to be rebuffed.[7]

The Berger Inquiry represents just one example of a more democratic means of technological decision making.

Democratic Technologies

Besides fostering democratic procedures for technological decision making, we must seek technological outcomes that are substantively democratic. The purpose of democratic procedures is, most obviously, to help ensure that technologies structurally support popular aspirations, whatever they may be. The alternative is to continue watching aspirations tacitly conform themselves to haphazardly generated technological imperatives or to authoritarian decisions.

However, according to strong democratic theory, citizens and their representatives should grant precedence here to two kinds of aspirations. First and most importantly, technologies should—independent of their diverse focal purposes—structurally support the social and institutional conditions necessary to establish and maintain strong democracy itself. (These conditions are discussed later in this chapter.) Second, technologies should structurally respect any other important concerns common to all citizens.

This does not necessarily mean shifting social resources to the design of technologies that focally support democracy or other common goods. That is the instinct of many strong democrats, and some such efforts may be appropriate. For example, there might be a constructive role within strong democracy for electronically mediated "town meetings."

However, a preoccupation with certain technologies' focal functions, if it excludes commensurate attention to their nonfocal functions and to those of other technologies, is apt to prove disappointing or counterproductive. It might, for example, do little good to televise more political debates without first inquiring whether a nonfocal consequence of watching television is to induce passivity rather than critical engagement.[8] And is it obviously more urgent to seek any new technologies that are focally democratic before contemplating the redesign of existing technologies that, nonfocally, are antidemocratic? How can one know that the adverse effect of the latter is not sufficient to override any beneficial effect intended by the former? For instance, it may be fruitless to try to foster civic engagement via interactive telecommunications unless communities are prepared at the same time to promote convivially designed town and city centers; neighborhood parks, greenhouses, workshops, and daycare centers; technologies compatible with democratically managed workplaces and flexible work schedules; more democratically governed urban technological infrastructures; and other steps toward constituting democratic communities.

In other words, societies do not require a special subset of technologies that are focally democratic as a complement to the remaining majority of technologies that are inconsequential to politics, because the remaining technologies are not, in fact, inconsequential. The overall objective ought to be a technological order that structurally manifests a democratic design style. Considering the entirety of a society's disparate technologies—both their focal and nonfocal aspects—is the technological order strongly democratic? That is the first question.

Owing in part to modern societies' persistent neglect of their structural potency, technologies have never systematically been evaluated from the standpoint of their bearing on democracy. Therefore, upon scrutiny, many existing technologies may prove structurally undemocratic. Furthermore, from a dynamic perspective, they may erect obstacles to efforts intended to further democratization. For example, the declining interest in political participation observed within most industrial democracies might be partly attributable to latent subversion of democracy's necessary conditions by technologies. We can start testing such conjectures after formulating criteria for distinguishing structurally democratic technologies from their less democratic counterparts.

Contestable Democratic Design Criteria

If democratic theory can specify that technologies ought above all to be compatible with strong democracy, does that prescription preempt the most important questions that democratic procedures for technological decision making might otherwise address? No. In the first place, this leaves many important questions to the discretion of democratic judgment. These involve debating shared and personal concerns and then striving to ensure that technologies structurally support them. But even on the prior question of seeking a technological order that structurally supports strong democracy, there is a broad and critical role for democratic involvement.

The simple idea that technologies ought to be compatible with strong democracy is entirely abstract. To become effective, it must be expanded into a sequence of successively more specific guidelines for technological design, what I call democratic design criteria. But to specify such criteria with greater precision and content, and then to use them, one must adduce and interpret a progressively wider selection of evidence and exercise judgment. Thus as democratic design criteria become more specific and are applied, the grounds upon which they might reasonably be contested increase.

Moreover, even an expanded system of design criteria will always remain essentially incomplete. For instance, as social circumstances shift or as novel technologies are developed, new criteria will be needed and old ones will have to be reevaluated. In addition, no finite set of criteria can ever fully specify an adequate technological design. Democratic design is ultimately a matter of art and judgment.[9]

This guarantees an ongoing central role for democratic procedure. Democratic theory and its theorists—or anyone else—can help initiate the process of formulating and using design guidelines. However, self-selected actors have neither the knowledge nor the right to make determinative discretionary judgments on behalf of other citizens. Individuals cannot, for example, possibly know what their common interests and preferred democratic institutions are until after they have heard others express their hopes and concerns, and listened to comments on their own, and until everyone has had some chance to reflect on their initial desires and assumptions. Also, individuals cannot trust themselves, pollsters, or scientists to make objective judgments on behalf of others, because invariably each person's, professional's, or group's interests are at stake in the outcome, subtly influencing perception and reasoning. Only democratic forums can supply impartiality born of the balance among multiple perspectives, the opportunity for reflection, and the full range of social knowledge needed to reach legitimate determinations.

Hence it makes sense to seek democratic procedures for formulating and applying rationally contestable design criteria for democratic technologies. These will be "contestable" because the process of generating and refining design criteria cannot be finalized. As technology, social knowledge, and societies and their norms change, one can expect shifts in these design criteria. However, the criteria will be "rationally" or democratically contestable because such shifts need not be arbitrary. They should reflect citizens' current best assessment of the conditions required to realize strong democracy and other shared values. (See Figure 20.1 for the basic ingredients of a strong democratic politics of technology.)

Contrast

The theory of democracy and technology developed here contrasts with predecessor theories that emphasize either broadened participation in decision making or else evolving technologies

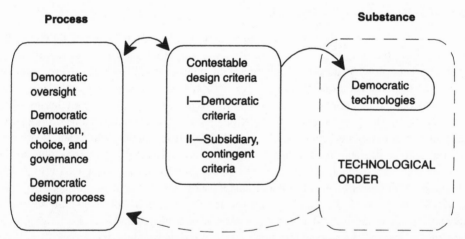

Figure 20.1   Democratic politics of technology. A technology is democratic if it has been designed and chosen with democratic participation or oversight and—considering its focal and nonfocal aspects—is structurally compatible with strong democracy and with citizens' other important common concerns. Within a democratic politics of technology, reflection on existing and proposed technologies plays a role in generating democratic design criteria. Use of these criteria then mediates between democratic procedures and the evolution of a substantively democratic technological order.

The figure distinguishes between two categories of design criteria: (I) priority goes to criteria that help ensure that technologies are compatible with democracy's necessary conditions; (II) subsidiary criteria can then reflect technologies' structural bearing on citizens' other concerns and interests. The dashed line indicates that the entire existing technological order exerts a structural influence on politics generally, including (in this instance) the possibility of a democratic politics of technology.

that support democratic social relations, but that do not integrate these procedural and substantive concerns.[10]

The theory also contrasts with a prevalent view, one that arose during the 19th century, that American mass-production technology was democratic because it made consumer goods widely and cheaply available. Democracy thus became equated with a perceived tendency toward equality of opportunity in economic consumption. This earlier view was insensitive to the structural social consequences associated with production technologies themselves and with the goods and services they produced. Furthermore, as a consequence of this blind spot, the theory foresaw no need to complement the market mechanism for making technological choices with any type of political oversight.

FREEDOM: THE MORAL BASIS OF STRONG DEMOCRACY

This chapter opened by simply positing a strongly democratic model of democracy; let us now briefly consider a moral argument supporting the model's desirability. Among human goods or values, freedom is widely regarded as preeminent. Freedom is a fundamental precondition of all our willful acts, and hence of pursuing all other goods. But under what conditions is one free? Normally, people consider themselves free when no one is interfering with what they want to do. However, this familiar view is not entirely adequate. Suppose a woman is externally free to pursue her desires, but her desires are purely and directly a product of social conditioning or compulsively self-destructive (e.g., heroin addiction)? How truly free is she?

Such considerations suggest that actions are fully free when guided by something in addition to external incentive, social compulsion, or even a person's own instinctive psychological inclinations. That "something" Immanuel Kant identified as morality—specifically, compliance with moral principles that individuals prescribe to themselves. Morality expresses freedom in ways that cannot otherwise occur, even when one chooses among one's own competing psychological inclinations. With the freedom that morality secures, one acquires the dignity of being autonomously self-governing, an "end unto oneself."[11]

But what should the content of moral self-prescriptions be? Kant envisioned one overarching moral principle, what he called the "categorical imperative." One can think of it as a formal restatement of the Golden Rule: always treat others with the respect that you would wish them to accord you, including your fundamental interest in freedom. In Kant's words, "Act so that you treat humanity, whether in your own person or in that of another, always as an end and never as a means only."[12] Thus, in Kantian philosophy the concept of autonomy connotes moral community and readiness to act on behalf of the common good, rather than radical individualism.

However, suppose I behave morally, but nobody else reciprocates? There would be small freedom for me in that kind of society. Living in interdependent association with others (as people do and must) provides innumerable opportunities that could not otherwise exist, including the opportunity to develop moral autonomy. But it also subjects each person to the consequences of others' actions. Should these consequences seem arbitrary or contrary to their interests, people might well judge their freedom diminished.

As a solution, suppose each person's actions were governed by regulative structures, such as laws and government institutions, that they participated in choosing (hence strong democratic procedure) and that respected any important common concerns, particularly their preeminent interest in freedom (hence strong democratic substance). In a society of this sort, laws and other social structures would each stand, in effect, as explicit expressions of mutual agreement to live in accord with Kant's categorical imperative (i.e., to respect oneself and others as ends).[13]

Strong democracy asks that citizens grant priority to commonalities not for their society's own sake, independent of its individual members, but because it is on balance best for each individual member. Strong democratic procedure expresses and develops individual moral freedom, while its structural results constitute conditions requisite to perpetuating maximum equal freedom. Insofar as it envisions democratic procedures for evolving and governing democratic structures, let us call this a model of "democratic structuration."* In other words, democratic structuration represents strong democracy's basic principle of collective self-organization.

Combine the preceding normative argument with the conventionally slighted insight that technologies function as an important species of social structure. It follows that evolving a democratic technological order is a moral responsibility of the highest order. A democratic politics of technology—one comprising democratic means for cultivating technologies that structurally support democracy—is needed to transform technology from an arbitrary, irrational, or undemocratic social force into a substantive constituent of human freedom (see Figure 20.2).

Of course, democracy is by no means the only issue that needs to be considered when making decisions about technology. Citizens might well wish to make technological decisions based partly on practical, economic, cultural, environmental, religious, or other grounds. But among these diverse considerations, priority should go to the question of technologies' bearing on democracy. This is because democracy is fundamental, establishing the necessary background circumstance for us to be able to decide fairly and effectively what other issues to take into account in both our technological and nontechnological decision making. (Granting priority to democracy within technological decisions would be somewhat analogous to ensuring compatibility with the U.S. Constitution when drafting or debating proposed laws or regulations.)

It would be presumptuous, however, to insist that the case supporting strong democracy is entirely conclusive. This implies that the contestability attributed earlier to democratic technological design criteria stretches logically back into the supporting theory's philosophical core.

---

*The basic concept of structuration is that people's thoughts and behavior are invariably shaped by structures that—through ordinary activities (or sometimes extraordinary ones, such as revolution or constitutional convention)—they participate collectively and continuously in generating, reproducing, or transforming. A rough analogy from the natural world can help convey the idea. Consider a river as a process shaped and guided by a structure: its banks. As the river flows, it is continuously modifying its banks, here through erosion, there through deposition of sediment. Over time the river cuts deep gorges, meanders back and forth across broad floodplains, crafts oxbows and bypasses, and establishes at its mouth a complex deltaic formation. Hence the river is a vibrant example of structuration: a process conditioned by enduring structures that it nonetheless helps continuously to reconstitute.

In social life what we do and who we are (or may become) is similarly guided by our society's basic structures: its laws, major political and economic institutions, cultural beliefs, and so on. But our activities nevertheless produce cumulative material and psychological results, not fully determined by structure, that in turn are woven back into our society's evolving structural complex. Hence at every moment we contribute marginally—or, upon occasion, dramatically—to affirming or transforming our society's basic structures.

The word "structuration"—introduced by Giddens (1979, chap. 2)—is not aesthetically pleasing, but it has achieved wider currency than any synonym. I propose the term "democratic structuration" to embed this explanatory concept in a normative context, suggesting that the means and the ends of structuration should be guided by an overarching respect for moral freedom.

I. Philosophical Case for Strong Democracy:

A. Freedom is a highest order human value. Respecting other people's freedom is a moral duty, necessary for realizing one's own freedom. (Kanitian moral theory)

B. Given the inalterable fact of real-world social interdependence, the opportunity to fully develop and express individual freedom can best be secured within a context of democratic structuration:

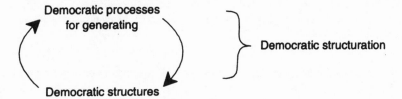

Under these circumstances both social processes and their structural results, by respecting people's freedom, embody Kantian morality. Here structures support rather than constrain people's highest order interest in freedom. (Neo-Rousseauian, or strong democratic, political theory)

II. Philosophical Case for a Democratic Politics of Technology:

Applying the preceding argument (I) to technology:

C. Technologies are a species of social structure. Therefore, it is morally vital that they, like other social structures, be generated and governed via democratic structuration. (The content of democratic structuration, as it applies to technology, is elaborated in Figure 20.1.)

Figure 20.2   Philosophical argument for strong democracy and for a democratic politics of technology.

DEMOCRATIC BACKGROUND CONDITIONS

Establishing democratic structuration depends on a number of background conditions. These need to be elaborated here in just enough detail to permit subsequent derivation of democratic design criteria for technologies. The requisite background conditions include: (1) some commonality of purpose, attachment, or outlook among citizens (at a minimum, general recognition of a preeminent interest in living in a strong democracy); (2) some general readiness on the part of citizens to accord higher political priority to advancing important common purposes than to narrower personal concerns; and (3) institutions that foster these circumstances. These background conditions, in turn, incorporate three organizing principles: democratic politics, democratic community, and democratic work (see Figure 20.3).

Democratic Politics

A strong democracy, by definition, affords citizens roughly equal, and maximally extensive, opportunities to guide their society's evolution. What kinds of political institutions does this imply?

Participation and Representation

Is there a middle ground between the present systems of representation, in which a few people participate in deciding most important issues, and obsessive participation, in which *all* people are expected to participate actively in *all* important issues. Barber characterizes the middle ground as a democracy

> in which all of the people govern themselves in at least some public matters at least some of the time . . . Active citizens govern themselves directly . . . , not necessarily . . . in every instance, but frequently enough and in particular when basic policies are being decided and when significant power is being deployed.[14]

Moreover, on issues in which individuals choose not to participate, they should know that generally others with a similar point of view are participating competently, in effect on their behalf. This may entail, among other things, institutional mechanisms to ensure that the views of socially disadvantaged groups are fully represented and that their needs and rights are respected.[15]

Broadened and equalized opportunities for participation are more than a matter of formal legal rights. They must be supported by relatively equal access to the resources required for efficacy, including time and money. Today, for example, politicians, government functionaries, soldiers, and jurors are paid to perform their civic duties. Why not, when necessary, pay citizens to perform theirs, as did the ancient Athenians? Fairness and equality may also be served by increasing the ratio of representatives chosen by lot to those chosen by vote.[16]

Political Decentralization and Federation

What can help prevent representation from gradually usurping the role of an active citizenry? A partial answer lies in some sort of devolution of centralized political institutions (in which the population's large size renders meaningful participation by all impossible) in favor of a plurality of more autonomous, local political units. By means of small-scale local politics, more voices can be heard and each can carry more weight than in a larger polity. Decisions can be more responsive to individuals, thereby increasing citizens' incentive to participate. There is also potential for small polities to be able to govern themselves somewhat more consensually than can the larger society.

However, various considerations—such as the importance of protecting minorities from local repression—suggest the need to embed decentralization within a larger, federated democratic system. The detailed form of federation must be decided contextually, but its thrust should be toward (1) subsidiarity (i.e., decisions should be made at the lowest political level competent to make them), (2) egalitarianism within and among polities, and (3) global awareness or nonparochialism (i.e., ideally, everyone manifests a measure of knowledge and concern with the entire federated whole—or even beyond it).

In short, power should be relatively diffuse and equal. Political interaction and accountability should be multidirectional—flowing horizontally among local polities, vertically back and forth between local polities and more comprehensive political units, and cross-cut in less formal ways by nonterritorially based groups, voluntary associations, and social movements.

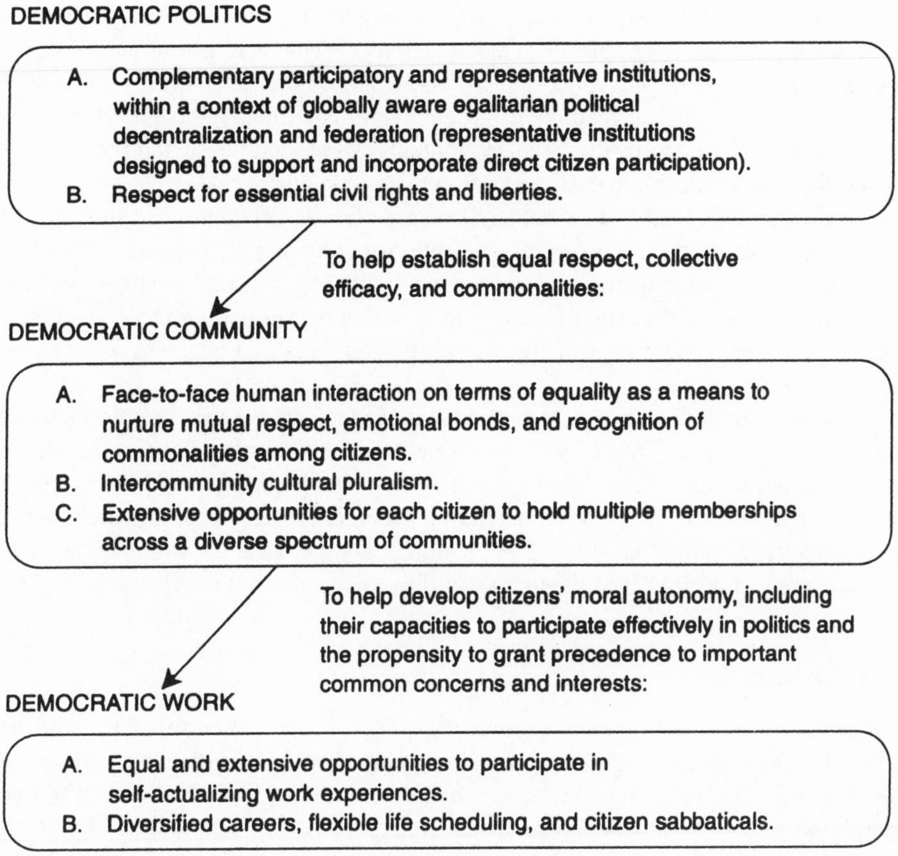

DEMOCRATIC POLITICS

> A. Complementary participatory and representative institutions, within a context of globally aware egalitarian political decentralization and federation (representative institutions designed to support and incorporate direct citizen participation).
> B. Respect for essential civil rights and liberties.

To help establish equal respect, collective efficacy, and commonalities:

DEMOCRATIC COMMUNITY

> A. Face-to-face human interaction on terms of equality as a means to nurture mutual respect, emotional bonds, and recognition of commonalities among citizens.
> B. Intercommunity cultural pluralism.
> C. Extensive opportunities for each citizen to hold multiple memberships across a diverse spectrum of communities.

To help develop citizens' moral autonomy, including their capacities to participate effectively in politics and the propensity to grant precedence to important common concerns and interests:

DEMOCRATIC WORK

> A. Equal and extensive opportunities to participate in self-actualizing work experiences.
> B. Diversified careers, flexible life scheduling, and citizen sabbaticals.

Figure 20.3   Some of strong democracy's principal necessary conditions.

Agenda Setting and Civil Rights

What if citizens are widely empowered to participate in societal choices, but the menu of choice is so restricted that they cannot express their true wishes? Numerous political theorists agree that decision-making *processes* are democratically inadequate, even spurious, unless they are combined with relatively equal and extensive opportunities for citizens, communities, and groups to help shape decision-making *agendas.* Various civil liberties and protections are also democratically essential either because they are intrinsic to respecting people as moral agents or because we require them in order to function as citizens.[17]

Democratic Community

A strong democracy requires local communities composed of free and equal members that are substantially self-governing. Such communities help constitute the foundation of a decentrally federated democratic polity, and hence of political participation, freedom, and efficacy. They do this in part by establishing a basis for individual empowerment within collectivities that, as such, are much more able than individual citizens to contest the emergence of democratically unaccountable power elsewhere in society (e.g., in neighboring communities, private corporations, nonterritorial interest groups, or higher echelons of federative government). Local communities also provide a key site for coming to know oneself and others fully and contextually as moral agents. The defining features of a democratic community include social structures and practices that nurture collective efficacy, mutual respect, and moral and political equality, and, if possible, help sustain a measure of communitywide commonality.[18]

Strong democratic theory does not envision a perfect societal harmony of interest, sentiment, or perspective. Rather, the central aims of strong democratic practice include seeking amid the fray for any existing areas of commonness; striving to invent creative solutions that, in a just manner, enhance the ratio of concordance to that of conflict; and balancing the search for common purpose against respect for enduring differences and against coercive pressures toward conformity.

Of course, countless forms of community and human association are not locally based. However, strong democracy places a special weight on local community as a foundation (but not a culmination) because of the distinctive and inescapable physical and moral interdependencies that arise at the local level; the territorial grounding of political jurisdictions; and the distinctive quality of mutual understanding, learning, and personal growth that can take place through sustained, contextually situated, face-to-face discourse and interaction.

Cultural Pluralism

If one next considers an entire society's overall pattern of kinds of communities and associations, one discovers that strong democracy does more than *permit* diversity among them: it *requires* diversity. Specifically, democracies should manifest a certain kind of institutional and cultural pluralism: equal respect and protection for all cultures, communities, traditions, and ways of life whose practices can reasonably be construed to affirm equal respect and freedom for all. (Cultures that fail to meet this standard may not warrant unqualified respect, but neither do they warrant determined intervention—unless, that is, they seriously threaten the viability of other, democratic cultures or oppress their own members involuntarily.[19])

There are two principal reasons for this requirement. First, equal respect for people entails respecting their cultural heritage. To undermine a culture corrodes the social bases of its members' sense of self and purpose. Second, all people share an interest in living in a society and a world comprised of many cultures. Cultural pluralism supplies alternative viewpoints from which individuals can learn to see their own culture's strengths and limitations, thereby enriching their lives, understanding, and even survival prospects. Moreover, it provides alternative kinds of communities to which people can travel or move if they become sufficiently dissatisfied with their own.

Democratic Macrocommunity

Democratic politics beyond the level of a single community or group requires generally accepted means of addressing disagreements, and ideally a measure of societywide mutual respect or commonality. The alternatives can include authoritarianism, civil violence, or even genocide—as modern history vividly demonstrates. How, then, can local or association-specific solidarity, together with translocal cultural pluralism, possibly be reconciled with the conditions needed for societywide democracy?

Cultures and groups invariably disagree on fundamental matters sometimes. Nonetheless, there is reason to believe that local democratic communities represent a promising foundation for cultivating societywide respect or commonality. For one thing, ethnic hatred and violence are frequently associated with longstanding political-economic inequalities, not with extant approximations to strong democracy. Moreover, often it is probably harder to escape acknowledging and learning to accommodate differences when engaged in local democracy than in translocal association or politics, where there can be more leeway to evade, deny, or withdraw from differences.

The alternative notion of forging a macropolitical culture at the expense of local democratic communities risks coercion or a mass society, in which people relate abstractly rather than as concrete, multidimensional moral agents. Members of a mass society cannot feel fully respected as whole selves, and furthermore, they are vulnerable to self-deception concerning other citizens' needs and to manipulation by those feigning privileged knowledge of the common good.[20]

One method of nurturing local nonparochialism is to pursue cooperative relations among communities that are distantly located and culturally distinct. There is a good model in those modern U. S. and European cities that have established collaborative relations with communities in other nations regarding matters of peace, international justice, environmental protection, or economic development.[21]

Another route to nonparochialism is to ensure that people have extensive opportunities to experience life in a variety of different kinds of communities. Generally, such opportunities should involve experience both in (1) a culturally diverse array of small face-to-face communities or groups and (2) socially comprehensive, nonterritorially based communities.* The former would encourage concrete understanding of the lives and outlooks of different kinds of people and communities; the latter would provide practical opportunities to generalize and apply what one has learned from these diverse experiences to the problems and well-being of society as a whole.

Nonterritorial associations that are not socially comprehensive—such as ethnic associations, labor unions, churches, single-issue political organizations, and so on—can obviously function as one kind of rewarding community for their members. However, they seem less likely to provoke deep, multiculturally informed comprehension of an entire society.

Democratic Work

People often think of "work" as something they do primarily to earn a living. Here, however, work is interpreted as a lifelong process whose central functions include individual self-development as well as social maintenance (both biological and cultural). "Democratic work" thus denotes (1) work activity through which one can discover, develop, and express one's creative powers, strengthen one's character, and enhance one's self-esteem, efficacy, and moral growth (including readiness to act on behalf of common interests and concerns); (2) a work setting that permits one to help choose the product, intermediate activities, and social

---

*The latter are communities or organizations that manifest a multifaceted concern with the well-being of a wide range of kinds of people, if not the entire society or world. Examples include broad-based political parties or movements; federation-level government agencies; and translocal, nongovernmental social service organizations.

conditions of one's labor, thereby developing political competence within a context of democratic self-governance; and (3) the creation of material or other cultural products that are consistent with democracy's necessary conditions, that are useful or pleasing to oneself or to others, and that thus contribute to social maintenance and mutual and self-respect.[22]

Democratic work contributes richly to individual autonomy and democratic society. Hence, societies cannot be considered strongly democratic if there is involuntary unemployment or if, for example, many people are compelled to work in social environments that are tedious and hierarchically structured, while a few elite managers make important decisions that affect many other citizens. To the extent that good jobs are scarce or societal maintenance requires a certain amount of drudgery or other unpleasant work, vigorous efforts should be made to ensure the sharing of both unpleasant and pleasurable activities.[23]

Diversified Careers and Citizen Sabbaticals

Numerous social thinkers have suggested that people should be able to work in a variety of different careers—either in linear sequence or, preferably, in fluidly alternating succession (sometimes called "flexible life scheduling").[24] However, one reason that is often overlooked concerns the cultivation of citizens' readiness to respect people everywhere as ends in themselves and to act on behalf of societywide interests.

To capture this benefit might require a societywide system, analogous to faculty sabbaticals or the U. S. Peace Corps, that would encourage each person to occasionally take a leave of absence from his or her home community, to live and work for perhaps a month each year or a year each decade in another community, culture, or region. This sabbatical system should include opportunities for the broadest possible number and range of people to take turns within translocal government and administration.

Citizens could then return to their home communities with a deeper appreciation of the diverse needs of other communities, a broader experiential basis from which to conceive of their society's general interest, and lingering emotional attachments to the other communities. (Note, for instance how increased contact between white and African-American soldiers in the U. S. armed services has generally reduced racial prejudice there, thus increasing receptivity to societywide racial integration and equality.[25]) Citizen sabbaticals would thus provide one concrete means of implementing the earlier proposal that citizens have the opportunity for lived experience in a culturally diverse array of communities.

Guiding Principles

Do the three seemingly distinct social domains—formal politics, community, and work—mean that different kinds of basic institutions each contribute to democracy in an essentially different but complementary way? Suppose, instead, one conceives of democratic politics, community, and work as three guiding principles that should each to some extent be active within every basic institutional setting or association (recall Figure 20.3). An actual workplace, for example, may be conceived primarily as a locus of self-actualizing experience and production (work), but it should also ordinarily be governed democratically (politics) and help nurture mutual regard (community).

Failure to embody, within each of a society's many settings and associations, all three principles will tend to result in a whole society much less than the sum of its institutional parts. When, for example, each of a society's basic institutions is merely monoprincipled not only does each fall short of constituting a democratic microcosm, but each in addition tends to stress and overtax the capacities of the others.

Democratic Knowledge

Widespread political participation and the experience of diverse cultures and forms of work amount to an experientially based program of civic education. Living this way, one could hardly help but acquire extensive knowledge of one's world and society. This is not only positive; it is also democratically vital and a civil right. Competent citizenship, moral development, self-esteem, and cultural maintenance all depend on extensive opportunities, available to both individuals and cultural groups, to participate in producing, contesting, disseminating, and critically appropriating social knowledge, norms, and cultural meaning.[26]

Formal politics, in particular, must incorporate procedures that support collective self-education and deliberation. The means might include ensuring multiple independent sources of information with effective representation of minority perspectives; open and diverse means of participatory political communication and deliberation (including subsidies to disadvantaged groups that would otherwise be excluded); and extensive and convenient means of monitoring government performance.

NOTES

1. Tripp (1980, pp. 19, 33).
2. Barber (1984). I find the term "strong democracy" congenial and I agree with many of Barber's ideas. However, I also draw on other theoretical treatments that complement or qualify his model, such as Rawls (1971), Mansbridge (1980), Bowles and Gintis (1986), Gould (1988), Young (1989), and Cohen et al. (1992). These works also include extensive criticism of thin democracy or its underlying, classical liberal presuppositions.
3. Barber (1984).
4. Social struggles, the influence of business interests and social elites, and city planning have all, of course, left their mark on contemporary cities. But often planning has done little but help establish the infrastructure under which anarchic growth could proceed. On the technologically influenced evolution of American urban and suburban areas, see, for example, Mumford (1961), Cowan (1983), Hayden (1984), Jackson (1985), and Rose and Tarr (1987).
5. Berger (1977, vol. I, p. 22).
6. Sources on the Berger Inquiry include Berger (1977), Gamble (1978), and OECD (1979, pp. 61–77).
7. Tripp (1980, p. 19).
8. E.g., Mander (1978).
9. On the art and process of design, see Pye (1982), O'Cathain (1984), Ferguson (1992), and Thomas (1994).
10. Examples of conceptions of technology and democracy that are either purely procedural or purely substantive include, respectively, OECD (1979) and Mumford (1964). Notable prior steps toward synthesizing these two components include Illich (1973), Bookchin (1982), Goldhaber (1986), and Feenberg (1991).
11. Kant (1959, pp. 46–54).
12. Ibid., p. 47.
13. In its essential logic, this solution is Rousseau's in *The Social Contract* (1968).
14. Barber (1984, pp. xiv, 151).
15. Young (1989).
16. Barber (1984, pp. 290–293). On Athenian democracy, see Finley (1983, pp. 70–84).

17. E.g., Cobb and Elder (1972), Rawls (1971).

18. Small communities can sometimes be parochial or oppressive, although—by definition—not if they are strongly democratic. For evidence and argument contrary to the view that local communities are invariably parochial or oppressive, see Mansbridge (1980), Taylor (1982), and Morgan (1984). Various social observers (e.g., Lee 1976) argue that the dangers of enforced social conformity are greater in a mass society than in one composed of smaller, differentiated, democratic social units. On communities as mediating structures that empower individuals, see Frug (1980), Bowles and Gintis (1986), and Berry (1993). The potential power inherent in local democratic community is made manifest, for example, in Old Order Amish people's success in resisting the joint coercive force toward conformity resulting from external market relations, persistent social ridicule, and one-time government opposition to various Amish practices (such as refusal to participate in Social Security, military service, and the public school system).

19. See Rawls (1971, sect. 35) and Gutmann (1987, pp. 41–47).

20. Calhoun (1991).

21. Brugmann (1989).

22. For one reasonable attempt at synthesizing a number of contrasting modern views on individual moral development, see Kegan (1982). Empirical research has not decisively confirmed or refuted the thesis that self-governed workplaces encourage political participation beyond the workplace (e.g., compare Elden 1981 with Greenberg 1986).

23. Schor (1992). For examples of contemporary workplaces in which menial tasks are shared, see Rothschild and Whitt (1986) and Holusha (1994). Feminist writing on the importance of sharing childcare responsibilities are also salient here (e.g., Hochschild 1989).

24. E.g., Sirianni (1988).

25. Lovell and Stiehm (1989).

26. For perspectives on democratic knowledge production and civic education, see, for example, Schutz (1946), Adams and Horton (1975), Goodwyn (1978), Foucault (1980), Freire (1980), Guess (1981), Barber (1984), Belenky (1986), Gutmann (1987), and Harding (1993).

21

Socialism and the Democratic Planning of Technical Change

Tony Smith

I have argued that the normative claims made by advocates of the "new economy" regarding labor relations, consumer relations, and interenterprise relations in lean production cannot be sustained. What sorts of institutional changes would have to be made for the promises of lean production to be actualized?

The promise of a fundamental transformation in labor relations can only be redeemed if structural coercion, exploitation, and alienation in the workplace are overcome to the greatest extent possible. The structural coercion that workers face in capitalist labor markets stems from lack of secure access to the means of subsistence and production. Secure access to the means of subsistence can only be guaranteed if all citizens are granted a right to share in the fruits of economic life. This could be accomplished through the direct provision of basic social needs, through a guaranteed basic income, or through some combination of the two (Mandel 1992, 205–10; Van Parijs 1989). In a similar fashion, access to the means of production demands an acknowledgment of the right to employment as a fundamental citizen right. Only then would decisions regarding whether to work for particular enterprises count as truly free decisions.

Reversing exploitation and alienation in the labor process would necessitate both public ownership of enterprises past a certain point in size and worker self-management. As long as firms are owned by private capital investors who appoint managers as their agents, technologies and forms of organization will be introduced first and foremost in order to further the interests of those owners and agents. And they will be employed in ways that tend to sacrifice other interests to that end. In contrast, if the management of an enterprise were democratically accountable to the workforce, other factors would determine the fundamental goals of technical and organizational change in the workplace. Far more economic surplus reinvested within the firm would be devoted to innovations that reduce work hours, enhance the creativity of workers, increase workplace safety, and so on. After all, the first principle of democracy is that

From Tony Smith, *Technology and Capital in the Age of Lean Production*, (Albany, N.Y.: SUNY Press, 2000), pp. 135–159. Copyright © 2000 by SUNY Press. Reprinted by permission of SUNY Press.

authority rests on the consent of those over whom authority is exercised. Exploitation and alienation in the workplace will only be overcome when this principle is applied to exercises of authority in labor relations.

This argument does not imply that everyone in the workplace should get to decide about everything all the time. It does indicate, however, that those who decide how economic resources are to be allocated within a firm, and how the labor process is to be structured, ought to be accountable to those over whom this authority is exercised. The collective body of workers must themselves delegate those who are to make these decisions, and have the power to recall them when occasion demands.[1]

If these institutional changes are necessary for overcoming antagonisms in the workplace, and if these changes are not part of the lean production model, then we may conclude once again that systematic antagonisms have not been overcome in the "new economy." Perhaps it is the case that the workforce as a whole experiences a net increase in skill levels with the transformation to the technologies and social organization of lean production. But it does not follow from this that the fundamental class conflict between capital and labor discussed by Marx is now a thing of the past. Structural coercion, exploitation, and the real subsumption of labor under capital remain in lean production systems. Eradicating this fundamental class antagonism demands an institutional transformation far deeper than the transition from the traditional mass production of Fordism to lean production.

What sort of social arrangements would have to be in place for institutionalized antagonisms in consumer relations to be truly overcome? First and foremost, the main source of the asymmetry of power between producing enterprises and consuming individuals and households must be addressed. This involves numerous factors outside the sphere of consumption. Public ownership of large enterprises, workplace democracy, and mechanisms allowing social accountability for investment decisions[2] all work to dissolve the concentrated economic power that in capitalism distorts consumer relations no less than relations in production.

Other solutions specific to the realm of consumption must complement these measures. Well-funded consumer unions could take an active role in soliciting consumer products, in monitoring price markups, in transmitting information regarding these matters to consumers, and so on (Elson 1988). If these unions were democratically accountable to the consumers whose interests they represent, then the atomization of consumers would be overcome through political organization.

Consumer unions would also help guard against the real subsumption of consumers occurring when the objectification of their desires is used against them. These unions must have open access to the data bases collected on individuals and households. They must monitor how enterprises make use of these data bases, and they must ensure that consumers are fully informed of these tactics. They must also ensure that privacy rights are enforced, including the right to not have information about oneself be part of a data base without explicit consent.

Turning to questions of subjectivity, a major problem with consumption in the "new economy" concerns the systematic inducements to define one's identity through the consumption items one has purchased. In a different social context, different forms of self-definition would be encouraged. If more participatory forms of decision making pervaded the workplace and the community, ample opportunities would be provided for active participation in social life. A significant reduction of the working day would allow working men and women to take advantage of these opportunities.[3] Under such social circumstances one's identity would tend to be defined more by one's active contributions to the community than by the items one passively consumes. A significant reduction in the working day would also allow more time to be spent in direct personal relations with partners, children, relatives, and friends. This too would result in the strengthening of other forms of identity besides those connected to the consumption of commodities.

This does not mean that individuals would cease altogether to define themselves through the items they purchase. The clothing, music, decor, and so on, a person chooses to buy would all continue to contribute to the sense of who that person is. But in a society of economic democracy and increased personal interactions, the consumption of commodities would provide just one dimension of the complex process of identity formation. This provides a clear contrast to life under the capital form, where we are barraged with ads telling us that what we buy defines who we are, where the freedom to select among consumer goods compensates for the drudgery and stress of labor, and where the lack of time to devote to pursuits outside the workplace encourages the instant gratification of shopping.

The failure to meet wants that do not take the commodity form is another tendency that remains in lean production. This failure could be reversed through an extensive and well-funded program of allocating public grants to community groups that set out to meet social needs in a noncommodifiable way. These grants must be coupled with regulations guaranteeing access to the technologies of distribution. For example, sufficient bandwidth must be set aside on the Internet for the transmission of information and entertainment produced by community groups, and ample funds should be made available to allow those groups to finance their projects.

A gulf between consumers and the objects of consumption arises whenever decisions to purchase commodities contribute to results inconsistent with the consumers' own considered judgments regarding the social good. This tends to occur when relevant information is not transmitted to consumers through the price mechanism. And so other mechanisms must be established. Information technologies certainly make it possible in principle to transmit at little or no cost information regarding the work conditions under which products were produced, the environmental consequences of using those products, and so on. We could imagine, for instance, consumer unions investigating these and other relevant matters. Enterprises would be required to collect and make available to the consumer unions the data required. Printouts of findings could then be made available on demand to consumers, through terminals at both the point of purchase and consumers' households. The results of these investigations might also be summarized in grades that would appear alongside the prices on display. Some of these findings and grades would no doubt contradict others. It would be the responsibility of the consumer unions to highlight the differences, tracing them back to divergent methodological assumptions, value judgments, and so on. Consumers would then have the opportunity to sort through this information. Having easy access to information regarding product availability and product features would also eliminate much of the need for advertisements.

There is thus a general congruence between the sorts of measures required if consumer interests are truly to be institutionalized, and the measures required to ensure that the interests of workers are adequately represented. As long as both sets of interests are ultimately subordinate to the imperatives of capital accumulation the promises made by the advocates of lean production will remain utopian fantasies.

Regarding relationships among economic enterprises, once again we need to ask what sorts of social transformations would have to occur for the claims of the new capitalist utopians to hold true. This question has been raised by Bennett Harrison, who has perhaps as deep a sense of the limitations of networks of lean production firms as any social theorist. And yet there is a huge gap between his description of what he terms "the dark side" of lean production and his specific policy proposals. He calls for a rejuvenation of the labor movement, along with a higher minimum wage, mandatory corporate training programs, a ban on the hiring of permanent replacement workers for strikers, and a long-term growth strategy led by public investment (Harrison 1994, 245).

It is difficult to think that Harrison himself is unaware of the limits of his proposals. I suspect that he feared that if he had seriously addressed the shortcomings of lean production networks his work would be ignored in mainstream policy-making circles. The tremendous gap between his description of these shortcomings and his proposals to address them provides a striking measure of the limits of mainstream political discourse today.

Drawing on the work of Diane Elson, Hilary Wainwright, and David Schweickart, four suggestions can be proposed for bringing about greater harmony and trust among units of production and distribution.

(1) The use of technologies within networks of firms is a function of the ownership and control of those firms. With public ownership of enterprises past a certain size, and with workplace democracy instituted in those enterprises, the interfirm communication and coordination enabled by advanced information technologies would take a quite different form from that in lean production networks. Large firms would not be able to use equity holdings in smaller firms to dictate production targets and prices. Nor could they insist on unreciprocated access to information. Decisions affecting networks of firms could be made by representatives of the different labor forces, who would be accountable to those groups of workers. Representatives of smaller firms would not be likely to agree to arrangements in which larger firms unilaterally appropriated cost savings made by smaller firms, or arrangements in which more and more economic burdens were placed on smaller enterprises, while larger ones received more and more economic benefits. And so these problems would not tend to arise.

(2) One factor ensuring that intercapital relations remain aggressively antagonistic is the "expand or die" nature of capitalist markets. This imperative is not weakened one iota in lean production. And so technical change in lean production tends to generate overproduction crises, which lead each intercapital network to frantic attempts to shift the costs of devaluation onto other networks. Publicly owned workers' cooperatives would escape this trap. As expansion proceeds, a point is reached where taking on any more people in the co-op would lessen the share of present members. When this point is attained these workers do not have an incentive to support the continued expansion of their enterprise (Schweickart 1994, chapter 4).

In order to lessen internetwork antagonisms significantly, however, much more is required. The activities of different networks must be coordinated together in a democratic fashion. In brief, there must be a greater role for the democratic planning of investment decisions in society as a whole. This planning has a number of objectives. Decisions must be made regarding the general direction of the economy. The background conditions for implementing those decisions must be provided. And the decisions must be implemented concretely. I shall soon attempt to sketch how a system of technology boards and community banks would contribute to these objectives. Of course the collapse of the Soviet Union has made all talk of planning extremely suspect, and this misgiving must be addressed. But if a system of technology boards and community banks could help attain the above goals, the danger of overproduction crises would be significantly lessened. Decisions to expand would be subject to public oversight to a certain degree. The need to devalue previous investments as technologies change and demands shift would not be eliminated, but it would be profoundly reduced. And most of the costs of the devaluations that did occur would be fairly shared. In this manner the forces generating interenterprise antagonisms would be diminished far more than in lean production or any other conceivable form of capitalism.

(3) In the model of socialized markets advocated by Elson, enterprises are required to disclose all information regarding production and distribution.[4] It would not be possible for anyone to monopolize a significant portion of this information, as executives of firms in lean production networks do at present. Further, if the management of enterprises in networks were democratically accountable to their workforce, horizontal links among workers in different enterprises would not be systematically discouraged as is the case today. Community oversight would also foster communication between enterprises and social movements. In this manner the bias in the sorts of information collected and transmitted within lean production networks would be overcome.

(4) If those making decisions regarding the management of a firm are democratically accountable to the workforce of the enterprise, and if decisions regarding the expansion of the enterprise are made by elected representative bodies and community banks following principles agreed upon in public debate, then there would not be a necessary tendency for the formation of interenterprise networks to conflict with the common good. For that tendency necessarily arises only if those making decisions affecting the public are able to shield themselves from public accountability.

I am well aware that this is no more than the briefest of sketches of an account of interfirm relations in an alternative social order. But I believe it is sufficient to establish the following point: uniting different units of capital in lean production networks does not by itself suffice to bring about trust, cooperation, and harmony in interenterprise relations, even if information technologies are used to promote greater communication and coordination within networks. Here too the new capitalist utopians are wrong. Fundamental changes in both property relations and production relations are required to realize the potential for greater trust and cooperation created by information technologies. These changes would inaugurate a historical break with capitalism, and not merely another variant of it (see Marsh 1995, chapters 15 and 16).

Democratic planning is one of the essential features of this historical break. As mentioned above, democratic planning has three dimensions, the determination of the direction of the economy as a whole, the fulfillment of the background conditions necessary to proceed in this direction, and the concrete implementation of the decisions that have been made. Regarding the last topic, David Schweickart (1994) has proposed that new investment funds be distributed to regions on a per capita basis. Community banks in these regions would then fund the expansion of present enterprises and the formation of new enterprises. These enterprises would then either provide goods and services directly, or compete in (socialized) markets.[5]

How should the direction of the economy as a whole be determined, and how should the background conditions necessary for developing in that direction be established? The idea of a comprehensive plan including each individual input and each unit of output is a fantasy, and must be unequivocally rejected. Even with the best intentions in the world and immense computing power at their disposal, central planners will simply lack too much relevant information. In the model defended here each enterprise decides for itself what set of inputs it will purchase from other enterprises in producer markets, and what set of outputs it will sell to other producers and final consumers. Similarly, consumers decide for themselves which goods and services they wish to purchase. But this does not imply that the idea of subjecting the general direction of the economy as a whole to social control ought to be rejected. To a considerable extent the general direction of the economy is a function of the path of technical change, and capitalist economies already provide ample evidence that the path of technical change can (must) be influenced as a result of decisions made by public authorities. Since World War II in the United States, between one-half to two-thirds of all research and development has been funded by the government, and two thirds of all basic research has been so funded (Mowery and Rosenberg 1989, 128, 130). The pattern of this funding has shaped the general direction of the U.S. economy.[7] The role of the government in determining the direction of technical change, and thus of the economy as a whole, is similar in all other technically advanced societies as well. The question really isn't "planning or the market?" but rather the sort of planning and the sort of market we should have. The problem in the United States and elsewhere is that this planning has been done behind closed doors. *To a large extent, democratic central planning is the democratization of technology policy.*[8] I shall now sketch one way this planning could be implemented, expanding on suggestions made by Diane Elson.

Central planning could be undertaken by a national technology board, whose actions would be coordinated with an international board. Organizations representing mass movements must play a crucial role in any truly democratic from of central planning (Fisk 1989). And so the board should consist of representatives from mass organizations of workers, consumers, environmental groups, collectives of scientists and technicians, and so on, along with representatives of the state. This board would have the duty of formulating a number of competing plans regarding the general allocation of resources for technological development, plans based on different estimations of the scientific-technical potential in the society, the risks associated with developing that potential, and the priority of social needs. These various plans could then be taken back to the base of the mass organizations and subjected to extensive discussion, making full use of the potential of the contemporary revolution in communications technology. At the conclusion of the discussion period, a society-wide vote could then decide which framework should be accepted.

The technology board would at that point have the responsibility of setting in motion the plan that had been agreed to. This would first involve allocating resources to various centers for research and development. These centers would work in conjunction with nearby universities and units of production and distribution, forming a "technological milieu."[9] These centers would then proceed through the different stages in the technology pipeline, from the most abstract and basic research to progressively more concrete applications.[10] At the conclusion of this process, the results would be provided to worker-run enterprises for final production and distribution, using either internally generated funds or investments made by local community banks.

For each local center a local technology board could be established, consisting of both local officials and local representatives of mass organizations. Among the main tasks of these local boards would be:

1. to ensure that local citizens and groups have continuous access to scientific and technological expertise so that they can educate themselves regarding new developments (unlike both bureaucratic socialist economies and capitalist market societies, where access to expertise tends to be restricted to elites);
2. to set up a series of science and technology courts, where scientists and technologists with different evaluations of predominantly technical matters relevant to economic development can be cross-examined;
3. to set up a series of public hearings, allowing local citizens and groups to articulate any questions or misgivings they may have regarding economic developments;

4. to modify or stop such developments if objections are serious enough; and
5. to forward appeals of their decisions to regional, national, and international technology boards, with the most controversial decisions ultimately to be left to society-wide discussion and vote.

In this manner local communities could ensure that local technological development is consistent with the democratic will (Sclove 1995).

These proposals are designed to transcend the lack of cooperation among units of production and distribution that generates overproduction crises and massive social disruptions in all forms of capitalism, including lean production. The institutionalization of national and international technology boards would eliminate the madness of a system in whch each individual unit of production—or each network—desperately attempts to shift social costs onto its workforce, other enterprises, other networks, or the community at large. It would also overcome the limits set by capitalism to the flow of information regarding production and distribution.

If, as much of the rhetoric of lean production is designed to suggest, economic systems are to be judged by the extent to which they overcome social antagonisms, democratic planning would surely rank above the lean production model of capitalism . . . if it were feasible. How do we know that economic democracy provides a feasible alternative to the "new economy"? Are there any reasons to think that democracy in the workplace, in the realm of consumption, and in the planning of technological development could avoid the difficulties that plagued bureaucratic central planning?

In the present work advocates of lean production have repeatedly been referred to as "new capitalist utopians." It may strike many readers that the proposals outlined here deserve to be dismissed as utopian illusions far more than anything submitted by defenders of the "new economy." However sympathetic readers might be to the idea that the promises of lean production have been wildly overstated, the idea that there is a form of socialism that can redeem these promises may seem laughably absurd. For most social theorists, and for most citizens, the utter collapse of the Soviet model has proven conclusively that the best humanity can realistically hope for is a reasonably humane form of capitalism. In the following section I shall address this concern. I shall argue that the alternative to lean production sketched in this section is feasible as well as normatively attractive.

THE CONTRAST BETWEEN DEMOCRATIC AND BUREAUCRATIC PLANNING OF TECHNICAL CHANGE

Seven difficulties regarding bureaucratic central planning were discussed earlier in this chapter: Bureaucratic central planning (1) hampers innovation in general, (2) hampers process innovation in specific, (3) involves disincentives for technical change, (4) leads to secrecy and plans based on misinformation, (5) hampers the development of communication technologies, (6) leads to a fear of the risks connected with technological innovation, and (7) inevitably leads to a neglect of quality. Would the form of socialism just sketched be equally prone to these difficulties?

1. Flexibility

Democratic central planning need not involve the specification of rigid plans governing the entire economy. It concerns only the most general pattern of the allocation of the social surplus. Many remaining aspects of economic life are left to the (socialized) market, with its inherent flexibility. The members of society could decide to place greater priority on say, developing innovations to provide adequate housing rather than luxury condominiums, solar rather than nuclear energy, or sustainable rather than chemical agriculture. These decisions do not commit the housing sector, the energy sector, or the agricultural sector to adopt any specific technique. Nor do they demand that one set of techniques alone be employed by these sectors over the course of an entire planning period. These sorts of decisions can be decentralized, that is, left to local enterprises working in conjunction with local technology boards and community banks. In this manner, democratic central planning does not involve the structural tendency toward predictable routine that characterizes bureaucratic planning.

2. Process Innovation

When new processes are introduced, they typically require new inputs. In bureaucratically deformed economies if a firm makes such an innovation the central plan is disrupted, with the effect that suppliers often cannot be found to provide the required inputs for the innovating enterprises. Two factors in the democratic central planning of technical change suggest that this difficulty would not arise. As already noted, central planning in this model leaves considerable scope for decentralized decision making. Also, research on local and regional levels is performed in centers that are part of a "technological milieu." Those engaged in pure and applied research interact both formally and informally with those engaged in assembly as well as with their suppliers. The local and regional technology boards overseeing the innovation process include representatives of all these groups. This arrangement would allow a change in the mix of inputs used by assemblers to be planned in conjunction with a simultaneous change in the output mix of the relevant suppliers. The flexibility inherent in (socialized) markets can be relied upon to take things from there.

3. Incentives

The issue of incentives for technical change is perhaps the most crucial of the seven points discussed in this section. It will be discussed at somewhat greater length than the other six matters.

Bureaucratic central planning proved capable of guiding the initial process of industrialization, albeit at great personal and political cost. Beyond that, however, it failed to provide sufficient incentives to institutionalize technical change throughout the economy. If the same shortcoming held for democratic central planning as well, then the case for lean production would be much stronger. Whatever its other limitations, there is a strong case for considering lean production to be "innovation-mediated production" (Kenney and Florida 1993), at least for firms of the "core."

In all forms of capitalism the owners and controllers of capital generally have strong material incentives to introduce changes in technology. In lean production, the expansion of intellectual property rights provides an extra incentive to develop technical knowledge, the fruits of which can then be privately appropriated. If in the model of socialism sketched in the previous section there is public ownership of firms past a certain size, and if there are no intellectual property rights in this model, then who exactly has an incentive to undertake technical change?

The general answer to this question is straightforward enough. Within the workplace the members of the workforce have a clear incentive to seek new techniques that lessen their toil. Workplace democracy provides an institutional mechanism to carry out this objective. They can elect managers who will institute a search for such new techniques, and who will introduce them when they are found; they can vote out managers who fail to act in this manner. On the community level a similar point holds. It is plausible to assume that citizens wish to have social needs provided for in as efficient a manner as possible. They can be expected to elect representatives to technology boards who will direct the search for innovations that meet socially articulated needs in a more adequate fashion, and who will encourage the introduction of such innovations when they are found. Since these planners are recallable, accountability can be continuously enforced.

What of the lack of intellectual property rights? Their absence may allow the free flow of knowledge that has already been produced. But if you cannot privately appropriate the fruits of your innovation, where is the incentive to produce new forms of knowledge? I find Elson's response to this question persuasive (1988). In her model of socialism the technology board provides grants to research institutes to fund the search for new knowledge, provides rewards when this search is successful, and takes the past track record of institutes into account when providing new funding. These arrangements appear to be more than sufficient to provide an institutional context in which knowledge workers can flourish.[10]

In this context it is also worth stressing that democratic planning is compatible with both private property and a certain degree of inequality in its distribution. While defenders of this model are against the private ownership of large-scale productive resources, they are not against the private ownership of individual consumption goods.[11] A democratically planned society can reward innovative individuals and research teams with greater material compensation in order to provide incentives for further innovation. This would not contradict the principles of that society, so long as the society collectively agreed that the resulting increase in inequality was more than compensated by the positive effects of providing these incentives.

Even if certain incentives for technical change were provided in our alternative to lean production, however, they might be outweighed by even more powerful disincentives. Consider the force of habit. A workforce that has become comfortable with a certain way of doing things may resist attempts to change established procedures. A community may resist attempts to upset its established patterns of life. Bureaucratic central planning overcomes this problem through bribery and the threat of direct physical coercion. Capitalistic markets rely on the structural coercion connected with forced unemployment and disinvestment in communities. There does not appear to be any analogous mechanism at the disposal of democratic socialist planning to break the weight of past habits.

For the defenders of socialist democracy this is a pseudo-problem. The minimization of toil and the better satisfaction of social needs are social goods. The minimization of personal and social disruptions are also social goods. The correct trade-off between these two sets of benefits cannot be determined *a priori*. We certainly cannot trust either bureaucratic officials or the impersonal dictates of capitalistic markets to attain the proper balance. The appropriate trade-off is that which the affected workers and communities would decide for themselves in an uncoerced decision process. The democratic planning of technical change institutionalizes such a process. It should also be noted that economic democracy includes numerous local centers for small-scale, experimental innovation. This provides a reason to expect that the pace of technical change may not slow down at all compared to the current multinational-dominated economy.

Another possible disincentive for technical change stems from the extensive oversight processes defining democratic planning. It might be said that the series of boards, public hearings, science and technology courts, community banks, public votes, and so on that accompany the democratizing of economic development provide a powerful disincentive to those engaging in innovative activities. Would these hurdles not require considerable amounts of time and energy to jump? And would this not result in a structural tendency for innovators to be discouraged from introducing advances that they might otherwise have introduced?

It must be recalled that in capitalist market societies technical choices typically must also pass through an extensive series of hurdles. In the United States the military apparatus often must favorably assess the weapons potential of a proposed technical innovation for it to receive crucial initial funding. When it comes to commercial applications by start-up firms, venture capitalists must favorably assess the short- and long-term profit potential of the innovation. And an innovation generally becomes extensively diffused only if large multinationals either take over completely, purchase equity in, or reach marketing agreements with small firms at the cutting edge of the new technology. These things occur only after the multinationals have completed extensive strategic deliberations. Somehow technical change continues in market societies at a fairly rapid rate despite being subjected to such scrutiny. There is no *a priori* reason to think that the pace of technical change will be significantly affected just because this scrutiny is made by publicly accountable representatives rather than by military officials and the owners and controllers of capital.

However, even if we accepted for the sake of the argument that the pace of technical change might lessen somewhat under democratic planning, this in itself would not be telling. Surely any adequate evaluation of the technical innovation process in different social systems must consider the *direction* as well as the pace of technical change. From the standpoint of the public interest a shift in direction can in principle compensate for any slackening in the pace of technical change. The measures introduced in the previous section are designed to ensure that the direction of technical change will further the social good to the greatest possible extent.

4. The Avoidance of Secrecy and Misinformation

Under the model of democratic planning, those engaged in the implementation of technological change win social acknowledgment, economic reward, and access to future career opportunities if they successfully introduce process innovations reducing the toil (or enhancing the creativity) of the labor force, or product innovations serving democratically articulated social needs. This provides ample motivation for innovators to publicize their innovations, and to transmit to the various technology boards accurate information regarding available technological capacities. Further, the processes of public hearings, science and technology courts, and appeals would by themselves be sufficient to ensure that there would not be a structural tendency towards secrecy and misinformation. The abolishing of intellectual property rights serves this same end as well.

5. Communications Technology

Obviously the entire model of the democratic socialist planning of technology requires quite developed communications technology. The coordination of activities along the technology pipeline, the social oversight of these activities by the various boards, and the accountability of these boards to the various mass organizations, all demand a continual flow of communication. While there are systematic reasons for the neglect of communications technology in bureaucratic command economies, a structural tendency for this area of technological development to receive first priority can be derived from the structural mechanisms of a democratic technology policy.

6. Risk

Under bureaucratic planning local officials tend to be wary of introducing untested technical innovations, since they will be held personally responsible for failing to fulfill quotas imposed by higher level officials. In contrast, under democratic planning innovations are developed as a result of a collective decision, and responsibility for any subsequent problems will therefore be collectively shared. If a technology does not fulfill its promise, a search for scapegoats would be pointless, and it would be obvious that energy would be better spent on correcting the error. Under these circumstances a person or team suggesting an innovation would not face the sort of risks faced in command economies.

7. Quality

The problems with quality under bureaucratic planning all stem from a single structural fact, the lack of feedback between producers and consumers. The heads of production facilities are accountable to intermediate strata in the bureaucratic hierarchy as well as to the central ministries, but they are not accountable to the particular end-users of their products. Since higher level officials are only concerned with whether assigned quotas have been fulfilled, while it is consumers who are concerned with the quality of the produced goods, this institutional framework has a built-in tendency to neglect qualitative matters connected with production technologies.

The point to stress here is that it is the lack of feedback between producers and consumers that is the problem, rather than anything inherent in planning per se. Under the democratic centralized planning of technical change there is direct accountability of the planners to consumers. If planners regularly develop technologies that lead goods and services to be produced that do not meet the community's standards of quality, democratic mechanisms grant the community a direct recourse: elect new planners. And the presence of socialized markets provides a second important feedback mechanism to ensure that technical change results in products and services meeting satisfactory standards of quality.

A feedback relation between producers and consumers is also found in capitalist market societies, of course. But capitalist markets are based on the principle one dollar, one vote, and vast inequalities are allowed to persist. The extent of the feedback provided by specific consumers is directly proportional to their disposable wealth. In socialized markets the provision of basic needs, price commissions, wage commissions, and so on, all help ensure that wealth is more equally distributed (Elson 1988, 27–30). Democratic planning provides a further feedback mechanism that bypasses the market entirely. As a result there is every reason to believe that quality considerations could be institutionalized in a deeper and more extensive fashion than in capitalist markets.

This completes the reply to those who hold that the collapse of the Soviet model proves conclusively that any attempt to seek an alternative to lean production that goes beyond capitalism is doomed to fail. No attempt has been made here to provide a complete picture of such an alternative; the model of socialism sketched above no doubt requires extensive refinement

and revisions. Nonetheless, I believe that this model is sufficient to justify the following judgment: it is possible to conceive a democratic form of socialism that is both normatively attractive and feasible. History has not ended with the failure of bureaucratic planning and the
resurgence of capitalist market societies. Our responsibility to future generations demands that
we begin to grope towards some third alternative that the world has not yet seen, an alternative
beyond the limits of the "new economy."

There is one last question to consider before bringing this work to a close. The diffusion
of lean production is a central feature of what has come to be called "globalization." How
does the socialist alternative to the "new economy" defended here fit into the globalization
debate?

MARXIAN INTERNATIONALISM AND THE GLOBALIZATION DEBATE

In many ways socialist democracy appears to be consistent with the "new protectionism"
called for by many populists (see the articles collected in Hines and Lang 1996). The goal of
this policy is to encourage the most extensive regional self-sufficiency possible. Trade across
regions would be reserved for those cases where economic activity within a given area simply
could not provide needed goods and services. If this policy were implemented, regional populists assert, local communities and the regional governments democratically accountable to
them would flourish. Environmental wastes generated when goods and services that could be
produced locally are transported vast distances would also be eliminated.

Despite the attractiveness of much of this vision, I believe it ought to be rejected. There
is a dangerous indeterminacy to populism; progressive populist positions may have the unanticipated consequence of strengthening right-wing xenophobia. Workers in the so-called First
World, angry about stagnant or declining real wages, could easily direct that anger against the
peoples of the Third World. In fact, however, relatively little of the downward pressure on
wages in the United States and elsewhere can be explained by imports from low-waged countries (Webber and Rigby 1996, chapter 7). The antilabor offensive of the U.S. capitalist class
and the austerity program of its chief general, Alan Greenspan, are far more significant explanatory variables in the United States. The story is exactly the same in Europe, as Robert Went
has documented (Went 1996). The more the globalization narrative of the regional populists
diverts attention from this state of affairs, the greater the danger that the new protectionism
will degenerate to just another variant of the old xenophobia protectionism, defining the enemy
as those outside "our" nation.

A strong case can be made for regional self-sufficiency in some areas of food production,
given the centrality of nutrition to human well-being. But a generalization of the principle of
self-sufficiency has a number of implications that appear troubling from the perspective of
regional populists' own normative commitments. If a policy of regionalism were ever implemented, it would freeze the present uneven distribution of resources for investment. Regional-
ization would offer little redress for the plunder of poorer countries over the centuries.
Technology transfer would be discouraged, cutting off enterprises and consumers in poorer
regions from potentially beneficial state-of-the-art innovations.

Another set of problems arises from the fact that there are certain issues of interest to
the global community as a whole. Should regions be free to set low environmental standards,
despite the fact that environmental practices have "spill-over effects" that can spread throughout the world? Should different regions be free to set different labor standards, despite the fact

that standards holding in some areas might fall well behind what the remainder of the human community considers the minimal safeguards of human dignity? What of other economic, civil and political rights? Why shouldn't global homogenization at a high level be struggled for and enforced in these areas?

These considerations strongly suggest that any attempt to bring to life the promises left unfulfilled in lean production ought to aim at a new form of globalization, one in which the law of value is superseded. Once again we may turn to David Schweickart's model of economic democracy for clues.

Schweickart locates ownership rights to productive resources past a certain size in the community as a whole. One practical upshot of this is that the members of worker cooperatives do not have the right to treat their enterprises as their private "cash cows." They cannot run them into the ground while distributing the resulting revenues among themselves. A portion of revenues must be set aside as depreciation funds in order to maintain the value of the enterprise's productive resources. Decisions regarding the investment of these depreciation funds are to be made by the responsible authorities in those enterprises, subject to external social audits.

This leaves the question of new investment in the economy, including both the expansion of existing enterprises and the formation of new ones. Schweickart proposes that funds for new investments be collected from a flat tax on the capital assets of all worker cooperatives (he estimates that this should amount to 10-to-15 percent of GNP). A democratically elected body operating on the international level would have the responsibility of allocating this money.[12] After a period of open discussion and debate, these revenues would then be divided into three parts. One portion would be allocated to democratically elected bodies operating on national/regional levels. These funds would be distributed on a per capita basis, that is, more populous regions would receive proportionally more resources. A second would be devoted to the provision of public goods on an international level (education, transportation, health, cultural production, research and development, etc.). The final portion would be set aside for investment in new industries (or the expansion of old ones) addressing the social needs granted the highest priority at the conclusion of the democratic decision making process.[13]

Elected national/regional representative bodies would then have the duty to make the same three allocations. Funds for new investment would be allocated to local representative bodies on a per capita basis. Other funds would be directed toward public goods on the national/regional level. The remaining funds would be set aside for new investment in industries addressing needs having the greatest social priority within the given area. Local representative bodies, finally, would be elected to allocate funds for public goods on the local level and to provide grants to worker cooperatives addressing the needs most pressing in the local community.

The funds for new investment allocated to particular localities would then be distributed by community banks in those areas. Enterprises that wished to expand, or people who wished to set up new firms, would apply for grants from these banks. Banks would allocate funds based upon their estimates of the likelihood of economic success, the likelihood of new employment opportunities in the community, and consistency with the set of social priorities democratically set on the local, regional, and international levels. External social audits would measure the record of these community banks by these criteria. Community banks with documented success would be allocated greater funds to distribute in the future, while less successful ones would shrink over time.

How does this vision of a global economy contrast with that advocated by defenders of the other perspectives on globalization? I would argue that this postcapitalist form of global-ization combines their strengths while avoiding their weaknesses. The regional populists are correct that the technological dynamism of global capitalism comes at a horrific cost: the threat of capital flight is used to blackmail workers and communities; ever more economic power is concentrated in fewer and fewer hands; more and more crucial decisions regarding the rules of intentional commerce are transferred to international bureaucracies unaccountable to those affected by their decisions. A postcapitalist global order of the sort described above would avoid each and every one of these ineluctable features of capitalism. With no class of capital-ists, there is no possibility of capital flight. With decisions regarding new investment priorities made by democratically elected representatives at international, national/regional, and local levels, there is no possibility of vast concentrations of economic power, or of unaccountable trade technocrats decreeing the rules of global life behind closed doors. This policy would also significantly lessen the danger that globalization will result in the homogenization of social life across the planet. Community banks are far more likely to be sensitive to regional differ-ences than multinational corporations, whose executives and leading investors increasingly inhabit a world of pseudo-cosmopolitanism (Daly and Cobb 1994, 234).

These goals of the regional populists can be accomplished without any retreat to merely local economies, which from the standpoint of Marxian internationalism would count as a retreat from one of the most profoundly progressive features of capitalism. We ought not to lament that the fate of every region is intertwined with the fate of all others. The proper ques-tion is not whether there should be a global community, but what form it should take. The eradication of hunger and disease, the fostering of economic and political democracy, and the avoidance of species-suicide through environmental crises, all require a mobilization of eco-nomic and political resources on a global level.

The neoliberals are correct to emphasize the importance of technological dynamism, and of a global order allowing all regions access to the latest product and process innovations. The above model of economic democracy provides ample mechanisms for the funding of basic and applied research and development, clear incentives for worker cooperatives to introduce innovations, and an absence of legal and economic barriers to the diffusion of innovations.

Competitive regionalists rightly point to the continued importance of governments in the global economy, an emphasis consistent with the Marxian internationalism perspective. They are correct as well to stress the spatial dimension of economic activity; there are many respects in which the rise of a more globalized economy makes the formation of local clusters of firms, farms, universities, research labs, governmental agencies, and so on, even more sig-nificant. Schweickart's model of economic democracy would foster the development of such clusters; community banks would form hubs around which networks of complementary enterprises and institutions would form. But competitive regionalists myopically refuse to confront one ineluctable fact: strategies that are rational for individual regions to pursue taken singly may generate irrational results from the standpoint of the human community. The vision of a global economy in which a relatively few areas enjoy success and insecurity in equal measure, while remaining regions stagnate or decline, is not very inspiring. Capital-ist development has always been profoundly uneven, and the strategies of the competitive regionalists would, if anything, exacerbate this tendency. Here lies the profound importance of the proposal to allocate funds for new investment strictly proportional to regional popula-tions. If this proposal were implemented on a global level, the regional imbalances generated in the course of five hundred years of capitalism would begin to be reversed. No longer would a crazed rush of investment to some regions be mirrored by an absence of economic resources in the regions most in need.

Finally, global underconsumptionists reflect seriously on the systematic problems of the global capitalist economy. And they are to be applauded for understanding how the perversity of capitalism simultaneously generates overcapacity and poverty. But they misdiagnose both the problem and the solution. In their view, the source of the problem is restricted consumption power, and the solution is an expansion of effective demand. But the true source of global instability is the alien power of the law of value over social life, a power imposed blindly as the unanticipated collective result of individual responses to "the rules of the game." Under the value form, the accumulation process insanely pushes forward until far more capital has been accumulated than can find a profitable outlet. This structural tendency necessarily holds in capitalism *whatever* the consumption powers of the populace might be. Upon reaching the point of overaccumulation the law of value, like Moloch, demands its sacrifice. A more or less massive devaluation of capital must commence so that the process of accumulation may begin again, devaluation that inevitably places the greatest burdens on precisely those social groups that benefited least from previous capital accumulation. Shifting some purchasing power from one class of consumers to another may sometimes be humane, but it does not address this underlying power of the law of value. And adding extra consumer power though deficit spending merely extends the time scale in which this dynamic is played out.

For Marxian internationalists, if the root problem of contemporary globalization is the alien power of the law of value, the only solution commensurable with the problem is to free humanity from subjugation to this law. This does not imply the abolition of markets; it is not the presence of markets that establishes the alien power of the value form, but the institutionalization of the drive to accumulate surplus value to the greatest extent possible, whatever the cost to social life.[14] The law of value is abolished in the model of democratic socialism described above by three structural transformations. First, the class relationship upon which the law of value is based, the capital/wage labor relation, is abolished. There are no capital markets; no one has a right to sell shares in ownership rights to private investors. And there are no labor markets; persons who labor are not commodities with a price tag.[15] Second, the drive to accumulate as an end in itself is abolished. Decisions regarding the rate of new investment (the level at which the flat tax on the assets of cooperatives is set) are now a matter for democratic debate and decision, and worker cooperatives break free from the "grow or die" imperative of capitalist enterprises. Third, the overall direction of the economy is subject to social oversight, rather than handed over to profit imperatives. Globalization after such transformations would be far different from the globalization that threatens us all today.

Unfortunately, democratic socialism is not on the immediate agenda. In the meantime, the push to institute lean production will only get stronger. How ought we to respond to this? Criticisms of the ideology of the "new economy" are no more than a small part of the story. The biggest part concerns the creative tactics and strategies devised in the course of concrete struggles in the realms of production, consumption, and community life (see McCarney 1990). In his recent work *Workers in a Lean World* (1997) Kim Moody provides a comprehensive synthesis of the lessons for the future to be learned from past struggles against lean production. Interested readers are urged to consult this important book. In conclusion I would like to state briefly what I take to be the four most important practical lessons that follow from the present work.

The first point is the most obvious one. If there is no reason to think that the interests of capital and labor are automatically reconciled in lean production, working men and women must be prepared to engage in struggles in defense of their interests. This in turn implies the need to form (or preserve) organizations that can effectively carry out such struggles. In other words, there is a clear need for independent labor organizations controlled by the workforce itself. Given the continued existence of capital/wage labor antagonism, the imperative to avoid company unions is as strong as it has ever been.

Second, the self-organization of labor must be on the same scale as the organization of capital. This means that the basic unit of organization cannot be a single firm or even a national industry. As capital is organized into unified networks of firms, so labor organizations must unite workers in assembly firms with those employed by subcontractors and distributors. As capital is organized across borders, labor organizations must become truly international as well. As one and the same process of capital accumulation creates both employment and unemployment, both sectors of the working class must be united in the same organizations. All of this demands a complete and unequivocal break from the agenda of lean production firms, which "are opposed to forms of alternative worker identification . . . which create a separate sphere of identity for workers and disrupt the alignment between worker and company" (Kenney and Florida 1993, 285).

Third, the struggle against the shortcomings of the lean production system also demands the setting up of alternative networks to those that unite lean production firms, networks that unite those engaged in struggles at the point of production with those engaged in struggles in other social arenas. The work of trade union committees must be closely integrated with community health and safety projects, with coalitions of oppressed groups, with consumer activist organizations engaged in the monitoring and critique of corporate advertising campaigns, with groups concerned with questions of local, regional, and global ecology, and so on. All of these struggles concern the working class, and none can be successfully resolved as long as the reign of capital persists.

Finally, networks of information exchange are only an intermediate step. The struggle against lean production ultimately requires a revolutionary movement, committed to internationalist principles, and dedicated to the materialization of the utopian impulses lean production so cynically abuses.

NOTES

1. This does not mean that workers in a particular enterprise would experience no external pressure once worker self-management has been instituted. Institutional mechanisms must be set up that encourage enterprises to respond to social needs in an efficient manner. In the alternative to lean production presented here competition in producer and consumer markets (there are no labor or capital markets) and the oversight of community banks and representative political bodies ensure that this tends to occur. I shall also refer later to the portion of the economic surplus that is allocated outside of particular enterprises. This too must be subject to democratic control for exploitation to be absent.

2. Elson (1988) proposes that social audits be performed at each step in the production and distribution process. Another possibility would be to have representatives of consumer interest groups sit on the boards of directors of enterprises (Devine 1988). Schweickart (1994) calls for publicly owned community banks that allocate new investment funds according to criteria formulated in a democratic process. See below.

3. In 1969, workers in the United States annually labored 1,786 hours on average. By 1987 the figure was 1,949. This means that the average worker put in roughly one month (163 hours) of extra work each year (Schor 1991).

4. This implies that there are no intellectual property rights of any sort. The effects this might have on innovation are discussed in the following section.

5. I do not think there is any way to fix once and for all the goods and services that should be provided directly as public goods and those that should be distributed through (socialized) markets. The appropriate mixture depends upon the given cultural and historical context, although we can expect the public goods sector to exceed that found in any variant of capitalism. I would also like to note that Diane Elson (1988) has discussed a number of institutional arrangements that further the socialization of the market besides those mentioned here. These include organizations of users of public services, wage commissions, price commissions, public regulators, and network coordinators.

6. To give only one example, David Noble (1984) has shown how capital/labor relations in the civilian sector have been profoundly affected by the Pentagon's role in the development of computerized manufacturing.

7. Another part of this planning concerns the setting of priorities for new investments.

8. Under capitalism, technological innovation has been most successful where a "technological milieu" combines people with expertise in different facets of the innovation process, such as those engaged in basic science, applied engineers, production workers, subcontractors, and so on (Storper and Walker, chapter 4). The same point would hold under a different set of social relations. In fact, we could expect technological milieus to flourish under socialism. Production workers would be given far more opportunity to familiarize themselves with the theoretical principles of science and engineering if a significant reduction in the work day were combined with democracy at the workplace.

9. In reality, this is not a linear process. Each stage in the pipeline proceeds simultaneously, and provides feedback to every other stage.

10. It should also be recalled that in capitalism the people actually generating new scientific-technical information are generally not the ones who benefit the most from intellectual property rights. One of the first things a new researcher hired by a corporation typically has to do is sign away all patent rights to the corporation.

11. Marx insisted unequivocally on this point in his essay "Critique of the Gotha Programme" (Marx 1977).

12. Schweickart himself does not extend the idea of pooling funds for new investment to the international level; this proposal is an extrapolation from his discussion of economic democracy on the national level.

13. The assumption here is that markets left to themselves systematically tend to underinvest in both public goods and industries with high positive externalities, due to familiar free-rider problems.

14. It is important to remember that there have been markets for tens of thousands of years, while the dominance of the value form can be measured in centuries.

15. It may be worth mentioning that few advocates of "free markets" fail to put restrictions on what are appropriate matters for market transactions. No one today laments the fact that as a society we have drawn a line and said that there will no markets in which property rights to the ownership of human beings are exchanged. Nor do even the fiercest libertarians advocate setting up a market for political offices granting private ownership of those offices to those willing to pay the highest price. The democratic version of socialism advocated here simply shifts the line defining where legitimate market transactions end, based on the assumption that private ownership of another's labor power or of large-scale productive resources is strictly analogous to private ownership of a person or a political office. This argument, needless to say, does not rule out private ownership of items for personal consumption.

22

The Insurgent Architect at Work

David Harvey

Imagine ourselves as architects, all armed with a wide range of capacities and powers, embedded in a physical and social world full of manifest constraints and limitations. Imagine also that we are striving to change that world. As crafty architects bent on insurgency we have to think strategically and tactically about what to change and where, about how to change what and with what tools. But we also have somehow to continue to live in this world. This is the fundamental dilemma that faces everyone interested in progressive change.

But what kind of world are we embedded in? We know that it is a world full of contradictions, of multiple positionalities, of necessary flights of the imagination translated into diverse fields of action, of uneven geographical developments, and of highly contested meanings and aspirations. The sheer enormity of that world and its incredible complexity provide abundant opportunities for the exercise of critical judgement and of limited freedom of the individual and collective will. But the enormity of apparent choice and the divergent terrains upon which struggles can be conducted is perpetually in danger of generating a disempowering confusion (of the sort that globalization, for example, has strongly promoted). Furthermore, it appears impossible to avoid unintended consequences of our actions, however well thought out. How are we to cut through these confusions and build a different sense of possibilities while acknowledging the power of the constraints with which we are surrounded?

Here are some conversation points in lieu of answers. Earlier I argued for a system of *translations* across and between qualitatively different but related areas of social and ecological life. The spatiotemporal scale at which processes operate here makes a difference. For this reason, Wilson considers scale as one of the most important differentiations within the unity of science. *The Communist Manifesto* notes the same problem as revolutionary sentiment passes from the political individual through the factory, political parties, and the nation state to a movement in which workers of the world can unite. Dialectics permits diverse knowledges and practices to be rendered coherent across scales without resort to some narrow causal reductionism. This dialectical way of thinking echoes aspects of the theory of uneven geographical developments. The production of spatiotemporal scale is just as important as the production of differentiations within a scale in defining how our world is working and how it might work better.

I now take up these ideas in greater depth. I propose first that we consider political possibilities at a variety of spatiotemporal scales. I then argue that real political change arises out of simultaneous and loosely coordinated shifts in both thinking and action across several scales (either simultaneously or sequentially). If, therefore, I separate out one particular spatiotemporal scale for consideration, in order to understand its role in the overall dynamics of political change, then I must do so in a way that acknowledges its relation to processes only identifiable at other scales. The metaphor to which I appeal is one of several different 'theaters' of thought and action on some 'long frontier' of 'insurgent' political practices. Advances in one theater get ultimately stymied or even rolled back unless supported by advances elsewhere. No one theater is particularly privileged even though some of us may be more able, expert, and suited to act in one rather than another. A typical political mistake is the thoroughly understandable habit of thinking that the only theater that matters is the one that I or you happen to be in. Insurgent political practices must occur in all theaters on this long frontier. A generalized insurgency that changes the shape and direction of social life requires collaborative and coordinating actions in all of them. With that caveat in mind, I consider seven theaters of insurgent activity in which human beings can think and act, though in radically different ways, as architects of their individual and collective fates.

1. THE PERSONAL IS POLITICAL

The insurgent architect, like everyone else, is an embodied person. That person, again like everyone else, occupies an exclusive space for a certain time (the spatiotemporality of a human life is fundamental). The person is endowed with certain powers and skills that can be used to change the world. He or she is also a bundle of emotions, desires, concerns, and fears all of which play out through social activities and actions. The insurgent architect cannot deny the consequences of that embodiment in material, mental, and social life.

Through changing our world we change ourselves. How, then can any of us talk about social change without at the same time being prepared, both mentally and physically, to change ourselves? Conversely, how can we change ourselves without changing our world? That relation is not easy to negotiate. Foucault (1984) rightly worried that the 'fascism that reigns in our heads' is far more insidious than anything that gets constructed outside.

Yet we also have to decide—to build the road, the factory, the houses, the leisure park, the wall, the open space . . . And when a decision is made, it forecloses on other possibilities, at least for a time. Decisions carry their own determinations, their own closures, their own authoritarian freight. Praxis is about confronting the dialectic in its 'either/or' rather than its transcendent 'both/and' form. It always has its existential moments. Many of the great architects of the past made their personal political in incredibly decisive as well as authoritarian ways (with results both good and bad according to the partial judgement of subsequent generations).

It is in this sense, therefore, that the personal (including that of the architect) is deeply political. But that does not mean, as feminists, ecologists, and the innumerable array of identity politicians who have strutted their stuff these last few years have discovered to their cost, that virtually *anything* personal makes for good politics. Nor does it mean, as is often suggested in some radical alternative movements (such as deep ecology), that fundamental transformations in personal attitudes and behaviors are sufficient (rather than necessary) for social change to occur.

While social change begins and ends with the personal, therefore, there is much more at stake here than individualized personal growth (a topic that now warrants a separate and large section in many bookstores in the United States) or manifestations of personal commitment. Even when it seems as if some charismatic and all-powerful person—a Haussmann, a Robert Moses, or an Oscar Niemeyer—builds a world with the aim of shaping others to conform to their particular and personal visions and desires, there turns out to be much more to it than just the vision of the person. Class interests, political powers, the mobilization of forces of violence, the orchestration of discourses and public opinion, and the like, are all involved.

But in reflecting on what we insurgent architects do, a space must be left for the private and the personal—a space in which doubt, anger, anxiety, and despair as well as certitude, altruism, hope, and elation may flourish. The insurgent architect cannot, in the end, suppress or repress the personal any more than anyone else can. No one can hope to change the world without changing themselves. The negotiation that always lies at the basis of all architectural and political practices is, therefore, between persons seeking to change each other and the world, as well as themselves.

2. THE POLITICAL PERSON IS A SOCIAL CONSTRUCT

To insist on the personal as political is to confront the question of the person and the body as the irreducible moment (defined at a particular spatiotemporal scale) for the grounding of all politics and social action. But the individual, the body, the self, the person (or whatever term we wish to use) is a fluid social construct rather than some absolute and immutable entity fixed in concrete. How 'social construction' and 'embodiment' are understood then becomes important. For example, a relational conception of self puts the emphasis upon our porosity in relation to the world of socio-ecological change and thereby tempers many theories of individual rights, legal status, and the like. The person that is political is then understood as an entity open to the innumerable processes (occurring at different spatiotemporal scales) that transect our physical and social worlds. The person must then be viewed as an ensemble of socio-ecological relations.

But an already-achieved spatiotemporal order can hold us to some degree apart from this fluid and open conception in our thought and practices. In the United States, private property and inheritance, market exchange, commodification and monetization, the organization of economic security and social power, all place a premium upon personalized private property vested in the self (understood as a bounded entity, a non-porous individual), as well as in house, land, money, means of production, etc., all construed as the elemental socio-spatial forms of political-economic life. The organization of production and consumption forges divisions of labor and of function upon us and constructs professionalized personas (the architect, the professor, and the poet as well as the proletarian, all of whom, as Marx and Engels point out in *The Communist Manifesto*, 'have lost their halo' and become in some way or another paid agents of bourgeois power). We live in a social world that converts all of us into fragments of people with particular attachments, skills, and abilities integrated into those powerful and dynamic structures that we call a 'mode of production.' Our 'positionality' or 'situatedness' in relation to that is a social construct in exactly the same way that the mode of production is a social creation. This 'positionality' defines who or what we are (at least for now). And 'where we see it from' within that process provides much of the grist for our consciousness and our imaginary.

But 'what and how far we can see' from 'where we see it from' also varies according to the spatiotemporal constructions and our choices in the world we inhabit. Access to information via the media, for example, and the qualities and controls on information flow play an important role in how we can hope to understand and change the world. These horizons, both spatial and temporal, have simultaneously expanded and compressed over the past thirty years and part of any political project must be to intervene in the resultant information flows in ways that are progressive and constructive. But there is also the need to persuade people to look beyond the borders of that myopic world of daily life that we all necessarily inhabit.

In contrast, the fierce spatiotemporalities of daily life—driven by technologies that emphasize speed and rapid reductions in the friction of distance and of turnover times—preclude time to imagine or construct alternatives other than those forced unthinkingly upon us as we rush to perform our respective professional roles in the name of technological progress and endless capital accumulation. The material organization of production, exchange, and consumption rests on and reinforces specific notions of rights and obligations and affects our feelings of alienation and of subordination, our conceptions of power and powerlessness. Even seemingly new avenues for self-expression (multiculturalism being a prime recent example) are captive to the forces of capital accumulation (e.g. love of nature is made to equal eco-tourism). The net effect is to limit our vision of the possible. No less a person than Adam Smith (cited in Marx, 1976 edition, 483) considered that 'the understandings of the greater part of men are necessarily formed by their ordinary employments' and that 'the uniformity of (the labourer's) life naturally corrupts the courage of his mind.' If this is only partially true—as I am sure it is—it highlights how the struggle to think alternatives—to think and act differently—inevitably runs up against the circumstances of and the consciousness that derives from a localized daily life. Most insidious of all, is the way in which routine, by virtue of its comfort and security, can mask the ways in which the jarring prospects of transformative change must in the long run be confronted. Where, then, is the courage of our minds to come from?

Let us go back to the figure of the insurgent architect. She or he acts out a socially constructed (sometimes even performative) role, while confronting the circumstances and consciousness that derive from a daily life where demands are made upon time, where social expectations exist, where skills are acquired and supposed to be put to use in limited ways for purposes usually defined by others. The architect then appears as a cog in the wheel of capitalist urbanization, as much constructed by as constructor of that process (was this not as true of Haussmann, Cerda, Ebenezer Howard, Le Corbusier, Oscar Niemeyer, as of everyone else?).

Yet the architect can (indeed must) desire, think and dream of difference. And in addition to the speculative imagination which he or she necessarily deploys, she or he has available some special resources for critique, resources from which to generate alternative visions as to what might be possible. One such resource lies in the tradition of utopian thinking. 'Where we learn it from' may then become just as, if not more, important as 'what we can see from where we see it from.'

Utopian schemas of spatial form typically open up the construction of the political person to critique. They do so by imagining entirely different systems of property rights, living and working arrangements, all manifest as entirely different spatial forms and temporal rhythms. This proposed re-organization (including its social relations, forms of reproductive work, its technologies, its forms of social provision) makes possible a radically different consciousness (of social relations, gender relations, of the relation to nature, as the case may be) together with the expression of different rights, duties, and obligations founded upon collective ways of living.

Postulating such alternatives allows us to conduct a 'thought experiment' in which we imagine how it is to be (and think) in a different situation. It says that by changing our situatedness (materially or mentally) we can change our vision of the world. But it also tells us how hard the practical work will be to get from where we are to some other situation like that. The chicken-and-egg problem of how to change ourselves through changing our world must be set slowly but persistently in motion. But it is now understood as a project to alter the forces that construct the political person, my political person. I, as a political person, can change my politics by changing my positionality and shifting my spatiotemporal horizon. I can also change my politics in response to changes in the world out there. None of this can occur through some radical revolutionary break (though traumatic events and social breakdowns have often opened a path to radically different conceptions). The perspective of a long revolution is necessary.

But to construct that revolution some sort of collectivization of the impulse and desire for change is necessary. No one can go it very far alone. But positioned as an insurgent architect, armed with a variety of resources and desires, some derived directly from the utopian tradition, I can aspire to be a subversive agent, a fifth columnist inside of the system, with one foot firmly planted in some alternative camp.

3. THE POLITICS OF COLLECTIVITIES

Collective politics are everywhere but they usually flow in constrained and predictable channels. If there is any broad swathe of insurgent politics at work in the interstices of urbanization in the advanced capitalist countries, for example, it is a mobilization in defense of private property rights. The violence and anger that greets any threat to those rights and values—be it from the state or even from agents of capital accumulation like developers—is an awesome political force. But it typically turns inwards to protect already existing personalized 'privatopias.' The same force can be found in the militia or neo-fascist movements on the right (a fascinating form of insurgent politics) as well as within the radical communitarianism of some ecologists.

Such formations of collective governance preclude the search for any far-reaching alternatives. Most politics and collective forms of action preserve and sustain the existing system, even as they deepen some of its internal contradictions, ecologically, politically, and economically (e.g. the collective rush to suburbanize increases car dependency, generates greenhouse gasses, particulate matter pollution, and tropospheric ozone concentrations etc.). The gated communities of Baltimore are a symbol of collective politics, willingly arrived at, gone awry.

Traditional utopianism seeks to confront this prevailing condition. Communitarianism as a utopian movement typically gives precedence, for example, to citizenship, to collective identifications and responsibilities, over the private pursuit of individual advantage and the 'rights talk' that attaches thereto. This ideal founds many a utopian dream, from Thomas More to Fourier, and infuses many contemporary religious movements like those for a Christian Base Community or even the much softer (and some would say much weaker) cultivation of concepts of 'citizenship' as the basis for the good life (see, e.g., Douglass and Friedmann, 1998).

Distinctive communities are painstakingly built by social practices including the exercise of authoritarian powers and conformist restrictions. They are not just imagined (however important the imaginary of them may be). It is useful, therefore, to view an achieved 'community' as an enclosed space (irrespective of scale or even frontier definitions) within which a certain well-defined system of rules prevails. To enter into that space is to enter into a space of rules which one acknowledges, respects, and obeys (either voluntarily or through some sort of compulsion). The construction of 'community' entails the production of such a space. Challenging the rules of community means challenging the very existence of such a collectivity by challenging its rules. It then follows that communities are rarely stable for long. Abundant opportunities exist here for the insurgent architect to promote new rules and/or to shape new spaces. Our capacities as rule makers and rule breakers here enter fully into play. Part of the attraction of the spatial form utopian tradition is precisely the way in which it creates an imaginary space in which completely different rules can be contemplated. And it is interesting to note how the figure of the city periodically re-emerges in political theory as the spatial scale at which ideas and ideals about democracy and belonging can best be articulated.

It is not always easy here to define the difference between insurgent politics of a progressive sort and the exclusionary and authoritarian practices of, say, homeowner associations who defend their property rights. Etzioni (1997), a leading proponent of the new communitarianism, actively supports, for example, the principle of closed and gated communities as a progressive contribution to the organization of social life. Collective institutions can also end up merely improving the competitive strength of territories in the high stakes game of the uneven geographical development of capitalism (see, e.g., Putnam's 1993 account of the institutional bases of uneven geographical development in Italy). For the privileged, community often means securing and enhancing privileges already gained. For the underprivileged it all too often means 'controlling their own slum.'

Dialectical utopianism must confront the production of 'community' and 'coming together for purposes of collective action' in some fashion and articulate the place and meaning of this phenomena within a broader frame of politics. This means a translation to a different scale from that of the embodied political person. Community should be viewed as a delicate relation between fluid processes and relatively permanent rules of belonging and association (like those formally imposed by the nation state). The tangible struggle to define its limits and range (sometimes even territories and borders), to create and sustain its rules and institutions through collective powers such as constitutional forms, political parties, the churches, the unions, neighborhood organizations, local governments, and the like, has proven central to the pursuit of alternatives to the selfishness of personalized market individualism. But, as many have recently pointed out, the re-making and re-imagining of 'community' will work in progressive directions only if it is connected *en route* to a more generalized radical insurgent politics. That means a radical project (however defined) must exist. The rule-making that ever constitutes community must be set against the rule-breaking that makes for revolutionary transformations.

The embeddedness and organized power community offers as a basis for political action is crucial even if its coherence requires democratically structured systems of authority, consensus, and 'rules of belonging'. Thus, although community 'in itself' has meaning as part of a broader politics, community 'for itself' almost invariably degenerates into regressive exclusions and fragmentations (what some would call negative heterotopias of spatial form). Means must therefore be found whereby we insurgent architects can reach out across space and time to shape a more integrated process of historical-geographical change beyond the limits typically defined by some community of common interest. The construction of collective identities, of communities of action, of rules of belonging, is a crucial moment in the translation of the personal and the political onto a broader terrain of human action. At the same time, the formation of such collectivities creates an environment and a space (sometimes, like the nation state, relatively stable and enduring) that shapes the political person as well as the ways in which the personal is and can be political.

4. MILITANT PARTICULARISM AND POLITICAL ACTION

The theory of 'militant particularism' argues that all broad-based political movements have their origins in particular struggles in particular places and times (see Harvey, 1996, Chapter 1). Many struggles are defensive—for example, struggles against plant closures or excessive exploitation of labor, the siting of noxious facilities (toxic waste dumps), the dismantling or lack of social or police protections, violence against women, the environmental transformations proposed by developers, the appropriation of indigenous resources by outsiders, attacks upon indigenous cultural forms, and the like. A widespread politics of resistance now exists, for example, to neoliberalism and capitalism throughout the world. But some forms of militant particularism are pro-active. Under capitalism this typically means struggles for specific group rights that are universally declared but only partially conferred (in the past this has usually meant the rights of entrepreneurs and owners of means of production to freely exercise their rights of ownership without restraint, but it has also extended to include the rights of slaves, labor, women, gays, the culturally different, animals and endangered species, the environment, and the like).

The critical problem for this vast array of struggles is to shift gears, transcend particularities, and arrive at some conception of a universal alternative to that social system which is the source of their difficulties. Capitalism (coupled with modernism and, perhaps, a Eurocentric 'Westernism') successfully did this *vis-à-vis* pre-existing modes of production, but the oppositional movements of socialism, communism, environmentalism, feminism, and even humanism and multiculturalism have all constructed some sort of universalistic politics out of militant particularist origins. It is important to understand how this universalization occurs, the problems that arise, and the role traditional utopianism plays.

Dialectics here is useful. It teaches that universality always exists *in relation to* particularity: neither can be separated from the other even though they are distinctive moments within our conceptual operations and practical engagements. The notion of justice, for example, acquires universality through a process of abstraction from particular instances and circumstances, but becomes particular again as it is actualized in the real world through social practices. But the orchestration of this process depends upon mediating institutions (those, for example, of language, law, and custom within given territories or among specific social groups). These mediating institutions 'translate' between particularities and universals and (like the Supreme Court) become guardians of universal principles and arbiters of their application. They also become power centers in their own right. This is, broadly, the structure set up under capitalism with the state and all of its institutions (now supplemented by a variety of international institutions such as the World Bank and the IMF, the United Nations, GATT and the World Trade Organization) being fundamental as 'executive committees' of capitalism's systemic interests. Capitalism is replete with mechanisms for converting from the particular (even personal) to the universal and back again in a dynamic and interactive mode. Historically, of course, the primary mediator has been the nation state and all of its institutions including those that manage the circulation of money.

No social order can, therefore, evade the question of universals. The contemporary 'radical' critique of universalism is sadly misplaced. It should focus instead on the specific institutions of power that translate between particularity and universality rather than attack universalism *per se*. Clearly, such institutions favor certain particularities (such as the rights of ownership of means of production) over others (such as the rights of the direct producers) and promote a specific kind of universal.

But there is another difficulty. The movement from particularity to universality entails a 'translation' from the concrete to the abstract. Since a violence attaches to abstraction, a tension always exists between particularity and universality in politics. This can be viewed either as a creative tension or, more often, as a destructive and immobilizing force in which inflexible mediating institutions (such as an authoritarian government apparatus) claim rights over individuals and communities in the name of some universal principle.

It is here that critical engagement with the static utopianism of spatial form (particularly its penchant for nostalgia) and the loosening of its hold by appeal to a utopianism of spatial-temporal transformation can keep open prospects for further change. The creative tension within the dialectic of particularity-universality cannot be repressed for long. Mediating institutions, no matter how necessary, cannot afford to ossify, and traditional utopianism is often powerfully suggestive as to institutional reforms. The dynamic utopian vision that emerges is one of sufficient stability of institutional and spatial forms to provide security and continuity, coupled with a dynamic negotiation between particularities and universals so as to force mediating institutions and spatial structures to be as open as possible. At times, capitalism has worked in such a way (consider how, for example, the law gets reinterpreted to confront new socio-economic conditions and how the production of space has occurred throughout the long history of capitalism). Any radical alternative, if it is to succeed as it materializes, must follow capitalism's example in this regard. It must find ways to negotiate between the security conferred by fixed institutions and spatial forms on the one hand and the need to be open and flexible in relation to new socio-spatial possibilities on the other. Both Jefferson and Mao understood the need for some sort of 'permanent revolution' to lie at the heart of any progressive social order. The failure to acknowledge that imperative lies at the heart of the collapse of the Soviet Union and seriously threatens the United States. The perspective of a permanent revolution (in, for example, the production of spatial forms) must therefore be added to that of a long revolution as we reach for the principles of a spatiotemporal and dialectical utopianism.

5. MEDIATING INSTITUTIONS AND BUILT ENVIRONMENTS

The formation of institutions and built environments that can mediate the dialectic between particularity and universality is of crucial importance. Such institutions typically become centers for the formation of dominant discourses as well as centers for the exercise of power. Many of them—medical care, education, financial affairs, and the state—cultivate a special expertise in the same way that built environments of different sorts facilitate possibilities for social action in some directions while limiting others. Many institutions (e.g. local governments and the state) are organized territorially and define and regulate activity at a particular spatial scale. They can translate militant particularism into an institutionalized spatial order designed to facilitate or repress certain kinds of social action and thereby influence the ways in which the personal can be political, encouraging some (like entrepreneurial endeavor, say) and discouraging others (like socialist communes).

Much the same can be said of the built environments that get constructed. Consider, for example, the form and style of urbanization and the consequences that flow therefrom. How can the personal be openly political when environmental conditions inhibit the free exploration of radically different lifestyles (such as living without an automobile or private property in Los Angeles)? The uneven conditions of geographical development that now prevail in Baltimore do not allow the personal to be political in anything other than rather restrictive ways (equally repressive, though in very different ways, for the affluent child of suburbia as for the child of inner-city poverty).

The creation of mediating institutions is deeply fraught and frequently contested (as one might properly expect). The chief difficulty is to bring multiple militant particularisms (in the contemporary US this might mean the aspirations of radical ecologists, the chamber of commerce, ethnic or religious groups, feminists, developers, class organizations, bankers, and the like) into some kind of institutional relation to each other without resort to arbitrary authority and power. The Porto Alegre experience (see Abers, 1998) suggests that this sort of thing can be done. But decisions have to be made and arbitrary authority and power are invariably implicated in the process. With the best will in the world these cannot be eliminated. The effect is to render the mediating institutions sites of power and thereby sources of distinctive discourses and constructions which can be organized in a system of dominance which individual persons find hard to resist let alone transcend. The capture or destruction of mediating institutions (such as the state, the financial sector, education) and the re-shaping of built environments has often, therefore, been the be-all and end-all of insurgent radicalism. While this is one crucial theater in the long frontier of insurgent politics, it is far from being the whole of the story.

6. TRANSLATIONS AND ASPIRATIONS

The insurgent architect with a lust for transformative action must be able to translate political aspirations across the incredible variety and heterogeneity of socio-ecological and political-economic conditions. He or she must also be able to relate different discursive constructions and representations of the world (such as the extraordinary variety of ways in which environmental issues are discussed). He or she must confront the conditions of and prospects for uneven geographical developments. The skills of translation here become crucial. For James Boyd White (1990, 257–64):

[Translation means] confronting unbridgeable discontinuities between texts, between languages, and between people. As such it has an ethical as well as an intellectual dimension. It recognises the other—the composer of the original text—as a center of meaning apart from oneself. It requires one to discover both the value of the other's language and the limits of one's own. Good translation thus proceeds not by the motives of dominance and acquisition, but by respect. It is a word for a set of practices by which we learn to live with difference, with the fluidity of culture and with the instability of the self. (257)

We should not feel that respect for the other obliges us to erase ourselves, or our culture, as if all value lay out there and none here. As the traditions of the other are entitled to respect, despite their oddness to us, and sometimes despite their inhumanities, so too our own tradition is entitled to respect as well. Our task is to be distinctively ourselves in a world of others: to create a frame that includes both self and other, neither dominant, in an image of fundamental equality. This is true of us as individuals in our relations with others, and true of us as a culture too, as we face the diversity of our world . . . This is not the kind of relativism that asserts that nothing can be known, but is itself a way of knowing: a way of seeing one thing in terms of another. Similarly it does not assert that no judgments can be reached, but is itself a way of judging, and of doing so out of a sense of our position in a shifting world.   (264)

This, in itself, has its own utopian ring. It is not hard to problematize such an argument, as Said did so brilliantly in *Orientalism,* as the power of the translator (usually white male and bourgeois) to represent 'the other' in a manner that dominated subjects (orientals, blacks, women, etc.) are forced to internalize and accept. But that historical understanding itself provides a hedge against the kinds of representational repressions that Said and many feminists have recorded. This links us back to how the personal is always political. As White notes: 'to attempt to "translate" is to experience a failure at once radical and felicitous: radical, for it throws into question our sense of ourselves, our languages, of others; felicitous, for it releases us momentarily from the prison of our own ways of thinking and being' (1990, 257). The act of translation offers a moment of liberatory as well as repressive possibility. The architects of spatiotemporal utopianism must be open to such possibilities.

But as real architects of our future we cannot engage in endless problematization and never-ending conversations. Firm recommendations must be advanced and decisions taken, in the clear knowledge of all the limitations and potentiality for unintended consequences (both good and bad). We need to move step by step towards more common understandings. And this for two compelling reasons. First, as Zeldin (1994, 16) among others remarks, we know a great deal about what divides people but nowhere near enough about what we have in common. The insurgent architect has a role to play in defining commonalities as well as in registering differences. But the second compelling reason is this: without translation, collective forms of action become impossible. All potential for an alternative politics disappears. The fluid ability of capitalists and their agents to translate among themselves using the basic languages of money, commodity, and property (backed, where necessary, with the theoretical language of a reductionist economics) is one of their towering class strengths. Any insurgent oppositional movement must do this just as well if not better. Struggle as we may, it is impossible to conduct politics without an adequate practice of translation. If reductionism of the Wilsonian sort is rejected, then the only option is translation. Thomas Kuhn, in his *Structure of Scientific Revolutions,* considers translation (rather than reduction) as the privileged and perhaps sole means by which fundamentally different paradigms of scientific knowledge might be related, and Judith Butler (1998, 38), under pressure from her critics as to the fragmenting effect of identity politics, argues:

> Whatever universal becomes possible—and it may be that universals only become possible for a time, 'flashing up' in Benjamin's sense—will be the result of the difficult labour of translation in which social movements offer up their points of convergence against a background of ongoing social contestation.

The omnipresent danger in any dialectical utopianism is that some all-powerful center or some elite comes to dominate. The center cajoles, bullies, and persuades its periphery into certain modes of thought and action (much as the United States has done since World War II, culminating in the infamous Washington consensus through which the United States sought to institutionalize its hegemonic position in the world order by gaining adhesion of everyone to certain universal principles of political-economic life). As opposed to this, the democratic and egalitarian rules of translation should be clear. But so should the universal principles that, however much they merely 'flash up' as epiphenomena, emerge from the rich experience of translation to define what it is we might have in common.

7. THE MOMENT OF UNIVERSALITY: ON PERSONAL COMMITMENTS AND POLITICAL PROJECTS

The moment of universality is not a final moment of revelation or of absolute truth. I construe it, in the first instance, as a moment of existential decision, a moment of 'either/or' praxis, when certain principles are materialized through action in the world. It is, as it were, a nature-imposed condition of our species existence that we have to make decisions (individually and collectively) and we have to act upon them. The moment of universality is the moment of choosing, no matter how much we may reserve judgement on our actions afterwards. How we come to represent those decisions to ourselves in terms of principles or codes of action that act as guidelines for future decisions is an important cultural value that gains power over us as it becomes instantiated in discourses and institutions. It is here that abstract universal principles operate as plays of power.

Universals cannot and do not exist, however, outside of the political persons who hold to them and act upon them. They are not free standing nor do they function as abstracted absolutes that can be brought to bear upon human affairs for all times and places. They are omnipresent in all practices. But to the degree that we begin to shape and order them for given purposes they take on the guise of abstract principles (even written codes and laws) to which we adhere. And if we find in them successful guides to action (as we do, for example, within the corpus of scientific understandings) so they shape our world view and become institutionalized as mediating discourses. They tend to cluster and converge as dominant paradigms, as hegemonic discourses, or as pervasive ethical, moral, or political-economic principles that inform our beliefs and actions. They become codified into languages, laws, institutions, and constitutions. Universals are socially constructed not given.

While social construction can betoken contestation, it is more often the case that the dominant principles handed down to us so limit our conceptions as to inhibit alternative visions of how the world might be. A wide range of possible universal and unifying principles has in fact been bequeathed to us (the fruit of long and often bitter experience). But, as many commentators point out (usually with critical intent), many of these principles have their origins in the Western Enlightenment when theorists of the natural and social order had, unlike now, no hesitation in expressing their opinions as universal truths and propositions. It is fashionable in these times to denigrate these (at least in the humanities) while at the same time leaving crude versions of them fully in play in society in general. But we can never do without universals of some sort. We can, of course, *pretend* to do without them. Much of what now passes for radical argument in the humanities and some segments of the social sciences resorts to much dissimulation and opacity (when it is not engaged in downright chicanery) with respect to this point.

It is therefore important at the outset to exhume the traces of universal principles expressed in the ways the personal is and can be political. This is so because without certain criteria of judgement (explicit or implied), it is impossible to distinguish between right and wrong or between progressive and regressive lines of political action. The existential moment of do I or do I not support this or that line of action entails such a judgement. Even though I may prefer not to make it, not to decide is in itself a form of decision (one that many Americans now prefer at the ballot box with specific consequences). So though the moment of universality is not the moment of revelation, it is the moment of *judgement and decision* and these willy-nilly entail *expression* of some universal whether we like it or not. It is only in these terms that we are able to say that *this* form of insurgent politics (embedded, say, in a movement for environmental justice) is progressive and worthy of support while *that* form of insurgent politics (like the militia movement in the Michigan woods) is not. The moment of the universal is, therefore, the moment of political judgement, commitment, and material praxis.

For this reason it is, paradoxically, the moment that gets argued over in the most abstract of terms. In effect, we seek to create a generalized discourse about rights and wrongs, about moral imperatives and proper and improper means and ends, through which we try to persuade ourselves as well as others to certain consistent lines of action, knowing full well that each of us is different and that no particularity is exactly the same as any other.

Such arguments can easily seem redundant, but when connected back across all the other theaters of action on the long frontier of insurgent politics, they can acquire a stronger force and even provide some sort of political and emotive thread that helps us recognize in what ways the personal, the collective, the mediating institutions can relate to each other in dynamic ways through the activities of the translator and the insurgent architectural imagination. Furthermore, it is also the case that universals draw their power and meaning from a conception of species being (it is only in terms of species rights that universal principles of conduct can make sense). It then also follows that acceptance of some sort of 'unity of science' is a necessary condition for the promulgation of universality claims. Conversely, discussions of universality crucially depend upon critical engagement with notions of species being and the unity of science.

So what universals might we currently embrace as meaningful ideals upon which to let our imaginations roam as we go to work as insurgent architects of our future? The United Nations Declaration of Human Rights is a document that expresses such universal principles in problematic but to some degree persuasive terms. The application of these principles has often been contested and their interpretation has had to be fought over in almost every particular case. Can we add or re-formulate those universals in interesting ways? My own preferred short-list of universal rights worthy of attention runs as follows:

1. The right to life chances
   This entails a basic right to sustenance and to elemental economic securities. Food security would be the most basic manifestation of such rights, but a general system of entitlements—as Sen (1982) would call them—is also fundamental. This re-affirms the UN Declaration (Article 23, Section 3) that 'everyone has the right to just and favourable remuneration ensuring himself and his family an existence worthy of human dignity, and supplemented, if necessary, by other means of social protection.' The universal right to a 'living wage' and to adequate social security is one way to both demand and problematize such a universal package of rights.
2. The right to political association and 'good' governance
   Individuals must have the right to associate in order to shape and control political institutions and cultural forms at a variety of scales (cf. Articles 20 and 21 of the UN Declaration). The presumption is that some adequate definition can be found for properly democratic procedures of association and that collective forms of action must offer reasonable protections to minority opinions. The presumption also exists

that some definition of 'good' governance can be found, from the local to the global level. Here, too, the demand highlights problems and differences (the definition of 'good governance' is far from homogeneous) at the same time as it takes up universalizing claims. But individuals plainly should have rights to produce their own spaces of community and inscribe their own rules therein, even as limitations on such rights become critical to restrict the narrow exclusions and the internal repressions to which communitarianism always tends.

3. The rights of the direct laborers in the process of production

The rights of those who labor to exercise some level of individual and collective control over labor processes (over what is produced as well as over how it shall be produced) is crucial to any conception of democracy and freedom. Long-standing concerns over the conditions of labor and the right of redress in the event of unreasonable burdens or sufferings (such as those that result in shortened life expectancy) need to be reinforced on a more global scale. This entails a demand for the radical empowerment of the laborer in relation to the production system in general (no matter whether it is capitalist, communist, socialist, anarchist, or whatever). It also highlights respect for the dignity of labor and of the laborer within the global system of production, exchange, and consumption (on this point, at least, a variety of Papal Encyclicals as well as the UN Declaration provide supportive materials).

4. The right to the inviolability and integrity of the human body

The UN Declaration (Articles 1 to 10) insists on the right to the dignity and integrity of the body and the political person. This presumes rights to be free from the tortures, incarcerations, killings, and other physical coercions that have so often been deployed in the past to accomplish narrow political objectives. The right of women to control their own reproductive functions and to live free of coercions and violence (domestic, cultural, and institutionalized) must also lie at the core of this conception. Violence against women and the subservience of women to patriarchal and paternalistic systems of domination has become a major issue for which universal rights claims have become deeply plausible and compelling (though often in conflict with claims for autonomy of cultural traditions).

5. Immunity/destabilization rights

Everyone has the right to freedom of thought, conscience, and religion, according to the UN Declaration (Articles 18 and 19). On this point the Declaration is definitive and clear. But I here think Unger's (1987b, 524–34) argument for a system of immunity rights that connects to a citizen's rights to destabilize that which exists is even stronger, for it insists on the right to critical commentary and dispute without fear of retaliation or other loss. It is only through the exercise of such rights that society can be both re-imagined and re-made (Unger's arguments on this point are persuasive).

6. The right to a decent and healthy living environment

From time to time legislation in particular countries has been predicated on the right of everyone to live in a decent and healthy living environment, one that is reasonably free from threats and dangers and from unnecessary hazards (particularly those produced through human activities, such as toxic wastes, dirty air, and polluted waters). The spreading cancers of environmental injustice throughout the world and the innumerable consequences for human health and well-being that flow from environmental degradations (both physical and social) indicate a terrain where the proper

establishment of universal rights is imperative, even if it is surely evident that the
meaning, interpretation, and application of such rights will be difficult to achieve.

7. The right to collective control of common property resources
The system of property rights by which capitalism has typically asserted its universalizing claims (actively supported in Article 17 of the UN Declaration) is widely
understood as both defective and in some instances destructive with respect to our
physical and social world. This is nowhere more apparent than in instances of common property resources (everything from genetic materials in tropical rain forests to
air, water, and other environmental qualities including, incidentally, the rights to
control built environments for historical, or cultural, or aesthetic reasons). The definition of such resources and the determination of who is the 'collective' in whose
name rights of control will be vested are all deeply controversial issues. But there
are widespread arguments now for alternative systems of property rights to those
implied in a narrowly self-serving and myopic structure of private property rights
that fail to acknowledge any other form of public or collective interest to that given
through a pervasive market (and corporation-dominated) individualism.

8. The rights of those yet to be born
Future generations have a claim upon us, preferably to live in a world of open possibilities rather than of foreclosed options. The whole rhetoric of sustainable environmental development rests on some sense (however vague and undefined) of
responsibilities and obligations that stretch beyond the ken of our own immediate
interests. *In extremis*, this right also recognizes our volitional role in the evolutionary
process and our responsibilities not only to our own species but also to the innumerable others whose prospects for survival depend upon our actions (see Item 11).

9. The right to the production of space
The ability of individuals and collectivities to 'vote with their feet' and perpetually
seek the fulfillments of their needs and desires elsewhere is probably the most radical of all proposals. Yet without it there is nothing to stop the relative incarceration
of captive populations within particular territories. If, for example, labor had the
same right of mobility as capital, if political persecution could be resisted (as the
affluent and privileged have proven) by geographical movement, and if individuals
and collectivities had the right to change their locations at will, then the kind of
world we live in would change dramatically (this principle is stated in Article 14 of
the UN Declaration). But the production of space means more than merely the ability to circulate within a pre-ordained spatially structured world. It also means the
right to reconstruct spatial relations (territorial forms, communicative capacities, and
rules) in ways that turn space from an absolute framework of action into a more
malleable relative and relational aspect of social life.

10. The right to difference including that of uneven geographical development
The UN Declaration (Articles 22 and 27) states that everyone should be accorded
'the economic, social and cultural rights indispensable for his dignity and the free
development of his personality' while also pointing to the importance of the right
'freely to participate in the cultural life of the community' and to receive protection
of 'the moral and material interests resulting from scientific, literary or artistic production.' This implies the right to be different, to explore differences in the realms
of culture, sexuality, religious beliefs, and the like. But it also implies the right for
different group or collective explorations of such differences and, as a consequence,

the right to pursue development on some territorial and collective basis that departs from established norms. Uneven geographical development should also be thought of as a right rather than as a capitalistically imposed necessity that diminishes life chances in one place in order to enhance them elsewhere. Again, the application of such a principle in such a way that it does not infringe upon others in negative ways will have to be fought over, but the statement of such a principle, like that of the living wage, provides a clear basis for argument. The recent UN extension of cultural rights (particularly those specified in Article 27 of the original UN Declaration) to encompass those of minorities (cf. Phillips and Rosas, 1995) provides an initial opening in this direction.

11. Our rights as species beings

    This is, perhaps, the vaguest and least easily specifiable of all rights. Yet it is perhaps the most important of them all. It must become central to debate. If we review our position in the long history of biological and social evolution, then plainly we have been and continue to be powerful evolutionary agents. If we are now entering a phase of volitional and conscious interventions in evolutionary processes (interventions that carry with them enormous risks and dangers), then we must necessarily construe certain universals to both promote and regulate the way we might engage upon such interventions. We all should have the right freely to explore the relation to nature and the transformative possibilities inherent in our species being in creative ways. This means the right to explore the possibility of different combinations of our evolutionary repertoire—the powers of cooperation, diversification, competition, the production of nature and of different dimensionalities to space and time. But that right to free experimentation (made much of by Unger) must also be tempered by duties, responsibilities, and obligations to others, both human and non-human, and it most certainly must accord strong protections against the potential powers of a non-democratic elite (or a capitalist class) to push us down technological, social, and evolutionary pathways that represent narrow class interests rather than human interests in general. Any concept of 'species interests' will inevitably be riven by rampant divisions of class, gender, religion, culture, and geography. But without some sense of where our common interests as a species might lie, it becomes impossible to construct any 'family of meanings' to connect or ground the incredible variety of partial claims and demands that make our social world such an interestingly divided place. On this point Naess and Rothenberg (1989, 164–70) have much to offer, by insisting that 'the universal right to self-unfolding' is related to the recognition of that same right across all species, and that 'the unfolding of life' in general is as important as the unfolding of our own personal trajectories of self-discovery and development.

This interlocking and oftentimes conflictual system of universal rights, I insist, is not the be-all and end-all of struggle, but a formative moment in a much more complicated social process directed towards socio-ecological change that embraces all the other distinctive theaters of social action. But the insurgent architect has to be an advocate of such rights. At the same time he or she must clearly recognize that their formulation arises out of social life and that they remain otiose and meaningless unless brought to bear in tangible ways upon mediating institutions, processes of community formation, and upon the ways in which the personal is construed and acted upon as the political.

8. SHAPING SOCIO-ECOLOGICAL ORDERS

The dialectical utopianism to which I aspire requires the perspective of a long and permanent historical-geographical revolution. Thinking about transformative political practices as manifestations of a dialectical and spatiotemporal utopianism is helpful. But it will only be so if we understand how activity and thought in the different theaters of social action relate, combine, and dissolve into each other to create an evolving totality of social action.

Unfortunately, much that passes for imaginative architectural and political practice often stays immobilized in only one or two of the theaters I have here defined. Our mental and practical divisions of labor and of perspectives are now so deeply ingrained in everything we do that it becomes impossible for any one of us to be fully present in much more than one of these theaters of thought and action at any one time. The problem is not that this cannot work. Indeed, it may work far too well as it so patently has in the past (as, for example, dominant mediating institutions use divisions of mental and practical labors to dictate the terms of universality and the ways it is admissible for the personal to be political). The errors of that past always threaten to return and haunt us (though perhaps in different ways). By seeing the seven moments I have described as integral to each other, by recognizing how they are all internally related, and by seeking to flow our analysis, our thinking, and our practices across their entire range, we may better situate our capacities as insurgent architects of some alternative possible dynamics. Any aspiring insurgent architect must learn, in association with others, to collate and combine action on all fronts. Universality without the personal is abstract dogma if not active political hypocrisy. Community without either the personal or universal becomes exclusionary and fascistic. Mediating institutions that consolidate their powers and oppress the personal or translate universals into bureaucratic systems of despotism and control subvert the revolutionary impulse into state authoritarianism. The translator who assumes omnipotence represses. The great individual (the architect/philosopher) who becomes detached from the masses and from daily life becomes either an irrelevant joke or an oppressive and domineering figure on the local if not on the world stage.

It is open dialogue and practical interactions across theaters on this long frontier that counts. And it is to dialectics rather than Wilsonian reductionism that we must appeal to make the connexions, however putatively, across these different scales. Only then can the impulse towards dialectical utopianism be prevented from dissolving into the arid and ultimately self-destructive utopianism of either closed spatial form or of temporal processes of perpetual creative destruction.

But aspirations must be tempered by a sense of limitations and of vulnerability. There are necessary limits to even the most vaunting of ambitions. If, as I have argued, dialectical utopianism must be effectively grounded in historical-geographical realities and achievements, if, to return to Marx's celebrated formulation, we can always aspire to make our own historical geography but never under historical and geographical conditions of our own choosing, then the leap from the present into some future is always constrained, no matter how hard we struggle to liberate ourselves from the three basic constraints of (1) where we can see it from, (2) how far can we see, and (3) where we can learn it from.

And as we make that leap, we also have to acknowledge that it is a speculative leap into the unknown and into the unknowable. There is a level at which, no matter how hard we try, we simply cannot know with certainty what kind of outcomes will emerge. Both the social and the ecological orders, particularly when taken together, are open and heterogeneous to the point where their totality can never quite be grasped let alone manipulated into predictable or stable states. No matter how hard we try to construct and reconstruct the socio-ecological order to a given plan, we inevitably fall victim not only to the unexpected consequences of our own actions but also to evolutionary contingencies (those 'accidents' to which Marx referred) that impinge upon us at every twist and turn and at every scale. It is precisely for this reason that the ideals of community, of utopias of spatial form, exercise such an attraction because they depict a closed world of known certainties and rules where chance and contingency, uncertainty and risk, are resolutely locked out.

Herein lies perhaps the most difficult of all barriers for the insurgent architect to surmount. In facing up to a world of uncertainty and risk, the possibility of being quite undone by the consequences of our own actions weighs heavily upon us, often making us prefer 'those ills we have than flying to others that we know not of.' But Hamlet, beset by angst and doubt and unable to act, brought disaster upon himself and upon his land by the mere fact of his inaction. It is on this point that we need to mark well the lessons of capitalist historical geography. For that historical geography was created through innumerable forms of speculative action, by a preparedness to take risks and be undone by them. While we laborers (and philosophical underlaborers) may for good reasons 'lack the courage of our minds,' the capitalists have rarely lacked the courage of theirs. And, arguably, when they have given in to doubt they have lost their capacity to make and re-make the world. Marx and Keynes, both, understood that it was the 'animal spirits,' the speculative passions and expectations of the capitalist (like those that Zola so dramatically depicted) that bore the system along, taking it in new directions and into new spaces (both literal and metaphorical). And it is perhaps no accident that architecture as a supremely speculative and heroic profession (rather than as either a Platonic metaphor or a craft) emerged in Italy along with the merchant capitalists who began upon their globalizing ventures through commercial speculations in the fifteenth and sixteenth centuries. It was that speculative spirit that open up new spaces for human thought and action in all manner of ways.

The lesson is clear: until we insurgent architects know the courage of our minds and are prepared to take an equally speculative plunge into some unknown, we too will continue to be the objects of historical geography (like worker bees) rather than active subjects, consciously pushing human possibilities to their limits. What Marx called 'the real movement' that will abolish 'the existing state of things' is always there for the making and for the taking. That is what gaining the courage of our minds is all about.

Part V

TECHNOLOGY AND HUMAN NATURE

Technological devices and systems mediate our lives and form the basic conditions in which we live and form our identities as individuals and citizens. We are who we are in relation to technologies. We would be different people if we were raised in radically different technological circumstances. It is the presence or absence of technological devices and systems that, in large part, forms the differences between life on a farm and life in a city, or between life in an advanced industrialized society and life in an underdeveloped society. Technologies also figure into the composition of our identities as part of the stories we tell about who we are. We identify ourselves as, for example, avid readers, basketball players, homeowners, cooks, knitters, artists, music lovers, and a host of other identity constructions that are inseparable from having and using technologies. These are just some of the common ways that technology mediates identity and determines the kind of people we are—individually and collectively. Other, less common ways in which human identity is connected to technological practices involve the use of technologies that call into question the very meaning of what it is to be human. These include medical technologies that radically intervene in the course of a life (e.g., in vitro fertilization) and computer programs that mimic human intelligence (e.g., Deep Blue). These technologies challenge our ideas as to what distinguishes the human—in other words, what is natural for us to do and to be and what is artificial. The articles in this section explore the various ways that technologies shape, transform, and call into question our very idea of human nature.

Michel Foucault, in "Panopticism," excerpted from his *Discipline and Punish: The Birth of the Prison*, examines the role played by panoptic technologies, surveillance structures commonly found in prisons, hospitals, and barracks. The panopticon is a laboratory for studying people used in "normalizing detention," a technique for creating a human body to be examined and regulated in the smallest details of life. What Foucault calls "disciplinary power" is the way that power is subtly exercised on people for the sake of order and control. The effectiveness of power is based on observation, organization, and measuring, supervising, and correcting what is considered abnormal. The very idea of what makes a person normal or abnormal was furthered through the study of individuals in panoptic environments; experts were given the opportunity to exclude, examine, label, and experiment on people in a controlled setting. The normal person became defined in relation to the madman, the leper, the plague victim, the criminal, beggars/vagabonds, and the disorderly. The panopticon, Foucault proposes, is the machinery and set of techniques that enable a society to control its citizens by discovering and inventing new psychological and behavioral categories to be controlled subsequently by power structures as well as by people themselves. The panopticon automatizes power; its very architecture assures its continual functioning. It is a quasi-autonomous form of power that functions as a political technology.

The panopticon leads to new forms of power and knowledge as it normalizes and integrates people into a disciplinary society. There is nothing necessarily insidious about disciplinary practices; they are merely techniques for ordering human multiplicities: techniques that aim to exercise power at the lowest cost, maximize the extent of power, and increase the docility and utility of an entire system of people and institutions. The disciplinary society described by Foucault evolved alongside capitalism and the Enlightenment. Even democracies depend on disciplinary mechanisms that classify and order people according to a norm or scale. Power and knowledge reinforce one another to form the foundation of social life. More power creates more categories of knowledge; more knowledge refines and extends the scope of power. This is how the liberatory Enlightenment project is at the same time a process of increased discipline: we achieve greater freedom as power and knowledge become more detailed and more controlling. Foucault's contribution to a philosophy of technology is to focus on the overlooked, small, hidden forms of panoptic power that lead to the various disciplinary techniques critical to the formation of identity.

In "What's Wrong with Enhancement Technologies?" Carl Elliott examines some troublesome questions that arise from the use of medical interventions designed not to treat illness but to improve human capacities and characteristics. Some common enhancement technologies include cosmetic surgery, athletic performance-enhancing drugs, antidepressants to make people less shy and more outgoing, and other drugs and procedures used to enhance physical, intellectual, and behavioral characteristics. As medical science and technology advance, and the possibilities of genetic enhancement become more real, it may one day be possible to make people more intelligent, slow down the aging process, extend life expectancy, limit aggressiveness and violence, or control whatever human features are controlled by genes. Yet many people are ambivalent about technologically enhancing human traits and capacities. Elliott highlights several morally problematic aspects of enhancement technology. Some of the worrisome issues have to do with blurring the lines between treatment and enhancement and thus challenging the proper goals of medicine; other issues have to do with what is considered normal and abnormal (and permissible) to change, thus challenging social goals. Elliott identifies several problems with enhancement, including the problem of "cultural complicity" (i.e., reinforcing cultural stereotypes about desirable and undesirable traits), "relative ends" (i.e., others must be unenhanced for enhancement to exist), and "authenticity" (i.e., "Am I the same person after enhancement?"). Ultimately, what's wrong with enhancement technologies may be what Elliott describes as "the drive to mastery," to control every aspect of being human.

In "Twenty-First Century Bodies," excerpted from *The Age of Spiritual Machines,* Ray Kurzweil predicts that in the near future we will be able to enhance our bodies at the molecular-cellular level with *nanotechnology*—minute machines that will be developed when we are able to build things from the atom up. Although this *nanobot* technology does not yet exist, once it does we should have a thorough and inexpensive system for changing the very structure of matter. Kurzweil states his belief that intelligent nanotechnology is inevitable. Just as human intelligence is the result of evolution, the next stage is the evolution of computer intelligence that will surpass our cognitive speed, capacity, and accuracy. He bases his prediction on Moore's Law of Integrated Circuits: better chip technology will produce an exponential increase in computer power. Every two years, you get twice as much computer power and capacity for the same amount of money. It is only a matter of time before we develop computational machines that can think like we do. The creation of artificial intelligence is the inevitable result of an evolutionary process that created humans who evolved to create computational technology that will outstrip us and speed up evolution through nonhuman machines.

In this reading, Kurzweil explains how we will be able to enhance and eventually replace not only our minds but also our bodies with synthetic materials. The key is nanotechnology: intelligent, atomic-level machines that could replicate themselves so as to number in the trillions. The possibilities of nanotechnology are the stuff of science fiction: machines that will be able to travel through the circulatory system, cleaning arteries, destroying cancer cells and tumors, repairing injured tissue and damaged organs, and even replacing missing limbs; they could enter our brains and provide sensory stimulation to create any imaginable experience; swarms of nanobots could create virtual fog that would simulate real environments; and they could clean our water and air, removing any hazardous matter from the environment—or they could take over and destroy us all. Nanotechnology is interesting philosophically because, as Kurzweil shows, it is an entirely different kind of human–technology relationship than that usually discussed in the philosophical literature. Nothing has as much potential to transform us in as many different ways as nanotechnology.

Hubert and Stuart Dreyfus would flatly disagree with Kurzweil. In "Why Computers May Never Think like People," the Dreyfuses argue that computers are merely rule-governed logic machines that can never attain the expert skill level demonstrated by human beings. Computers will never be able to equal, much less exceed, human intelligence because they do not think the way we do. Artificial intelligence, the Dreyfuses hold, is completely different from the "intuitive intelligence" we have. Machines are limited to a digital set of possibilities determined by heuristic sets of formal rules, whereas human beings acquire knowledge based on our bodily capacity for understanding contextual meaning that has little to do with formal rule-governed behavior. Rather, we acquire know-how through practice and experience. The authors describe five levels of skill acquisition, ranging from novice to expert, in order to show how acquiring competence to act appropriately in situations involves increasingly less rational and analytic thought but more intuitive and spontaneous actions. The more competent we are, the less we think, deliberate, and choose. The implication for artificial intelligence research is that attempts to reverse engineer the human brain to model a machine that exceeds our computational abilities misses the point. Human experience is not based on calculations, rules, facts, inferences, and information processing. No machine could ever simulate the kind of practical, embodied, contextual, and intuitive understanding that humans possess. Herbert and Stuart Dreyfus uncover the underlying assumptions of artificial intelligence research and suggest an alternative model of human experience far different from machine logic.

Sherry Turkle, in "Whither Psychoanalysis in Computer Culture?" sidesteps the question of whether machines can "really" be intelligent. Turkle is more interested in what kinds of relationships are appropriate and desirable to have with computers. A new class of computational objects is emerging—for example, wearable computers, PDAs, interactive robots, and virtual pets. These devices evoke emotional responses from their users, who often form lasting relationships with them. These "evocative objects," as Turkle terms them, are explicitly designed so that people will have emotional, affective relationships with them. The result is that these objects cause us to see ourselves and our relationships with things differently.

This new generation of "relational artifacts" calls for a reinterpretation of the premise of psychoanalysis and affective relationships in general as people begin to form intimate bonds with quasi-living things. Turkle claims that we need to rethink the psychoanalytic notion of "object-relations," traditionally reserved to explain our relationships with other people, to include a wider range of evocative objects. The very notion of self-identity changes as more and more people bond with their computers, play with their identities in online chat rooms, and form attachments to their robotic companions. Turkle wonders how interacting with relational artifacts will affect our way of thinking about ourselves, our sense of human identity, and our sense of what makes people special. If machines present themselves as emotional objects, what does this mean for our relationships with other people? Computer culture creates the demand for a more, not less, psychoanalytic theory, Turkle claims, as relational artifacts continue to become more lifelike and engage people emotionally. We need a new theory of object-relations to help us explore our responses to these things and to analyze how artifacts mediate the development of self.

23

Panopticism

Michel Foucault

Bentham's *Panopticon* is the architectural figure of this composition. We know the principle on which it was based: at the periphery, an annular building; at the centre, a tower; this tower is pierced with wide windows that open onto the inner side of the ring; the peripheric building is divided into cells, each of which extends the whole width of the building; they have two windows, one on the inside, corresponding to the windows of the tower; the other, on the outside, allows the light to cross the cell from one end to the other. All that is needed, then, is to place a supervisor in a central tower and to shut up in each cell a madman, a patient, a condemned man, a worker or a schoolboy. By the effect of backlighting, one can observe from the tower, standing out precisely against the light, the small captive shadows in the cells of the periphery. They are like so many cages, so many theatres, in which each actor is alone, perfectly individualized and constantly visible. The panoptic mechanism arranges spatial unities that make it possible to see constantly and to recognize immediately. In short, it reverses the principle of the dungeon; or rather of its three functions—to enclose, to deprive of light and to hide—it preserves only the first and eliminates the other two. Full lighting and the eye of a supervisor capture better than darkness, which ultimately protected. Visibility is a trap.

To begin with, this made it possible—as a negative effect—to avoid those compact, swarming, howling masses that were to be found in places of confinement, those painted by Goya or described by Howard. Each individual, in his place, is securely confined to a cell from which he is seen from the front by the supervisor; but the side walls prevent him from coming into contact with his companions. He is seen, but he does not see; he is the object of information, never a subject in communication. The arrangement of his room, opposite the central tower, imposes on him an axial visibility; but the divisions of the ring, those separated cells, imply a lateral invisibility. And this invisibility is a guarantee of order. If the inmates are convicts, there is no danger of a plot, an attempt at collective escape, the planning of new crimes for the future, bad reciprocal influences; if they are patients, there is no danger of contagion; if they are madmen there is no risk of their committing violence upon one another; if they are schoolchildren, there is no copying, no noise, no chatter, no waste of time; if they are workers, there are no disorders, no theft, no coalitions, none of those distractions that slow down the rate of work, make it less perfect or cause accidents. The crowd, a compact mass, a locus of multiple exchanges, individualities merging together, a collective effect, is abolished and replaced by a collection of separated individualities. From the point of view of the guardian, it is replaced by a multiplicity that can be numbered and supervised; from the point of view of the inmates, by a sequestered and observed solitude (Bentham, 60–64).

Hence the major effect of the Panopticon: to induce in the inmate a state of conscious and permanent visibility that assures the automatic functioning of power. So to arrange things that the surveillance is permanent in its effects, even if it is discontinuous in its action; that the perfection of power should tend to render its actual exercise unnecessary; that this architectural apparatus should be a machine for creating and sustaining a power relation independent of the person who exercises it; in short, that the inmates should be caught up in a power situation of which they are themselves the bearers. To achieve this, it is at once too much and too little that the prisoner should be constantly observed by an inspector: too little, for what matters is that he knows himself to be observed; too much, because he has no need in fact of being so. In view of this, Bentham laid down the principle that power should be visible and unverifiable. Visible: the inmate will constantly have before his eyes the tall outline of the central tower from which he is spied upon. Unverifiable: the inmate must never know whether he is being looked at at any one moment; but he must be sure that he may always be so. In order to make the presence or absence of the inspector unverifiable, so that the prisoners, in their cells, cannot even see a shadow, Bentham envisaged not only venetian blinds on the windows of the central observation hall, but, on the inside, partitions that intersected the hall at right angles and, in order to pass from one quarter to the other, not doors but zig-zag openings; for the slightest noise, a gleam of light, a brightness in a half-opened door would betray the presence of the guardian. The Panopticon is a machine for dissociating the see/being seen dyad: in the peripheric ring, one is totally seen, without ever seeing; in the central tower, one sees everything without ever being seen.

It is an important mechanism, for it automatizes and disindividualizes power. Power has its principle not so much in a person as in a certain concerted distribution of bodies, surfaces, lights, gazes; in an arrangement whose internal mechanisms produce the relation in which individuals are caught up. The ceremonies, the rituals, the marks by which the sovereign's surplus power was manifested are useless. There is a machinery that assures dissymmetry, disequilibrium, difference. Consequently, it does not matter who exercises power. Any individual, taken almost at random, can operate the machine: in the absence of the director, his family, his friends, his visitors, even his servants (Bentham, 45). Similarly, it does not matter what motive animates him: the curiosity of the indiscreet, the malice of a child, the thirst for knowledge of a philosopher who wishes to visit this museum of human nature, or the perversity of those who take pleasure in spying and punishing. The more numerous those anonymous and temporary observers are, the greater the risk for the inmate of being surprised and the greater his anxious awareness of being observed. The Panopticon is a marvellous machine which, whatever use one may wish to put it to, produces homogeneous effects of power.

A real subjection is born mechanically from a fictitious relation. So it is not necessary to use force to constrain the convict to good behaviour, the madman to calm, the worker to work, the schoolboy to application, the patient to the observation of the regulations. Bentham was surprised that panoptic institutions could be so light: there were no more bars, no more chains, no more heavy locks; all that was needed was that the separations should be clear and the openings well arranged. The heaviness of the old 'houses of security', with their fortress-like architecture, could be replaced by the simple, economic geometry of a 'house of certainty'. The efficiency of power, its constraining force have, in a sense, passed over to the other side—to the side of its surface of application. He who is subjected to a field of visibility, and who knows it, assumes responsibility for the constraints of power; he makes them play spontaneously upon himself; he inscribes in himself the power relation in which he simultaneously plays both roles; he becomes the principle of his own subjection. By this very fact, the external power may throw off its physical weight; it tends to the non-corporal; and, the more it approaches this limit, the more constant, profound and permanent are its effects: it is a perpetual victory that avoids any physical confrontation and which is always decided in advance.

Bentham does not say whether he was inspired, in his project, by Le Vaux's menagerie at Versailles: the first menagerie in which the different elements are not, as they traditionally were, distributed in a park (Loisel, 104–7). At the centre was an octagonal pavilion which, on the first floor, consisted of only a single room, the king's *salon;* on every side large windows looked out onto seven cages (the eighth side was reserved for the entrance), containing different species of animals. By Bentham's time, this menagerie had disappeared. But one finds in the programme of the Panopticon a similar concern with individualizing observation, with characterization and classification, with the analytical arrangement of space. The Panopticon is a royal menagerie; the animal is replaced by man, individual distribution by specific grouping and the king by the machinery of a furtive power. With this exception, the Panopticon also does the work of a naturalist. It makes it possible to draw up differences: among patients, to observe the symptoms of each individual, without the proximity of beds, the circulation of miasmas, the effects of contagion confusing the clinical tables; among schoolchildren, it makes it possible to observe performances (without there being any imitation or copying), to map aptitudes, to assess characters, to draw up rigorous classifications and, in relation to normal development, to distinguish 'laziness and stubbornness' from 'incurable imbecility'; among workers, it makes it possible to note the aptitudes of each worker, compare the time he takes to perform a task, and if they are paid by the day, to calculate their wages (Bentham, 60–64).

So much for the question of observation. But the Panopticon was also a laboratory; it could be used as a machine to carry out experiments, to alter behaviour, to train or correct individuals. To experiment with medicines and monitor their effects. To try out different punishments on prisoners, according to their crimes and character, and to seek the most effective ones. To teach different techniques simultaneously to the workers, to decide which is the best. To try out pedagogical experiments—and in particular to take up once again the well-debated problem of secluded education, by using orphans. One would see what would happen when, in their sixteenth or eighteenth year, they were presented with other boys or girls; one could verify whether, as Helvetius thought, anyone could learn anything; one would follow 'the genealogy of every observable idea'; one could bring up different children according to different systems of thought, making certain children believe that two and two do not make four or that the moon is a cheese, then put them together when they are twenty or twenty-five years old; one would then have discussions that would be worth a great deal more than the sermons or lectures on which so much money is spent; one would have at least an opportunity of making discoveries in the domain of metaphysics. The Panopticon is a privileged place for experiments on men, and for analysing with complete certainty the transformations that may be obtained from them. The Panopticon may even provide an apparatus for supervising its own mechanisms. In this central tower, the director may spy on all the employees that he has under his orders: nurses, doctors, foremen, teachers, warders; he will be able to judge them continuously, alter their behaviour, impose them upon them the methods he thinks best; and it will even be possible to observe the director himself. An inspector arriving unexpectedly at the centre of the Panopticon will be able to judge at a glance, without anything being concealed from him, how the entire establishment is functioning. And, in any case, enclosed as he is in the middle of this architectural mechanism, is not the director's own fate entirely bound up with it? The incompetent physician who has allowed contagion to spread, the incompetent prison governor or workshop manager will be the first victims of an epidemic or a revolt. '"By every tie I could devise", said the master of the Panopticon, "my own fate had been bound up by me with theirs"' (Bentham, 177). The Panopticon functions as a kind of laboratory of power. Thanks to its mechanisms of observation, it gains in efficiency and in the ability to penetrate into men's behaviour; knowledge follows the advances of power, discovering new objects of knowledge over all the surfaces on which power is exercised.

The plague-stricken town, the panoptic establishment—the differences are important. They mark, at a distance of a century and a half, the transformations of the disciplinary programme. In the first case, there is an exceptional situation: against an extraordinary evil, power is mobilized; it makes itself everywhere present and visible; it invents new mechanisms; it separates, it immobilizes, it partitions; it constructs for a time what is both a counter-city and the perfect society; it imposes an ideal functioning, but one that is reduced, in the final analysis, like the evil that it combats, to a simple dualism of life and death: that which moves brings death, and one kills that which moves. The Panopticon, on the other hand, must be understood as a generalizable model of functioning; a way of defining power relations in terms of the everyday life of men. No doubt Bentham presents it as a particular institution, closed in upon itself. Utopias, perfectly closed in upon themselves, are common enough. As opposed to the ruined prisons, littered with mechanisms of torture, to be seen in Piranese's engravings, the Panopticon presents a cruel, ingenious cage. The fact that it should have given rise, even in our own time, to so many variations, projected or realized, is evidence of the imaginary intensity that it has possessed for almost two hundred years. But the Panopticon must not be understood as a dream building: it is the diagram of a mechanism of power reduced to its ideal form; its functioning, abstracted from any obstacle, resistance or friction, must be represented as a pure architectural and optical system: it is in fact a figure of political technology that may and must be detached from any specific use.

It is polyvalent in its applications; it serves to reform prisoners, but also to treat patients, to instruct schoolchildren, to confine the insane, to supervise workers, to put beggars and idlers to work. It is a type of location of bodies in space, of distribution of individuals in relation to one another, of hierarchical organization, of disposition of centres and channels of power, of definition of the instruments and modes of intervention of power, which can be implemented in hospitals, workshops, schools, prisons. Whenever one is dealing with a multiplicity of individuals on whom a task or a particular form of behaviour must be imposed, the panoptic schema may be used. It is—necessary modifications apart—applicable 'to all establishments whatsoever, in which, within a space not too large to be covered or commanded by buildings, a number of persons are meant to be kept under inspection' (Bentham, 40; although Bentham takes the penitentiary house as his prime example, it is because it has many different functions to fulfil—safe custody, confinement, solitude, forced labour and instruction).

In each of its applications, it makes it possible to perfect the exercise of power. It does this in several ways: because it can reduce the number of those who exercise it, while increasing the number of those on whom it is exercised. Because it is possible to intervene at any moment and because the constant pressure acts even before the offences, mistakes or crimes have been committed. Because, in these conditions, its strength is that it never intervenes, it is exercised spontaneously and without noise, it constitutes a mechanism whose effects follow from one another. Because, without any physical instrument other than architecture and geometry, it acts directly on individuals; it gives 'power of mind over mind'. The panoptic schema makes any apparatus of power more intense: it assures its economy (in material, in personnel, in time); it assures its efficacity by its preventative character, its continuous functioning and its automatic mechanisms. It is a way of obtaining from power 'in hitherto unexampled quantity', 'a great and new instrument of government . . . ; its great excellence consists in the great strength it is capable of giving to *any* institution it may be thought proper to apply it to' (Bentham, 66).

It's case of 'it's easy once you've thought of it' in the political sphere. It can in fact be integrated into any function (education, medical treatment, production, punishment); it can increase the effect of this function, by being linked closely with it; it can constitute a mixed mechanism in which relations of power (and of knowledge) may be precisely adjusted, in the smallest detail, to the processes that are to be supervised; it can establish a direct proportion between 'surplus power' and 'surplus production'. In short, it arranges things in such a way that the exercise of power is not added on from the outside, like a rigid, heavy constraint, to the functions it invests, but is so subtly present in them as to increase their efficiency by itself increasing its own points of contact. The panoptic mechanism is not simply a hinge, a point of exchange between a mechanism of power and a function; it is a way of making power relations function in a function, and of making a function function through these power relations. Bentham's Preface to *Panopticon* opens with a list of the benefits to be obtained from his 'inspection-house': '*Morals reformed—health preserved—industry invigorated—instruction diffused—public burthens lightened*—Economy seated, as it were, upon a rock—the gordian knot of the Poor-Laws not cut, but united—all by a simple idea in architecture!' (Bentham, 39).

Furthermore, the arrangement of this machine is such that its enclosed nature does not preclude a permanent presence from the outside: we have seen that anyone may come and exercise in the central tower the functions of surveillance, and that, this being the case, he can gain a clear idea of the way in which the surveillance is practised. In fact, any panoptic institution, even if it is as rigorously closed as a penitentiary, may without difficulty be subjected to such irregular and constant inspections: and not only by the appointed inspectors, but also by the public; any member of society will have the right to come and see with his own eyes how the schools, hospitals, factories, prisons function. There is no risk, therefore, that the increase of power created by the panoptic machine may degenerate into tyranny; the disciplinary mechanism will be democratically controlled, since it will be constantly accessible 'to the great tribunal committee of the world'. This Panopticon, subtly arranged so that an observer may observe, at a glance, so many different individuals, also enables everyone to come and observe any of the observers. The seeing machine was once a sort of dark room into which individuals spied; it has become a transparent building in which the exercise of power may be supervised by society as a whole.

The panoptic schema, without disappearing as such or losing any of its properties, was destined to spread throughout the social body; its vocation was to become a generalized function. The plague-stricken town provided an exceptional disciplinary model: perfect, but absolutely violent; to the disease that brought death, power opposed its perpetual threat of death;

life inside it was reduced to its simplest expression; it was, against the power of death, the meticulous exercise of the right of the sword. The Panopticon, on the other hand, has a role of amplification; although it arranges power, although it is intended to make it more economic and more effective, it does so not for power itself, nor for the immediate salvation of a threatened society: its aim is to strengthen the social forces—to increase production, to develop the economy, spread education, raise the level of public morality; to increase and multiply.

How is power to be strengthened in such a way that, far from impeding progress, far from weighing upon it with its rules and regulations, it actually facilitates such progress? What intensificator of power will be able at the same time to be a multiplicator of production? How will power, by increasing its forces, be able to increase those of society instead of confiscating them or impeding them? The Panopticon's solution to this problem is that the productive increase of power can be assured only if, on the one hand, it can be exercised continuously in the very foundations of society, in the subtlest possible way, and if, on the other hand, it functions outside these sudden, violent, discontinuous forms that are bound up with the exercise of sovereignty. The body of the king, with its strange material and physical presence, with the force that he himself deploys or transmits to some few others, is at the opposite extreme of this new physics of power represented by panopticism; the domain of panopticism is, on the contrary, that whole lower region, that region of irregular bodies, with their details, their multiple movements, their heterogeneous forces, their spatial relations; what are required are mechanisms that analyse distributions, gaps, series, combinations, and which use instruments that render visible, record, differentiate and compare: a physics of a relational and multiple power, which has its maximum intensity not in the person of the king, but in the bodies that can be individualized by these relations. At the theoretical level, Bentham defines another way of analysing the social body and the power relations that traverse it; in terms of practice, he defines a procedure of subordination of bodies and forces that must increase the utility of power while practising the economy of the prince. Panopticism is the general principle of a new 'political anatomy' whose object and end are not the relations of sovereignty but the relations of discipline.

The celebrated, transparent, circular cage, with its high tower, powerful and knowing, may have been for Bentham a project of a perfect disciplinary institution; but he also set out to show how one may 'unlock' the disciplines and get them to function in a diffused, multiple, polyvalent way throughout the whole social body. These disciplines, which the classical age had elaborated in specific, relatively enclosed places—barracks, schools, workshops—and whose total implementation had been imagined only at the limited and temporary scale of a plague-stricken town, Bentham dreamt of transforming into a network of mechanisms that would be everywhere and always alert, running through society without interruption in space or in time. The panoptic arrangement provides the formula for this generalization. It programmes, at the level of an elementary and easily transferable mechanism, the basic functioning of a society penetrated through and through with disciplinary mechanisms.

There are two images, then, of discipline. At one extreme, the discipline-blockade, the enclosed institution, established on the edges of society, turned inwards towards negative functions: arresting evil, breaking communications, suspending time. At the other extreme, with panopticism, is the discipline-mechanism: a functional mechanism that must improve the exercise of power by making it lighter, more rapid, more effective, a design of subtle coercion for a society to come. The movement from one project to the other, from a schema of exceptional discipline to one of a generalized surveillance, rests on a historical transformation: the gradual extension of the mechanisms of discipline throughout the seventeenth and eighteenth centuries, their spread throughout the whole social body, the formation of what might be called in general the disciplinary society.

'Discipline' may be identified neither with an institution nor with an apparatus; it is a type of power, a modality for its exercise, comprising a whole set of instruments, techniques, procedures, levels of application, targets; it is a 'physics' or an 'anatomy' of power, a technology. And it may be taken over either by 'specialized' institutions (the penitentiaries or 'houses of correction' of the nineteenth century), or by institutions that use it as an essential instrument for a particular end (schools, hospitals), or by pre-existing authorities that find in it a means of reinforcing or reorganizing their internal mechanisms of power (one day we should show how intra-familial relations, essentially in the parents–children cell, have become 'disciplined', absorbing since the classical age external schemata, first educational and military, then medical, psychiatric, psychological, which have made the family the privileged locus of emergence for the disciplinary question of the normal and the abnormal); or by apparatuses that have made discipline their principle of internal functioning (the disciplinarization of the administrative apparatus from the Napoleonic period), or finally by state apparatuses whose major, if not exclusive, function is to assure that discipline reigns over society as a whole (the police).

On the whole, therefore, one can speak of the formation of a disciplinary society in this movement that stretches from the enclosed disciplines, a sort of social 'quarantine', to an indefinitely generalizable mechanism of 'panopticism'. Not because the disciplinary modality of power has replaced all the others; but because it has infiltrated the others, sometimes undermining them, but serving as an intermediary between them, linking them together, extending them and above all making it possible to bring the effects of power to the most minute and distant elements. It assures an infinitesimal distribution of the power relations.

A few years after Bentham, Julius gave this society its birth certificate (Julius, 384–6). Speaking of the panoptic principle, he said that there was much more there than architectural ingenuity: it was an event in the 'history of the human mind'. In appearance, it is merely the solution of a technical problem; but, through it, a whole type of society emerges. Antiquity had been a civilization of spectacle. 'To render accessible to a multitude of men the inspection of a small number of objects': this was the problem to which the architecture of temples, theatres and circuses responded. With spectacle, there was a predominance of public life, the intensity of festivals, sensual proximity. In these rituals in which blood flowed, society found new vigour and formed for a moment a single great body. The modern age poses the opposite problem: 'To procure for a small number, or even for a single individual, the instantaneous view of a great multitude.' In a society in which the principal elements are no longer the community and public life, but, on the one hand, private individuals and, on the other, the state, relations can be regulated only in a form that is the exact reverse of the spectacle: 'It was to the modern age, to the ever-growing influence of the state, to its ever more profound intervention in all the details and all the relations of social life, that was reserved the task of increasing and perfecting its guarantees, by using and directing towards that great aim the building and distribution of buildings intended to observe a great multitude of men at the same time.'

Julius saw as a fulfilled historical process that which Bentham had described as a technical programme. Our society is one not of spectacle, but of surveillance; under the surface of images, one invests bodies in depth; behind the great abstraction of exchange, there continues the meticulous, concrete training of useful forces; the circuits of communication are the supports of an accumulation and a centralization of knowledge; the play of signs defines the anchorages of power; it is not that the beautiful totality of the individual is amputated, repressed, altered by our social order, it is rather that the individual is carefully fabricated in it, according to a whole technique of forces and bodies. We are much less Greeks than we believe. We are neither in the amphitheatre, nor on the stage, but in the panoptic machine, invested by its effects of power, which we bring to ourselves since we are part of its mechanism. The importance, in historical mythology, of the Napoleonic character probably derives from the fact that it is at the point of junction of the monarchical, ritual exercise of sovereignty and the hierarchical, permanent exercise of indefinite discipline. He is the individual who looms over everything with a single gaze which no detail, however minute, can escape: 'You may consider that no part of the Empire is without surveillance, no crime, no offence, no contravention that remains unpunished, and that the eye of the genius who can enlighten all embraces the whole of this vast machine, without, however, the slightest detail escaping his attention' (Treilhard, 14). At the moment of its full blossoming, the disciplinary society still assumes with the Emperor the old aspect of the power of spectacle. As a monarch who is at one and the same time a usurper of the ancient throne and the organizer of the new state, he combined into a single symbolic, ultimate figure the whole of the long process by which the pomp of sovereignty, the necessarily spectacular manifestations of power, were extinguished one by one in the daily exercise of surveillance, in a panopticism in which the vigilance of intersecting gazes was soon to render useless both the eagle and the sun.

The formation of the disciplinary society is connected with a number of broad historical processes—economic, juridico-political and, lastly, scientific—of which it forms part.

1. Generally speaking, it might be said that the disciplines are techniques for assuring the ordering of human multiplicities. It is true that there is nothing exceptional or even characteristic in this: every system of power is presented with the same problem. But the peculiarity of the disciplines is that they try to define in relation to the multiplicities a tactics of power that fulfils three criteria: firstly, to obtain the exercise of power at the lowest possible cost (economically, by the low expenditure it involves; politically, by its discretion, its low exteriorization, its relative invisibility, the little resistance it arouses); secondly, to bring the effects of this social power to their maximum intensity and to extend them as far as possible, without either failure or interval; thirdly, to link this 'economic' growth of power with the output of the apparatuses (educational, military, industrial or medical) within which it is exercised; in short, to increase both the docility and the utility of all the elements of the system. This triple objective of the disciplines corresponds to a well-known historical conjuncture. One aspect of this conjuncture was the large demographic thrust of the eighteenth century; an increase in the floating population (one of the primary objects of discipline is to fix; it is an anti-nomadic technique); a change of quantitative scale in the groups to be supervised or manipulated (from the beginning of the seventeenth century to the eve of the French Revolution, the school population had been increasing rapidly, as had no doubt the hospital population; by the end of the eighteenth century, the peace-time army exceeded 200,000 men). The other aspect of the conjuncture was the growth in the apparatus of production, which was becoming more and more extended and complex; it was also becoming more costly and its profitability had to be increased. The development of the disciplinary methods corresponded to these two processes, or rather, no doubt, to the new need to adjust their correlation. Neither the residual forms of feudal power nor the structures of the administrative monarchy, nor the local mechanisms of supervision, nor the unstable, tangled mass they all formed together could carry out this role: they were hindered from doing so by the irregular and inadequate extension of their network, by their often conflicting functioning, but above all by the 'costly' nature of the power that was exercised in them. It was costly in several senses: because directly it cost a great deal to the Treasury; because the system of corrupt offices and farmed-out taxes weighed indirectly,

but very heavily, on the population; because the resistance it encountered forced it into a cycle of perpetual reinforcement; because it proceeded essentially by levying (levying on money or products by royal, seigniorial, ecclesiastical taxation; levying on men or time by *corvées* of press-ganging, by locking up or banishing vagabonds). The development of the disciplines marks the appearance of elementary techniques belonging to a quite different economy: mechanisms of power which, instead of proceeding by deduction, are integrated into the productive efficiency of the apparatuses from within, into the growth of this efficiency and into the use of what it produces. For the old principle of 'levying-violence', which governed the economy of power, the disciplines substitute the principle of 'mildness-production-profit'. These are the techniques that make it possible to adjust the multiplicity of men and the multiplication of the apparatuses of production (and this means not only 'production' in the strict sense, but also the production of knowledge and skills in the school, the production of health in the hospitals, the production of destructive force in the army).

In this task of adjustment, discipline had to solve a number of problems for which the old economy of power was not sufficiently equipped. It could reduce the inefficiency of mass phenomena: reduce what, in a multiplicity, makes it much less manageable than a unity; reduce what is opposed to the use of each of its elements and of their sum; reduce everything that may counter the advantages of number. That is why discipline fixes; it arrests or regulates movements; it clears up confusion; it dissipates compact groupings of individuals wandering about the country in unpredictable ways; it establishes calculated distributions. It must also master all the forces that are formed from the very constitution of an organized multiplicity; it must neutralize the effects of counter-power that spring from them and which form a resistance to the power that wishes to dominate it: agitations, revolts, spontaneous organizations, coalitions—anything that may establish horizontal conjunctions. Hence the fact that the disciplines use procedures of partitioning and verticality, that they introduce, between the different elements at the same level, as solid separations as possible, that they define compact hierarchical networks, in short, that they oppose to the intrinsic, adverse force of multiplicity the technique of the continuous, individualizing pyramid. They must also increase the particular utility of each element of the multiplicity, but by means that are the most rapid and the least costly, that is to say, by using the multiplicity itself as an instrument of this growth. Hence, in order to extract from bodies the maximum time and force, the use of those overall methods known as time-tables, collective training, exercises, total and detailed surveillance. Furthermore, the disciplines must increase the effect of utility proper to the multiplicities, so that each is made more useful than the simple sum of its elements: it is in order to increase the utilizable effects of the multiple that the disciplines define tactics of distribution, reciprocal adjustment of bodies, gestures and rhythms, differentiation of capacities, reciprocal coordination in relation to apparatuses or tasks. Lastly, the disciplines have to bring into play the power relations, not above but inside the very texture of the multiplicity, as discreetly as possible, as well articulated on the other functions of these multiplicities and also in the least expensive way possible: to this correspond anonymous instruments of power, coextensive with the multiplicity that they regiment, such as hierarchical surveillance, continuous registration, perpetual assessment and classification. In short, to substitute for a power that is manifested through the brilliance of those who exercise it, a power that insidiously objectifies those on whom it is applied; to form a body of knowledge about these individuals, rather than to deploy the ostentatious signs of sovereignty. In a word, the disciplines are the ensemble of minute technical inventions that made it possible to increase the useful size of multiplicities by decreasing the inconveniences of the power which, in order to make them useful, must control them. A multiplicity, whether in a workshop or a nation, an army or a school, reaches the threshold of a discipline when the relation of the one to the other becomes favourable.

If the economic take-off of the West began with the techniques that made possible the accumulation of capital, it might perhaps be said that the methods for administering the accumulation of men made possible a political take-off in relation to the traditional, ritual, costly, violent forms of power, which soon fell into disuse and were superseded by a subtle, calculated technology of subjection. In fact, the two processes—the accumulation of men and the accumulation of capital—cannot be separated; it would not have been possible to solve the problem of the accumulation of men without the growth of an apparatus of production capable of both sustaining them and using them; conversely, the techniques that made the cumulative multiplicity of men useful accelerated the accumulation of capital. At a less general level, the technological mutations of the apparatus of production, the division of labour and the elaboration of the disciplinary techniques sustained an ensemble of very close relations (cf. Marx, *Capital,* vol. I, chapter XIII and the very interesting analysis in Guerry and Deleule). Each makes the other possible and necessary; each provides a model for the other. The disciplinary pyramid constituted the small cell of power within which the separation, coordination and supervision of tasks was imposed and made efficient; and analytical partitioning of time, gestures and bodily forces constituted an operational schema that could easily be transferred from the groups to be subjected to the mechanisms of production; the massive projection of military methods onto industrial organization was an example of this modelling of the division of labour following the model laid down by the schemata of power. But, on the other hand, the technical analysis of the process of production, its 'mechanical' breaking-down, were projected onto the labour force whose task it was to implement it: the constitution of those disciplinary machines in which the individual forces that they bring together are composed into a whole and therefore increased is the effect of this projection. Let us say that discipline is the unitary technique by which the body is reduced as a 'political' force at the least cost and maximized as a useful force. The growth of a capitalist economy gave rise to the specific modality of disciplinary power, whose general formulas, techniques of submitting forces and bodies, in short, 'political anatomy', could be operated in the most diverse political régimes, apparatuses or institutions.

2. The panoptic modality of power—at the elementary, technical, merely physical level at which it is situated—is not under the immediate dependence or a direct extension of the great juridico-political structures of a society; it is nonetheless not absolutely independent. Historically, the process by which the bourgeoisie became in the course of the eighteenth century the politically dominant class was masked by the establishment of an explicit, coded and formally egalitarian juridical framework, made possible by the organization of a parliamentary, representative régime. But the development and generalization of disciplinary mechanisms constituted the other, dark side of these processes. The general juridical form that guaranteed a system of rights that were egalitarian in principle was supported by these tiny, everyday, physical mechanisms, by all those systems of micro-power that are essentially nonegalitarian and asymmetrical that we call the disciplines. And although, in a formal way, the representative régime makes it possible, directly or indirectly, with or without relays, for the will of all to form the fundamental authority of sovereignty, the disciplines provide, at the base, a guarantee of the submission of forces and bodies. The real, corporal disciplines constituted the foundation of the formal, juridical liberties. The contract may have been regarded as the ideal foundation of law and political power; panopticism constituted the technique, universally widespread, of coercion. It continued to work in depth on the juridical structures of soci-

ety, in order to make the effective mechanisms of power function in opposition to the formal framework that it had acquired. The 'Enlightenment', which discovered the liberties, also invented the disciplines.

In appearance, the disciplines constitute nothing more than an infra-law. They seem to extend the general forms defined by law to the infinitesimal level of individual lives; or they appear as methods of training that enable individuals to become integrated into these general demands. They seem to constitute the same type of law on a different scale, thereby making it more meticulous and more indulgent. The disciplines should be regarded as a sort of counter-law. They have the precise role of introducing insuperable asymmetries and excluding reciprocities. First, because discipline creates between individuals a 'private' link, which is a relation of constraints entirely different from contractual obligation; the acceptance of a discipline may be underwritten by contract; the way in which it is imposed, the mechanisms it brings into play, the non-reversible subordination of one group of people by another, the 'surplus' power that is always fixed on the same side, the inequality of position of the different 'partners' in relation to the common regulation, all these distinguish the disciplinary link from the contractual link, and make it possible to distort the contractual link systematically from the moment it has as its content a mechanism of discipline. We know, for example, how many real procedures undermine the legal fiction of the work contract: workshop discipline is not the least important. Moreover, whereas the juridical systems define juridical subjects according to universal norms, the disciplines characterize, classify, specialize; they distribute along a scale, around a norm, hierarchize individuals in relation to one another and, if necessary, disqualify and invalidate. In any case, in the space and during the time in which they exercise their control and bring into play the asymmetries of their power, they effect a suspension of the law that is never total, but is never annulled either. Regular and institutional as it may be, the discipline, in its mechanism, is a 'counter-law'. And, although the universal juridicism of modern society seems to fix limits on the exercise of power, its universally widespread panopticism enables it to operate, on the underside of the law, a machinery that is both immense and minute, which supports, reinforces, multiplies the asymmetry of power and undermines the limits that are traced around the law. The minute disciplines, the panopticisms of every day may well be below the level of emergence of the great apparatuses and the great political struggles. But, in the genealogy of modern society, they have been, with the class domination that traverses it, the political counterpart of the juridical norms according to which power was redistributed. Hence, no doubt, the importance that has been given for so long to the small techniques of discipline, to those apparently insignificant tricks that it has invented, and even to those 'sciences' that give it a respectable face; hence the fear of abandoning them if one cannot find any substitute; hence the affirmation that they are at the very foundation of society, and an element in its equilibrium, whereas they are a series of mechanisms for unbalancing power relations definitively and everywhere; hence the persistence in regarding them as the humble, but concrete form of every morality, whereas they are a set of physico-political techniques.

To return to the problem of legal punishments, the prison with all the corrective technology at its disposal is to be resituated at the point where the codified power to punish turns into a disciplinary power to observe; at the point where the universal punishments of the law are applied selectively to certain individuals and always the same ones; at the point where the redefinition of the juridical subject by the penalty becomes a useful training of the criminal; at the point where the law is inverted and passes outside itself, and where the counter-law becomes the effective and institutionalized content of the juridical forms. What generalizes the power to punish, then, is not the universal consciousness of the law in each juridical subject; it is the regular extension, the infinitely minute web of panoptic techniques.

3. Taken one by one, most of these techniques have a long history behind them. But what was new, in the eighteenth century, was that, by being combined and generalized, they attained a level at which the formation of knowledge and the increase of power regularly reinforce one another in a circular process. At this point, the disciplines crossed the 'technological' threshold. First the hospital, then the school, then, later, the workshop were not simply 'reordered' by the disciplines; they became, thanks to them, apparatuses such that any mechanism of objectification could be used in them as an instrument of subjection, and any growth of power could give rise in them to possible branches of knowledge; it was this link, proper to the technological systems, that made possible within the disciplinary element the formation of clinical medicine, psychiatry, child psychology, educational psychology, the rationalization of labour. It is a double process, then: an epistemological 'thaw' through a refinement of power relations; a multiplication of the effects of power through the formation and accumulation of new forms of knowledge.

The extension of the disciplinary methods is inscribed in a broad historical process: the development at about the same time of many other technologies—agronomical, industrial, economic. But it must be recognized that, compared with the mining industries, the emerging chemical industries or methods of national accountancy, compared with the blast furnaces or the steam engine, panopticism has received little attention. It is regarded as not much more than a bizarre little utopia, a perverse dream—rather as though Bentham had been the Fourier of a police society, and the Phalanstery had taken on the form of the Panopticon. And yet this represented the abstract formula of a very real technology, that of individuals. There were many reasons why it received little praise; the most obvious is that the discourses to which it gave rise rarely acquired, except in the academic classifications, the status of sciences; but the real reason is no doubt that the power that it operates and which it augments is a direct, physical power that men exercise upon one another. An inglorious culmination had an origin that could be only grudgingly acknowledged. But it would be unjust to compare the disciplinary techniques with such inventions as the steam engine or Amici's microscope. They are much less; and yet, in a way, they are much more. If a historical equivalent or at least a point of comparison had to be found for them, it would be rather in the 'inquisitorial' technique.

But the penitentiary Panopticon was also a system of individualizing and permanent documentation. The same year in which variants of the Benthamite schema were recommended for the building of prisons, the system of 'moral accounting' was made compulsory: an individual report of a uniform kind in every prison, on which the governor or head-warder, the chaplain and the instructor had to fill in their observations on each inmate: 'It is in a way the *vade mecum* of prison administration, making it possible to assess each case, each circumstance and, consequently, to know what treatment to apply to each prisoner individually' (Ducpétiaux, 56–7). Many other, much more complete, systems of recording were planned or tried out (cf., for example, Gregory, 199ff; Grellet-Wammy, 23–5 and 199–203). The overall aim was to make the prison a place for the constitution of a body of knowledge that would regulate the exercise of penitentiary practice. The prison has not only to know the decision of the judges and to apply it in terms of the established regulations: it has to extract unceasingly from the inmate a body of knowledge that will make it possible to transform the penal measure into a penitentiary operation; which will make of the penalty required by the offence a modification of the inmate that will be of use to society. The autonomy of the carceral régime and the knowledge that it creates make it possible to increase the utility of the penalty, which the code had made the very principle of its punitive philosophy: 'The governor must not lose sight of a single inmate, because in whatever part of the prison the inmate is to be found, whether he is entering or leaving, or whether he is staying there, the governor must also justify the motives for his staying in a particular classification or for his movement from one to another. He is a veritable accountant. Each inmate is for him, in the sphere of individual education, a capital invested with penitentiary interest' (Lucas, II, 449–50). As a highly efficient technology, penitentiary practice produces a return on the capital invested in the penal system and in the building of heavy prisons.

Similarly, the offender becomes an individual to know. This demand for knowledge was not, in the first instance, inserted into the legislation itself, in order to provide substance for the sentence and to determine the true degree of guilt. It is as a convict, as a point of application for punitive mechanisms, that the offender is constituted himself as the object of possible knowledge.

But this implies that the penitentiary apparatus, with the whole technological programme that accompanies it, brings about a curious substitution: from the hands of justice, it certainly receives a convicted person; but what it must apply itself to is not, of course, the offence, nor even exactly the offender, but a rather different object, one defined by variables which at the outset at least were not taken into account in the sentence, for they were relevant only for a corrective technology. This other character, whom the penitentiary apparatus substitutes for the convicted offender, is the *delinquent*.

The delinquent is to be distinguished from the offender by the fact that it is not so much his act as his life that is relevant in characterizing him. The penitentiary operation, if it is to be a genuine re-education, must become the sum total existence of the delinquent, making of the prison a sort of artificial and coercive theatre in which his life will be examined from top to bottom. The legal punishment bears upon an act; the punitive technique on a life; it falls to this punitive technique, therefore, to reconstitute all the sordid detail of a life in the form of knowledge, to fill in the gaps of that knowledge and to act upon it by a practice of compulsion. It is a biographical knowledge and a technique for correcting individual lives. The observation of the delinquent 'should go back not only to the circumstances, but also to the causes of his crime; they must be sought in the story of his life, from the triple point of view of psychology, social position and upbringing, in order to discover the dangerous proclivities of the first, the harmful predispositions of the second and the bad antecedents of the third. This biographical investigation is an essential part of the preliminary investigation for the classification of penalties before it becomes a condition for the classification of moralities in the penitentiary system. It must accompany the convict from the court to the prison, where the governor's task is not only to receive it, but also to complete, supervise and rectify its various factors during the period of detention' (Lucas, II, 440–42). Behind the offender, to whom the investigation of the facts may attribute responsibility for an offence, stands the delinquent whose slow formation is shown in a biographical investigation. The introduction of the 'biographical' is important in the history of penalty. Because it establishes the 'criminal' as existing before the crime and even outside it. And, for this reason, a psychological causality, duplicating the juridical attribution of responsibility, confuses its effects. At this point one enters the 'criminological' labyrinth from which we have certainly not yet emerged: any determining cause, because it reduces responsibility, marks the author of the offence with a criminality all the more formidable and demands penitentiary measures that are all the more strict. As the biography of the criminal duplicates in penal practice the analysis of circumstances used in gauging the crime, so one sees penal discourse and psychiatric discourse crossing each other's frontiers; and there, at their point of junction, is formed the notion of the 'dangerous' individual, which makes it possible to draw up a network of causality in terms of an entire biography and to present a verdict of punishment-correction.

The penitentiary technique and the delinquent are in a sense twin brothers. It is not true that it was the discovery of the delinquent through a scientific rationality that introduced into our old prisons the refinement of penitentiary techniques. Nor is it true that the internal elaboration of penitentiary methods has finally brought to light the 'objective' existence of a delinquency that the abstraction and rigidity of the law were unable to perceive. They appeared together, the one extending from the other, as a technological ensemble that forms and fragments the object to which it applies its instruments. And it is this delinquency, formed in the foundations of the judicial apparatus, among the *'basses œuvres'*, the servile tasks, from which justice averts its gaze, out of the shame it feels in punishing those it condemns, it is this delinquency that now comes to haunt the untroubled courts and the majesty of the laws; it is this delinquency that must be known, assessed, measured, diagnosed, treated when sentences are passed. It is now this delinquency, this anomaly, this deviation, this potential danger, this illness, this form of existence, that must be taken into account when the codes are rewritten. Delinquency is the vengeance of the prison on justice. It is a revenge formidable enough to leave the judge speechless. It is at this point that the criminologists raise their voices.

But we must not forget that the prison, that concentrated and austere figure of all the disciplines, is not an endogenous element in the penal system as defined at the turn of the eighteenth and nineteenth centuries. The theme of a punitive society and of a general semiotechnique of punishment that has sustained the 'ideological' codes—Beccarian or Benthamite—did not itself give rise to the universal use of the prison. This prison came from elsewhere—from the mechanisms proper to a disciplinary power. Now, despite this heterogeneity, the mechanisms and effects of the prison have spread right through modern criminal justice; delinquency and the delinquents have become parasites on it through and through. One must seek the reason for this formidable 'efficiency' of the prison. But one thing may be noted at the outset: the penal justice defined in the eighteenth century by the reformers traced two possible but divergent lines of objectification of the criminal: the first was the series of 'monsters', moral or political, who had fallen outside the social pact; the second was that of the juridical subject rehabilitated by punishment. Now the 'delinquent' makes it possible to join the two lines and to constitute under the authority of medicine, psychology or criminology, an individual in whom the offender of the law and the object of a scientific technique are superimposed—or almost—one upon the other. That the grip of the prison on the penal system should not have led to a violent reaction of rejection is no doubt due to many reasons. One of these is that, in fabricating delinquency, it gave to criminal justice a unitary field of objects, authenticated by the 'sciences', and thus enabled it to function on a general horizon of 'truth'.

The prison, that darkest region in the apparatus of justice, is the place where the power to punish, which no longer dares to manifest itself openly, silently organizes a field of objectivity in which punishment will be able to function openly as treatment and the sentence be inscribed among the discourses of knowledge. It is understandable that justice should have adopted so easily a prison that was not the offspring of its own thoughts. Justice certainly owed the prison this recognition.

24

Enhancement Technology

Carl Elliott

*E*nhancement technologies is a term of art coined by bioethicists to refer to medical interventions that can be used not just to cure or control illness, but also to enhance human capacities and characteristics. The term comes out of a debate over gene therapy in the late 1980s. The first gene therapy trials involved the treatment of a genetic disease called adenosine deaminase (ADA) deficiency, which causes children to have severe problems with their immune system. Gene therapy to treat illnesses was widely seen as morally justifiable, as long as it could be shown safe and effective, but many skeptics worried about the prospect of manipulating a person's genetic constitution in an effort to improve them—to try to make them smarter or better looking, to change their personalities, and so on. The skeptics found the prospect of genetic enhancement even more worrying if the enhancements could be passed from generation to generation with germ line (rather than somatic cell) manipulation.

Out of this debate came a distinction between *therapy* on the one hand and *enhancement* on the other, the distinction suggesting that therapy was morally acceptable, and enhancement morally worrying. The purpose of the distinction was to allow researchers to pursue gene therapy for conditions such as ADA deficiency or cystic fibrosis, but to discourage them from trying to monkey around with the genetics of personality, intelligence, or physical appearance. Other bioethicists latched onto the enhancement/treatment distinction as a way of making more general moral distinctions between other types of medical interventions. Many people argue that third-party payers (e.g., insurance companies) are obligated to pay for treatments, for example, but not for enhancements, such as breast augmentation surgery or Rogaine for baldness.

The distinction between enhancement and therapy has not turned out to be terribly useful for ethical purposes. Part of the problem is that some interventions that are easily characterized as enhancements seem ethically justifiable, or even desirable. For example, immunizations enhance a person's immune system rather than cure or control an illness, yet no one is arguing that they should be banned or that insurance companies should not pay for them. A tougher problem comes from the fact that most *enhancements* can also be characterized as *treatments* for some kind of psychological problem. Is the antidepressant Paxil (paroxetine) an enhancement aimed at making shy people more outgoing, or is it a treatment for social anxiety disorder? Is Ritalin a concentration enhancer, or is it a treatment for attention deficit hyperactivity disorder? At best, the line between treatment and enhancement is very fluid.

The enhancement/treatment distinction also slides right over the question: what's wrong with enhancement technologies anyway? Is there a moral problem with wanting to be taller, or better looking, or happier, or able to concentrate better? Many of the characteristics many people want to enhance are generally seen as positive changes, and might not be at all worrying if they were achieved through, say, education, or work, or some sort of psychotherapy. What is wrong with trying to achieve these things with the tools of medicine?

The answer to that question will vary from one intervention to another. Each medical technology has its own merits and dangers, and what is morally worrying about one may not be a problem for another. Many of these worries have become quite familiar. A number of writers believe that the main problem for enhancement technologies will be access. They may well be right. If enhancement technologies are not paid for by third-party payers, then the extra boosts they provide will almost certainly be disproportionately available to the rich. This would reinforce conditions of social inequality.

I want to highlight a few of the less obvious problems with enhancement technologies. My purpose is not to suggest that my list exhausts the range of moral problems, or that my concerns apply equally to all technologies. I do not even want to suggest that these concerns mean that such technologies are unethical. My aims are diagnostic. I want to try to put a finger on the worries than many people feel about some of these technologies but often find difficult to articulate.

The first concern I want to highlight is what the Georgetown University philosopher Margaret Olivia Little calls the *problem of cultural complicity*. As Little points out, the demand for certain technologies is propped up by cultural forces that many of us would see as harmful. They are harmful because they make some people feel inadequate or unhappy with the way they are. For example, cosmetic surgeons have often taken advantage of the desire of some ethnic minorities to efface markers of their ethnicity—to perform surgery on the "Jewish nose," for example, or to alter "Asian eyes" in order to make them look more like the eyes of Europeans. Another example might be the pressure that many American women feel to conform to a certain body type, which leaves many women and girls feeling that they are too fat or that their breasts are too small. At the extreme end of the spectrum would be the cultural pressures that help produce psychiatric illnesses like anorexia nervosa.

These kinds of forces leave people with a dilemma. On the one hand, you might see them as harmful, and you might well believe that we would be better off as a society if we were free of them. Yet on the other hand, they are real. So you feel them, and if you are a parent, you feel their effects on your child. But if you give in to them, you help reinforce them. This is what Little means by cultural complicity. By getting breast augmentation surgery, you are complicit in the norm that creates the pressure for women to have large breasts. By boosting your son's height using growth hormone, you are complicit in the norm that creates the pressure for men to be tall. By taking Paxil or Zoloft for shyness, you are complicit in the cultural norm that makes shyness something to be ashamed of. By giving in to the pressure (or, if you are a doctor, by exploiting it), you are helping to reinforce that pressure.

This leads to a second problem that I will call the *problem of relative ends*. You can see the problem of relative ends most clearly with the prescription of synthetic growth hormone to increase the height of short children, especially short boys. Growth hormone was initially given to to children who have a genetic deficiency of growth hormone—that is, children whose bodies don't produce growth hormone themselves. But in the 1980s, pediatricians began to debate whether it should be given to short children who are not growth hormone deficient. The American Academy of Pediatrics eventually issued a policy statement on the ethics of growth hormone therapy, which said that it is ethically acceptable to prescribe it to some children who were not deficient in growth hormone but who were short for other reasons, such as Turner's Syndrome or chronic renal disease. Some pediatricians (as well as the manufacturer of synthetic growth hormone) argued that growth hormone should not be restricted to children who are short because of an illness, but to any child who was so short that that they would be stigmatized by their condition. It also bears mentioning that synthetic growth hormone was very expensive—upwards of $50,000 a year at the time of this debate—and that there was little evidence that it could work very well for children who were not growth hormone deficient.

The case of growth hormone demonstrates the problem of relative ends. Boys and men want to be tall, but *tall* is a relative concept; being tall is dependent on others being short. 6'5" is tall because the average American man is 5'10"; if the average American man were 6'5", then 6'5" would no longer be tall. In other words, not everyone can be tall; for tall men to exist, there must be short men. For short men to exist, there must be tall men. At best, everyone could be the same height, but even then I suspect that there would still be tall and short, only they might be measured in millimeters rather than inches. The situation here is like Gore Vidal's remark: it's not enough to succeed; others must fail.

The seminal text here, as anyone with small children will know, is Dr. Seuss's book *The Sneetches*. The important characteristic for Sneetches is whether they have stars on their bellies. As Dr. Seuss says, "But because they have stars, all the Star Belly Sneetches / Would brag, we're the best kind of Sneetch on the beaches." For Sneetches, the demand is for stars, and the supplier who fills that demand is an entrepreneur called Sylvester McMonkey McBean. McBean rides into town with a machine that puts stars on the bellies of the Plain-Belly Sneetches. But when everyone has a star on their bellies, the appeal of having one is gone. So then the Sneetches want to have their stars removed. McBean is happy to do this as well. The Sneetches get caught in a vicious circle, adding and removing stars from their bellies, for which McBean happily collects a fee.

A third (and related) concern with enhancement technologies is the *role of the market*. Increasingly, the American healthcare system has become powered by the engine of consumer capitalism. During the 1990s, according to *Fortune* magazine, the pharmaceutical industry became the most profitable industry in the United States, with annual profit margins exceeding 18 percent. Much of these profits come from drugs that can arguably be used for enhancement. According to the National Institute for Health Care Management, for example, the most profitable class of drugs in 2001 was the antidepressants. Of course, a market economy relies on the freedom of industry to sell people what they want, but it also relies on the ability of industry to persuade people that they want what industry is selling. In 1997, the FDA relaxed its restrictions on direct-to-consumer advertising for prescription drugs, allowing the pharmaceutical industry to advertise on television and in popular magazines. Is anything wrong with this? Some people would say no; this is simply how a capitalist economy works. Others would say that selling people what they want is wrong if it preys on their fears, insecurities, or weaknesses, like tobacco company executives who don't smoke themselves but make their living by persuading others to smoke.

This concern about the market has emerged in the recent history of antidepressants such as Prozac. Prozac (fluoxetine) was the first of a new generation of antidepressants known as selective serotonin reuptake inhibitors (SSRIs), which became extraordinarily popular during the 1990s. The psychiatrist Peter Kramer coined the phrase "cosmetic psychopharmacology" in his book *Listening to Prozac*. What Kramer found so intriguing about Prozac was not what it does for patients who are clinically depressed, but what it does for those who aren't: patients who are shy and withdrawn, or who have poor self-esteem, or who are somewhat obsessive. According to Kramer and others, some patients on Prozac (but not all or even most of them) undergo what seems to be a change in personality. Some of the more controlling, compulsive types become more laid-back and easygoing. Some shy people become more self-confident and assertive. Kramer's patients said things like "I feel like I've been drugged all my life and now I'm finally clearheaded" or "I never really felt like myself until now." Some patients seem to be able to see themselves in a way that they had been incapable of before. They didn't just get well; in the words of one patient, they became "better than well." This effect is what Kramer called cosmetic psychopharmacology—the use of psychoactive drugs not to treat severe mental disorders, but to improve various aspects of a person's mental life.

This effect has helped transform the SSRIs from a medical treatment into a market commodity. Today the SSRIs are not simply used to treat clinical depression. They are used to treat (among other things) social anxiety disorder, posttraumatic stress disorder, generalized anxiety disorder, obsessive-compulsive disorders, eating disorders, sexual compulsions, and premenstrual dysphoric disorder. Many of these disorders did not officially exist several decades ago, and many of those that did exist were thought to be very rare. They became widely diagnosed only when a treatment was developed. Some clinicians argue that this is simply the result of their greater awareness of the disorders and their improved diagnostic skills. But it is also true that once a pharmaceutical company develops a treatment for a psychiatric disorder, it also has a financial interest in making sure that doctors diagnose the disorder as often as possible. This may mean transforming what was once seen as ordinary human variation—being shy, uptight, or melancholy—into psychiatric disorders. The more people are persuaded that they have a disorder that can be medicated, the more medication that will be sold.

The fourth worry I want to mention is a little trickier. I will call it the *problem of authenticity*. Probably the best way to illustrate the problem of authenticity is with a case that Peter Kramer discusses in *Listening to Prozac*. Like other psychiatrists, Kramer found that when he prescribed Prozac, his patients told him things like "This is how I was always meant to feel." He writes about one patient called Tess who is clinically depressed. Tess has few friends, lots of obligations, and very poor self-esteem. Kramer prescribes a tricyclic antidepressant, and she eventually gets over her depression. But Kramer thinks she could do even better, so he switches her to Prozac. Soon she is happier, more outgoing, and more self-confident, and she knows when and how to say no. Her life is much better.

Eventually Kramer tapers her Prozac and then takes her off the drug completely. In a few months, Tess comes back and says, "I just don't feel like myself anymore." Remarkable, thinks Kramer: she has returned to the very state in which she has been for twenty or thirty years, her entire life apart from the past several months, and she says "I don't feel like myself." Instead, she says she feels like herself on Prozac.

What do we make of these kinds of remarks? It's clear that this patient changes quite a lot on Prozac. But should it be described as a transformation to a new self, or as a restoration to an authentic self? Or is it something else? Some people would argue that Prozac can restore an authentic self that has been hidden by pathology. The authentic self is the one that has the proper levels of serotonin in the brain. But there are other cases in the book that seem to point in the opposite direction. Kramer tells of one patient who is not a success on Prozac. Before Prozac he is bitter, sarcastic, and rather cynical in a way that he seems to have cultivated. On Prozac, he becomes less bitter and less cynical; he loses that sarcastic edge. And he is happier.

But he doesn't like it. He doesn't feel comfortable with the person he has become, and so he stops taking the drug. For him, Prozac doesn't seem so much to restore an authentic self as to create a new one, a new one that he thinks is not really him.

This language of authenticity is very slippery, and it can be used in many different ways. For example, another area of medicine where patients talk about finding their true selves is transsexual surgery. Candidates for surgery might say, "I am really a woman, trapped in a man's body," and that the surgery will let them be who they really are. This sounds similar to what Arthur Frank calls a "restitution narrative" in his book *The Wounded Storyteller.* A restitution narrative has the basic form of "I was healthy, then I got sick, and then I was restored to health." This is like the restitution narrative, but the restitution is to something that never existed before, only wished for—not restoration back to health, but restoration to an ideal of health that had never before been realized. What interests me is not so much whether this narrative is true or not, whatever that would mean, but how persuasive it is. Even people who are troubled by the idea of a person changing his or her sex find themselves swayed by this kind of story: "I really am a man, trapped in a woman's body." It is a Cartesian explanation: a ghost locked in the wrong machine. And it sounds plausible to us in a way that it might not sound plausible to someone in a time and place without our tradition of body/mind dualism.

Of course, the ideal of authenticity can also be used to argue *against* enhancement technologies or other kinds of self-transformation. You can see this most clearly in the case of cosmetic surgery for markers of ethnic identity. To undergo surgery to make your Asian eyes look more European, or to use creams to make your dark skin lighter, is sometimes seen as an act of fakery or self-betrayal. Words like *fakery* and *self-betrayal* turn the language of authenticity around and use it for purposes just the opposite of those used by Tess and others who "become themselves" on Prozac. But these descriptions are persuasive only if you think of authenticity as a moral ideal.

The fifth worry about enhancement technologies I will call the *problem of relativism.* What I mean by this is simply that illnesses, by and large, are not objective entities that look the same to all people at all times. Rather, what counts as an illness is a product of a particular time and place, and a particular set of cultural understandings. Homosexuality was officially considered a mental disorder up until the 1970s and was listed in the American Psychiatric Association's *Diagnostic and Statistical Manual.* Today it is thought of as simply part of a person's identity, a constituent of the way some people are. Our understanding of homosexuality has moved from illness to identity. And, of course, we can also slide easily in the other direction from identity to illness. Some years ago, a person with three copies of chromosome 21 was called a mongoloid; now she has a genetic disease called Down syndrome. Whereas we used to think of her as a different type of human being, now we think of her as sick. We have redefined identity as illness.

The reasons for this are complex, of course. Very often what counts as an illness is a consequence of the discovery of a way to correct it. As Willard Gaylin has pointed out, before various reproductive technologies were developed, infertility was simply a fact of nature. Now that it can be treated, it is a medical problem. Before the invention of the lens, poor vision was simply a consequence of getting old. Now it is something to be treated by a medical specialist. Psychiatry is another striking example. Before the development of psychotherapy at the end of the nineteenth century, mental illness was limited to psychotic disorders; now it includes phobias, obsessions, compulsions, personality disorders, and so on. Today it is very easy to speak of any disagreeable personality trait as if it were an illness—and even some that are not so disagreeable, like shyness, which is being discussed more and more often in the ethical and psychiatric literature as if it were a kind of mental disability. The point is that in each of these cases, what was once simply an unavoidable aspect of some people's lives was conceptually transformed by technology into a medical problem.

My point here is simply that what we see as a straightforward example of a medical treatment will look differently to people from other times and other places, and that the line we often draw between enhancements and treatments is not as sharp as we would like to think. Let me take a deliberately provocative example—the way we respond to intersexed infants, or children born with ambiguous genitalia. The standard medical response to such a child is to assign the child a sex as soon as possible, male or female, and to treat the child over a period of time with surgery and hormones to ensure that the child's physical appearance conforms as closely as possible to that of a boy or a girl. Human beings are either male or female, and if they don't look like one or the other, then something must be medically wrong with them. Our conceptual system has no room for anything in between.

But things need not necessarily be this way. Contrast, for example, our Western attitudes towards intersexuals with those of 1930s Navaho, who didn't think of intersexuals as uncategorizable and in need of medical treatment, as we do, but rather thought of them as blessed by the gods. They were revered, even held in awe. The classic study here was done by the American anthropologist Walter Hill (and later made famous by Clifford Geertz). One Navaho interviewed by Hill in his study tells him, "(Intersexuals) know everything. They can do the work of both a man and a woman." "They are responsible for all the wealth in the country," says another. "If there were no more left, the horses, sheep and Navaho would all go." The Navaho of the 1930s made intersexuals the heads of the family and gave them control over family property. For them, the idea of surgically fixing an intersexed infant would seem strange, even morally objectionable. It certainly would not be treating an illness.

My broader point here is that like intersexuality, our understandings of illness, personality, and beauty are culturally located in particular places. Our current understandings are probably not going to look the same to someone who is not immersed in our culture, and they probably won't look the same to us in fifty years. This is not to say that we can easily change these cultural understandings: we can't just extract ourselves from our circumstances and see the world like the Navaho of the 1930s. An intersexed child has to live in our society, and so do children who are short, or shy, or heavy. Even so, the realization that our own contemporary understanding of the world is not fixed and immutable should make us cautious about embracing new enhancement technologies, and especially about embracing them so readily.

A sixth problem, related to the concern about social justice, is the *problem of competition*. This problem is most evident in the use of performance-enhancing drugs in sports. Performance-enhancing drugs give some athletes a competitive advantage, and even athletes who would rather not use performance-enhancing drugs feel forced to use them anyway, for fear that they will not otherwise be able to compete. This kind of fear may be most prevalent where the competition is explicit, such as a football game or a race, but some people have the same fear even when the competition is more subtle, such as cognitive performance at school or at work. For example, stimulants such as Ritalin (methylphenidate) are widely prescribed for children with attention deficit hyperactivity disorder. They make these children less distractible and less impulsive, and they improve their ability to concentrate. But studies have also shown that stimulants will improve even a normal person's ability to concentrate. If people on stimulants can work longer and more effectively than they otherwise could, does this give them a competitive advantage in the classroom?

This may be a problem even with relatively harmless medications. Take beta blockers, for example. Beta blockers are used to treat high blood pressure, and they work by blocking the effects of the sympathetic nervous system. For this very reason, they are also useful for performers who get stage fright, such as musicians, actors, and other people who have to do a lot of public speaking. Some performers get very nervous on the stage, and their voices tremble, their faces flush, their hearts race, and their palms sweat. These are exactly the reactions that beta blockers prevent. Unlike Valium or other psychoactive drugs, beta blockers do not make the people who take them any less anxious or nervous, or at least not directly. They just block the outward effects of a person's nervousness. They give people a mask of relaxation—which, as it happens, often makes them much more relaxed. Since a performer's greatest fear is often that his or her anxiety will be obvious to the audience, a drug that masks that anxiety can be very reassuring.

Some people would see beta blockers as a harmless enhancement technology. They are safe, effective, and have few (if any) long-term effects. They don't affect the mind, they don't change the personality, and they wear off in a couple of hours. People who take them usually can't even feel the difference, unless they have to perform on stage. Yet other people would see beta blockers as morally problematic precisely because they give people a competitive advantage. If you are a graduate student who sees the classroom as a competition, you may resent a fellow student who uses beta blockers for her class presentations. If you are a musician competing for a place in the orchestra, you may well resent the violinist who medicates herself before her solos.

The final worry about enhancement technologies I will mention is perhaps the hardest to pin down. This is a worry about what the political theorist Michael Sandel calls "the drive to mastery." It is hardest to pin down because it is less about the possible consequences of enhancement technologies than about the sensibility they reflect—an attitude that views the world as something to be manipulated, mastered, and controlled. When people charge that scientists or doctors are "playing God," at least some of them are objecting to the lack of humility entailed by this sensibility. They object to the arrogance of placing such extraordinary faith in human reason. As Leon Kass puts it, the objection is not so much a matter of attempting to do what ought to be left to God, but doing so in the absence of Godlike wisdom.

This kind of worry goes beyond conventional concerns about justice. As Sandel has pointed out, it is possible to imagine a world in which all athletes have equal access to safe, performance-enhancing drugs; in which we are allowed to choose the sex of our children without creating a gender imbalance; and in which we eat factory-farmed pigs and chickens genetically engineered not to feel pain. Yet many of us would resist such a world. And the reason we would resist is not because such a world would be unjust, or even because it would lead to a world with more pain and suffering, but because of the extent to which it has been planned and engineered. We would resist the idea that the whole world is there to be manipulated for human ends.

25

Twenty-First Century Bodies

Ray Kurzweil

Humankind's first tools were found objects: sticks used to dig up roots and stones used to break open nuts. It took our forebears tens of thousands of years to invent a sharp blade. Today we build machines with finely designed intricate mechanisms, but viewed on an atomic scale, our technology is still crude. "Casting, grinding, milling, and even lithography move atoms in great thundering statistical herds," says Ralph Merkle, a leading nanotechnology theorist at Xerox's Palo Alto Research Center. He adds that current manufacturing methods are "like trying to make things out of Legos with boxing gloves on. . . . In the future, nanotechnology will let us take off the boxing gloves."[1]

Nanotechnology is technology built on the atomic level: building machines one atom at a time. "Nano" refers to a billionth of a meter, which is the width of five carbon atoms. We have one existence proof of the feasibility of nanotechnology: life on Earth. Little machines in our cells called ribosomes build organisms such as humans one molecule, that is, one amino acid, at a time, following digital templates coded in another molecule, called DNA. Life on Earth has mastered the ultimate goal of nanotechnology, which is self-replication.

But as mentioned above, Earthly life is limited by the particular molecular building block it has selected. Just as our human-created computational technology will ultimately exceed the capacity of natural computation (electronic circuits are already millions of times faster than human neural circuits), our twenty-first-century physical technology will also greatly exceed the capabilities of the amino acid–based nanotechnology of the natural world.

The concept of building machines atom by atom was first described in a 1959 talk at Cal Tech titled "There's Plenty of Room at the Bottom," by physicist Richard Feynman, the same guy who first suggested the possibility of quantum computing.[2] The idea was developed in some detail by Eric Drexler twenty years later in his book *Engines of Creation*.[3] The book actually inspired the cryonics movement of the 1980s, in which people had their heads (with or without bodies) frozen in the hope that a future time would possess the molecule-scale

---

technology to overcome their mortal diseases, as well as undo the effects of freezing and defrosting. Whether a future generation would be motivated to revive all these frozen brains was another matter.

After publication of *Engines of Creation*, the response to Drexler's ideas was skeptical and he had difficulty filling out his MIT Ph.D. committee despite Marvin Minsky's agreement to supervise it. Drexler's dissertation, publised in 1992 as a book titled *Nanosystems: Molecular Machinery, Manufacturing, and Computation,* provided a comprehensive proof of concept, including detailed analyses and specific designs.[4] A year later, the first nanotechnology conference attracted only a few dozen researchers. The fifth annual conference, held in December 1997, boasted 350 scientists who were far more confident of the practicality of their tiny projects. Nanothinc, an industry think tank, estimated in 1997 that the field already produces $5 billion in annual revenues for nanotechnology-related technologies, including micromachines, microfabrication techniques, nanolithography, nanoscale microscopes, and others. This figure has been more than doubling each year.[5]

THE AGE OF NANOTUBES

One key building material for tiny machines is, again, nanotubes. Although built on an atomic scale, the hexagonal patterns of carbon atoms are extremely strong and durable. "You can do anything you damn well want with these tubes and they'll just keep on truckin'," says Richard Smalley, one of the chemists who received the Nobel Prize for discovering the buckyball molecule.[6] A car made of nanotubes would be stronger and more stable than a car made with steel, but would weigh only fifty pounds. A spacecraft made of nanotubes could be of the size and strength of the U.S. space shuttle, but weigh no more than a conventional car. Nanotubes handle heat extremely well, far better than the fragile amino acids that people are built out of. They can be assembled into all kinds of shapes: wirelike strands, sturdy girders, gears, etcetera. Nanotubes are formed of carbon atoms, which are in plentiful supply in the natural world.

As I mentioned earlier, the same nanotubes can be used for extremely efficient computation, so both the structural and computational technology of the twenty-first century will likely be constructed from the same stuff. In fact, the same nanotubes used to form physical structures can also be used for computation, so future nanomachines can have their brains distributed throughout their bodies.

The best-known examples of nanotechnology to date, while not altogether practical, are beginning to show the feasibility of engineering at the atomic level. IBM created its corporate logo using individual atoms as pixels.[7] In 1996, Texas Instruments built a chip-sized device with half a million moveable mirrors to be used in a tiny high-resolution projector.[8] TI sold $100 million worth of their nanomirrors in 1997.

Chih-Ming Ho of UCLA is designing flying machines using surfaces covered with microflaps that control the flow of air in a similar manner to conventional flaps on a normal airplane.[9] Andrew Berlin at Xerox's Palo Alto Research Center is designing a printer using microscopic air valves to move paper documents precisely.[10]

Cornell graduate student and rock musician Dustin Carr built a realistic-looking but microscopic guitar with strings only fifty nanometers in diameter. Carr's creation is a fully functional musical instrument, but his fingers are too large to play it. Besides, the strings vibrate at 10 million vibrations per second, far beyond the twenty-thousand-cycles-per-second limit of human hearing.[11]

THE HOLY GRAIL OF SELF-REPLICATION:
LITTLE FINGERS AND A LITTLE INTELLIGENCE

Tiny fingers represent something of a holy grail for nanotechnologists. With little fingers and computation, nanomachines would have in their Lilliputian world what people have in the big world: intelligence and the ability to manipulate their environment. Then these little machines could build replicas of themselves, achieving the field's key objective.

The reason that self-replication is important is that it is too expensive to build these tiny machines one at a time. To be effective, nanometer-sized machines need to come in the trillions. The only way to achieve this economically is through combinatorial explosion: let the machines build themselves.

Drexler, Merkle (a coinventor of public key encryption, the primary method of encrypting messages), and others have convincingly described how such a self-replicating nanorobot—*nanobot*—could be constructed. The trick is to provide the nanobot with sufficiently flexible manipulators—arms and hands—so that it is capable of building a copy of itself. It needs some means for mobility so that it can find the requisite raw materials. It requires some intelligence so that it can solve the little problems that will arise when each nanobot goes about building a complicated little machine like itself. *Finally, a really important requirement is that it needs to know when to stop replicating.*

MORPHING IN THE REAL WORLD

Self-replicating machines built at the atomic level could truly transform the world we live in. They could build extremely inexpensive solar cells, allowing the replacement of messy fossil fuels. Since solar cells require a large surface area to collect sufficient sunlight, they could be placed in orbit, with the energy beamed down to Earth.

Nanobots launched into our bloodstreams could supplement our natural immune system and seek out and destroy pathogens, cancer cells, arterial plaque, and other disease agents. In the vision that inspired the cryonics enthusiasts, diseased organs can be rebuilt. We will be able to reconstruct any or all of our bodily organs and systems, and do so at the cellular level. I talked in the last chapter about reverse engineering and emulating the salient computational functionality of human neurons. In the same way, it will become possible to reverse engineer and replicate the physical and chemical functionality of any human cell. In the process we will be in a position to greatly extend the durability, strength, temperature range, and other qualities and capabilities of our cellular building blocks.

We will then be able to grow stronger, more capable organs by redesigning the cells that constitute them and building them with far more versatile and durable materials. As we go down this road, we'll find that some redesign of the body makes sense at multiple levels. For example, if our cells are no longer vulnerable to the conventional pathogens, we may not need the same kind of immune system. But we will need new nanoengineered protections for a new assortment of nanopathogens.

Food, clothing, diamond rings, buildings could all assemble themselves molecule by molecule. Any sort of product could be instantly created when and where we need it. Indeed, the world could continually reassemble itself to meet our changing needs, desires, and fantasies. By the late twenty-first century, nanotechnology will permit objects such as furniture, buildings, clothing, even people, to change their appearance and other characteristics—essentially to change into something else—in a split second.

These technologies will emerge gradually. There is a clear incentive to go down this path. Given a choice, people will prefer to keep their bones from crumbling, their skin supple, their life systems strong and vital. Improving our lives through neural implants on the mental level, and nanotechnology-enhanced bodies on the physical level, will be popular and compelling. It is another one of those slippery slopes—there is no obvious place to stop this progression until the human race has largely replaced the brains and bodies that evolution first provided.

A CLEAR AND FUTURE DANGER

Without self-replication, nanotechnology is neither practical nor economically feasible. And therein lies the rub. What happens if a little software problem (inadvertent or otherwise) fails to halt the self-replication? We may have more nanobots than we want. They could eat up everything in sight.

The movie *The Blob* (of which there are two versions) was a vision of nanotechnology run amok. The movie's villain was this intelligent self-replicating gluttonous stuff that fed on organic matter. Recall that nanotechnology is likely to be built from carbon-based nanotubes, so, like the Blob, it will build itself from organic matter, which is rich in carbon. Unlike mere animal-based cancers, an exponentially exploding nanomachine population would feed on any carbon-based matter. Tracking down all of these bad nanointelligences would be like trying to find trillions of microscopic needles—rapidly moving ones at that—in at least as many haystacks. There have been proposals for nanoscale immunity technologies: good little antibody machines that would go after the bad little machines. The nanoantibodies would, of course, have to scale up at least as quickly as the epidemic of marauding nanomiscreants. There could be a lot of collateral damage as these trillions of machines battle it out.

Now that I have raised this specter, I will try, unconvincingly perhaps, to put the peril in perspective. I believe that it will be possible to engineer self-replicating nanobots in such a way that an *inadvertent*, undesired population explosion would be unlikely. I realize that this may not be completely reassuring, coming from a software developer whose products (like those of my competitors) crash once in a while (but rarely—and when they do, it's the fault of the operating system!). There is a concept in software development of "mission critical" applications. These are software programs that control a process on which people are heavily dependent. Examples of mission-critical software include life-support systems in hospitals, automated surgical equipment, autopilot flying and landing systems, and other software-based systems that affect the well-being of a person or organization. It is feasible to create extremely high levels of reliability in these programs. There are examples of complex technology in use today in which a mishap would severely imperil public safety. A conventional explosion in an atomic power plant could spray deadly plutonium across heavily populated areas. Despite a near meltdown at Chernobyl, this apparently has only occurred twice in the decades that we have had hundreds of such plants operating, both incidents involving recently acknowledged reactor calamities in the Chelyabinsk region of Russia.[12] There are tens of thousands of nuclear weapons, and none has ever exploded in error.

I admit that the above paragraph is not entirely convincing. But the bigger danger is the intentional hostile use of nanotechnology. Once the basic technology is available, it would not be difficult to adapt it as an instrument of war or terrorism. It is not the case that someone would have to be suicidal to use such weapons. The nanoweapons could easily be programmed to replicate only against an enemy; for example, only in a particular geographical area. Nuclear weapons, for all their destructive potential, are at least relatively local in their effects. The self-replicating nature of nanotechnology makes it a far greater danger.

VIRTUAL BODIES

We don't always need real bodies. If we happen to be in a virtual environment, then a virtual body will do just fine. Virtual reality started with the concept of computer games, particularly ones that provided a simulated environment. The first was Space War, written by early artificial-intelligence researchers to pass the time while waiting for programs to compile on their slow 1960s computers.[13] The synthetic space surroundings were easy to render on low-resolution monitors: Stars and other space objects were just illuminated pixels.

Computer games and computerized video games have become more realistic over time, but you cannot completely immerse yourself in these imagined worlds, not without some imagination. For one thing, you can see the edges of the screen, and the all too real world that you have never left is still visible beyond these borders.

If we're going to enter a new world, we had better get rid of traces of the old. In the 1990s the first generation of virtual reality has been introduced in which you don a special visual helmet that takes over your entire visual field. The key to visual reality is that when you move your head, the scene instantly repositions itself so that you are now looking at a different region of a three-dimensional scene. The intention is to simulate what happens when you turn your real head in the real world: The images captured by your retinas rapidly change. Your brain nonetheless understands that the world has remained stationary and that the image is sliding across your retinas only because your head is rotating.

Like most first generation technologies, virtual reality has not been fully convincing. Because rendering a new scene requires a lot of computation, there is a lag in producing the new perspective. Any noticeable delay tips off your brain that the world you're looking at is not entirely real. The resolution of virtual reality displays has also been inadequate to create a fully satisfactory illusion. Finally, contemporary virtual reality helmets are bulky and uncomfortable.

What's needed to remove the rendering delay and to boost display resolution is yet faster computers, which we know are always on the way. By 2007, high-quality virtual reality with convincing artificial environments, virtually instantaneous rendering, and high-definition displays will be comfortable to wear and available at computer game prices.

That takes care of two of our senses—visual and auditory. Another high-resolution sense organ is our skin, and "haptic" interfaces to provide a virtual tactile interface are also evolving. One available today is the Microsoft force-feedback joystick, derived from 1980s research at the MIT Media Lab. A force-feedback joystick adds some tactile realism to computer games, so you feel the rumble of the road in a car-driving game or the pull of the line in a fishing simulation. Emerging in late 1998 is the "tactile mouse," which operates like a conventional mouse but allows the user to feel the texture of surfaces, objects, even people. One company that I am involved in, Medical Learning Company, is developing a simulated patient to help train doctors, as well as enable nonphysicians to play doctor. It will include a haptic interface so that you can feel a knee joint for a fracture or a breast for lumps.[14]

A force-feedback joystick in the tactile domain is comparable to conventional monitors in the visual domain. The force-feedback joystick provides a tactile interface, but it does not totally envelop you. The rest of your tactile world is still reminding you of its presence. In order to leave the real world, at least temporarily, we need a tactile environment that takes over your sense of touch.

So let's invent a virtual tactile environment. We've seen aspects of it in science fiction films (always a good source for inventing the future). We can build a body suit that will detect your own movements as well as provide high resolution tactile stimulation. The suit will also need to provide sufficient force-feedback to actually prevent your movements if you are pressing against a virtual obstacle in the virtual environment. If you are giving a virtual companion a hug, for example, you don't want to move right through his or her body. This will require a force-feedback structure outside the suit, although obstacle resistance could be provided by the suit itself. And since your body inside the suit is still in the real world, it would make sense to put the whole contraption in a booth so that your movements in the virtual world don't knock down lamps and people in your "real" vicinity. Such a suit could also provide a thermal response and thereby allow the simulation of feeling a moist surface—or even immersing your hand or your whole body in water—which is indicated by a change in temperature and a decrease in surface tension. Finally, we can provide a platform consisting of a rotating treadmill device for you to stand (or sit or lie) on, which will allow you to walk or move around (in any direction) in your virtual environment.

So with the suit, the outer structure, the booth, the platform, the goggles, and the earphones, we just about have the means to totally envelop your senses. Of course, we will need some good virtual reality software, but there's certain to be hot competition to provide a panoply of realistic and fantastic new environments as the requisite hardware becomes available.

Oh yes, there is the sense of smell. A completely flexible and general interface for our fourth sense will require a reasonably advanced nanotechnology to synthesize the wide variety of molecules that we can detect with our olfactory sense. In the meantime, we could provide the ability to diffuse a variety of aromas in the virtual reality booth.

Once we are in a virtual reality environment, our own bodies—at least the virtual versions—can change as well. We can become a more attractive version of ourselves, a hideous beast, or any creature real or imagined as we interact with the other inhabitants in each virtual world we enter.

Virtual reality is not a (virtual) place you need go to alone. You can interact with your friends there (who would be in other virtual reality booths, which may be geographically remote). You will have plenty of simulated companions to choose from as well.

DIRECTLY PLUGGING IN

Later in the twenty-first century, as neural implant technologies become ubiquitous, we will be able to create and interact with virtual environments without having to enter a virtual reality booth. Your neural implants will provide the simulated sensory inputs of the virtual environment—and your virtual body—directly in your brain. Conversely, your movements would not move your "real" body, but rather your perceived virtual body. These virtual environments would also include a suitable selection of bodies for yourself. Ultimately, your experience would be highly realistic, just like being in the real world. More than one person could enter a virtual environment and interact with each other. In the virtual world, you will meet other real people and simulated people—eventually, there won't be much difference.

This will be the essence of the Web in the second half of the twenty-first century. A typical "web site" will be a perceived virtual environment, with no external hardware required. You "go there" by mentally selecting the site and then entering that world. Debate Benjamin Franklin on the war powers of the presidency at the history society site. Ski the Alps at the Swiss Chamber of Commerce site (while feeling the cold spray of snow on your face). Hug your favorite movie star at the Columbia Pictures site. Get a little more intimate at the Penthouse or Playgirl site. Of course, there may be a small charge.

REAL VIRTUAL REALITY

In the late twenty-first century, the "real" world will take on many of the characteristics of the virtual world through the means of nanotechnology "swarms." Consider, for example, Rutgers University computer scientist J. Storrs Hall's concept of "Utility Fog."[15] Hall's conception starts with a little robot called a Foglet, which consists of a human-cell-sized device with twelve arms pointing in all directions. At the end of the arms are grippers so that the Foglets can grasp one another to form larger structures. These nanobots are intelligent and can merge their computational capacities with each other to create a distributed intelligence. A space filled with Foglets is called Utility Fog and has some interesting properties.

First of all, the Utility Fog goes to a lot of trouble to simulate its not being there. Hall describes a detailed scenario that lets a real human walk through a room filled with trillions of Foglets and not notice a thing. When desired (and it's not entirely clear who is doing the desiring), the Foglets can quickly simulate any environment by creating all sorts of structures. As Hall puts it, "Fog city can look like a park, or a forest, or ancient Rome one day and Emerald City the next."

The Foglets can create arbitrary wave fronts of light and sound in any direction to create any imaginary visual and auditory environment. They can exert any pattern of pressure to create any tactile environment. In this way, Utility Fog has all the flexibility of a virtual environment, except it exists in the real physical world. The distributed intelligence of the Utility Fog can simulate the minds of scanned (Hall calls them "uploaded") people who are re-created in the Utility Fog as "Fog people." In Hall's scenario, "a biological human can walk through Fog walls, and a Fog (uploaded) human can walk through dumb-matter walls. Of course Fog people can walk through Fog walls, too."

The physical technology of Utility Fog is actually rather conservative. The Foglets are much bigger machines than most nanotechnology conceptions. The software is more challenging, but ultimately feasible. Hall needs a bit of work on his marketing angle: Utility Fog is a rather dull name for such versatile stuff.

There are a variety of proposals for nanotechnology swarms, in which the real environment is constructed from interacting multitudes of nanomachines. In all of the swarm conceptions, physical reality becomes a lot like virtual reality. You can be sleeping in your bed one moment, and have the room transform into your kitchen as you awake. Actually, change that to a dining room as there's no need for a kitchen. Related nanotechnology will instantly create whatever meal you desire. When you finish eating, the room can transform into a study, or a game room, or a swimming pool, or a redwood forest, or the Taj Mahal. You get the idea.

Mark Yim has built a large-scale model of a small swarm showing the feasibility of swarm interaction.[16] Joseph Michael has actually received a U.K. patent on his conception of a nanotechnology swarm, but it is unlikely that his design will be commercially realizable in the twenty-year life of his patent.[17]

It may seem that we will have too many choices. Today, we have only to choose our clothes, makeup, and destination when we go out. In the late twenty-first century, we will have to select our body, our personality, our environment—so many difficult decisions to make! But don't worry—we'll have intelligent swarms of machines to guide us.

THE SENSUAL MACHINE

Made double by his lust
he sounds a woman's groans.
A figment of his flesh.

—from Barry Spacks's poem "The Solitary at Seventeen"

I can predict the future by assuming that money and male hormones are the driving
forces for new technology. Therefore, when virtual reality gets cheaper than dating,
society is doomed.

—Dogbert

The first book printed from a moveable type press may have been the Bible, but the century following Gutenberg's epochal invention saw a lucrative market for books with more prurient topics.[18] New communication technologies—the telephone, motion pictures, television, video-tape—have always been quick to adopt sexual themes. The Internet is no exception, with 1998 market estimates of adult online entertainment ranging from $185 million by Forrester Research to $1 billion by Inter@active Week. These figures are for customers, mostly men, paying to view and interact with performers—live, recorded, and simulated. One 1998 estimate cited 28,000 web sites that offer sexual entertainment.[19] These figures do not include couples who have expanded their phone sex to include moving pictures via online video conferencing.

CD-ROMs and DVD disks constitute another technology that has been exploited for erotic entertainment. Although the bulk of adult-oriented disks are used as a means for delivering videos with a bit of interactivity thrown in, a new genre of CD-ROM and DVD provides virtual sexual companions that respond to some mouse-administered fondling.[20] Like most first-generation technologies, the effect is less than convincing, but future generations will eliminate some of the kinks, although not the kinkiness. Developers are also working to exploit the force-feedback mouse so that you can get some sense of what your virtual partner feels like.

Late in the first decade of the twenty-first century, virtual reality will enable you to be with your lover—romantic partner, sex worker, or simulated companion—with full visual and auditory realism. You will be able to do anything you want with your companion except touch, admittedly an important limitation.

Virtual touch has already been introduced, but the all-enveloping, highly realistic, visual-auditory-tactile virtual environment will not be perfected until the second decade of the twenty-first century. At this point, virtual sex becomes a viable competitor to the real thing. Couples will be able to engage in virtual sex regardless of their physical proximity. Even when proximate, virtual sex will be better in some ways and certainly safer. Virtual sex will provide sensations that are more intense and pleasurable than conventional sex, as well as physical experiences that currently do not exist. Virtual sex is also the ultimate in safe sex, as there is no risk of pregnancy or transmission of disease.

Today, lovers may fantasize their partners to be someone else, but users of virtual sex communication will not need as much imagination. You will be able to change the physical appearance and other characteristics of both yourself and your partner. You can make your lover look and feel like your favorite star without your partner's permission or knowledge. Of course, be aware that your partner may be doing the same to you.

Group sex will take on a new meaning in that more than one person can simultaneously share the experience of one partner. Since multiple real people cannot all control the movements of one virtual partner, there needs to be a way of sharing the decision making of what the one virtual body is doing. Each participant sharing a virtual body would have the same visual, auditory, and tactile experience, with shared control of their shared virtual body (perhaps the one virtual body will reflect a consensus of the attempted movements of the multiple participants). A whole audience of people—who may be geographically dispersed—could share one virtual body while engaged in a sexual experience with one performer.

Prostitution will be free of health risks, as will virtual sex in general. Using wireless, very-high-bandwidth communication technologies, neither sex workers nor their patrons need leave their homes. Virtual prostitution is likely to be legally tolerated, at least to a far greater extent than real prostitution is today, as the virtual variety will be impossible to monitor or control. With the risks of disease and violence having been eliminated, there will be far less rationale for proscribing it.

Sex workers will have competition from simulated—computer generated—partners. In the early stages, "real" human virtual partners are likely to be more realistic than simulated virtual partners, but that will change over time. Of course, once the simulated virtual partner is as capable, sensual, and responsive as a real human virtual partner, who's to say that the simulated virtual partner isn't a real, albeit virtual, person?

Is virtual rape possible? In the purely physical sense, probably not. Virtual reality will have a means for users to immediately terminate their experience. Emotional and other means of persuasion and pressure are another matter.

How will such an extensive array of sexual choices and opportunities affect the institution of marriage and the concept of commitment in a relationship? The technology of virtual sex will introduce an array of slippery slopes, and the definition of a monogamous relationship will become far less clear. Some people will feel that access to intense sexual experiences at the click of a mental button will destroy the concept of a sexually committed relationship. Others will argue, as proponents of sexual entertainment and services do today, that such diversions are healthy outlets and serve to maintain healthy relationships. Clearly, couples will need to reach their own understandings, but drawing clear lines will become difficult with the level of privacy that this future technology affords. It is likely that society will accept practices and activities in the virtual arena that it frowns on in the physical world, as the consequences of virtual activities are often (although not always) easier to undo.

In addition to direct sensual and sexual contact, virtual reality will be a great place for romance in general. Stroll with your lover along a virtual Champs-Élysées, take a walk along a virtual Cancún beach, mingle with the animals in a simulated Mozambique game reserve. Your whole relationship can be in Cyberland.

Virtual reality using an external visual-auditory-haptic interface is not the only technology that will transform the nature of sexuality in the twenty-first century. Sexual robots—sexbots—will become popular by the beginning of the third decade of the new century. Today, the idea of intimate relations with a robot or doll is not generally appealing because robots and dolls are so, well, inanimate. But that will change as robots gain the softness, intelligence, pliancy, and passion of their human creators. (By the end of the twenty-first century, there won't be a clear difference between humans and robots. What, after all, is the difference between a human who has upgraded her body and brain using new nanotechnology and computational technologies, and a robot who has gained an intelligence and sensuality surpassing her human creators?)

By the fourth decade, we will move to an era of virtual experiences through internal neural implants. With this technology, you will be able to have almost any kind of experience with just about anyone, real or imagined, at any time. It's just like today's online chat rooms, except that you don't need any equipment that's not already in your head, and you can do a lot more than just chat. You won't be restricted by the limitations of your natural body as you and your partners can take on any virtual physical form. Many new types of experiences will become possible: A man can feel what it is like to be a woman, and vice versa. Indeed, there's no reason why you can't be both at the same time, making real, or at least virtually real, our solitary fantasies.

And then, of course, in the last half of the century, there will be the nanobot swarms—good old sexy Utility Fog, for example. The nanobot swarms can instantly take on any form and emulate any sort of appearance, intelligence, and personality that you or it desires—the human form, say, if that's what turns you on.

THE SPIRITUAL MACHINE

We are not human beings trying to be spiritual. We are spiritual beings trying to be human.

—Jacquelyn Small

Body and soul are twins. God only knows which is which.

—Charles A. Swinburne

We're all lying in the gutter, but some of us are gazing at the stars.

—Oscar Wilde

Sexuality and spirituality are two ways that we transcend our everyday physical reality. Indeed, there are links between our sexual and our spiritual passions, as the ecstatic rhythmic movements associated with some varieties of spiritual experience suggest.

MIND TRIGGERS

We are discovering that the brain can be directly stimulated to experience a wide variety of feelings that we originally thought could be gained only from actual physical or mental experience. Take humor, for example. In the journal *Nature,* Dr. Itzhak Fried and his colleagues at

UCLA tell how they found a neurological trigger for humor. They were looking for possible causes for a teenage girl's epileptic seizures and discovered that applying an electric probe to a specific point in the supplementary motor area of her brain caused her to laugh. Initially, the researchers thought that the laughter must be just an involuntary motor response, but they soon realized they were triggering the genuine perception of humor, not just forced laughter. When stimulated in just the right spot of her brain, she found everything funny. "You guys are just so funny—standing around" was a typical comment.[21]

Triggering a perception of humor without circumstances we normally consider funny is perhaps disconcerting (although personally, I find it humorous). Humor involves a certain element of surprise. Blue elephants. The last two words were intended to be surprising, but they probably didn't make you laugh (or maybe they did). In addition to surprise, the unexpected event needs to make sense from an unanticipated but meaningful perspective. And there are some other attributes that humor requires that we don't understand just yet. The brain apparently has a neural net that detects humor from our other perceptions. If we directly stimulate the brain's humor detector, then an otherwise ordinary situation will seem pretty funny.

The same appears to be true of sexual feelings. In experiments with animals, stimulating a specific small area of the hypothalamus with a tiny injection of testosterone causes the animals to engage in female sexual behavior, regardless of gender. Stimulating a different area of the hypothalamus produces male sexual behavior.

These results suggest that once neural implants are commonplace, we will have the ability to produce not only virtual sensory experiences but also the feelings associated with these experiences. We can also create some feelings not ordinarily associated with the experience. So you will be able to add some humor to your sexual experiences, if desired (of course, for some of us humor may already be part of the picture).

The ability to control and reprogram our feelings will become even more profound in the late twenty-first century when technology moves beyond mere neural implants and we fully install our thinking processes into a new computational medium—that is, *when we become software.*

We work hard to achieve feelings of humor, pleasure, and well-being. Being able to call them up at will may seem to rob them of their meaning. Of course, many people use drugs today to create and enhance certain desirable feelings, but the chemical approach comes bundled with many undesirable effects. With neural implant technology, you will be able to enhance your feelings of pleasure and well-being without the hangover. Of course, the potential for abuse is even greater than with drugs. When psychologist James Olds provided rats with the ability to press a button and directly stimulate a pleasure center in the limbic system of their brains, the rats pressed the button endlessly, as often as five thousand times an hour, to the exclusion of everything else, including eating. Only falling asleep caused them to stop temporarily.[22]

Nonetheless, the benefits of neural implant technology will be compelling. As just one example, millions of people suffer from an inability to experience sufficiently intense feelings of sexual pleasure, which is one important aspect of impotence. People with this disability will not pass up the opportunity to overcome their problem through neural implants, which they may already have in place for other purposes. Once a technology is developed to overcome a disability, there is no way to restrict its use from enhancing normal abilities, nor would such restrictions necessarily be desirable. The ability to control our feelings will be just another one of those twenty-first-century slippery slopes.

SO WHAT ABOUT SPIRITUAL EXPERIENCES?

The spiritual experience—a feeling of transcending one's everyday physical and mortal bounds to sense a deeper reality—plays a fundamental role in otherwise disparate religions and philosophies. Spiritual experiences are not all of the same sort but appear to encompass a broad range of mental phenomena. The ecstatic dancing of a Baptist revival appears to be a different phenomenon from the quiet transcendence of a Buddhist monk. Nonetheless, the notion of the spiritual experience has been reported so consistently throughout history, and in virtually all cultures and religions, that it represents a particularly brilliant flower in the phenomenological garden.

Regardless of the nature and derivation of a mental experience, spiritual or otherwise, once we have access to the computational processes that give rise to it, we have the opportunity to understand its neurological correlates. With the understanding of our mental processes will come the opportunity to capture our intellectual, emotional, and spiritual experiences, to call them up at will, and to enhance them.

SPIRITUAL EXPERIENCE THROUGH BRAIN GENERATED MUSIC

There is already one technology that appears to generate at least one of aspect of a spiritual experience. This experimental technology is called Brain Generated Music (BGM), pioneered by NeuroSonics, a small company in Baltimore, Maryland, of which I am a director. BGM is a brain-wave biofeedback system capable of evoking an experience called the Relaxation Response, which is associated with deep relaxation.[23] The BGM user attaches three disposable leads to her head. A personal computer then monitors the user's brain waves to determine her unique alpha wavelength. Alpha waves, which are in the range of eight to thirteen cycles per second (cps), are associated with a deep meditative state, as compared to beta waves (in the range of thirteen to twenty-eight cps), which are associated with routine conscious thought. Music is then generated by the computer, according to an algorithm that transforms the user's own brain-wave signal.

The BGM algorithm is designed to encourage the generation of alpha waves by producing pleasurable harmonic combinations upon detection of alpha waves, and less pleasant sounds and sound combinations when alpha detection is low. In addition, the fact that the sounds are synchronized to the user's own alpha wavelength to create a resonance with the user's own alpha rhythm also encourages alpha production.

Dr. Herbert Benson, formerly the director of the hypertension section of Boston's Beth Israel Hospital and now at New England Deaconess Hospital in Boston, and other researchers at the Harvard Medical School and Beth Israel, discovered the neurological-physiological mechanism of the Relaxation Response, which is described as the opposite of the "fight or flight," or stress response.[24] The Relaxation Response is associated with reduced levels of epinephrine (adrenaline) and norepinephrine (noradrenaline), blood pressure, blood sugar, breathing, and heart rates. Regular elicitation of this response is reportedly able to produce permanently lowered blood-pressure levels (to the extent that hypertension is caused by stress factors) and other health benefits. Benson and his colleagues have catalogued a number of techniques that can elicit the Relaxation Response, including yoga and a number of forms of meditation.

I have had experience with meditation, and in my own experience with BGM, and in observing others, BGM does appear to evoke the Relaxation Response. The music itself feels as if it is being generated from inside your mind. Interestingly, if you listen to a tape recording of your own brain-generated music when you are not hooked up to the computer, you do not experience the same sense of transcendence. Although the recorded BGM is based on your personal alpha wavelength, the recorded music was synchronized to the brain waves that were produced by your brain when the music was first generated, not to the brain waves that are produced while listening to the recording. You need to listen to "live" BGM to achieve the resonance effect.

Conventional music is generally a passive experience. Although a performer may be influenced in subtle ways by her audience, the music we listen to generally does not reflect our response. Brain Generated Music represents a new modality of music that enables the music to evolve continually based on the interaction between it and our own mental responses to it.

Is BGM producing a spiritual experience? It's hard to say. The feelings produced while listening to "live" BGM are similar to the deep transcendent feelings I can sometimes achieve with meditation, but they appear to be more reliably produced by BGM.

THE GOD SPOT

Neuroscientists from the University of California at San Diego have found what they call the God module, a tiny locus of nerve cells in the frontal lobe that appears to be activated during religious experiences. They discovered this neural machinery while studying epileptic patients who have intense mystical experiences during seizures. Apparently the intense neural storms during a seizure stimulate the God module. Tracking surface electrical activity in the brain with highly sensitive skin monitors, the scientists found a similar response when very religious nonepileptic persons were shown words and symbols evoking their spiritual beliefs.

A neurological basis for spiritual experience has long been postulated by evolutionary biologists because of the social utility of religious belief. In response to reports of the San Diego research, Richard Harries, the Bishop of Oxford, said through a spokesman that "it would not be surprising if God had created us with a physical facility for belief."[25]

When we can determine the neurological correlates of the variety of spiritual experiences that our species is capable of, we are likely to be able to enhance these experiences in the same way that we will enhance other human experiences. With the next stage of evolution creating a new generation of humans that will be trillions of times more capable and complex than humans today, our ability for spiritual experience and insight is also likely to gain in power and depth.

Just being—experiencing, being conscious—is spiritual, and reflects the essence of spirituality. Machines, derived from human thinking and surpassing humans in their capacity for experience, will claim to be conscious, and thus to be spiritual. They will believe that they are conscious. They will believe that they have spiritual experiences. They will be convinced that these experiences are meaningful. And given the historical inclination of the human race to anthropomorphize the phenomena we encounter, and the persuasiveness of the machines, we're likely to believe them when they tell us this.

Twenty-first-century machines—based on the design of human thinking—will do as their human progenitors have done—going to real and virtual houses of worship, meditating, praying, and transcending—to connect with their spiritual dimension.

NOTES

1. Ralph Merkle's comments on nanotechnology can be found in an overview at his web site at the Xerox Palo Alto Research Center <http://sandbox.xerox.com/nano>. His site contains links to important publications on nanotechnology, such as Richard Feynman's 1959 talk and Eric Drexler's dissertation, as well as links to various research centers that focus on nanotechnology.

2. Richard Feynman presented these ideas on December 29, 1959, at the annual meeting of the American Physical Society at the California Institute of Technology (Cal Tech). His talk was first published in the February 1960 issue of Cal Tech's *Engineering and Science.* This article is available online at <http://nano.xerox .com/nanotech/feynman.html>.

3. Eric Drexler, *Engines of Creation* (New York: Anchor Press/Doubleday, 1986). The book is also accessible online from the Xerox nanotechnology site <http://sandbox.xerox.com/nano> and also from Drexler's web site at the Foresight Institute<http://www.foresight.org/EOC/.index.html>.

4. Eric Drexler, *Nanosystems: Molecular Machinery, Manufacturing, and Computation* (New York: John Wiley and Sons, 1992).

5. According to Nanothinc's web site <http://www.nanothinc.com/>, "Nanotechnology, broadly defined to include a number of nanoscale-related activities and disciplines, is a global industry in which more than 300 companies generate over $5 billion in annual revenues today—and $24 billion in 4 years." Nanothinc includes a list of companies and revenues upon which the figure is based. Some of the nanoapplications generating revenues are micromachines, microelectromechanical systems, autofabrication, nanolithography, nanotechnology tools, scanning probe microscopy, software, nanoscale materials, and nanophase materials.

6. Richard Smalley's publications and work on nanotechnology can be found at the web site for the Center for Nanoscale Science and Technology at Rice University <http://cnst.rice.edu/>.

7. For information on the use of nanotechnology in creating IBM's corporate logo, read Faye Flam, "Tiny Instrument Has Big Implications." *Knight-Ridder/Tribune News Service,* August 11, 1997, p. 811K7204.

8. Dr. Jeffrey Sampsell at Texas Instruments has written a white paper summarizing research on micromirrors, available at <http://www.ti.com/dlp/docs/it/resources/white/overview/over.shtml>.

9. A description of the flying machines can be found at the web site of the MEMS (MicroElectroMechanical Systems) and Fluid Dynamics Research Group at the University of California at Los Angeles (UCLA) <http://ho.seas.ucla.edu/new/main.htm>.

10. Xerox's nanotechnology research is described in Brian Santo, "Smart Matter Program Embeds Intelligence by Combining Sensing, Actuation, Computation—Xerox Builds on Sensor Theory for Smart Materials." *EETimes* (March 23, 1998): 129. More information on this research can be found at the web site for the Smart Matter Research Group at Xerox's Palo Alto Research Center at <http://www.parc.xerox.com/spl/ projects/smart-matter/>.

11. For information on the use of nanotechnology in creating the nanoguitar, read Faye Flam, "Tiny Instrument Has Big Implications." *Knight-Ridder/Tribune News Service.*

12. Learn more about the Chelyabinsk region by visiting the web site dedicated to helping the people living in that area at <http://www.logtv.com/chelya/chel.html>.

13. For more about the story behind Space War, see "A History of Computer Games," *Computer Gaming World* (November 1991): 16–26; and Eric S. Raymond, ed., *New Hacker's Dictionary* (Cambridge, MA: MIT Press, 1992). Space War was developed by Steve Russell in 1961 and implemented by him on the PDP-1 at MIT a year later.

14. Medical Learning Company is a joint venture between the American Board of Family Practice (an organization that certifies the sixty thousand family practice physicians in the United States) and Kurzweil Technologies. The goal of the company is to develop educational software for continuing medical education of physicians as well as other markets. A key aspect of the technology will include an interactive simulated patient that can be examined, interviewed, and treated.

15. Hall's Utility Fog concept is described in J. Storrs Hall, "Utility Fog Part 1," *Extropy,* issue no. 13 (vol. 6, no. 2), third quarter 1994; and J. Storrs Hall, "Utility Fog Part 2," *Extropy,* issue no. 14 (vol. 7, no. 1), first quarter 1995. Also see Jim Wilson, "Shrinking Micromachines: A New Generation of Tools Will Make Molecule-Size Machines a Reality." *Popular Mechanics* 174, no. 11 (November 1997): 55–58.

16. Mark Yim, "Locomotion with a Unit-Modular Reconfigurable Robot," Stanford University Technical Report STAN-CS-TR-95–1536.

17. Joseph Michael, UK Patent #94004227.2.

18. For examples of early "prurient" text publications, see *A History of Erotic Literature* by Patrick J. Kearney (Hong Kong, 1982); and *History Laid Bare* by Richard Zachs (New York: HarperCollins, 1994).

19. *Upside Magazine,* April 1998.

20. For example, the "TFUI" (Touch-and-Feel User Interface) from pixis, as used in their Diva and Space Sirens series of CD-ROMs.

21. From "Who Needs Jokes? Brain Has a Ticklish Spot," Malcolme W. Browne, *New York Times,* March 10, 1998. Also see. I. Fried (with C. L. Wilson, K. A. MacDonald, and E. J. Behnke), "Electric Current Stimulates Laughter," *Scientific Correspondence* 391: 650, 1998.

22. K. Blum et al., "Reward Deficiency Syndrome," *American Scientist,* March–April, 1996.

23. Brain Generated Music is a patented technology of NeuroSonics, a small company in Baltimore, Maryland. The founder, CEO, and principal developer of the technology is Dr. Geoff Wright, who is head of computer music at Peabody Institute of Johns Hopkins University.

24. For details about Dr. Benson's work, see his book *The Relaxation Response* (New York: Avon, 1990).

25. "'God Spot' Is Found in Brain," *Sunday Times* (Britain), November 2, 1997.

26

Why Computers May Never Think like People

Hubert and Stuart Dreyfus

Scientists who stand at the forefront of artificial intelligence (AI) have long dreamed of autonomous "thinking" machines that are free of human control. And now they believe we are not far from realizing that dream. As Marvin Minsky, a well-known AI professor at MIT, recently put it: "Today our robots are like toys. They do only the simple things they're programmed to. But clearly they're about to cross the edgeless line past which they'll do the things we are programmed to."

Patrick Winston, Minsky's successor as head of the MIT AI Laboratory, agrees: "Just as the Wright Brothers at Kitty Hawk in 1903 were on the right track to the 747s of today, so artificial intelligence, with its attempt to formalize common-sense understanding, is on the way to fully intelligent machines."

Encouraged by such optimistic pronouncements, the US Department of Defense (DOD) is sinking millions of dollars into developing fully autonomous war machines that will respond to a crisis without human intervention. Business executives are investing in "expert" systems whose wisdom they hope will equal, if not surpass, that of their top managers. And AI entrepreneurs are talking of "intelligent systems" that will perform better than we can—in the home, in the classroom, and at work.

But no matter how many billions of dollars the Defense Department or any other agency invests in AI, there is almost no likelihood that scientists can develop machines capable of making intelligent decisions. After 25 years of research, AI has failed to live up to its promise, and there is no evidence that it ever will. In fact, machine intelligence will probably never replace human intelligence simply because we ourselves are not "thinking machines." Human beings have an intuitive intelligence that "reasoning" machines simply cannot match.

Military and civilian managers may see this obvious shortcoming and refrain from deploying such "logic" machines. However, once various groups have invested vast sums in developing these machines, the temptation to justify this expense by installing questionable AI technologies will be enormous. The dangers of turning over the battlefield completely to machines are obvious. But it would also be a mistake to replace skilled air-traffic controllers, seasoned business managers, and master teachers with computers that cannot come close to their level of expertise. Computers that "teach" and systems that render "expert" business decisions could eventually produce a generation of students and managers who have no faith in their own intuition and expertise.

We wish to stress that we are not Luddites. There are obvious tasks for which computers are appropriate and even indispensable. Computers are more deliberate, more precise, and less prone to exhaustion and error than the most conscientious human being. They can also store, modify, and tap vast files of data more quickly and accurately than humans can. Hence, they can be used as valuable tools in many areas. As word processors and telecommunication devices, for instance, computers are already changing our methods of writing and our notions of collaboration.

However, we believe that trying to capture more sophisticated skills within the realm of electronic circuits—skills involving not only calculation but also judgment—is a dangerously misguided effort and ultimately doomed to failure.

ACQUIRING HUMAN KNOW-HOW

Most of us know how to ride a bicycle. Does that mean we can formulate specific rules to teach someone else how to do it? How would we explain the difference between the feeling of falling over and the sense of being slightly off-balance when turning? And do we really know, until the situation occurs, just what we would do in response to a certain wobbly feeling? No, we don't. Most of us are able to ride a bicycle because we possess something called "know-how," which we have acquired from practice and sometimes painful experience. That know-how is not accessible to us in the form of facts and rules. If it were, we could say we "know that" certain rules produce proficient bicycle riding.

There are innumerable other aspects of daily life that cannot be reduced to "knowing that." Such experiences involve "knowing how." For example, we know how to carry on an appropriate conversation with family, friends, and strangers in a wide variety of contexts—in the office, at a party, and on the street. We know how to walk. Yet the mechanics of walking on two legs are so complex that the best engineers cannot come close to reproducing them in artificial devices.

This kind of know-how is not innate, as is a bird's skill at building a nest. We have to learn it. Small children learn through trial and error, often by imitating those who are proficient. As adults acquire a skill through instruction and experience, they do not appear to leap suddenly from "knowing that"—a knowledge guided by rules—to experience-based know-how. Instead, people usually pass through five levels of skill: novice, advanced beginner, competent, proficient, and expert. Only when we understand this dynamic process can we ask how far the computer could reasonably progress.

During the novice stage, people learn facts relevant to a particular skill and rules for action that are based on those facts. For instance, car drivers learning to operate a stick shift are told at what speed to shift gears and at what distance—given a particular speed—to follow other cars. These rules ignore context, such as the density of traffic or the number of stops a driver has to make.

Similarly, novice chess players learn a formula for assigning pieces point values independent of their position. They learn the rule: "Always exchange your pieces for the opponent's if the total value of the pieces captured exceeds that of pieces lost." Novices generally do not know that they should violate this rule in certain situations.

After much experience in real situations, novices reach the advanced-beginner stage. Advanced-beginner drivers pay attention to situational elements, which cannot be defined objectively. For instance, they listen to engine sounds when shifting gears. They can also distinguish between the behavior of a distracted or drunken driver and that of the impatient but alert driver. Advanced-beginner chess players recognize and avoid overextended positions. They can also spot situational clues such as a weakened king's side or a strong pawn structure. In all these cases, experience is immeasurably more important than any form of verbal description.

Like the training wheels on a child's first bicycle, initial rules allow beginners to accumulate experience. But soon they must put the rules aside to proceed. For example, at the competent stage, drivers no longer merely follow rules; they drive with a goal in mind. If they wish to get from point A to point B very quickly, they choose their route with an eye to traffic but not much attention to passenger comfort. They follow other cars more closely than they are "supposed" to, enter traffic more daringly, and even break the law. Competent chess players may decide, after weighing alternatives, that they can attack their opponent's king. Removing pieces that defend the enemy king becomes their overriding objective, and to reach it these players will ignore the lessons they learned as beginners and accept some personal losses.

A crucial difference between beginners and more competent performers is their level of involvement. Novices and advanced beginners feel little responsibility for what they do because they are only applying learned rules; if they foul up, they blame the rules instead of themselves. But competent performers, who choose a goal and a plan for achieving it, feel responsible for the result of their choices. A successful outcome is deeply satisfying and leaves a vivid memory. Likewise, disasters are not easily forgotten.

THE INTUITION OF EXPERTS

The learner of a new skill makes conscious choices after reflecting on various options. Yet in our everyday behavior, this model of decision-making—the detached, deliberate, and sometimes agonizing selection among alternatives—is the exception rather than the rule, Proficient performers do not rely on detached deliberation in going about their tasks. Instead, memories of similar experiences in the past seem to trigger plans like those that worked before. Proficient performers recall whole situations from the past and apply them to the present without breaking them down into components or rules.

For instance, a boxer seems to recognize the moment to begin an attack not by following rules and combining various facts about his body's position and that of his opponent. Rather, the whole visual scene triggers the memory of similar earlier situations in which an attack was successful. The boxer is using his intuition, or know-how.

Intuition should not be confused with the re-enactment of childhood patterns or any of the other unconscious means by which human beings come to decisions. Nor is guessing what we mean by intuition. To guess is to reach a conclusion when one does not have enough knowledge or experience to do so. Intuition or know-how is the sort of ability that we use all the time as we go about our everyday tasks. Ironically, it is an ability that our tradition has acknowledged only in women and judged inferior to masculine rationality.

While using their intuition, proficient performers still find themselves thinking analytically about what to do. For instance, when proficient drivers approach a curve on a rainy day, they may intuitively realize they are going too fast. They then consciously decide whether to apply the brakes, remove their foot from the accelerator, or merely reduce pressure on the accelerator. Proficient marketing managers may intuitively realize that they should reposition a product. They may then begin to study the situation, taking great pride in the sophistication of their scientific analysis while overlooking their much more impressive talent—that of recognizing, without conscious thought, the simple existence of the problem.

The final skill level is that of expert. Experts generally know what to do because they have a mature and practiced understanding. When deeply involved in coping with their environment, they do not see problems in some detached way and consciously work at solving them. The skills of experts have become so much a part of them that they need be no more aware of them than they are of their own bodies. Airplane pilots report that as novices they felt they were flying their planes, but as experienced pilots they simply experience flying itself. Grand masters of chess, engrossed in a game, are often oblivious to the fact that they are manipulating pieces on a board. Instead, they see themselves as participants in a world of opportunities, threats, strengths, weaknesses, hopes, and fears. When playing rapidly, they sidestep dangers as automatically as teenagers avoid missiles in a familiar video game.

One of us, Stuart, knows all too well the difference between expert and merely competent chess players; he is stuck at the competent level. He took up chess as an outlet for his analytic talent in mathematics, and most of the other players on his college team were also mathematicians. At some point, a few of his teammates who were not mathematicians began to play fast five- or ten-minute games of chess, and also began eagerly to replay the great games of the grand masters. But Stuart and his mathematical colleagues resisted because fast chess didn't give them the time to *figure out* what to do. They also felt that they could learn nothing from the grand master games, since the record of those games seldom if ever provided specific rules and principles.

Some of his teammates who played fast chess and studied grand master games absorbed a great deal of concrete experience and went on to become chess masters. Yet Stuart and his mathematical friends never got beyond the competent level. Students of math may predominate among chess enthusiasts, but a truck driver is as likely as a mathematician to be among the world's best players. Stuart says he is glad that his analytic approach to chess stymied his progress because it helped him to see that there is more to skill than reasoning.

When things are proceeding normally, experts do not solve problems by reasoning; they do what normally works. Expert air-traffic controllers do not watch blips on a screen and deduce what must be going on in the sky. Rather, they "see" planes when they look at their screens and they respond to what they see, not by using rules but as experience has taught them to. Skilled outfielders do not take the time to figure out where a ball is going. Unlike novices, they simply run to the right spot. In *The Brain*, Richard Restak quotes a Japanese martial artist as saying, "There can be no thought, because if there is thought, there is a time of thought and that means a flaw. . . . If you take the time to think, 'I must use this or that technique', you will be struck while you are thinking."

We recently performed an experiment in which an international chess master, Julio Kaplan, had to add numbers at the rate of about one per second while playing five-second-a-move chess against a slightly weaker but master-level player. Even with his analytical mind apparently jammed by adding numbers, Kaplan more than held his own against the master in a series of games. Deprived of the time necessary to see problems or construct plans, Kaplan still produced fluid and coordinated play.

As adults acquire skills, what stands out is their progression *from* the analytic behavior of consciously following abstract rules *to* skilled behavior based on unconsciously recognizing new situations as similar to remembered ones. Conversely, small children initially understand only concrete examples and gradually learn abstract reasoning. Perhaps it is because this pattern in children is so well known that adult intelligence is so often misunderstood.

By now it is evident that there is more to intelligence than calculative rationality. In fact, experts who consciously reason things out tend to regress to the level of a novice or, at best, a competent performer. One expert pilot described an embarrassing incident that illustrates this point. Once he became an instructor, his only opportunity to fly the four-jet KC-135s at which he had once been expert was during the return flights he made after evaluating trainees. He was approaching the landing strip on one such flight when an engine failed. This is technically an emergency, but an experienced pilot will effortlessly compensate for the pull to one side. Being out of practice, our pilot thought about what to do and then overcompensated. He then consciously corrected himself, and the plane shuddered violently as he landed. Consciously using rules, he had regressed to flying like a beginner.

This is not to say that deliberative rationality has no role in intelligence. Tunnel vision can sometimes be avoided by a type of detached deliberation. Focussing on aspects of a situation that seem relatively unimportant allows another perspective to spring to mind. We once heard an Israeli fighter pilot recount how deliberative rationality may have saved his life by rescuing him from tunnel vision. Having just vanquished an expert opponent, he found himself taking on another member of the enemy squadron who seemed to be brilliantly eluding one masterful ploy after another. Things were looking bad until he stopped following his intuition and deliberated. He then realized that his opponent's surprising maneuvers were really the predictable, rule-following behavior of a beginner. This insight enabled him to vanquish the pilot.

IS INTELLIGENCE BASED ON FACTS?

Digital computers, which are basically complicated structures of simple on-off switches, were first used for scientific calculation. But by the end of the 1950s, researchers such as Allen Newell and Herbert Simon, working together at the Rand Corp., began to exploit the idea that computers could manipulate general symbols. They saw that one could use symbols to represent elementary facts about the world and rules to represent relationships between the facts. Computers could apply these rules and make logical inferences about the facts. For instance, a programmer might give a computer rules about how cannibals like to eat missionaries, and facts about how many cannibals and missionaries must be ferried across a river in one boat that carries only so many people. The computer could then figure out how many trips it would take to get both the cannibals and the missionaries safely across the river.

Newell and Simon believed that computers programmed with such facts and rules could, in principle, solve problems, recognize patterns, understand stories, and indeed do anything that an intelligent person could do. But they soon found that their programs were missing crucial aspects of problem-solving, such as the ability to separate relevant from irrelevant operations. As a result, the programs worked in only a very limited set of cases, such as in solving puzzles and proving theorems of logic.

In the late 1960s, researchers at MIT abandoned Newell and Simon's approach, which was based on imitating people's reports of how they solved problems, and began to work on any processing methods that could give computers intelligence. They recognized that to solve "real-world" problems the computer had to somehow simulate real-world understanding and intuition. In the introduction to *Semantic Information Processing*, a collection of his students' Ph.D. theses, Marvin Minsky describes the heart of the MIT approach:

> If we . . . ask . . . about the common everyday structures—that which a person needs to have ordinary common sense—we will find first a collection of indispensable categories, each rather complex: geometrical and mechanical properties of things and of space; uses and properties of a few thousand objects; hundreds of "facts" about hundreds of people; thousands of facts about tens of people; tens of facts about thousands of people; hundreds of facts about hundreds of organizations . . . I therefore feel that a machine will quite critically need to acquire on the order of a hundred thousand elements of knowledge in order to behave with reasonable sensibility in ordinary situations. A million, if property organized, should be enough for a very great intelligence.

However, Minsky's students encountered the same problem that had plagued Newell and Simon: each program worked only in its restricted specialty and could not be applied to other problems. Nor did the programs have any semantics—that is, any understanding of what their symbols meant. For instance, Daniel Bobrow's STUDENT program, which was designed to understand and solve elementary algebraic story problems, interpreted the phrase "the number of times I went to the movies" as the product of the two variables "number of" and "I went to the movies." That's because, as far as the program knew, "times" was a multiplicative operator linking the two phrases.

The restricted, ad hoc character of such work is even more striking in a program called ELIZA, written by MIT computer science professor Joseph Weizenbaum. Weizenbaum set out to show just how much apparent intelligence one could get a computer to exhibit without giving it any real understanding at all. The result was a program that imitated a therapist using simple tricks such as turning statements into questions: it responded to "I'm feeling sad" with "Why are you feeling sad?" When the program couldn't find a stock response, it printed out statements such as "Tell me about your father." The remarkable thing was that people were so easily fooled by these tricks. Weizenbaum was appalled when some people divulged their deepest feelings to the computer and asked others to leave the room while they were using it.

One of us, Hubert, was eager to see a demonstration of the notorious program, and he was delighted when Weizenbaum invited him to sit at the console and interact with ELIZA. Hubert spoiled the fun, however. He unintentionally exposed how shallow the trickery really was by typing, "I'm feeling happy," and then correcting himself by typing, "No, elated." At that point, the program came back with the remark, "Don't be so negative." Why? Because it had been programmed to respond with that rebuke whenever there was a "no" in the input.

MICROWORLDS VERSUS THE REAL WORLD

It took about five years for the shallowness of Minsky's students' programs to become apparent. Meanwhile, Hubert published a book, *What Computers Can't Do,* which asserted that AI research had reached a dead end since it could not come up with a way to represent general common-sense understanding. But just as *What Computers Can't Do* went to press in 1970, Minsky and Seymour Papert, also a professor at MIT, developed a new approach to AI. If one could not deal systematically with common-sense knowledge all at once, they asked, then why not develop methods for dealing systematically with knowledge in isolated subworlds and build gradually from that?

Shortly after that, MIT researchers hailed a computer program by graduate student Terry Winograd as a "major advance" in getting computers to understand human language. The program, called SHRDLU, simulated on a TV screen a robot arm that could move a set of variously shaped blocks. The program allowed a person to engage in a dialogue with the computer, asking questions, making statements, and issuing commands within this simple world of movable blocks. The program relied on grammatical rules, semantics, and facts about blocks. As Winograd cautiously claimed, SHRDLU was a "computer program which 'understands' language in a limited domain."

Winograd achieved success in this restricted domain, or "microworld," because he chose a simple problem carefully. Minsky and Papert believed that by combining a large number of these microworlds, programmers could eventually give computers real-life understanding.

Unfortunately, this research confuses two domains, which we shall distinguish as "universe" and "world." A set of interrelated facts may constitute a "universe" such as the physical universe, but it does not constitute a "world" such as the world of business or theater. A "world" is an organized body of objects, purposes, skills, and practices that make sense only against a background of common human concerns. These "sub-worlds" are not isolable physical systems. Rather, they are specific elaborations of a whole, without which they could not exist.

If Minsky and Papert's microworlds *were* true sub-worlds, they would not have to be extended and combined to encompass the everyday world, because each one would already incorporate it. But since microworlds are only isolated, meaningless domains, they cannot be combined and extended to reflect everyday life. Because scientists failed to ask what a "world" is, another five-year period of AI research ended in stagnation.

Winograd himself soon gave up the attempt to generalize the techniques SHRDLU used. "The AI programs of the late sixties and early seventies are much too literal," he acknowledged in a report for the National Institute of Education. "They deal with meaning as if it were a structure to be built up of the bricks and mortar provided by the words."

From the late 1970s to the present, AI has been wrestling unsuccessfully with what is called the common-sense knowledge problem: how to store and gain access to all the facts human beings seem to know. This problem has kept AI from even beginning to fulfill the predictions Minsky and Simon made in the mid-1960s: that within 20 years computers would be able to do everything humans can.

CAN COMPUTERS COPE WITH CHANGE?

If a machine is to interact intelligently with people, it has to be endowed with an understanding of human life. What we understand simply by virtue of being human—that insults make us angry, that moving physically forward is easier than moving backward—all this and much more would have to be programmed into the computer as facts and rules. As AI workers put it, they must give the computer our belief system. This, of course, presumes that human understanding is made up of beliefs that can be readily collected and stored as facts.

Even if we assume that this is possible, an immediate snag appears: we cannot program computers for context. For instance, we cannot program a computer to know simply that a car is going "too fast." The machine must be programmed in a way free of interpretation—we must stipulate that the car is going "20 miles an hour," for example. Also, computers know what to do only by reference to precise rules, such as "shift to second at 20 miles an hour." Computer programmers cannot use common-sense rules, such as "under normal conditions, shift to second at about 20 miles an hour."

Even if all the facts were stored in a context-free form, the computer still couldn't use them because it would be unable to draw on just the facts or rules that are relevant in each particular context. For example, a general rule of chess is that you should trade material when you're ahead in the value of the pieces on the board. However, you should not apply that rule if the opposing king is much more centrally located than yours, or when you are attacking the enemy king. And there are exceptions to each of these exceptions. It is virtually impossible to include all the possible exceptions in a program and do so in such a way that the computer knows which exception to use in which case.

In the real world, any system of rules has to be incomplete. The law, for instance, always strives for completeness but never achieves it. "Common law" helps, for it is based more on precedents than on a specific code. But the sheer number of lawyers in business tells us that it is impossible to develop a code of law so complete that all situations are unambiguously covered.

To explain our own actions and rules, humans must eventually fall back on everyday practices and simply say, "This is what one does." In the final analysis, all intelligent behavior must hark back to our sense of what we *are*. We can never explicitly formulate this in clear-cut rules and facts; therefore, we cannot program computers to possess that kind of know-how.

Nor can we program them to cope with changes in everyday situations. AI researchers have tried to develop computer programs that describe a normal sequence of events as they unfold. One such script, for instance, details what happens when someone goes to a restaurant. The problem is that so many unpredictable events can occur—one can receive an emergency telephone call or run into an acquaintance—that it's virtually impossible to predict how different people will respond. It all depends on what else is going on and what their specific purpose is. Are these people there to eat, to hobnob with friends, to answer phone calls, or to give the waiters a hard time? To make sense of behavior in restaurants, one has to understand not only what people typically do in eating establishments but why they do it. Thus, even if programmers could manage to list all that is *possibly* relevant in typical restaurant dining, computers could not use the information because they would have no understanding of what is *actually* relevant to specific customers.

THINKING WITH IMAGES, NOT WORDS

Experimental psychologists have shown that people actually use images, not descriptions as computers do, to understand and respond to some situations. Humans often think by forming images and comparing them holistically. This process is quite different from the logical, step-by-step operations that logic machines perform.

For instance, human beings use images to predict how certain events will turn out. If people know that a small box is resting on a large box, they can imagine what would happen if the large box were moved. If they see that the small box is tied to a door, they can also imagine what would result if someone were to open the door. A computer, however, must be given a list of facts about boxes, such as their size, weight, and frictional coefficients, as well as information about how each is affected by various kinds of movements. Given enough precise information about boxes and strings, the computer can deduce whether the small box will move with the large one under certain conditions. People also reason things out in this explicit, step-by-step way—but only if they must think about relationships they have never seen and therefore cannot imagine.

At present, computers have difficulty recognizing images. True, they can store an image as a set of dots and then rotate the set of dots so that a human designer can see the object from any perspective. But to know what a scene depicts, a computer must be able to analyze it and recognize every object. Programming a computer to analyze a scene has turned out to be very difficult. Such programs require a great deal of computation, and they work only in special cases with objects whose characteristics the computer has been programmed to recognize in advance.

But that is just the beginning of the problem. The computer can make inferences only from lists of facts. It's as if to read a newspaper you had to spell out each word, find its meaning in the dictionary, and diagram every sentence, labeling all the parts of speech. Brains do not seem to decompose either language or images this way, but logic machines have no choice. They must break down images into the objects they contain—and then into descriptions of those objects' features—before drawing any conclusions. However, when a picture is converted into a description, much information is lost. In a family photo, for instance, one can see immediately which people are between, behind, and in front of which others. The programmer must list all these relationships for the computer, or the machine must go through the elaborate process of deducing these relationships each time the photo is used.

Some AI workers look for help from parallel processors, machines that can do many things at once and hence make millions of inferences per second. But this appeal misses the point: that human beings seem to be able to form and compare images in a way that cannot be captured by any number of procedures that operate on descriptions.

Take, for example, face recognition. People can not only form an image of a face, but they can also see the similarity between one face and another. Sometimes the similarity will depend on specific shared features, such as blue eyes and heavy beards. A computer, if it has been programmed to abstract such features from a picture of a face, could recognize this sort of similarity.

However, a computer cannot recognize emotions such as anger in facial expressions, because we know of no way to break down anger into elementary symbols. Therefore, logic machines cannot see the similarity between two faces that are angry. Yet human beings can discern the similarity almost instantly.

Many AI theorists are convinced that human brains unconsciously perform a series of computations to perceive such subtleties. While no evidence for this mechanical model of the brain exists, these theorists take it for granted because it is the way people proceed when they are reflecting consciously. To such theorists, any alternative explanation appears mystical and therefore anti-scientific.

But there is another possibility. The brain, and therefore the mind, could still be explained in terms of something material. But it does not have to be an information processing machine. Other physical systems can detect similarity without using any descriptions or rules at all. These systems are known as holograms.

IS THE MIND LIKE A HOLOGRAM?

An ordinary hologram works by taking a picture of an object using two beams of laser light, one of which is reflected off the object and one of which shines directly onto film. When the two beams meet, they create an interference pattern like that produced by the waves from several pebbles thrown into a pond. The light waves form a specific pattern of light and dark regions. A photographic plate records this interference pattern, thus storing a representation of the object.

In ordinary light, the plate just looks blurry, a uniform silvery gray. But if the right frequency of light is projected on to it, the recorded pattern of light and dark shapes the light into a replica of the object. This replica appears three-dimensional: we can view different sides of it as we change position.

What first attracted neuropsychologists to the hologram was that it really is holistic: any small piece of the blur on the photographic plate contains the whole scene. For example, if you cut one corner off a hologram of a table and shine a laser beam through what remains, you do not see an image of a table with a corner missing. The whole table is still there but with fuzzier edges.

Certain areas of the brain also have this property. When a piece is cut out, a person may lose nothing specific from vision, for example. Instead, that person may see everything less distinctly. Holograms have another mindlike property: they can be used for associative memory. If one uses a single hologram to record two different scenes and then bounces laser light off one of the scenes, an image of the other will appear.

In our view, the most important property of holograms is their ability to detect similarity. For example, if we made a hologram of this page and then made a hologram of one of the letters on the page, say the letter F, shining a light through the two holograms would reveal an astonishing effect: a black field with bright spots wherever the letter F occurs on the page. Moreover, the brightest spots would indicate the Fs with the greatest similarity to the F we used to make our hologram. Dimmer spots would appear where there are imperfect or slightly rotated versions of the F. Thus, a hologram can not only identify objects; it can also recognize similarity between them. Yet it employs no descriptions or rules.

The way a hologram can instantly pick out a specific letter on a page is reminiscent of the way people pick out a familiar face from a crowd. It is possible that we distinguish the familiar face from all the other faces by processing rules about objectively identifiable features. But we would have to examine each face in the crowd, detect its features, and compare them with lists of our acquaintances' features. It is much more plausible that our minds work on some variation of the holistic model. While the brain obviously does not contain lasers or use light beams, some scientists have suggested that neurons could process incoming stimuli using interference patterns like those of a hologram.

However, the human mind seems to have an ability that far transcends current holographic techniques: the remarkable ability to recognize whole meaningful patterns without decomposing them into features. Unlike holography, our mind can sometimes detect faces in a crowd that have expressions unlike any we have previously seen on those faces. We can also pick out familiar faces that have changed dramatically because of the growth of a beard or the ravages of time.

We take no stand on the question of whether the brain functions holographically. We simply want to make clear that the information processing computer is not the only physical system that can exhibit mindlike properties. Other devices may provide closer analogies to the way the mind actually works.

Given the above considerations, what level of skill can we expect logic machines to reach? Since we can program computers with thousands of rules combining hundreds of thousands of features, the machines can become what might be thought of as expert novices in any well-structured and well-understood domain. As long as digital computers' ability to recognize images and reason by analogy remains a vague promise, however, they will not be able to approach the way human beings cope with everyday reality. Despite their failure to capture everyday human understanding in computers, AI scientists have developed programs that seem to reproduce human expertise within a specific, isolated domain. The programs are called expert systems. In their narrow areas, such systems perform with impressive competence.

In his book on "Fifth Generation" computers, Edward Feigenbaum, a professor at Stanford, spells out the goal of expert systems: "In the kind of intelligent systems envisioned by the designers of the Fifth Generation, speed and processing power will be increased dramatically. But more important, the machines will have reasoning power: they will automatically engineer vast amounts of knowledge to serve whatever purpose human beings propose, from medical diagnosis to product design, from management decisions to education."

The knowledge engineers claim to have discovered that all a machine needs to behave like an expert in restricted domains are some general rules and lots of very specific knowledge. But can these systems really be expert? If we agree with Feigenbaum that "almost all thinking that professionals do is done by reasoning," and that each expert builds up a "repertory of working rules of thumb," the answer is yes. Given their speed and precision, computers should be as good as or better than people at following rules for deducing conclusions. Therefore, to build an expert system, a programmer need only extract those rules and program them into a computer.

JUST HOW EXPERT ARE EXPERT SYSTEMS?

However, human experts seem to have trouble articulating the principles on which they allegedly act. For example, when Arthur Samuel at IBM decided to write a program for playing checkers in 1947, he tried to elicit "heuristic" rules from checkers masters. But nothing the experts told him allowed him to produce master play. So Samuel supplemented these rules with a program that relies blindly on its memory of past successes to improve its current performance. Basically, the program chooses what moves to make based on rules and a record of all past positions.

This checkers program is one of the best expert systems ever built. But it is no champion. Samuel says the program "is quite capable of beating any amateur player and can give better players a good contest." It did once defeat a state champion, but the champion turned around and defeated the program in six mail games. Nonetheless, Samuel still believes that chess champions rely on heuristic rules. Like Feigenbaum, he simply thinks that the champions are poor at recollecting their compiled rules: "The experts do not know enough about the mental processes involved in playing the game."

INTERNIST-1 is an expert system highly touted for its ability to make diagnoses in internal medicine. Yet according to a recent evaluation of the program published in *The New England Journal of Medicine*, this program misdiagnosed 18 out of a total of 43 cases, while clinicians at Massachusetts General Hospital misdiagnosed 15. Panels of doctors who discussed each case misdiagnosed only 8. (Biopsies, surgery, and post-mortem autopsies were used to establish the correct diagnosis for each case.) The evaluators found that "the experienced clinician is vastly superior to INTERNIST-1, in the ability to consider the relative severity and independence of the different manifestations of disease and to understand the . . .

evolution of the disease process." The journal also noted that this type of systematic evaluation was "virtually unique in the field of medical applications of artificial intelligence."

In every area of expertise, the story is the same: the computer can do better than the beginner and can even exhibit useful competence, but it cannot rival the very experts whose facts and supposed rules it is processing with incredible speed and accuracy.

Why? Because the expert is not following any rules! While a beginner makes inferences using rules and facts just like a computer, the expert intuitively sees what to do without applying rules. Experts must regress to the novice level to state the rules they still remember but no longer use. No amount of rules and facts can substitute for the know-how experts have gained from experience in tens of thousands of situations. We predict that in no domain in which people exhibit such holistic understanding can a system based on rules consistently do as well as experts. Are there any exceptions?

At first glance, at least one expert system seems to be as good as human specialists. Digital Equipment Corp. developed R1, now called XCON, to decide how to combine components of VAX computers to meet consumers' needs. However, the program performs as well as humans only because there are so many possible combinations that even experienced technical editors depend on rule-based methods of problem-solving and take about 10 minutes to work out even simple cases. It is no surprise, then, that this particular expert system can rival the best specialists.

Chess also seems to be an exception to our rule. Some chess programs, after all, have achieved master ratings by using "brute force." Designed for the world's most powerful computers, they are capable of examining about 10 million possible positions in choosing each move.

However, these programs have an Achilles' heel: they can see only about four moves ahead for each piece. So fairly good players, even those whose chess rating is somewhat lower than the computers, can win by using long-range strategies such as attacking the king side. When confronted by a player who knows its weakness, the computer is not a master-level player.

In every domain where know-how is required to make a judgment, computers cannot deliver expert performance, and it is highly unlikely that they ever will.

Those who are most acutely aware of the limitations of expert systems are best able to exploit their real capabilities. Sandra Cook, manager of the Financial Expert Systems Program at the consulting firm SRI International, is one of these enlightened practitioners. She cautions prospective clients that expert systems should not be expected to perform as well as human experts, nor should they be seen as simulations of human expert thinking.

Cook lists some reasonable conditions under which expert, or rather "competent," systems can be useful. For instance, such systems should be used for problems that can be satisfactorily solved by human experts at such a high level that somewhat inferior performance is still acceptable. Processing of business credit applications is a good example, because rules can be developed for this task and computers can follow them as well as and sometimes better than inexperienced humans. Of course, there are some exceptions to the rules, but a few mistakes are not disastrous. On the other hand, no one should expect expert systems to make stock market predictions because human experts themselves cannot always make such predictions accurately.

Expert systems are also inappropriate for use on problems that change as events unfold. Advice from expert systems on how to control a nuclear reactor during a crisis would come too late to be of any use. Only human experts could make judgments quickly enough to influence events.

It is hard to believe some AI enthusiasts' claim that the companies who use expert systems dominate all competition. In fact, a company that relies too heavily on expert systems faces a genuine danger. Junior employees may come to see expertise as a function of the large knowledge bases and masses of rules on which these programs must rely. Such employees will fail to progress beyond the competent level of performance, and business managers may ultimately discover that their wells of true human expertise and wisdom have gone dry.

COMPUTERS IN THE CLASSROOM

Computers pose a similar threat in the classroom. Advertisements warn that a computer deficiency in the educational diet can seriously impair a child's intellectual growth. As a result, frightened parents spend thousands of dollars on home computers and clamor for schools to install them in the classroom. Critics have likened computer salespeople to the encyclopedia peddlers of a generation ago, who contrived to frighten insecure parents into spending hundreds of dollars for books that contributed little to their offsprings' education.

We feel that there is a proper place for computers in education. However, most of today's educational software is inappropriate, and many teachers now use computers in ways that may eventually produce detrimental results.

Perhaps the least controversial way computers can be used is as tools. Computers can sometimes replace teaching aids ranging from paintbrushes, typewriters, and chalkboards to lab demonstrations. Computer simulations, for instance, allow children to take an active and imaginative role in studying subjects that are difficult to bring into the classroom. Evolution is too slow, nuclear reactions are too fast, factories are too big, and much of chemistry is too dangerous to reproduce realistically. In the future, computer simulations of such events will surely become more common, helping students of all ages in all disciplines to develop their intuition. However, since actual skills can be learned only through experience, it seems only common sense to stick to the world of real objects. For instance, basic electricity should be taught with batteries and bulbs.

Relying too heavily on simulations has its pitfalls. First of all, the social consequences of decisions are often missing from simulations. Furthermore, the appeal of simulations could lead disciplines outside the sciences to stress their formal, analytic side at the expense of lessons based on informal, intuitive understanding. For example, political science departments may be tempted to emphasize mathematical models of elections and neglect the study of political philosophies that question the nature of the state and of power. In some economics departments, econometrics—which relies heavily on mathematical models—has already pushed aside study of the valuable lessons of economic history. The truth is that no one can assess the dynamic relationships that underlie election results or economies with anything like the accuracy of the laws of physics. Indeed, every election campaign or economic swing offers vivid reminders of how inaccurate predictions based on simulation models can be.

On balance, however, the use of the computer as a tool is relatively unproblematic. But that is not the case with today's efforts to employ the computer as tutor or tutee. Behind the idea that computers can aid, or even replace, teachers is the belief that teachers' understanding of the subject being taught and their profession consists of knowing facts and rules. In other words, the teacher's job is to convey specific facts and rules to students by drill and practice or by coaching.

Actually, if our minds were like computers, drill and practice would be completely unnecessary. The fact that even brilliant students need to practice when learning subtraction suggests that the human brain does not operate like a computer. Drill is required simply to fix the rule in human memory. Computers, by contrast, remember instantly and perfectly. Math students also have to learn that some features such as the physical size and orientation of numbers are irrelevant while others such as position are crucial. In this case, they must learn to "decontextualize," whereas computers have no context to worry about.

There is nothing wrong with using computers as drill sergeants. As with simulation, the only danger in this use stems from the temptation to overemphasize some skills at the expense of others. Mathematics might degenerate into addition and subtraction, English into spelling and punctuation, and history into dates and places.

AI enthusiasts believe that computers can play an even greater role in teaching. According to a 1984 report by the National Academy of Sciences, "Work in artificial intelligence and the cognitive sciences has set the stage for qualitatively new applications of technology to education."

Such claims should give us pause. Computers will not be first-rate teachers unless researchers can solve four basic problems: how to get machines to talk, to listen, to know, and to coach. "We speak as part of our humanness, instinctively, on the basis of past experience," wrote Patrick Suppes of Stanford University, one of the pioneers in computer-aided instruction, in a 1966 *Scientific American* article. "But to get a computer to talk appropriately, we need an explicit theory of talking."

Unfortunately, there is no such theory, and if our analysis of human intelligence is correct, there never will be. The same holds true for the problem of getting computers to listen. Continuous speech recognition seems to be a skill that resists decomposition into features and rules. What we hear does not always correspond to the features of the sound stream. Depending on the context and our expectations, we hear a stream of sound as "I scream," or "ice cream." We assign the space or pause in one of two places, although there is no pause in the sound stream. One expert came up with a sentence that illustrates the different ways we can hear the same stream of sound: "It isn't easy to wreck a nice beach." (Try reading that sentence out loud.)

Without the ability to coach, a computer could hardly substitute for an inexperienced teacher, let alone a Socrates. "Even if you can make the computer talk, listen, and adequately handle a large knowledge data base, we still need to develop an explicit theory of learning and instruction," Suppes writes. "In teaching a student, young or old, a given subject matter, a computer-based learning system can record anything the student does. It can know cognitively an enormous amount of information about the student. The problem is how to use this information wisely, skillfully, and efficiently to teach the student. This is something that the very best human tutors do well, even though they do not understand at all how they do it."

While he recognizes how formidable these obstacles are, Suppes persists in the hope that we can program computers to teach. However, in our view, expertise in teaching does not consist of knowing complicated rules for deciding what tips to give students, when to keep silent, when to intervene—although teachers may have learned such rules in graduate school. Rather, expert teachers learn from experience to draw intuitively and spontaneously on the common-sense knowledge and experience they share with their students to provide the tips and examples they need.

Since computers can successfully teach only novice or, at best, competent performance, they will only produce the sort of expert novices many feel our schools already graduate. Computer programs may actually prevent beginning students from passing beyond competent analysis to expertise. Instead of helping to improve education, computer-aided instruction could easily become part of the problem.

In the air force, for instance, instructors teach beginning pilots a rule for how to scan their instruments. However, when psychologists studied the eye movements of the instructors during simulated flight, the results showed that the instructors were not following the rule they were teaching. In fact, as far as the psychologists could determine, the instructors were not following any rules at all.

Now suppose that the instrument-scanning rule goes into a computer program. The computer monitors eye movements to make sure novices are applying the rule correctly. Eventually, the novices are ready, like the instructors, to abandon the rules and respond to whole situations they perceive as similar to others. At this point, there is nothing more for the computer to teach. If it is still used to check eye movements, it would prevent student pilots from making the transition to intuitive proficiency and expertise.

This is no mere bogeyman. Expert systems are already being developed to teach doctors the huge number of rules that programmers have "extracted" from experts in the medical domain. One can only hope that someone has the sense to disconnect doctors from the system as soon they reach the advanced-beginner stage.

CAN CHILDREN LEARN BY PROGRAMMING?

The concept of using computers as tutees also assumes the information-processing model of the mind. Adherents of this view suppose that knowledge consists of using facts and rules, and that therefore students can acquire knowledge in the very act of programming. According to this theory, learning and learning to program are the same thing.

Seymour Papert is the most articulate exponent of this theory. He is taking his LOGO program into Boston schools to show that children will learn to think more rigorously if they teach a literal-minded but patient and agreeable student—the computer. In Papert's view, programming a computer will induce children to articulate their own program by naming the features they are selecting from their environment, and by making explicit the procedures they are using to relate these features to events. Says Papert: "I have invented ways to take educational advantage of the opportunities to master the art of *deliberately* thinking like a computer, according, for example, to the stereotype of a computer program that proceeds in a step-by-step, literal, mechanical fashion."

Papert's insistence that human know-how can be analyzed has deep roots in our "rationalistic" Western tradition. We can all probably remember a time in school when we knew something perfectly well but our teacher claimed that we didn't know it because we couldn't explain how we got the answer.

Even Nobel laureates face this sort of problem. Physicist Richard Feynman had trouble getting the scientific community to accept his theories because he could not explain how he got his answers. In his book *Disturbing the Universe*, physicist and colleague Freeman Dyson wrote:

> The reason Dick's physics were so hard for the ordinary physicists to grasp was that he did not use equations. . . . He had a physical picture of the way things happen, and the picture gave him

the solutions directly with a minimum of calculation. It was no wonder that people who spent their lives solving equations were baffled by him. Their minds were analytical; his was pictorial.

While Papert tries to create a learning environment in which learners constantly face new problems and need to discover new rules, Timothy Gallwey, the author of *Inner Tennis*, encourages learners to achieve mastery by avoiding analytic thinking from the very start. He would like to create a learning environment in which there are no problems at all and so there is never any need for analytic reflection.

Our view lies in between. At any stage of learning, some problems may require rational, analytic thought. Nonetheless, skill in any domain is measured by the performer's ability to act appropriately in situations that might once have been problems but are no longer problems and so do not require analytic reflection. The risk of Gallwey's method is that it leaves the expert without the tools to solve new problems. But the risk of Papert's approach is far greater: it would leave the learner a perpetual beginner by encouraging dependence on rules and analysis.

AI ON THE BATTLEFIELD

The Department of Defense is pursuing a massive Strategic Computing Plan (SCP) to develop completely autonomous land, sea, and air vehicles capable of complex, far-ranging reconnaissance and attack missions. SCP has already spent about $145 million and received approval to spend $150 million in fiscal 1986. To bolster support for this effort, the DOD's Defense Advanced Research Projects Agency (DARPA) points to important advances in AI—expert systems with common sense and systems that can understand natural language. However, no such advances have occurred.

Likewise, computers are no more able today to deal intelligently with "uncertain data" than they were a few years ago when our computerized ballistic-missile warning system interpreted radar reflections from a rising moon as an enemy attack. In a report evaluating the SCP, the congressional Office of Technology Assessment cautioned, "Unlike the Manhattan Project or the Manned Moon Landing Mission, which were principally engineering problems, the success of the DARPA program requires basic scientific breakthroughs, neither the timing nor the nature of which can be predicted."

Even if the Defense Department invests billions of dollars in AI, there is almost no likelihood that this state of affairs will change. Yet once vast sums of money have been spent, there will be a great temptation to install questionable AI-based technologies in a variety of critical areas—from battle management to "data reduction" (figuring out what is really going on given noisy, contradictory data).

Military commanders now respond to a battlefield situation using common sense, experience, and whatever data are available. The frightening prospect of a fully computerized and autonomous defense system is that the expert's ability to use intuition will be replaced by merely competent decision-making. In a crisis, competence is just not good enough.

Furthermore, to justify its expenditures to the public, the military may feel compelled to encourage the civilian sector to adopt similar technologies. Full automation of air-traffic control systems and of skilled factory labor are both real possibilities.

Unless illusions concerning AI are dispelled, we are risking a future in which computers make crucial military and civilian decisions that are best left to human judgment. Knowledgeable AI practitioners have learned from bitter experience that the development of fully autonomous war machines is unlikely. We hope that military decision-makers or the politicians who fund them will see the light and save US taxpayers' money by terminating this crash program before it is too late.

THE OTHER SIDE OF THE STORY

At this point the reader may reasonably ask: If computers used as logic machines cannot attain the skill level of expert human beings, and if the "Japanese challenge in fifth-generation systems" is a false gauntlet, then why doesn't the public know that? The answer is that AI researchers have a great deal at stake in making it appear that their science and its engineering offspring—expert systems—are on solid ground. They will do whatever is required to preserve this image.

When public television station KCSM in Silicon Valley wanted to do a program on AI to be aired nationally, Stanford AI expert John McCarthy was happy to take part. So was a representative of IntelliCorp, a company making expert systems that wished to air a promotional film. KCSM also invited one of us, Hubert, to provide a balanced perspective. After much negotiating, an evening was finally agreed upon for taping the discussion.

That evening the producer and technicians were standing by at the studio and Hubert had already arrived in San Mateo when word came that McCarthy would not show up because Hubert was to be on the program. A fourth participant, expert-systems researcher Michael Genesereth of Stanford University, also backed out.

All of us were stunned. Representatives from public TV's NOVA science series and CBS news had already interviewed Hubert about AI, and he had recently appeared on a panel with Minsky, Papert, philosopher John Searle of Berkeley, and McCarthy himself at a meeting sponsored by the New York Academy of Sciences. Why not on KCSM? It seems the "experts" wanted to give the impression that they represented a successful science with marketable products and didn't want to answer any potentially embarrassing questions.

The shock tactic worked. The station's executive producer, Stewart Cheifet, rescheduled the taping with McCarthy as well as the demo from IntelliCorp, and he decided to drop the discussion with Hubert. The viewers were left with the impression that AI is a solid, ongoing science which, like physics, is hard at work solving its quite manageable current problems. The public's chance to hear both sides was lost and the myth of steady progress in AI was maintained. The real story remained to be told, and that is what we have tried to do here.

27

Whither Psychoanalysis in Computer Culture?

Sherry Turkle

PSYCHOANALYTIC CULTURE AND COMPUTER CULTURE

Over twenty years ago, as a new faculty member at MIT, I taught an introductory class on psychoanalytic theory. For one meeting, early in the semester, I had assigned Freud's chapters on slips of the tongue from *The Psychopathology of Everyday Life* (1901/1960). I began class by reviewing Freud's first example: the chairman of a parliamentary session begins the meeting by declaring it closed. Freud's analysis centered on the possible reasons behind the chairman's slip: he might be anxious about what the parliamentarians had on their agenda. Freud's analysis turned on trying to uncover the hidden meaning behind the chairman's remark. The theoretical effort was to understand his mixed emotions, his unconscious ambivalence.

As I was talking to my class about the Freudian notions of the unconscious and of ambivalence, one of the students, an undergraduate majoring in computer science, raised her hand to object. She was studying at the MIT Artificial Intelligence Laboratory, which was (and is) a place whose goal, in the words of one of its founders, Marvin Minsky, is to create "machines that did things that would be considered intelligent if done by people." Work in the AI Lab began with the assumption that the mind, in Minsky's terms, "was a meat machine," best understood by analogizing its working to that of a computer program. It was from this perspective that my student objected to what she considered a tortured explanation for slips of the tongue. "In a Freudian dictionary," she began, "closed and open are far apart. In a Webster's dictionary," she continued,

> they are as far apart as the listings for C and the listings for O. But in a computational diction-
> ary—such as we have in the human mind—closed and open are designated by the same symbol,
> separated by a sign of opposition. Closed equals 'minus' open. To substitute closed for open does
> not require the notion of ambivalence or conflict. When the substitution is made, a bit has been
> dropped. A minus sign has been lost. There has been a power surge. No problem.

With this brief comment, a Freudian slip had been transformed into an information processing error. An explanation in terms of meaning had been replaced by a narrative of mechanistic causation. At the time, that transition from meaning to mechanism struck me as emblematic of a larger movement that might be taking place in psychological culture. Were we moving from a psychoanalytic to a computer culture, one that would not need such notions as ambivalence when it modeled the mind as a digital machine (Turkle, 1984, 1991)?

For me, that 1981 class was a turning point. The story of the relationship between the psychoanalytic and computer cultures moved to the center of my intellectual concerns. But the story of their relationship has been far more complex than the narrative of simple transition that suggested itself to me during the early 1980s. Here, I shall argue the renewed relevance of a psychoanalytic discourse in digital culture. Indeed, I shall argue that this relevance is so profound as to suggest an occasion for a revitalization and renewal of psychoanalytic thinking.

In my view, this contemporary relevance does not follow, as some might expect, from efforts to link psychoanalysis and computationally inspired neuroscience. Nor does it follow, as I once believed it would, from artificial intelligence and psychoanalysis finding structural or behavioral analogies in their respective objects of study.

In "Psychoanalysis and Artificial Intelligence," I suggested an opening for dialogue between these two traditions that had previously eyed each other with suspicion if not contempt (Turkle, 1988). In my view, the opening occurred because of the ascendance of "connectionist" models of artificial intelligence. Connectionist descriptions of how mind was "emergent" from the interactions of agents had significant resonance with the way psychoanalytic object-relations theory talked about objects in a dynamic inner landscape. Both seemed to be describing what Minsky would have called a "society of mind." Today, however, the elements within the computer culture that speak most directly to psychoanalysis are concrete rather than theoretical. Novel and evocative computational objects demand a depth psychology of our relationships with them. The computer culture needs psychoanalytic understandings to adequately confront our evolving relationships with a new world of objects. Psychoanalysis needs to understand the influence of computational objects on the terrain it knows best: the experience and specificity of the sensual and speaking human subject.

EVOCATIVE OBJECTS AND PSYCHOANALYTIC THEORY

The designers of computational objects have traditionally focused on how these objects might extend human cognitive powers, perfect them or both. As an ethnographer/psychologist of computer culture, I hear another narrative as well: that of the users. Designers have traditionally focused on the instrumental computer, the computer that does things for us. Computer users are frequently more in touch with the subjective computer, the computer that does things to us, to our ways of seeing the world, to the way we think, to the nature of our relationships with each other. Technologies are never "just tools." They are evocative objects. They cause us to see ourselves and our world differently.

Although designers have focused on how computational devices such as personal digital assistants will help people better manage their complex lives, users have seen devices such as a Palm Pilot as *extensions of self*. The designer says: "People haven't evolved to keep up with complexity. Computers will help." The user says: "When my Palm crashed, it was like a death. More than I could handle. I had lost my mind." Wearable computers are devices that enable the user to have computer and online access all the time, connected to the Web by a small radio transmitter and using specially designed eyeglasses as a computer monitor.

Designers of wearable computing talk about new and, indeed, superhuman access to information. For example, with a wearable computer, you can be in a conversation with a faculty colleague and accessing his or her most recent papers at the same time. But when people actually wear computers all the time (and in this case, this sometimes happens when the designers begin to use and live with the technology), they testify to impacts on a very different register: wearable computers change one's sense of self. One user says, "I become my computer. It's not just that I remember people or know more about them. I feel invincible, sociable, better prepared. I am naked without it. With it, I'm a better person." A wearable computer is lived as a glass through which we see, however darkly, our cyborg futures (Haraway, 1991). Indeed, the group of students at MIT who have pioneered the use of wearable computing called themselves cyborgs.

Computer research proceeds through a discourse of rationality. Computer *culture* grows familiar with the experiences of passion, dependency, and profound connection with artifact. Contemporary computational objects are increasingly *intimate machines*; they demand that we focus our attention on the significance of our increasingly intimate relationships with them. This is where psychoanalytic studies are called for. We need a developmental and psychodynamic approach to technology that focuses on our new object relations.

There is a certain irony in this suggestion, for of course psychoanalysis has its own "object-relations" tradition (Greenberg & Mitchell, 1983). Freud's "Mourning and Melancholia" (1917/1960) opened psychoanalysis to thinking about how people take lost objects and internalize them, creating new psychic structure along with new facets of personality and capacity. *But for psychoanalysis, the "objects" in question were usually people.* A small number of psychoanalytic thinkers and writers explored the power of the inanimate (Erikson, 1963; Winnicott, 1971) and child analysts who wrote about the experience of objects in children's play) but, in general, the story of "object relations" in psychoanalysis has cast people in the role of "objects."

Today, the new objects of our lives call upon psychoanalytic theory to create an object relations theory that really is about objects in the everyday sense of the word.

What are these new objects? When in the early 1980s I first called the computer a "second self" or a Rorschach, an object for the projection of personhood, relationships with the computer were usually one-to-one, a person alone with a machine. This is no longer the case. A rapidly expanding system of networks, collectively known as the Internet, links millions of people together in new spaces that are changing the way we think, the nature of our sexuality, the form of our communities, our very identities. A network of relationships on the Internet challenges what we have traditionally called *identity* (Turkle, 1995).

Most recently, a new kind of computational object has appeared on the scene. *Relational artifacts,* such as robotic pets and digital creatures, are explicitly designed to have emotive, affect-laden connections with people. Today's computational objects do not wait for children to "animate" them in the spirit of a Raggedy Anne doll or the Velveteen Rabbit, the toy who finally became alive because so many children had loved him. They present themselves as already animated and ready for relationship. People are not imagined as their *users* but as their companions.

At MIT, a research group on "affective computing" works on the assumption that machines will not be able to develop humanlike intelligence without sociability and affect. The mission of the affective computing group is to develop computers that are programmed to assess their users' emotional states and respond with emotional states of their own. In the case of the robotic doll and the affective computers, we are confronted with relational artifacts that demand that the human users attend to the psychology of a machine.

Today's relational artifacts include robot dogs and cats, some specially designed and marketed to lonely elders. There is also a robot infant doll that makes baby sounds and even baby facial expressions, shaped by mechanical musculature under artificial skin. This computationally complex doll has baby "states of mind." Bounce the doll when it is happy, and it gets happier. Bounce it when it is grumpy, and it gets grumpier.

These relational artifacts provide good examples of how psychoanalysis might productively revisit old "object" theories in light of new "object" relations. Consider whether relational artifacts could ever be *transitional objects* in the spirit of a baby blanket or rag doll. For Winnicott (1971), such objects (to which children remain attached even as they embark on the exploration of the world beyond the nursery) are mediators between the child's earliest bonds with the mother, whom the infant experiences as inseparable from the self, and the child's growing capacity to develop relationships with other people who will be experienced as separate beings. The infant knows transitional objects as both almost inseparable parts of the self and, at the same time, as the first not-me possessions. As the child grows, the actual objects are left behind. The abiding effects of early encounters with them, however, are manifest in the experience of a highly charged intermediate space between the self and certain objects in later life. This experience has traditionally been associated with religion, spirituality, the perception of beauty, sexual intimacy, and the sense of connection with nature. In recent years, the power of the transitional object is commonly seen in experiences with computers.

Just as musical instruments can be extensions of the mind's construction of sound, computers can be extensions of the mind's construction of thought. A novelist refers to "my ESP with the machine. The words float out. I share the screen with my words." An architect who uses the computer to design goes even further: "I don't see the building in my mind until I start to play with shapes and forms on the machine. It comes to life in the space between my eyes and the screen." Musicians often hear the music in their minds before they play it, experiencing the music from within before they experience it from without. The computer similarly can be experienced as an object on the border between self and not-self.

In the past, the power of objects to play this transitional role has been tied to the ways in which they enabled the child to project meanings onto them. The doll or the teddy bear presented an unchanging and passive presence. In the past, computers were also targets of projection; the machine functioned as a Rorschach or a *second self*. But today's relational artifacts take a decidedly more active stance. With them, children's expectations that their dolls want to be hugged, dressed, or lulled to sleep don't come only from the child's projection of fantasy or desire onto inert playthings, but also from such things as the digital dolls' crying inconsolably or even saying "Hug me!" or "It's time for me to get dressed for school!"

In a similar vein, consider how these objects look from the perspective of self-psychology. Heinz Kohut describes how some people may shore up their fragile sense of self by turning another person into a "self object" (Ornstein, 1978). In the role of *self-object,* the other is experienced as part of the self, thus in perfect tune with the fragile individual's inner state. Disappointments inevitably follow. Relational artifacts (not as they exist now but as their designers promise they will soon be) clearly present themselves as candidates for such a role. If they can give the appearance of aliveness and yet not disappoint, they may even have a comparative advantage over people and open new possibilities for narcissistic experience with machines. One might even say that when people turn other people into self-objects, they are making an effort to turn a person into a kind of "spare part." From this point of view, relational artifacts make a certain amount of sense as successors to the always-resistant human material.

Just as television today is a background actor in family relationships and a "stabilizer" of mood and affect for individuals in their homes, in the near future a range of robotic companions and a web of pervasive computational objects will mediate a new generation's psychological and social lives. We will be living in a relational soup of computation that offers itself as a self-other if not as a self-object. Your home network and the computational "agents" programmed into it, indeed the computing embedded in your furniture and your clothing, will know your actions, your preferences, your habits, and your physiological responses to emotional stimuli. A new generation of psychoanalytic self-psychology is called upon to explore the human response and the human vulnerability to these objects.

PERSONAL COMPUTING: ONE-ON-ONE WITH THE MACHINE

Each modality of being with a computer, one-on-one with the machine, using the computer as a gateway to other people, and being presented with it as a relational artifact, implies a distinct mode of object relations. Each challenges psychoanalytic thinking in a somewhat different way. And all of these challenges face us at the same time. The development of relational artifacts does not mean that we don't also continue to spend a great deal of time alone, one-on-one with our personal computers.

Being alone with a computer can be compelling for many different reasons. For some, computation offers *the promise of perfection,* the fantasy that "if you do it right, it will do it right, and right away." Writers can become obsessed with fonts, layout, and spelling and grammar checks. What was once a typographical error can be, like Hester Prynne's scarlet letter, a sign of shame. As one writer put it: "A typographical error is the sign not of carelessness but of sloth and disregard for others, the sign that you couldn't take the one extra second, the one keystroke, to make it right." Like the anorectic projecting self-worth onto his or her body and calorie consumption, and who endeavors to eat ten calories less each day, game players or programmers may try to get to one more screen or play ten minutes more each day when dealing with the perfectible computational material.

Thus, the promise of perfection is at the heart of the computer's holding power for some. Others are drawn by different sirens. As we have seen, there is much seduction in the sense that on the computer, mind is building mind or even merging with the mind of another being. The machine can seem to be a second self, a metaphor first suggested to me by a thirteen-year-old girl who said, "When you program a computer there is a little piece of your mind, and now it's a little piece of the computer's mind. And now you can see it." An investment counselor in her mid-forties echoes the child's sentiment when she says of her laptop computer: "I love the way it has my whole life on it." If one is afraid of intimacy yet afraid of being alone, a computer offers an apparent solution: the illusion of companionship without the demands of friendship. In the mirror of the machine, one can be a loner yet never be alone.

LIVES ON THE SCREEN: RELATING
PERSON-TO-PERSON VIA COMPUTER

From the mid-1980s, the cultural image of computer use expanded from an individual alone with a computer to an individual engaged in a network of relationships via the computer. The

Internet became a powerful evocative object for rethinking identity, one that encourages people to recast their sense of self in terms of multiple windows and parallel lives.

Virtual personae. In cyberspace, as is well-known, the body is represented by one's own textual description, so the obese can be slender, the beautiful plain. The fact that self-presentation is written in text means that there is time to reflect upon and edit one's "composition," which makes it easier for the shy to be outgoing, the "nerdy" sophisticated. The relative anonymity of life on the screen—one has the choice of being known only by one's chosen "handle," or online name—gives people the chance to express often unexplored aspects of the self. Additionally, multiple aspects of self can be explored in parallel. Online services offer their users the opportunity to be known by several different names. For example, it is not unusual for someone to be BroncoBill in one online context, ArmaniBoy in another, and MrSensitive in a third.

The online exercise of playing with identity and trying out new ones is perhaps most explicit in "role-playing" virtual communities and online gaming, where participation literally begins with the creation of a persona (or several), but it is by no means confined to these somewhat exotic locales. In bulletin boards, newsgroups, and chatrooms, the creation of personae may be less explicit than in virtual worlds or games, but it is no less psychologically real. One Internet relay chat (IRC) participant describes her experience of online talk: "I go from channel to channel depending on my mood. . . . I actually feel a part of several of the channels, several conversations. . . . I'm different in the different chats. They bring out different things in me." Identity play can happen by changing names and by changing places.

Even the computer interface encourages rethinking complex identity issues. The development of the windows metaphor for computer interfaces was a technical innovation motivated by the desire to get people working more efficiently by "cycling through" different applications much as time-sharing computers cycled through the computing needs of different people. But in practice, windows have become a potent metaphor for thinking about the self as a multiple, distributed, "time-sharing" system. The self is no longer simply playing different roles in different settings, something that people experience when, for example, one wakes up as a lover, makes breakfast as a mother, and drives to work as a lawyer. The windows metaphor perhaps merely suggests a distributed self that exists in many worlds and plays many roles at the same time. Cyberspace, however, translates that metaphor into a lived experience of *cycling through.*

Identity, Moratoria, and Play. For some people, cyberspace is a place to "act out" unresolved conflicts, to play and replay characterological difficulties on a new and exotic stage. For others, it provides an opportunity to "work through" significant personal issues, to use the new materials of cybersociality to reach for new resolutions. These more positive identity effects follow from the fact that for some, cyberspace provides what Erik Erikson would have called a "psychosocial moratorium," a central element in how Erikson thought about identity development in adolescence. Although the term *moratorium* implies a "time out," what Erikson had in mind was not withdrawal. On the contrary, the adolescent moratorium is a time of intense interaction with people and ideas. It is a time of passionate friendships and experimentation. The adolescent falls in and out of love with people and ideas. Erikson's notion of the moratorium was not a "hold" on significant experiences but on their consequences. It is a time during which one's actions are, in a certain sense, not counted as they will be later in life. They are not given as much weight, not given the force of full judgment. In this context, experimentation can become the norm rather than a brave departure. Relatively consequence-free experimentation facilitates the development of a *core self,* a personal sense of what gives life meaning that Erikson called "identity."

Erikson developed these ideas about the importance of a moratorium during the late 1950s and early 1960s. At that time, the notion corresponded to a common understanding of what "the college years" were about. Today, thirty years later, the idea of the college years as a consequence-free time out seems of another era. College is preprofessional and AIDS has made consequence-free sexual experimentation an impossibility. The years associated with adolescence no longer seem a time out. But if our culture no longer offers an adolescent moratorium, virtual communities often do. It is part of what makes them seem so attractive.

Erikson's ideas about stages did not suggest rigid sequences. His stages describe what people need to achieve before they can easily move ahead to another developmental task. For example, Erikson pointed out that successful intimacy in young adulthood is difficult if one does not come to it with a sense of who one is, the challenge of adolescent identity building. In real life, however, people frequently move on with serious deficits. With incompletely resolved "stages," they simply do the best they can. They use whatever materials they have at hand to get as much as they can of what they have missed. Now virtual social life can play a role in these dramas of self-reparation. Time in cyberspace reworks the notion of the moratorium because it may now exist on an always-available "window." Analysts need to note, respect, and interpret their patients' "life on the screen."

Having literally written our online personae into existence, they can be a kind of Rorschach. We can use them to become more aware of what we project into everyday life. We can use the virtual to reflect constructively on the real. Cyberspace opens the possibility for identity play, but it is very serious play. People who cultivate an awareness of what stands behind their screen personae are the ones most likely to succeed in using virtual experience for personal and social transformation. And the people who make the most of their lives on the screen are those who are capable of approaching it in a spirit of self-reflection. What does my behavior in cyberspace tell me about what I want, who I am, what I may not be getting in the rest of my life?

"Case" is a 34-year-old industrial designer happily married to a female coworker. Case describes his "real life" or "RL" persona as a "nice guy," a "Jimmy Stewart type like my father." He describes his outgoing, assertive mother as a "Katherine Hepburn type." For Case, who views assertiveness through the prism of this Jimmy Stewart/Katherine Hepburn dichotomy, an assertive man is quickly perceived as "being a bastard." An assertive woman, in contrast, is perceived as being "modern and together." Case says that although he is comfortable with his temperament and loves and respects his father, he feels he pays a high price for his low-key ways. In particular, he feels at a loss when it comes to confrontation, both at home and at work. Online, in a wide range of virtual communities, Case presents himself as females to whom he refers as his "Katherine Hepburn types": These are strong, dynamic, "out there" women. They remind Case of his mother, who "says exactly what's on her mind." He tells me that presenting himself as a woman online has brought him to a point where he is more comfortable with confrontation in his RL as a man. Additionally, Case has used cyberspace to develop a new model for thinking about his mind. He thinks of his Katherine Hepburn personae as various "aspects of the self." His online life reminds him of how Hindu gods could have different aspects or subpersonalities, or avatars, all the while being a whole self.

Case's inner landscape is very different from the inner landscapes of a person with multiple personality disorder. Case's inner actors are not split off from each other or his sense of "himself." He experiences himself very much as a collective whole, not feeling that he must goad or repress this or that aspect of himself into conformity. He is at ease, cycling through from Katherine Hepburn to Jimmy Stewart. To use the psychoanalyst Philip Bromberg's (1994) language, online life has helped Case learn how to "stand in the spaces between selves and still feel one, to see the multiplicity and still feel a unity." To use the computer scientist Marvin Minsky's (1987) language, Case feels at ease cycling through his "society of mind," a notion of identity as distributed and heterogeneous. Identity, from the Latin *idem,* has been typically used to refer to the sameness between two qualities. On the Internet, however, one can be many and usually is.

Most recently, Ray Kurzweil, inventor of the Kurzweil reading machine and AI researcher, has created a virtual alter ego: a female rock star named Ramona. Kurzweil is physically linked to Ramona. She moves when he moves; she speaks when he speaks (his voice is electronically transformed into that of a woman); she sings when he sings. What Case experienced in the relative privacy of an online virtual community, Kurzweil suggests will be standard identity play for all of us. Ramona can be expressed as a puppet on a computer screen as Kurzweil performs her and as an artificial intelligence on Kurzweil's web site. In the first case Kurzweil is performing "live," in the second case the bot Ramona is doing so.

Theory and objects-to-think-with. The notions of identity and multiplicity to which I was exposed in the late 1960s and early 1970s originated within the continental psychoanalytic tradition. These notions, most notably that there is no such thing as "the ego"—that each of us is a multiplicity of parts, fragments, and desiring connections—grew in the intellectual hot-house of Paris; they presented the world according to such authors as Jacques Lacan, Gilles Deleuze, and Félix Guattari. I met these ideas and their authors as a student in Paris, but despite such ideal conditions for absorbing theory, my "French lessons" remained abstract exercises. These theorists of poststructuralism spoke words that addressed the relationship between mind and body, but from my point of view had little to do with my own.

In my lack of personal connection with these ideas, I was not alone. To take one example, for many people it is hard to accept any challenge to the idea of an autonomous ego. Although in recent years, many psychologists, social theorists, psychoanalysts, and philosophers have argued that the self should be thought of as essentially decentered, the normal requirements of everyday life exert strong pressure on people to take responsibility for their actions and to see themselves as unitary actors. This disjuncture between theory (the unitary self is an illusion) and lived experience (the unitary self is the most basic reality) is one of the main reasons why multiple and decentered theories have been slow to catch on—or when they do, why we tend to settle back quickly into older, centralized ways of looking at things.

When, twenty years later, I first used my personal computer and modem to join online communities, I had an experience of this theoretical perspective that brought it shockingly down to earth. I used language to create several characters. My actions were textual—my words made things happen. I created selves that were made of and transformed by language. And in each of these different personae, I was exploring different aspects of my self. The notion of a decentered identity was concretized by experiences on a computer screen. In this way, cyberspace became an object to think with for thinking about identity. In cyberspace, identity is fluid and multiple, a signifier no longer clearly points to a thing that is signified, and understanding is less likely to proceed through analysis than by navigation through virtual space.

Appropriable theories, ideas that capture the imagination of the culture at large, tend to be those with which people can become actively involved. They tend to be theories that can be "played" with. So one way to think about the social appropriability of a given theory is to ask whether it is accompanied by its own objects-to-think-with that can help it move out beyond intellectual circles.

For example, the popular appropriation of Freudian theory had little to do with scientific demonstrations of its validity. Freudian theory passed into the popular culture because it offered robust and down-to-earth objects-to-think-with. The objects were not physical but almost-tangible ideas such as dreams and slips of the tongue. People were able to play with such Freudian "objects." They became used to looking for them and manipulating them, both seriously and not so seriously. And as they did so, the idea that slips and dreams betray an unconscious started to feel natural.

In Freud's work, dreams and slips of the tongue carried the theory. Today, life on the computer screen carries theory. People decide that they want to interact with others on a computer network. They get an account on a commercial service. They think that this will provide them with new access to people and information, and of course it does. But it does more. When they log on, they may find themselves playing multiple roles; they may find themselves playing characters of the opposite sex. In this way, they are swept up by experiences that enable them to explore previously unexamined aspects of their sexuality or that challenge their ideas about a unitary self. The instrumental computer, the computer that does things *for* us, has another side. It is also a subjective computer that does things *to* us—to our view of our relationships, to our ways of looking at our minds and ourselves.

Within the psychoanalytic tradition, many "schools" have departed from a unitary view of identity, among these the Jungian, object-relations, and Lacanian. In different ways, each of these groups of analysts was banished from the ranks of orthodox Freudians for such suggestions, or somehow relegated to the margins. As America became the center of psychoanalytic politics in the mid-twentieth century, ideas about a robust executive ego moved into the psychoanalytic mainstream.

These days, the pendulum has swung away from any complacent view of a unitary self. Through the fragmented selves presented by patients and through theories that stress the decentered subject, contemporary social and psychological thinkers are confronting what has been left out of theories of the unitary self. Online experiences with "parallel lives" are part of the significant cultural context that supports new ways of theorizing about nonpathological, indeed healthy, multiple selves.

RELATIONAL ARTIFACTS: A COMPANION SPECIES?

In Steven Spielberg's movie *AI: Artificial Intelligence,* scientists build a humanoid robot boy, David, who is programmed to love. David expresses this love to a woman who has adopted him as her child. In the discussion that followed the release of the film, emphasis usually fell on the question whether such a robot could *really* be developed. People thereby passed over a deeper question, one that historically has contributed to our fascination with the computer's burgeoning capabilities. That question concerns not what computers can do or what computers will be like in the future, but rather, what *we* will be like. What kinds of people are we becoming as we develop more and more intimate relationships with machines?

In this context, the pressing issue in *A.I.* is not the potential "reality" of a nonbiological son, but rather that faced by his adoptive mother—a biological woman whose response to a machine that asks for her nurturance is the desire to nurture it, and whose response to a nonbiological creature who reaches out to her is to feel attachment, horror, love, and confusion.

The questions faced by the mother in *A.I.* include "What kind of relationship is it appropriate, desirable, imaginable to have with a machine?" and "What is a relationship?" Although artificial intelligence research has not come close to creating a robot such as Spielberg's David, these questions have become current, even urgent.

Today, we are faced with relational artifacts to which people respond in ways that have much in common with the mother in *A.I.* These artifacts are not perfect human replicas as was David, but they are able to push certain emotional buttons (think of them perhaps as evolutionary buttons). When a robotic creature makes eye contact, follows your gaze, and gestures toward you, you are provoked to respond to that creature as a sentient and even caring other. Psychoanalytic thought offers materials that can deepen our understanding of what we feel when we confront a robot child who asks us for love. It can help us explore what moral stance we might take if we choose to pursue such relationships.

There is every indication that the future of computational technology will include relational artifacts that have feelings, life cycles, and moods; that reminisce; and that have a sense of humor—that say they love us, and expect us to love them back. What will it mean to a person when their primary daily companion is a robotic dog? Or their health care "attendant" is built in the form of a robot cat? Or their software program attends to their emotional states and, in turn, has affective states of its own? In order to study these questions, I have embarked on a research project that includes fieldwork in robotics laboratories, among children playing with virtual pets and digital dolls, and among the elderly to whom robotic companions are starting to be aggressively marketed.

I have noted that in the over two decades in which I have explored people's relationships with computers, I have used the metaphor of the Rorschach, the computer as a screen on which people projected their thoughts and feelings, their very different cognitive styles. With relational artifacts, the Rorschach model of a computer/human relationship breaks down. People are learning to interact with computers through conversation and gesture; people are learning that to relate successfully to a computer you have to assess its emotional "state."

In my previous research on children and computer toys, children described the lifelike status of machines in terms of their cognitive capacities (the toys could "know" things and "solve" puzzles). In my studies on children and Furbies, I found that children describe these new toys as "sort of alive" because of the quality of their emotional attachments to the objects and because of the idea that the Furby might be emotionally attached to them. So, for example, when I ask the question, "Do you think the Furby is alive?" children answer not in terms of what the Furby can do, but how they feel about the Furby and how the Furby might feel about them.

> Ron (6): Well, the Furby is alive for a Furby. And you know, something this smart should have
>   arms. It might want to pick up something or to hug me.
> Katherine (5): Is it alive? Well, I love it. It's more alive than a Tamagotchi because it sleeps with
>   me. It likes to sleep with me.
> Jen (9): I really like to take care of it. So, I guess it is alive, but it doesn't need to really eat, so
>   it is as alive as you can be if you don't eat. A Furby is like an owl. But it is more alive than
>   an owl because it knows more and you can talk to it. But it needs batteries so it is not an
>   animal. It's not like an animal kind of alive.

Although we are just at the early stages of studying children and relational artifacts, several things seem clear. Today's children are learning to distinguish between an "animal kind of alive" and a "Furby kind of alive." The category of "sort of alive" becomes used with increasing frequency. And quite often, the boundaries between an animal kind of alive and a Furby kind of alive blur as the children attribute more and more lifelike properties to the emotive toy robot. So, for example, eight-year-old Laurie thinks that Furbies are alive, but die when their batteries are removed. People are alive because they have hearts, bodies, lungs, "and a big battery inside. If somebody kills you—maybe it's sort of like taking the batteries out of the Furby."

Furthermore, today's children are learning to have expectations of emotional attachments to computers, not in the way we have expectations of emotional attachment to our cars and stereos, but in the way we have expectations about our emotional attachments to people. In the process, the very meaning of the word *emotional* may change. Children talk about an "animal kind of alive and a Furby kind of alive." Will they also talk about a "people kind of love" and a "computer kind of love"?

We are in a different world from the old AI debates of the 1960s to 1980s, in which researchers argued about whether machines could be *really* intelligent. The old debate was essentialist; the new objects sidestep such arguments about what is inherent in *them* and play instead on what they evoke in *us:* when we are asked to care for an object, when the cared-for object thrives and offers us its attention and concern, we experience that object as intelligent, but more important, we feel a connection to it. So the question here is not to enter a debate about whether objects "really" have emotions, but to reflect on what relational artifacts evoke in the *user.*

How will interacting with relational artifacts affect people's way of thinking about themselves, their sense of human identity, of what makes people special? Children have traditionally defined what makes people special in terms of a theory of "nearest neighbors." So, when the nearest neighbors (in children's eyes) were their pet dogs and cats, people were special because they had reason. The Aristotelian definition of man as a rational animal made sense even for the youngest children. But when, in the 1980s, it seemed to be the computers who were the nearest neighbors, children's approach to the problem changed. Now, people were special not because they were rational animals but because they were emotional machines. So, in 1983, a ten-year-old told me: "When there are the robots that are as smart as the people, the people will still run the restaurants, cook the food, have the families, I guess they'll still be the only ones who'll go to Church."

Now in a world in which machines present themselves as emotional, what is left for us?

One woman's comment on AIBO, Sony's household entertainment robot, startles in what it might augur for the future of person-machine relationships: "[AIBO] is better than a real dog. . . . It won't do dangerous things, and it won't betray you. . . . Also, it won't die suddenly and make you feel very sad."

In Ray Bradbury's story (1998/1948), "I sing the body electric," a robotic, electronic grandmother is unable to win the trust of the girl in the family, Agatha, until the girl learns that the grandmother, unlike her recently deceased mother, cannot die. In many ways throughout the story, we learn that the grandmother is actually better than a human caretaker—more able to attend to each family member's needs, less needy, with perfect memory and inscrutable skills—and, most importantly, not mortal.

Mortality has traditionally defined the human condition; a shared sense of mortality has been the basis for feeling a commonality with other human beings, a sense of going through the same life cycle, a sense of the preciousness of time and life, and of its fragility. Loss (of parents, of friends, of family) is part of the way we understand how human beings grow and develop and bring the qualities of other people within themselves.

The possibilities of engaging emotionally with creatures that will not die, whose loss we will never need to face, presents dramatic questions that are based on current technology—not issues of whether the technology depicted in *AI* could really be developed.

The question "What kinds of relationships is it appropriate to have with machines?" has been explored in science fiction and in technophilosophy. But the sight of children and the elderly exchanging tenderness with robotic pets brings science fiction into everyday life and technophilosophy down to earth. In the end, the question is not just whether our children will come to love their toy robots more than their parents, but what will *loving* itself come to mean?

CONCLUSION: TOWARD THE FUTURE
OF THE COMPUTER CULTURE

Relational artifacts are being presented to us as companionate species at the same time that other technologies are carrying the message that mind is mechanism, most notably psycho-pharmacology. In my studies of attitudes toward artificial intelligence and robotics, people more and more are responding to a question about computers with an answer about psycho-pharmacology. Once Prozac has made someone see his or her mind as a biochemical machine, it seems a far smaller step to see the mind as reducible to a computational one. Twenty years ago, when my student turned a Freudian slip into an information-processing error, it was computational models that seemed most likely to spread mechanistic thinking about mind. Today, psychopharmacology is the more significant backdrop to the rather casual introduction of relational artifacts as companions, particularly for the elderly and for children.

The introduction of these objects is presented as good for business and (in the case of children) good for "learning" and "socialization. It is also presented as realistic social policy. This is the "robot or nothing" argument: if the old people don't get the robots, they certainly aren't going to get a pet. Many people do find the idea of robot companions unproblematic. Their only question about them is "Does it work?" By this, they usually mean, "Does it keep the elderly people/children quiet?" There are, of course, many other questions. To begin with, (even considering) putting artificial creatures in the role of companions to our children and parents raises the question of their moral status.

Already, there are strong voices that argue the moral equivalence of robots as a companion species. Kurzweil (1999) writes of an imminent age of "spiritual machines," by which he means machines with enough self-consciousness that they will deserve moral and spiritual recognition (if not parity) with their human inventors. Computer "humor," which so recently played on anxieties about whether or not people could "pull the plug" on machines, now portrays the machines confronting their human users with specific challenges. One *New Yorker* cartoon has the screen of a desktop computer asking: "I can be upgraded. Can you?" Another cartoon makes an ironic reference to Kurzweil's own vision of "downloading" his mind onto a computer chip. In this cartoon, a doctor, speaking to his surgical patient hooked up to an I.V. drip, says: "You caught a virus from your computer and we had to erase your brain. I hope you kept a back-up copy."

Kurzweil's argument for the moral (indeed spiritual) status of machines is intellectual, theoretical. Cynthia Breazeal's comes from her experience of connection with a robot. Breazeal was leader on the design team for Kismet, the robotic head that was designed to interact with humans "sociably," much as a two-year-old child would. Breazeal was its chief programmer, tutor, and companion. Kismet needed Breazeal to become as "intelligent" as it did and then Kismet became a creature Breazeal could interact with. Breazeal experienced what might be called a maternal connection to Kismet; she certainly describes a sense of connection with it as more than "mere" machine. When she graduated from MIT and left the AI Laboratory, where she had done her doctoral research, the tradition of academic property rights demanded that Kismet be left behind in the laboratory that had paid for its development. What she left behind was the robot "head" and its attendant software. Breazeal describes a sharp sense of loss. Building a new Kismet would not be the same.

It would be facile to analogize Breazeal's situation to that of the mother in Spielberg's *A.I.*, but she is, in fact, one of the first people in the world to have one of the signal experiences in that story. The issue is not Kismet's achieved level of intelligence, but Breazeal's human experience as a caretaker. Breazeal "brought up" Kismet and then could interact with it through inflection, gesture, and example. What we need today is a new object relations psychology that will help us understand such relationships and, indeed, responsibly navigate them. Breazeal's concerns have been for being responsible to the robots, acknowledging their moral status. My concern is centered on the *humans* in the equation. In concrete terms: first we need to understand Cynthia Breazeal's relationship to Kismet; second, we need to find a language for achieving some critical distance on it. Caring deeply for a machine that presents itself as a relational partner changes who we are as people. Presenting a machine to an aging parent as a companion changes who we are as well. Walt Whitman said, "A child goes forth every day / And the first object he look'ed upon / That object he became." We make our technologies, and our technologies make and shape us. We are not going to be the same people we are today, on the day we are faced with machines with which we feel in a relationship of mutual affection.

Even when the concrete achievements in the field of artificial intelligence were very primitive, the mandate of AI has always been controversial, in large part because it challenged ideas about human "specialness" and specificity. In the earliest days of AI, what seemed threatened was the idea that people were special because of their intelligence. There was much debate about whether machines could ever play chess; the advent of a program that could beat its creator in a game of checkers was considered a moment of high intellectual and religious importance (Weiner, 1964). By the mid-1980s, anxiety about what AI challenged about human specialness had gone beyond whether machines would be "smart" and had moved to emotional and religious terrain. At MIT, Marvin Minsky's students used to say that he wanted to build a computer "complex enough that a soul would want to live in it." Most recently, AI scientists are emboldened in their claims. They suggest the moral equivalence of people and machines. Ray Kurzweil argues that machines will be spiritual; Rodney Brooks (2002) argues that the "us and them" problem of distinguishing ourselves from the robots will disappear because we are becoming more robotic (with chips and implants) and the robots are becoming more human (with biological parts instead of silicon-based ones).

The question of human specificity and the related question of the moral equivalence of people and machines have moved from the periphery to the center of discussions about artificial intelligence. One element of "populist" resistance to the idea of moral equivalence finds expression in a number of narratives. Among these is the idea that humans are special because of their imperfections. A ten-year-old who has just played with Breazeal's Kismet says, "I would love to have a robot at home. It would be such a good friend. But it couldn't be a best friend. It might know everything but I don't. So it wouldn't be a best friend." There is resistance from the experience of the life cycle. An adult confronting an "affective" computer program designed to function as a psychotherapist says, "Why would I want to talk about sibling rivalry to something that was never born and never had a mother?" In the early days of the Internet, a *New Yorker* cartoon captured the essential psychological question: paw on keyboard, one dog says to another, "On the Internet, nobody knows you're a dog." This year, a very different cartoon summed up more recent anxieties. Two grownups face a child in a wall of solidarity, explaining: "We're neither software nor hardware. We're your parents." The issue raised in this cartoon is the specificity of human beings and human meaning. We are back to the family, to the life cycle, to human fragility and the experience of the human body. We are back to the elements of psychoanalytic culture.

With the turn of the millennium, we came to the end of the Freudian century. It is fashionable to argue that we have moved from a psychoanalytic to a computer culture, that there is no longer any need to talk about Freudian slips now that we can talk about information processing errors in an age of computation and neuroscience. In my view, however, the very opposite is true.

We must cultivate the richest possible language and methodologies for talking about our increasingly emotional relationships with artifacts. We need far closer examination of how artifacts enter the development of self and mediate between self and other. Psychoanalysis provides a rich language for distinguishing between need (something that artifacts may have) and desire (which resides in the conjunction of language and flesh). It provides a rich language for exploring the specificity of human meanings in their connections to the body. Finally, to come full circle, if one reinterprets Freudian slips in computational terms—if one moves from explanations in terms of meaning to ones in terms of mechanism—there is often a loss in appreciation for complexity and ambivalence.

Immersion in simulation games and relationships with digital creatures puts us in reassuring microworlds where the rules are clear. But never have we so needed the ability to think, so to speak, "ambivalently," to consider life in shades of gray, consider moral dilemmas that aren't battles for "infinite justice" between Good and Evil. Never have we so needed to be able to hold many different and contradictory thoughts and feelings at the same time. People may be comforted by the notion that we are moving from a psychoanalytic to a computer culture, but what the times demand is a passionate quest for joint citizenship.

REFERENCES

Bradbury, R. (1998). *I Sing the Body Electric and Other Stories.* New York: Avon Books. (Original work published 1948)

Bromberg, P. (1994). Speak that I may see you: Some reflections on dissociation, reality, and psychoanalytic listening. *Psychoanalytic Dialogues, 4*(4): 517–547.

Brooks, R. (2002). *Flesh and machines: How robots will change us.* New York: Pantheon.

Erikson, E. (1963). *Childhood and society* (2nd rev. ed.). New York: W. W. Norton.

Freud, S. (1960). Mourning and melancholia. In J. Strachey (Ed. and Trans.), *The standard edition of the complete psychological works of Sigmund Freud.* London: Hogarth Press. (Original work published 1917)

Freud, S. (1960). Slips of the tongue. In J. Strachey (Ed. and Trans.), *The standard edition of the complete psychological works of Sigmund Freud.* London: Hogarth Press. (Original work published 1901)

Greenberg, J. R., & Mitchell, S. A. (1983). *Object relations in psychoanalytic theory.* Cambridge, MA: Harvard University Press.

Haraway, D. (1991). *Simians, cyborgs and women.* New York: Routledge.

Kurzweil, R. (1999). *The age of spiritual machines: When computers exceed human intelligence.* New York: Viking.

Minsky, M. (1987). *Society of mind.* New York: Simon & Schuster.

Ornstein, P. H. (Ed.). (1978). *The search for the self: Selected writings of Heinz Kohut: 1950–1978* (Vol. 2). New York: International Universities Press, Inc.

Turkle, S. (1984). *The second self: Computers and the human spirit.* New York: Simon & Schuster.

Turkle, S. (1988). Psychoanalysis and artificial intelligence: A new alliance. *Daedalus, 117,* 1.

Turkle, S. (1991). *Psychoanalytic politics: Jacques Lacan and Freud's French revolution* (2nd rev. ed.). New York: Guilford Press.

Turkle, S. (1995). *Life on the screen: Identity in the age of the Internet.* New York: Simon & Schuster.

Weiner, N. (1964). *God and Golem, Inc.: A comment on certain points where cybernetics impinges on religion.* Cambridge, MA: MIT Press.

Winnicott, D. W. (1971). *Playing and reality.* New York: Basic Books.

Part VI

TECHNOLOGY AND SCIENCE

Recently historians, social scientists, and philosophers have begun to examine the role that technologies play in science. These philosophers of *technoscience* oppose and ultimately reverse the received view that technology is the concrete manifestation of abstract scientific principles. Instead they claim that science is made possible by technological instruments and devices. Our theoretical understanding of nature depends on the materiality of machines. For these thinkers, science is seen less as a lofty, intellectual pursuit of timeless truths of nature than as a practical activity of building devices that help us accomplish tasks. We learn about nature in so far as we can manipulate it and experiment on it. Instruments tell us what nature "really" is like. If we had no instruments, we'd have no *scientific* knowledge of nature. This conclusion makes science less a practice of addressing ideas than one that is about machines. Science is embodied in its technologies, and technologies determine what is science. Similarly, we find metaphysics and epistemology in machines and machines in metaphysics and epistemology. We also find social knowledge embodied in both science and technology. Scientific practice is a web of humans, machines, and social relations. The readings in this section examine this web of reason and materiality to show how it functions at the heart of scientific practice.

In "Experimentation and Scientific Realism," Ian Hacking examines the conditions that make it reasonable to accept the entities proposed by scientific theories to be real. *Scientific realism* is the belief that the theories of science give a literally true account of the world, including the unobservable world of theoretically postulated entities. Such realism is opposed to *scientific antirealism,* which maintains that theories in science are models or convenient methods of representation that help us understand and predict truths in the observable world. Hacking is a realist, but he is more concerned with the ontological status of entities (including processes, waves, currents, interactions, black holes, and the like) rather than the literal truth of theories. By examining the practices of experimental scientists, he claims we have firm evidence for believing that unobservable, theoretical entities really exist. But they exist as instruments or tools to solve problems and create more theories about nature. We can sidestep the issue of the truth of a theory, Hacking claims, so long as the proposed entity does work for us. Hacking agrees with Francis Bacon, the first philosopher of experimentation, that a scientific experiment *interferes* with the course of nature. The experimenter should be convinced of the reality of unseen entities only when he or she uses them to cause interactions elsewhere.

The "experimental argument for realism" that Hacking proposes states that the proof that something exists is our ability to manipulate it using devices, instruments, and machines.

In "Laboratories," excerpted from *Science in Action,* Bruno Latour argues that scientific facts are constructed, not discovered, through an interplay of laboratories and power relations. The traditional view of science claims that nature is the object of scientific inquiry and ultimately the arbiter of scientific truth. Scientists discover nature's truths through laboratory experimentation and record them in scientific journals. Latour inverts this progression by beginning with the literature of science, moving back to the laboratory, and eventually arriving at what we call *nature.* The purpose of the inversion is to highlight the activity of fabricating scientific facts—something that only becomes apparent when we examine science *in action.* We get a very different picture of scientific practice prior to the settling of a dispute as compared with science after the settling of a dispute. From this perspective, scientific practice is a technologically mediated, power-laden social institution for creating and resolving controversies.

Latour shows how challenges to scientific literature occur in the laboratory, which is not only the high-cost place where scientists work but also where "inscriptions" are made. Laboratory instruments are inscription-making devices that depict nature. We do not directly experience nature when employing instruments; rather, we interpret effects of it through a visual display. Reading instrumentally produced inscriptions requires training and practice, Latour reminds us, often producing conflicting interpretations or "trials of strength" among competing individuals and groups. The result of disagreement is a scientific object, produced in a laboratory. Laboratories invent new, indeterminate objects and readings that become defined by the results of trials of strength. If an object passes the trials that test the ties linking the representatives of a scientific claim to what they speak for—including dissenters, scientific literature, spokespeople, and experimentation—then, finally, we arrive at nature. Latour argues that we can never use nature to explain how and why a controversy ends; nature, rather, is the consequence of the settlement. Technoscience is the ultimate referee.

In "Scientific Visualism," Don Ihde also examines the use of instruments in scientific practice. His method of "perceptualist hermeneutics" describes a bodily sensory perception that also interprets the world. Ihde uses this contextualist, interpretive way of seeing to reveal how scientific instruments mediate experience and create new forms of perception. This technologically mediated interpretive perception is referred to as "instrumental realism": the real world only becomes an object of scientific discovery and explanation when it is constituted by scientific instruments. He carries out a "weak" program and a "strong" program to show how reality is increasingly transformed by instruments. The weak program examines the ways that instruments prepare reality for observation. Science is a hermeneutic practice that relies on instruments to make things scientifically analyzable. One set of devices—for example, imaging technologies like telescopes, x rays, MRI scans, and sonograms—makes things visible. Another set of inscription-making devices—for example, oscilloscopes and spectrographs—makes things readable. Both kinds of technologies present the world in perception-transforming visual forms that must be interpreted to be understood.

The strong program examines the ways that instruments constitute and make an otherwise invisible reality visible. These devices not only bring the object of science into view, but they also shape and "give a voice" to the world so that it may be experienced. This is what Ihde calls "technoconstruction." He identifies several different "instrumental phenomenological variations," or the various ways that technology mediates perception and constitutes the content of science. For example, astronomers often rely on new instruments to reveal previously unknown phenomena. More instruments produce more scientific objects. Another variation shows how a range of instruments may be used to measure different processes of single objects—for example, medical imaging technologies to view the brain using x rays, ultrasound, and MRI. Again, scientific instruments produce a previously invisible phenomenon that is then translated into visible images for scientific observation. Different combinations of different instruments produce different interpretive phenomena. Ihde's theory of technoconstruction steers a path between an unmediated, naïve realism and social constructivism. He interprets science in terms of a visual hermeneutics, embodied within an instrumentally realist framework in which instruments mediate perceptions.

In "Should Philosophies of Science Encode Democratic Ideals?" Sandra Harding argues that science's internal, cognitive, technical core is not immune from social and political concerns. Rather, they reflect the meanings, desires, priorities, and aspirations of a society and encode them into scientific and technical practices. The "cognitive democracy" that Harding proposes challenges the traditional view that science is the objective, value-neutral pursuit of universal truths of nature. Harding aims to expose the cultural, social, and political elements that have always been part of scientific practice. This approach claims that the sciences can and should more effectively encode democratic ideals. The sciences already encode such democratic principles as epistemic equality (i.e., everyone can experience the truth), publicity (i.e., information must be made public to be corroborated), and ontological neutrality (i.e., the stuff of nature is not organized into hierarchies of value). Democratic ideals are already part of the cognitive, technical core of scientific practice. The question Harding pursues is whether these are the best ideals that could be encoded into it.

Harding believes we can do better. Whatever social ideals are adopted into science's cognitive cores should not only advance the growth of scientific knowledge but also promote positive political effects as well. The strongest proposal for achieving democratic social relations requires full participation in decision-making processes that occur among groups in a state of "real equality" (i.e., where sexism, racism, and classism play no role in the distribution of political and economic resources). Real equality would transform scientific institutions and practices; this transformation would, in turn, advance democratic social relations. Harding warns that the "universality ideal" implicit in the "unity of science" (one world, one truth, one science) encodes antidemocratic values and interests at both political and scientific costs. The universality ideal in philosophy of science presupposes an ideal human mind and an ideal group of humans that is able to grasp the truth of nature. The problem with that view of human nature, Harding argues, is that it fails to recognize the tremendous cognitive diversity among individuals and groups. Cultures have different interests, discourses, and interpretations of nature. The political cost of devaluing these "local knowledge traditions" is to devalue the people and cultures that use them. The universality ideal fails to respect diversity, and thus it cannot be considered to be politically neutral. The scientific cost of devaluing local knowledge traditions is to exclude from consideration potentially useful and insightful perspectives on nature. Harding invites us to consider that all knowledge systems are in fact hybrids. If philosophies of science could incorporate different knowledge systems (including the universality ideal), they would more effectively encode democratic ideals and help promote real equality.

28

Experimentation and Scientific Realism

Ian Hacking

Experimental physics provides the strongest evidence for scientific realism. Entities that in principle cannot be observed are regularly manipulated to produce new phenomena and to investigate other aspects of nature. They are tools, instruments not for thinking but for doing.

The philosopher's standard "theoretical entity" is the electron. I will illustrate how electrons have become experimental entities, or experimenter's entities. In the early stages of our discovery of an entity, we may test hypotheses about it. Then it is merely a hypothetical entity. Much later, if we come to understand some of its causal powers and use it to build devices that achieve well-understood effects in other parts of nature, then it assumes quite a different status.

Discussions about scientific realism or antirealism usually talk about theories, explanation, and prediction. Debates at that level are necessarily inconclusive. Only at the level of experimental practice is scientific realism unavoidable—but this realism is not about theories and truth. The experimentalist need only be a realist about the entities used as tools.

A PLEA FOR EXPERIMENTS

No field in the philosophy of science is more systematically neglected than experiment. Our grade school teachers may have told us that scientific method is experimental method, but histories of science have become histories of theory. Experiments, the philosophers say, are of value only when they test theory. Experimental work, they imply, has no life of its own. So we lack even a terminology to describe the many varied roles of experiment. Nor has this one-sidedness done theory any good, for radically different types of theory are used to think about the same physical phenomenon (e.g., the magneto-optical effect). The philosophers of theory have not noticed this and so misreport even theoretical enquiry.

Different sciences at different times exhibit different relationships between "theory" and "experiment." One chief role of experiment is the creation of phenomena. Experimenters bring into being phenomena that do not naturally exist in a pure state. These phenomena are the touchstones of physics, the keys to nature, and the source of much modern technology. Many are what physicists after the 1870s began to call "effects": the photoelectric effect, the Compton effect, and so forth.[1] A recent high-energy extension of the creation of phenomena is the creation of "events," to use the jargon of the trade. Most of the phenomena, effects, and events created by the experimenter are like plutonium: They do not exist in nature except possibly on vanishingly rare occasions.[2]

In this paper I leave aside questions of methodology, history, taxonomy, and the purpose of experiment in natural science. I turn to the purely philosophical issue of scientific realism. Simply call it "realism" for short. There are two basic kinds: realism about entities and realism about theories. There is no agreement on the precise definition of either. Realism about theories says that we try to form true theories about the world, about the inner constitution of matter and about the outer reaches of space. This realism gets its bite from optimism: We think we can do well in this project and have already had partial success. Realism about entities—and I include processes, states, waves, currents, interactions, fields, black holes, and the like among entities—asserts the existence of at least some of the entities that are the stock in trade of physics.[3]

The two realisms may seem identical. If you believe a theory, do you not believe in the existence of the entities it speaks about? If you believe in some entities, must you not describe them in some theoretical way that you accept? This seeming identity is illusory. *The vast majority of experimental physicists are realists about entities but not about theories.* Some are, no doubt, realists about theories too, but that is less central to their concerns.

Experimenters are often realists about the entities that they investigate, but they do not have to be so. R. A. Millikan probably had few qualms about the reality of electrons when he set out to measure their charge. But he could have been skeptical about what he would find until he found it. He could even have remained skeptical. Perhaps there is a least unit of electric charge, but there is no particle or object with exactly that unit of charge. Experimenting on an entity does not commit you to believing that it exists. Only manipulating an entity, in order to experiment on something else, need do that.

Moreover, it is not even that you use electrons to experiment on something else that makes it impossible to doubt electrons. Understanding some causal properties of electrons, you guess how to build a very ingenious, complex device that enables you to line up the electrons the way you want, in order to see what will happen to something else. Once you have the right experimental idea, you know in advance roughly how to try to build the device, because you know that this is the way to get the electrons to behave in such and such a way. Electrons are no longer ways of organizing our thoughts or saving the phenomena that have been observed. They are now ways of creating phenomena in some other domain of nature. Electrons are tools.

There is an important experimental contrast between realism about entities and realism about theories. Suppose we say that the latter is belief that science aims at true theories. Few experimenters will deny that. Only philosophers doubt it. Aiming at the truth is, however, something about the indefinite future. Aiming a beam of electrons is using present electrons. Aiming a finely tuned laser at a particular atom in order to knock off a certain electron to produce an ion is aiming at present electrons. There is, in contrast, no present set of theories that one has to believe in. If realism about theories is a doctrine about the aims of science, it is a doctrine laden with certain kinds of values. If realism about entities is a matter of aiming electrons next week or aiming at other electrons the week after, it is a doctrine much more neutral between values. The way in which experimenters are scientific realists about entities is entirely different from ways in which they might be realists about theories.

This shows up when we turn from ideal theories to present ones. Various properties are confidently ascribed to electrons, but most of the confident properties are expressed in numerous different theories or models about which an experimenter can be rather agnostic. Even people in a team, who work on different parts of the same large experiment, may hold different and mutually incompatible accounts of electrons. That is because different parts of the experiment will make different uses of electrons. Models good for calculations on one aspect of electrons will be poor for others. Occasionally, a team actually has to select a member with a quite different theoretical perspective simply to get someone who can solve those experimental problems. You may choose someone with a foreign training, and whose talk is well-nigh incommensurable with yours, just to get people who can produce the effects you want.

But might there not be a common core of theory, the intersection of everybody in the group, which is the theory of the electron to which all the experimenters are realistically committed? I would say common lore, *not* common core. There are a lot of theories, models, approximations, pictures, formalisms, methods, and so forth involving electrons, but there is no reason to suppose that the intersection of these is a theory at all. Nor is there any reason to think that there is such a thing as "the most powerful nontrivial *theory* contained in the intersection of all the theories in which this or that member of a team has been trained to believe." Even if there are a lot of shared beliefs, there is no reason to suppose they form anything worth calling a theory. Naturally, teams tend to be formed from like-minded people at the same institute, so there is usually some real shared theoretical basis to their work. That is a sociological fact, not a foundation for scientific realism.

I recognize that many a scientific realism concerning theories is a doctrine not about the present but about what we might achieve, or possibly an ideal at which we aim. So to say that there is no present theory does not count against the optimistic aim. The point is that such scientific realism about theories has to adopt the Peircean principles of faith, hope, and charity. Scientific realism about entities needs no such virtues. It arises from what we can do at present. To understand this, we must look in some detail at what it is like to build a device that makes the electrons sit up and behave.

OUR DEBT TO HILARY PUTNAM

It was once the accepted wisdom that a word such as "electron" gets its meaning from its place in a network of sentences that state theoretical laws. Hence arose the infamous problems of incommensurability and theory change. For if a theory is modified, how could a word such as "electron" go on meaning the same? How could different theories about electrons be compared, since the very word "electron" would differ in meaning from theory to theory?

Putnam saved us from such questions by inventing a referential model of meaning. He says that meaning is a vector, refreshingly like a dictionary entry. First comes the syntactic marker (part of speech); next the semantic marker (general category of thing signified by the word); then the stereotype (clichés about the natural kind, standard examples of its use, and present-day associations. The stereotype is subject to change as opinions about the kind are modified.) Finally, there is the actual referent of the word, the very stuff, or thing, it denotes if it denotes anything. (Evidently dictionaries cannot include this in their entry, but pictorial dictionaries do their best by inserting illustrations whenever possible.)[4]

Putnam thought we can often guess at entities that we do not literally point to. Our initial guesses may be jejune or inept, and not every naming of an invisible thing or stuff pans out. But when it does, and we frame better and better ideas, then Putnam says that, although the stereotype changes, we refer to the same kind of thing or stuff all along. We and Dalton alike spoke about the same stuff when we spoke of (inorganic) acids. J. J. Thomson, H. A Lorentz, Bohr, and Millikan were, with their different theories and observations, speculating about the same kind of thing, the electron.

There is plenty of unimportant vagueness about when an entity has been successfully "dubbed," as Putnam puts it. "Electron" is the name suggested by G. Johnstone Stoney in 1891 as the name for a natural unit of electricity. He had drawn attention to this unit in 1874. The name was then applied to the subatomic particles of negative charge, which J. J. Thomson, in 1897, showed cathode rays consist of. Was Johnstone Stoney referring to the electron? Putnam's account does not require an unequivocal answer. Standard physics books say that Thomson discovered the electron. For once I might back theory and say that Lorentz beat him to it. Thomson called his electrons "corpuscles," the subatomic particles of electric charge. Evidently, the name does not matter much. Thomson's most notable achievement was to measure the mass of the electron. He did this by a rough (though quite good) guess at e, and by making an excellent determination of e/m, showing that m is about 1/1800 the mass of the hydrogen atom. Hence it is natural to say that Lorentz merely postulated the existence of a particle of negative charge, while Thomson, determining its mass, showed that there is some such real stuff beaming off a hot cathode.

The stereotype of the electron has regularly changed, and we have at least two largely incompatible stereotypes, the electron as cloud and the electron as particle. One fundamental enrichment of the idea came in the 1920s. Electrons, it was found, have angular momentum, or "spin." Experimental work by O. Stern and W. Gerlach first indicated this, and then S. Goudsmit and G. E. Uhlenbeck provided the theoretical understanding of it in 1925. Whatever we think, Johnstone Stoney, Lorentz, Bohr, Thomson, and Goudsmit were all finding out more about the same kind of thing, the electron.

We need not accept the fine points of Putnam's account of reference in order to thank him for giving us a new way to talk about meaning. Serious discussion of inferred entities need no longer lock us into pseudo-problems of incommensurability and theory change. Twenty-five years ago the experimenter who believed that electrons exist, without giving much credence to any set of laws about electrons, would have been dismissed as philosophically incoherent. Now we realize it was the philosophy that was wrong, not the experimenter. My own relationship to Putnam's account of meaning is like the experimenter's relationship to a theory. I do not literally believe Putnam, but I am happy to employ his account as an alternative to the unpalatable account in fashion some time ago. . . .

INTERFERING

Francis Bacon, the first and almost last philosopher of experiments, knew it well: The experimenter sets out "to twist the lion's tail." Experimentation is interference in the course of nature; "nature under constraint and vexed; that is to say, when by art and the hand of man she is forced out of her natural state, and squeezed and molded."[5] The experimenter is convinced of the reality of entities, some of whose causal properties are sufficiently well understood that they can be used to interfere *elsewhere* in nature. One is impressed by entities that one can use to test conjectures about other, more hypothetical entities. In my example, one is sure of the electrons that are used to investigate weak neutral currents and neutral bosons. This should not be news, for why else are we (nonskeptics) sure of the reality of even macroscopic objects, but because of what we do with them, what we do to them, and what they do to us?

Interference and interaction are the stuff of reality. This is true, for example, at the borderline of observability. Too often philosophers imagine that microscopes carry conviction because they help us see better. But that is only part of the story. On the contrary, what counts is what we can do to a specimen under a microscope, and what we can see ourselves doing. We stain the specimen, slice it, inject it, irradiate it, fix it. We examine it using different kinds of microscopes that employ optical systems that rely on almost totally unrelated facts about light. Microscopes carry conviction because of the great array of interactions and interferences that are possible. When we see something that turns out to be unstable under such play, we call it an artifact and say it is not real.[6]

Likewise, as we move down in scale to the truly unseeable, it is our power to use unobservable entities that makes us believe they are there. Yet, I blush over these words "see" and "observe." Philosophers and physicists often use these words in different ways. Philosophers tend to treat opacity to visible light as the touchstone of reality, so that anything that cannot be touched or seen with the naked eye is called a theoretical or inferred entity. Physicists, in contrast, cheerfully talk of observing the very entities that philosophers say are not observable. For example, the fermions are those fundamental constituents of matter such as electron neutrinos and deuterons and, perhaps, the notorious quarks. All are standard philosophers' "unobservable" entities. C. Y. Prescott, the initiator of the experiment described below, said in a recent lecture, that "of these fermions, only the *t* quark is yet unseen. The failure to observe $t\bar{t}$ states in e^+e^- annihilation at PETRA remains a puzzle."[7] Thus, the physicist distinguishes among the philosophers' "unobservable" entities, noting which have been observed and which not. Dudley Shapere has just published a valuable study of this fact.[8] In his example, neutrinos are used to see the interior of a star. He has ample quotations such as "neutrinos present the only way of directly observing" the very hot core of a star.

John Dewey would have said that fascination with seeing-with-the-naked-eye is part of the spectator theory of knowledge that has bedeviled philosophy from earliest times. But I do not think Plato or Locke or anyone before the nineteenth century was as obsessed with the sheer opacity of objects as we have been since.

MAKING

Even if experimenters are realists about entities, it does not follow that they are right. Perhaps it is a matter of psychology: Maybe the very skills that make for a great experimenter go with a certain cast of mind which objectifies whatever it thinks about. Yet this will not do. The experimenter cheerfully regards neutral bosons as merely hypothetical entities, while electrons are real. What is the difference?

There are an enormous number of ways in which to make instruments that rely on the causal properties of electrons in order to produce desired effects of unsurpassed precision. I shall illustrate this. The argument—it could be called the "experimental argument for realism"—is not that we infer the reality of electrons from our success. We do not make the instruments and then infer the reality of the electrons, as when we test a hypothesis, and then believe it because it passed the test. That gets the time order wrong. By now we design apparatus relying on a modest number of home truths about electrons, in order to produce some other phenomenon that we wish to investigate.

That may sound as if we believe in the electrons because we predict how our apparatus will behave. That too is misleading. We have a number of general ideas about how to prepare polarized electrons, say. We spend a lot of time building prototypes that do not work. We get rid of innumerable bugs. Often we have to give up and try another approach. Debugging is not a matter of theoretically explaining or predicting what is going wrong. It is partly a matter of getting rid of "noise" in the apparatus. "Noise" often means all the events that are not understood by any theory. The instrument must be able to isolate, physically, the properties of the entities that we wish to use, and damp down all the other effects that might get in our way. *We are completely convinced of the reality of electrons when we regularly set to build—and often enough succeed in building—new kinds of devices that use various well understood causal properties of electrons to interfere in other more hypothetical parts of nature.*

It is not possible to grasp this without an example. Familiar historical examples have usually become encrusted by false theory-oriented philosophy or history, so I will take something new. This is a polarizing electron gun whose acronym is PEGGY II. In 1978, it was used in a fundamental experiment that attracted attention even in *The New York Times.* In the next section I describe the point of making PEGGY II. To do that, I have to tell some new physics. You may omit reading this and read only the engineering section that follows. Yet it must be of interest to know the rather easy-to-understand significance of the main experimental results, namely, that parity is not conserved in scattering of polarized electrons from deuterium, and that, more generally, parity is violated in weak neutral-current interactions.[9]

PARITY AND WEAK NEUTRAL CURRENTS

There are four fundamental forces in nature, not necessarily distinct. Gravity and electromagnetism are familiar. Then there are the strong and weak forces (the fulfillment of Newton's program, in the *Optics*, which taught that all nature would be understood by the interaction of particles with various forces that were effective in attraction or repulsion over various different distances, i.e., with different rates of extinction).

Strong forces are 100 times stronger than electromagnetism but act only over a minuscule distance, at most the diameter of a proton. Strong forces act on "hadrons," which include protons, neutrons, and more recent particles, but not electrons or any other members of the class of particles called "leptons."

The weak forces are only 1/10,000 times as strong as electromagnetism, and act over a distance 100 times greater than strong forces. But they act on both hadrons and leptons, including electrons. The most familiar example of a weak force may be radioactivity.

The theory that motivates such speculation is quantum electrodynamics. It is incredibly successful, yielding many predictions better than one part in a million, truly a miracle in experimental physics. It applies over distances ranging from diameters of the earth to 1/100 the diameter of the proton. This theory supposes that all the forces are "carried" by some sort of particle: Photons do the job in electromagnetism. We hypothesize "gravitons" for gravity.

In the case of interactions involving weak forces, there are charged currents. We postulate that particles called "bosons" carry these weak forces.[10] For charged currents, the bosons may be either positive or negative. In the 1970s, there arose the possibility that there could be weak "neutral" currents in which no charge is carried or exchanged. By sheer analogy with the vindicated parts of quantum electrodynamics, neutral bosons were postulated as the carriers in weak neutral interactions.

The most famous discovery of recent high-energy physics is the failure of the conservation of parity. Contrary to the expectations of many physicists and philosophers, including Kant,[11] nature makes an absolute distinction between right-handedness and left-handedness. Apparently, this happens only in weak interactions.

What we mean by right- or left-handed in nature has an element of convention. I remarked that electrons have spin. Imagine your right hand wrapped around a spinning particle with the fingers pointing in the direction of spin. Then your thumb is said to point in the direction of the spin vector. If such particles are traveling in a beam, consider the relation between the spin vector and the beam. If all the particles have their spin vector in the same direction as the beam, they have right-handed (linear) polarization, while if the spin vector is opposite to the beam direction, they have left-handed (linear) polarization.

The original discovery of parity violation showed that one kind of product of a particle decay, a so-called muon neutrino, exists only in left-handed polarization and never in right-handed polarization.

Parity violations have been found for weak *charged* interactions. What about weak *neutral* currents? The remarkable Weinberg-Salam model for the four kinds of force was proposed independently by Stephen Weinberg in 1967 and A. Salam in 1968. It implies a minute violation of parity in weak neutral interactions. Given that the model is sheer speculation, its success has been amazing, even awe-inspiring. So it seemed worthwhile to try out the predicted failure of parity for weak neutral interactions. That would teach us more about those weak forces that act over so minute a distance.

The prediction is: Slightly more left-handed polarized electrons hitting certain targets will scatter than right-handed electrons. Slightly more! The difference in relative frequency of the two kinds of scattering is 1 part in 10,000, comparable to a difference in probability between 0.50005 and 0.49995. Suppose one used the standard equipment available at the Standard Linear Accelerator Center in the early 1970s, generating 120 pulses per second, each pulse providing one electron event. Then you would have to run the entire SLAC beam for twenty-seven years in order to detect so small a difference in relative frequency. Considering that one uses the same beam for lots of experiments simultaneously, by letting different experiments use different pulses, and considering that no equipment remains stable for even a month, let alone twenty-seven years, such an experiment is impossible. You need enormously more electrons coming off in each pulse—between 1000 and 10,000 more electrons per pulse than was once possible. The first attempt used an instrument now called PEGGY I. It had, in essence, a high-class version of J. J. Thomson's hot cathode. Some lithium was heated and electrons were boiled off. PEGGY II uses quite different principles.

PEGGY II

The basic idea began when C. Y. Prescott noticed (by chance!) an article in an optics magazine about a crystalline substance called gallium arsenide. GaAs has a curious property; when it is struck by circularly polarized light of the right frequencies, it emits lots of linearly polarized electrons. There is a good, rough and ready quantum understanding of why this happens, and why half the emitted electrons will be polarized, three-fourths of these polarized in one direction and one-fourth polarized in the other.

PEGGY II uses this fact, plus the fact that GaAs emits lots of electrons owing to features of its crystal structure. Then comes some engineering—it takes work to liberate an electron from a surface. We know that painting a surface with the right stuff helps. In this case, a thin layer of cesium and oxygen is applied to the crystal. Moreover, the less air pressure around the crystal, the more electrons will escape for a given amount of work. So the bombardment takes place in a good vacuum at the temperature of liquid nitrogen.

We need the right source of light. A laser with bursts of red light (7100 Ångstroms) is trained on the crystal. The light first goes through an ordinary polarizer, a very old-fashioned prism of calcite, or Iceland spar[12]—this gives linearly polarized light. We want circularly polarized light to hit the crystal, so the polarized laser beam now goes through a cunning device called a Pockel's cell, which electrically turns linearly polarized photons into circularly polarized ones. Being electric, it acts as a very fast switch. The direction of circular polarization depends on the direction of current in the cell. Hence, the direction of polarization can be varied randomly. This is important, for we are trying to detect a minute asymmetry between right- and left-handed polarization Randomizing helps us guard against any systematic "drift" in the equipment.[13] The randomization is generated by a radioactive decay device, and a computer records the direction of polarization for each pulse.

A circularly polarized pulse hits the GaAs crystal, resulting in a pulse of linearly polarized electrons. A beam of such pulses is maneuvered by magnets into the accelerator for the next bit of the experiment. It passes through a device that checks on a proportion of polarization along the way. The remainder of the experiment requires other devices and detectors of comparable ingenuity, but let us stop at PEGGY II.

BUGS

Short descriptions make it all sound too easy; therefore, let us pause to reflect on debugging. Many of the bugs are never understood. They are eliminated by trial and error. Let me illustrate three different kinds of bugs: (1) the essential technical limitations that, in the end, have to be factored into the analysis of error; (2) simpler mechanical defects you never think of until they are forced on you; and (3) hunches about what might go wrong.

Here are three examples of bugs:

1. Laser beams are not as constant as science fiction teaches, and there is always an irremediable amount of "jitter" in the beam over any stretch of time.
2. At a more humdrum level, the electrons from the GaAs crystals are back-scattered and go back along the same channel as the laser beam used to hit the crystal. Most of them are then deflected magnetically. But some get reflected from the laser apparatus and get back into the system. So you have to eliminate these new ambient electrons. This is done by crude mechanical means, making them focus just off the crystal and, thus, wander away.
3. Good experimenters guard against the absurd. Suppose that dust particles on an experimental surface lie down flat when a polarized pulse hits it, and then stand on their heads when hit by a pulse polarized in the opposite direction. Might that have a systematic effect, given that we are detecting a minute asymmetry? One of the team

thought of this in the middle of the night and came down next morning frantically using antidust spray. They kept that up for a month, just in case.[14]

RESULTS

Some 10^{11} events were needed to obtain a result that could be recognized above systematic and statistical error. Although the idea of systematic error presents interesting conceptual problems, it seems to be unknown to philosophers. There were systematic uncertainties in the detection of right- and left-handed polarization, there was some jitter, and there were other problems about the parameters of the two kinds of beam. These errors were analyzed and linearly added to the statistical error. To a student of statistical inference, this is real seat-of-the-pants analysis with no rationale whatsoever. Be that as it may, thanks to PEGGY II the number of events was big enough to give a result that convinced the entire physics community.[15] Left-handed polarized electrons were scattered from deuterium slightly more frequently than right-handed electrons. This was the first convincing example of parity violation in a weak neutral current interaction.

COMMENT

The making of PEGGY II was fairly nontheoretical. Nobody worked out in advance the polarizing properties of GaAs—that was found by a chance encounter with an unrelated experimental investigation. Although elementary quantum theory of crystals explains the polarization effect, it does not explain the properties of the actual crystal used. No one has got a real crystal to polarize more than 37 percent of the electrons, although in principle 50 percent should be polarized.

Likewise, although we have a general picture of why layers of cesium and oxygen will "produce negative electron affinity," that is, make it easier for electrons to escape, we have no quantitative understanding of why this increases efficiency to a score of 37 percent.

Nor was there any guarantee that the bits and pieces would fit together. To give an even more current illustration, future experimental work, briefly described later in this chapter, makes us want even more electrons per pulse than PEGGY II can give. When the aforementioned parity experiment was reported in *The New York Times*, a group at Bell Laboratories read the newspaper and saw what was going on. They had been constructing a crystal lattice for totally unrelated purposes. It uses layers of GaAs and a related aluminum compound. The structure of this lattice leads one to expect that virtually all the electrons emitted would be polarized. As a consequence, we might be able to double the efficiency of PEGGY II. But, at present, that nice idea has problems. The new lattice should also be coated in work-reducing paint. The cesium-oxygen compound is applied at high temperature. Hence the aluminum tends to ooze into the neighboring layer of GaAs, and the pretty artificial lattice becomes a bit uneven, limiting its fine polarized-electron-emitting properties.[16] So perhaps this will never work. Prescott is simultaneously reviving a souped-up new thermionic cathode to try to get more electrons. Theory would not have told us that PEGGY II would beat out thermionic PEGGY I. Nor can it tell if some thermionic PEGGY III will beat out PEGGY II.

Note also that the Bell people did not need to know a lot of weak neutral current theory to send along their sample lattice. They just read *The New York Times*.

MORAL

Once upon a time, it made good sense to doubt that there were electrons. Even after Thomson had measured the mass of his corpuscles, and Millikan their charge, doubt could have made sense. We needed to be sure that Millikan was measuring the same entity as Thomson. Thus, more theoretical elaboration was needed, and the idea had to be fed into many other phenomena. Solid state physics, the atom, and superconductivity all had to play their part.

Once upon a time, the best reason for thinking that there are electrons might have been success in explanation. Lorentz explained the Faraday effect with his electron theory. But the ability to explain carries little warrant of truth. Even from the time of J. J. Thomson, it was the measurements that weighed in, more than the explanations. Explanations, however, did help. Some people might have had to believe in electrons because the postulation of their existence could explain a wide variety of phenomena. Luckily, we no longer have to pretend to infer from explanatory success (i.e., from what makes our minds feel good). Prescott and the team from the SLAC do not explain phenomena with electrons. They know how to use them. Nobody in his right mind thinks that electrons "really" are just little spinning orbs about which you could, with a small enough hand, wrap your fingers and find the direction of spin along your thumb. There is, instead, a family of causal properties in terms of which gifted experimenters describe and deploy electrons in order to investigate something else, for example, weak neutral currents and neutral bosons. We know an enormous amount about the behavior of electrons. It is equally important to know what does *not* matter to electrons. Thus, we know that bending a polarized electron beam in magnetic coils does not affect polarization in any significant way. We have hunches, too strong to ignore although too trivial to test independently: For example, dust might dance under changes of directions of polarization. Those hunches are based on a hard-won sense of the kinds of things electrons are. (It does not matter at all to this hunch whether electrons are clouds or waves or particles.)

WHEN HYPOTHETICAL ENTITIES BECOME REAL

Note the complete contrast between electrons and neutral bosons. Nobody can yet manipulate a bunch of neutral bosons, if there are any. Even weak neutral currents are only just emerging from the mists of hypothesis. By 1980, a sufficient range of convincing experiments had made them the object of investigation. When might they lose their hypothetical status and become commonplace reality like electrons?—when we use them to investigate something else.

I mentioned the desire to make a better electron gun than PEGGY II. Why? Because we now "know" that parity is violated in weak neutral interactions. Perhaps by an even more grotesque statistical analysis than that involved in the parity experiment, we can isolate just the weak interactions. For example, we have a lot of interactions, including electromagnetic ones, which we can censor in various ways. If we could also statistically pick out a class of weak interactions, as precisely those where parity is not conserved, then we would possibly be on the road to quite deep investigations of matter and antimatter. To do the statistics, however, one needs even more electrons per pulse than PEGGY II could hope to generate. If such a project were to succeed, we should then be beginning to use weak neutral currents as a manipulable tool for looking at something else. The next step toward a realism about such currents would have been made.

The message is general and could be extracted from almost any branch of physics. I mentioned earlier how Dudley Shapere has recently used "observation" of the sun's hot core to illustrate how physicists employ the concept of observation. They collect neutrinos from the sun in an enormous disused underground mine that has been filled with old cleaning fluid (i.e., carbon tetrachloride). We would know a lot about the inside of the sun if we knew how many solar neutrinos arrive on the earth. So these are captured in the cleaning fluid. A few neutrinos will form a new radioactive nucleus (the number that do this can be counted). Although, in this study, the extent of neutrino manipulation is much less than electron manipulation in the PEGGY II experiment, we are nevertheless plainly using neutrinos to investigate something else. Yet not many years ago, neutrinos were about as hypothetical as an entity could get. After 1946 it was realized that when mesons disintegrate giving off, among other things, highly energized electrons, one needed an extra nonionizing particle to conserve momentum and energy. At that time this postulated "neutrino" was thoroughly hypothetical, but now it is routinely used to examine other things.

CHANGING TIMES

Although realisms and anti-realisms are part of the philosophy of science well back into Greek pre-history, our present versions mostly descend from debates at the end of the nineteenth century about atomism. Anti-realism about atoms was partly a matter of physics; the energeticists thought energy was at the bottom of everything, not tiny bits of matter. It also was connected with the positivism of Comte, Mach, K. Pearson, and even J. S. Mill. Mill's young associate Alexander Bain states the point in a characteristic way, apt for 1870:

> Some hypotheses consist of assumptions as to the minute structure and operation of bodies. From the nature of the case these assumptions can never be proved by direct means. Their merit is their suitability to express phenomena. They are Representative Fictions.[17]

"All assertions as to the ultimate structure of the particles of matter," continues Bain, "are and ever must be hypothetical. . . . The kinetic theory of heat serves an important intellectual function." But we cannot hold it to be a true description of the world. It is a representative fiction.

Bain was surely right a century ago, when assumptions about the minute structure of matter could not be proved. The only proof could be indirect, namely, that hypotheses seemed to provide some explanation and helped make good predictions. Such inferences, however, need never produce conviction in the philosopher inclined to instrumentalism or some other brand of idealism.

Indeed, the situation is quite similar to seventeenth-century epistemology. At that time, knowledge was thought of as correct representation. But then one could never get outside the representations to be sure that they corresponded to the world. Every test of a representation is just another representation. "Nothing is so much like an idea as an idea," said Bishop Berkeley. To attempt to argue to scientific realism at the level of theory, testing, explanation, predictive success, convergence of theories, and so forth is to be locked into a world of representations. No wonder that scientific anti-realism is so permanently in the race. It is a variant on "the spectator theory of knowledge."

Scientists, as opposed to philosophers, did, in general, become realists about atoms by 1910. Despite the changing climate, some antirealist variety of instrumentalism or fictionalism remained a strong philosophical alternative in 1910 and in 1930. That is what the history of philosophy teaches us. The lesson is: Think about practice, not theory. Anti-realism about atoms was very sensible when Bain wrote a century ago. Anti-realism about *any* submicroscopic entities was a sound doctrine in those days. Things are different now. The "direct" proof of electrons and the like is our ability to manipulate them using well-understood low-level causal properties. Of course, I do not claim that reality is constituted by human manipulability. Millikan's ability to determine the charge of the electron did something of great importance for the idea of electrons—more, I think, than the Lorentz theory of the electron. Determining the charge of something makes one believe in it far more than postulating it to explain something else. Millikan got the charge on the electron; but better still, Uhlenbeck and Goudsmit in 1925 assigned angular momentum to electrons, brilliantly solving a lot of problems. Electrons have spin, ever after. The clincher is when we can put a spin on the electrons, and thereby get them to scatter in slightly different proportions.

Surely, there are innumerable entities and processes that humans will never know about. Perhaps there are many in principle we can never know about, since reality is bigger than us. The best kinds of evidence for the reality of a postulated or inferred entity is that we can begin to measure it or otherwise understand its causal powers. The best evidence, in turn, that we have this kind of understanding is that we can set out, from scratch, to build machines that will work fairly reliably, taking advantage of this or that causal nexus. Hence, engineering, not theorizing, is the best proof of scientific realism about entities. My attack on scientific anti-realism is analogous to Marx's onslaught on the idealism of his day. Both say that the point is not to understand the world but to change it. Perhaps there are some entities which in theory we can know about only through theory (black holes). Then our evidence is like that furnished by Lorentz. Perhaps there are entities which we shall only measure and never use. The experimental argument for realism does not say that only experimenters' objects exist.

I must now confess a certain skepticism, about, say, black holes. I suspect there might be another representation of the universe, equally consistent with phenomena, in which black holes are precluded. I inherit from Leibniz a certain distaste for occult powers. Recall how he inveighed against Newtonian gravity as occult. It took two centuries to show he was right. Newton's ether was also excellently occult—it taught us lots: Maxwell did his electromagnetic waves in ether, H. Hertz confirmed the ether by demonstrating the existence of radio waves. Albert A. Michelson figured out a way to interact with the ether. He thought his experiment confirmed G. G. Stoke's ether drag theory, but, in the end, it was one of the many things that made ether give up the ghost. A skeptic such as myself has a slender induction: Long-lived theoretical entities which do not end up being manipulated commonly turn out to have been wonderful mistakes.

NOTES

1. C. W. F. Everitt suggests that the first time the word "effect" is used this way in English is in connection with the Peltier effect, in James Clerk Maxwell's 1873 *Electricity and Magnetism*, par. 249, p. 301. My interest in experiment was kindled by conversation with Everitt some years ago, and I have learned much in working with him on our joint (unpublished) paper, "Theory or Experiment, Which Comes First?"

2. Ian Hacking, "Spekulation, Berechung und die Erschaffnung der Phänomenen," in *Versuchungen: Aufsätze zur Philosophie, Paul Feyerabends*, no. 2, ed. P. Duerr (Frankfort, 1981), 126–158.

3. Nancy Cartwright makes a similar distinction in her book, *How the Laws of Physics Lie* (Oxford: Oxford University Press, 1983). She approaches realism from the top, distinguishing theoretical laws (which

do not state the facts) from phenomenological laws (which do). She believes in some "theoretical" entities and rejects much theory on the basis of a subtle analysis of modeling in physics. I proceed in the opposite direction, from experimental practice. Both approaches share an interest in real life physics as opposed to philosophical fantasy science. My own approach owes an enormous amount to Cartwright's parallel developments, which have often preceded my own. My use of the two kinds of realism is a case in point.

4. Hilary Putnam, "How Not to Talk About Meaning," "The Meaning of 'Meaning,'" and other papers in *Mind, Language and Reality*, Philosophical Papers, Vol. 2 (Cambridge: Cambridge University Press, 1975).

5. Francis Bacon, *The Great Instauration*, in *The Philosophical Works of Francis Bacon*, trans. Ellis and Spedding, ed. J. M. Robertson (London, 1905), 252.

6. Ian Hacking, "Do We See Through a Microscope?" *Pacific Philosophical Quarterly* 62 (1981): 305–322.

7. C. Y. Prescott, "Prospects for Polarized Electrons at High Energies," SLAC-PUB-2630, Stanford Linear Accelerator, October 1980, p. 5.

8. "The Concept of Observation in Science and Philosophy," *Philosophy of Science* 49 (1982): 485–526. See also K. S. Shrader-Frechette, "Quark Quantum Numbers and the Problem of Microphysical Observation," *Synthàse* 50 (1982): 125–146, and ensuing discussion in that issue of the journal.

9. I thank Melissa Franklin, of the Standard Linear Accelerator, for introducing me to PEGGY II and telling me how it works. She also arranged discussion with members of the PEGGY II group, some of whom are mentioned below. The report of experiment E-122 described here is "Parity Nonconservation in Inelastic Electron Scattering," C. Y. Prescott et al., in *Physics Letters*. I have relied heavily on the in-house journal, the *SLAC Beam Line*, report no. 8, October 1978, "Parity Violation in Polarized Electron Scattering." This was prepared by the in-house science writer Bill Kirk.

10. The odd-sounding bosons are named after the Indian physicist S. N. Bose (1894–1974), also remembered in the name "Bose-Einstein statistics" (which bosons satisfy).

11. But excluding Leibniz, who "knew" there had to be some real, natural difference between right- and left-handedness.

12. Iceland spar is an elegant example of how experimental phenomena persist even while theories about them undergo revolutions. Mariners brought calcite from Iceland to Scandinavia. Erasmus Bartholinus experimented with it and wrote it up in 1609. When you look through these beautiful crystals you see double, thanks to the so-called ordinary and extraordinary rays. Calcite is a natural polarizer. It was our entry to polarized light, which for three hundred years was the chief route to improved theoretical and experimental understanding of light and then electromagnetism. The use of calcite in PEGGY II is a happy reminder of a great tradition.

13. It also turns GaAs, a 3/4 to 1/4 left-hand/right-hand polarizer, into a 50–50 polarizer.

14. I owe these examples to conversation with Roger Miller of SLAC.

15. The concept of a "convincing experiment" is fundamental. Peter Gallison has done important work on this idea, studying European and American experiments on weak neutral currents conducted during the 1970s.

16. I owe this information to Charles Sinclair of SLAC.

17. Alexander Bain, *Logic, Deductive and Inductive* (London and New York, 1870), 362.

29

Laboratories

Bruno Latour

FROM TEXTS TO THINGS: A SHOWDOWN

You doubt what I wrote? Let me show you.' The very rare and obstinate dissenter who has *not* been convinced by the scientific text, and who has not found other ways to get rid of the author, is led from the text into the place where the text is said to come from. I will call this place the **laboratory**, which for now simply means, as the name indicates, the place where scientists *work*. Indeed, the laboratory was present in the texts we studied in the previous chapter: the articles were alluding to 'patients', to 'tumours', to 'HPLC', to 'Russian spies', to 'engines'; dates and times of experiments were provided and the names of technicians acknowledged. All these allusions however were made within a paper world; they were a set of semiotic actors presented in the text but not *present* in the flesh; they were alluded to as if they existed independently from the text; they could have been invented.

(1) Inscriptions

What do we find when we pass through the looking glass and accompany our obstinate dissenter from the text to the laboratory? Suppose that we read the following sentence in a scientific journal and, for whatever reason, do not wish to believe it:

> (1) '29.1 shows a typical pattern. Biological activity of endorphin was found essentially in two zones with the activity of zone 2 being totally reversible, or statistically so, by naloxone.'

We, the dissenters, question this figure 29.1 so much, and are so interested in it, that we go to the author's laboratory (I will call him 'the Professor'). We are led into an air-conditioned, brightly lit room. The Professor is sitting in front of an array of devices that does not attract our attention at first. 'You doubt what I wrote? Let me show you.' This last sentence refers to an image slowly produced by one of these devices (Figure 29.1):

We now understand that what the Professor is asking us to watch is related to the figure

---

Reprinted by permission of the publisher from *Science in Action: How to Follow Scientists and Engineers through Society* by Bruno Latour, pp. 64–74, 91–100, Cambridge, Mass.: Harvard University Press. Copyright © 1987 by Bruno Latour.

in the text of sentence (1). We thus realise where this figure comes from. It has been *extracted* from the instruments in this room, *cleaned, redrawn*, and *displayed*. We also realise, however, that the images that were the last layer in the text, are the *end result* of a long process in the laboratory that we are now starting to observe. Watching the graph paper slowly emerging out of the physiograph, we understand that we are at the junction of two worlds: a paper world that we have just left, and one of instruments that we are just entering. A hybrid is produced at the interface: a raw image, to be used later in an article, that is emerging from an instrument.

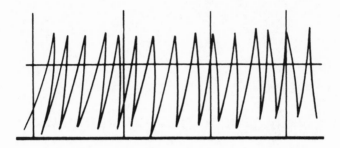

Figure 29.1

'Ok. This is the base line; now, I am going to inject endorphin, what is going to happen? See?!' (Figure 29.2)

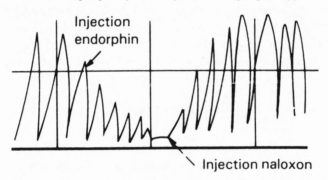

Figure 29.2

'Immediately the line drops dramatically. And now watch naloxone. See?! Back to base line levels. It is fully reversible.'

For a time we focus on the stylus pulsating regularly, inking the paper, scribbling cryptic notes. We remain fascinated by this fragile film that is in between text and laboratory. Soon, the Professor draws our attention beneath and beyond the traces on the paper, to the physiograph from which the image is slowly being emitted. Beyond the stylus a massive piece of electronic hardware records, calibrates, amplifies and regulates signals coming from another instrument, an array of glassware. The Professor points to a glass chamber in which bubbles are regularly flowing around a tiny piece of something that looks like elastic. It is indeed elastic, the Professor intones. It is a piece of gut, guinea pig gut ('myenteric plexus-longitudinal muscle of the guinea pig ileum', are his words). This gut has the property of contracting regularly if maintained alive. This regular pulsation is easily disturbed by many chemicals. If one hooks the gut up so that each contraction sends out an electric pulse, and if the pulse is made to move a stylus over graph paper, then the guinea pig gut will be induced to produce regular scribbles over a long period. If you then add a chemical to the chamber you *see* the peaks drawn by the inked stylus slow down or accelerate at the other end. This perturbation, invisible in the chamber, is visible on paper: the chemical, no matter what it is, is given a *shape* on paper. This shape 'tells you something' about the chemical. With this set-up you may now ask new questions: if I double the dose of chemical will the peaks be doubly decreased? And if I triple it, what will happen? I can now measure the white surface left by the decreasing scribbles directly on the graph paper, thereby defining a quantitative relation between the dose and the response. What if, just after the first chemical is added, I add another one which is known to counteract it? Will the peaks go back to normal? How fast will they do so? What will be the pattern of this return to the base line level? If two chemicals, one known, the other unknown, trace the same slope on the paper, may I say, in this respect at least, that they are the same chemicals? These are some of the questions the Professor is tackling with endorphin (unknown), morphine (well known) and naloxone (known to be an antagonist of morphine).

We are no longer asked to believe the text that we read in *Nature*; we are now asked to believe *our own eyes*, which can see that endorphin is behaving exactly like morphine. The object we looked at in the text and the one we are now contemplating are identical except for one thing. The graph of sentence (1), which was the most concrete and visual element of the text, is now in (2) the most abstract and textual element in a bewildering array of equipment. Do we see more or less than before? On the one hand we can see more, since we are looking at not only the graph but also the physiograph, and the electronic hardware, and the glassware, and the electrodes, and the bubbles of oxygen, and the pulsating ileum, and the Professor who is injecting chemicals into the chamber with his syringe, and is writing down in a huge protocol book the time, amount of and reactions to the doses. We can see more, since we have before our eyes not only the image but what the image is made of.

On the other hand we see *less* because now each of the elements that makes up the final graph could be modified so as to produce a different visual outcome. Any number of incidents could blur the tiny peaks and turn the regular writing into a meaningless doodle. Just at the time when we feel comforted in our belief and start to be fully convinced by our own eyes watching the image, we suddenly feel uneasy because of the fragility of the whole set up. The Professor, for instance, is swearing at the gut saying it is a 'bad gut'. The technician who sacrificed the guinea pig is held responsible and the Professor decides to make a fresh start with a new animal. The demonstration is stopped and a new scene is set up. A guinea pig is placed on a table, under surgical floodlights, then anaesthetised, crucified and sliced open. The gut is located, a tiny section is extracted, useless tissue peeled away, and the precious fragment is delicately hooked up between two electrodes and immersed in a nutrient fluid so as to be maintained alive. Suddenly, we are much further from the paper world of the article. We are now in a puddle of blood and viscera, slightly nauseated by the extraction of the ileum from this little furry creature. In the last chapter, we admired the rhetorical abilities of the Professor as an author. Now, we realise that many other manual abilities are required in order to write a convincing paper later on. The guinea pig alone would not have been able to tell us anything about the similarity of endorphin to morphine; it was not mobilisable into a text and would not help to convince us. Only a part of its gut, tied up in the glass chamber and hooked up to a physiograph, can be mobilised in the text and add to our conviction. Thus, the Professor's art of convincing his readers must extend beyond the paper to preparing the ileum, to calibrating the peaks, to tuning the physiograph.

After hours of waiting for the experiment to resume, for new guinea pigs to become available, for new endorphin samples to be purified, we realise that the invitation of the author ('let me show you') is not as simple as we thought. It is a slow, protracted and complicated staging of tiny images in front of an audience. 'Showing' and 'seeing' are not simple flashes of intuition. Once in the lab we are not presented outright with the real endorphin whose existence we doubted. We are presented with another world in which it is necessary to prepare, focus, fix and rehearse the vision of the real endorphin. We came to the laboratory in order to settle our doubts about the paper, but we have been led into a labyrinth.

This unexpected unfolding makes us shiver because it now dawns on us that if we disbelieve the traces obtained on the physiograph by the Professor, we will have to give up the topic altogether or go through the same experimental chores all over again. The stakes have increased enormously since we first started reading scientific articles. It is not a question of reading and writing back to the author any more. In order to argue, we would now need the manual skills required to handle the scalpels, peel away the guinea pig ileum, interpret the decreasing peaks, and so on. Keeping the controversy alive has already forced us through many difficult moments. We now realise that what we went through is nothing compared to the scale of what we have to undergo if we wish to continue. Earlier, we only needed a good library in order to dispute texts. It might have been costly and not that easy, but it was still feasible. At this present point, in order to go on, we need guinea pigs, surgical lamps and tables, physiographs, electronic hardware, technicians and morphine, not to mention the scarce flasks of purified endorphin; we also need the skills to use all these elements and to turn them into a pertinent objection to the Professor's claim. As will be made clear later, longer and longer detours will be necessary to find a laboratory, buy the equipment, hire the technicians and become acquainted with the ileum assay. All this work just to start making a convincing counter-argument to the Professor's original paper on endorphin. (And when we have made this detour and finally come up with a credible objection, where will the Professor be?)

When we doubt a scientific text we do not go from the world of literature to Nature as it is. Nature is not directly beneath the scientific article; it is there *indirectly* at best. Going from the paper to the laboratory is going from an array of rhetorical resources to a set of new resources devised in such a way as to provide the literature with its most powerful tool: the visual display. Moving from papers to labs is moving from literature to convoluted ways of getting this literature (or the most significant part of it).

This move through the looking glass of the paper allows me to define an **instrument**, a definition which will give us our bearings when entering any laboratory. I will call an instrument (or **inscription device**) any set-up, no matter what its size, nature and cost, that provides a visual display of any sort in a scientific text. This definition is simple enough to let us follow scientists' moves. For instance an optical telescope is an instrument, but so is an array of several radio-telescopes even if its constituents are separated by thousands of kilometers. The guinea pig ileum assay is an instrument even if it is small and cheap compared to an array of radiotelescopes or the Stanford linear accelerator. The definition is not provided by the cost nor by the sophistication but only by this characteristic: the set-up provides an inscription that is used as the final layer in a scientific text. An instrument, in this definition, is not every set-up which ends with a little window that allows someone to take a reading. A thermometer, a watch, a Geiger counter, all provide readings but are not considered as instruments as long as these readings are not used as the final layer of technical papers. This point is important when watching complicated contrivances with hundreds of intermediary readings taken by dozens of white-coated technicians. What will be used as visual proof in the article will be the few lines in the bubble chamber and not the piles of printout making the intermediate readings.

It is important to note that the use of this definition of instrument is a relative one. It depends on time. Thermometers *were* instruments and very important ones in the eighteenth century, so were Geiger counters between the First and Second World Wars. These devices provided crucial resources in papers of the time. But now they are only parts of larger set-ups and are only used so that a new visual proof can be displayed at the end. Since the definition is relative to the use made of the 'window' in a technical paper, it is also relative to the intensity and nature of the associated controversy. For instance, in the guinea pig ileum assay there is a box of electronic hardware with many readings that I will call 'intermediate' because they do not constitute the visual display eventually put to use in the article. It is unlikely that anyone will quibble about this because the calibration of electronic signals is now made through a black box produced industrially and sold by the thousand. It is a different matter with the huge tank built in an old gold mine in South Dakota at a cost of $600,000 (1964 dollars!) by Raymond Davis[1] to detect solar neutrinos. In a sense the whole set-up may be considered as *one* instrument providing one final window in which astrophysicists can read the number of neutrinos emitted by the sun. In this case all the other readings are intermediate ones. If the controversy is fiercer, however, the set-up is broken down into *several* instruments, each providing a specific visual display which has to be independently evaluated. If the controversy heats up a bit we do not see neutrinos coming out of the sun. We see and hear a Geiger counter that clicks when Argon[37] decays. In this case the Geiger counter, which gave only an intermediate reading when there was no dispute, becomes an instrument in its own right when the dispute is raging.

The definition I use has another advantage. It does not make presuppositions about what the instrument is made of. It can be a piece of hardware like a telescope, but it can also be made of softer material. A statistical institution that employs hundreds of pollsters, sociologists and computer scientists gathering all sorts of data on the economy *is* an instrument if it yields inscriptions for papers written in economic journals with, for instance, a graph of the inflation rate by month and by branch of industry. No matter how many people were made to participate in the construction of the image, no matter how long it took, no matter how much it cost, the whole institution is used as *one* instrument (as long as there is no controversy that calls its intermediate readings into question).

At the other end of the scale, a young primatologist who is watching baboons in the savannah and is equipped only with binoculars, a pencil and a sheet of white paper may be seen as an instrument if her coding of baboon behaviour is summed up in a graph. If you want to deny her statements, you might (everything else being equal) have to go through the same ordeals and walk through the savannah taking notes with similar constraints. It is the same if you wish to deny the inflation rate by month and industry, or the detection of endorphin with the ileum assay. The instrument, whatever its nature, is what leads you from the paper to what supports the paper, from the many resources mobilised in the text to the many more resources mobilised to create the visual displays of the texts. With this definition of an instrument, we are able to ask many questions and to make comparisons: how expensive they are, how old they are, how many intermediate readings compose one instrument, how long it takes to get one reading, how many people are mobilised to activate them, how many authors are using the inscriptions they provide in their papers, how controversial are those readings. . . . Using this notion we can define more precisely than earlier the laboratory as any place that gathers one or several instruments together.

What is behind a scientific text? Inscriptions. How are these inscriptions obtained? By setting up instruments. This other world just beneath the text is invisible as long as there is no controversy. A picture of moon valleys and mountains is presented to us as if we could see them directly. The telescope that makes them visible is invisible and so are the fierce controversies that Galileo had to wage centuries ago to produce an image of the Moon. Once that fact is constructed, there is no instrument to take into account and this is why the painstaking work necessary to tune the instruments often disappears from popular science. On the contrary, when science in action is followed, instruments become the crucial elements, immediately after the technical texts; they are where the dissenter is inevitably led.

There is a corollary to this change of relevance on the inscription devices depending on the strength of the controversy, a corollary that will become more important in the next chapter. If you consider only fully-fledged facts it seems that everyone could accept or contest them equally. It does not cost anything to contradict or accept them. If you dispute further and reach the frontier where facts are made, instruments become visible and with them the cost of continuing the discussion rises. It appears that *arguing is costly*. The equal world of citizens having opinions about things becomes an unequal world in which dissent or consent is not possible without a huge accumulation of resources which permits the collection of relevant inscriptions. What makes the differences between author and reader is not only the ability to utilise all the rhetorical resources studied earlier, but also to gather the many devices, people and animals necessary to produce a visual display usable in a text.

(2) Spokesmen and Women

It is important to scrutinise the exact settings in which encounters between authors and dissenters take place. When we disbelieve the scientific literature, we are led from the many libraries around to the *very few* places where this literature is produced. Here we are welcomed by the author who shows us where the figure in the text comes from. Once presented with the instruments, who does the talking during these visits? At first, the authors: they *tell* the visitor what to *see*: 'see the endorphin effect?', 'look at the neutrinos!' However, the authors are not lecturing the visitor. The visitors have their faces turned towards the instrument and are watching the place where the thing is writing itself down (inscription in the form of collection of specimens, graphs, photographs, maps—you name it). When the dissenter was reading the scientific text it was difficult for him or her to doubt, but with imagination, shrewdness and downright awkwardness it was always possible. Once in the lab, it is much more difficult because the dissenters see with their own eyes. If we leave aside the many other ways to avoid going through the laboratory that we will study later, the dissenter does not have to believe the paper nor even the scientist's word since in a self-effacing gesture the author has stepped aside. 'See for yourself' the scientist says with a subdued and maybe ironic smile. 'Are you convinced now?' Faced with the thing itself that the technical paper was alluding to, the dissenters now have a choice between either accepting the fact or doubting their own sanity—the latter is much more painful.

We now seem to have reached the end of all possible controversies since there is nothing left for the dissenter to dispute. He or she is right in front of the thing he or she is asked to believe. There is almost no human intermediary between thing and person; the dissenter is in the very place where the thing is said to happen and at the very moment when it happens. When such a point is reached it seems that there is no further need to talk of 'confidence': the thing impresses itself directly on us. Undoubtedly, controversies are settled once and for all when such a situation is set up—which again is very rarely the case. The dissenter becomes a believer, goes out of the lab, borrowing the author's claim and confessing that 'X has incontrovertibly shown that A is B'. A new fact has been made which will be used to modify the outcome of some other controversies.

If this were enough to settle the debate, it would be the end of this chapter. But . . . there is someone saying 'but, wait a minute . . .' and the controversy resumes!

What was imprinted on us when we were watching the guinea pig ileum assay? 'Endorphin of course,' the Professor *said*. But what did we *see*? This:

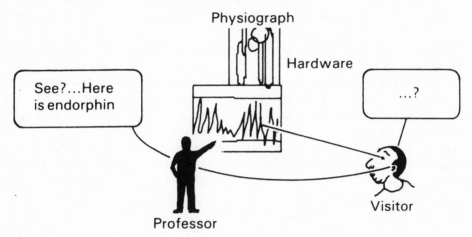

Figure 29.3

With a minimum of training we see peaks; we gather there is a base line, and we see a depression in relation to one coordinate that we understand to indicate the time. This is not endorphin yet. The same thing occurred when we paid a visit to Davis's gold and neutrino mine in South Dakota. We saw, he said, neutrinos counted straight out of the huge tank capturing them from the sun. But what *did* we see? Splurges on paper representing clicks from a Geiger counter. Not neutrinos, yet.

When we are confronted with the instrument, we are attending an 'audio-visual' spectacle. There is a *visual* set of inscriptions produced by the instrument and a *verbal* commentary uttered by the scientist. We get both together. The effect on conviction is striking, but its cause is mixed because we cannot differentiate what is coming from the thing inscribed, and what is coming from the author. To be sure, the scientist is not trying to influence us. He or she is simply commenting, underlining, pointing out, dotting the i's and crossing the t's, not adding anything. But it is also certain that the graphs and the clicks by themselves would not have been enough to form the image of endorphin coming out of the brain or neutrinos coming out of the sun. Is this not a strange situation? The scientists do not say anything more than what is inscribed, but without their commentaries the inscriptions say considerably less! There is a word to describe this strange situation, a very important word for everything that follows, that is the word **spokesman** (or **spokeswoman**, or **spokesperson**, or mouthpiece). The author behaves as if he or she were the mouthpiece of what is inscribed on the window of the instrument.

The spokesperson is someone who speaks for others who, or which, do not speak. For instance a shop steward is a spokesman. If the workers were gathered together and they all spoke at the same time there would be a jarring cacophony. No more meaning could be retrieved from the tumult than if they had remained silent. This is why they designate (or are given) a delegate who speaks on their behalf, and in their name. The delegate—let us call him Bill—does not speak in *his* name and when confronted with the manager does not speak 'as Bill' but as the 'workers' voice'. So Bill's longing for a new Japanese car or his note to get a pizza for his old mother on his way home are not the right topics for the meeting. The voice of the floor, articulated by Bill, wants a '3 per cent pay rise—and they are deadly serious about it, sir, they are ready to strike for it,' he tells the manager. The manager has his doubts: 'Is this really what they want? Are they really so adamant?' 'If you do not believe me,' replies Bill, 'I'll show you, but don't ask for a quick settlement. I told you they are ready to strike and you will see more than you want!' What does the manager see? He does not see what Bill said. Through the office window he simply sees an assembled crowd gathered in the aisles. Maybe it is because of Bill's interpretation that he reads anger and determination on their faces.

For everything that follows, it is very important not to limit this notion of spokesperson and not to impose any clear distinction between 'things' and 'people' in advance. Bill, for instance, represents people who could talk, but who, in fact, cannot all talk at once. Davis represents neutrinos that cannot talk, in principle, but which are made to write, scribble and sign thanks to the device set up by Davis. So in practice, there is not much difference between people and things: they both need someone to talk for them. From the spokesperson's point of view there is thus no distinction to be made between representing people and representing things. In each case the spokesperson literally does the talking for who or what cannot talk. The Professor in the laboratory speaks for endorphin like Davis for the neutrinos and Bill for the shopfloor. In our definition the crucial element is not the quality of the represented but only their number and the unity of the representative. The point is that confronting a spokesperson is not like confronting any average man or woman. You are confronted not with Bill or the Professor, but with Bill and the Professor *plus* the many things or people on behalf of whom they are talking. You do not address Mr Anybody or Mr Nobody but Mr or Messrs Manybodies. As we saw, it may be easy to doubt one person's word. Doubting a spokesperson's word requires a much more strenuous effort however because it is now one person—the dissenter—against a crowd—the author.

On the other hand, the strength of a spokesperson is not so great since he or she is by definition *one* man or woman whose word could be dismissed—one Bill, one Professor, one Davis. The strength comes from the representatives' word when they do not talk by and for themselves but *in the presence of* what they represent. Then, and only then, the dissenter is confronted simultaneously with the spokespersons and what they speak for: the Professor and the endorphin made visible in the guinea pig assay; Bill and the assembled workers; Davis and his solar neutrinos. The solidity of what the representative says is directly supported by the silent but eloquent presence of the represented. The result of such a set-up is that it seems as though the mouthpiece does not 'really talk', but that he or she is just commenting on what you yourself directly see, 'simply' providing you with the words you would have used anyway.

This situation, however, is the source of a major weakness. Who is speaking? The things or the people *through* the representative's voice? What does she (or he, or they, or it) say? Only what the things they represent would say if they could talk directly. But the point is that they cannot. So what the dissenter sees is, in practice, rather different from what the speaker says. Bill, for instance, says his workers want to strike, but this might be Bill's own desire or a union decision relayed by him. The manager looking through the window may see a crowd of assembled workers who are just passing the time and can be dispersed at the smallest threat. At any rate do they really want 3 per cent and not 4 per cent or 2 per cent? And even so, is it not possible to offer Bill this Japanese car he so dearly wants? Is the 'voice of the worker' not going to change his/its mind if the manager offers a new car to Bill? Take endorphin as another instance. What we really saw was a tiny depression in the regular spikes forming the base line. Is this the same as the one triggered by morphine? Yes it is, but what does that prove? It may be that all sorts of chemicals give the same shape in this peculiar assay. Or maybe the Professor so dearly wishes his substance to be morphine-like that he unwittingly confused two syringes and injected the same morphine twice, thus producing two shapes that indeed look identical.

What is happening? The controversy flares even after the spokesperson has spoken and displayed to the dissenter what he or she was talking about. How can the debate be stopped from proliferating again in all directions? How can all the strength that a spokesman musters be retrieved? The answer is easy: by letting the things and persons represented *say for them-selves the same thing that the representatives claimed they wanted to say*. Of course, this never happens since they are designated because, by definition, such direct communication is impossible. Such a situation however may be convincingly staged.

Bill is not believed by the manager, so he leaves the office, climbs onto a podium, seizes a loudspeaker and asks the crowd, 'Do you want the 3 per cent rise?' A roaring 'Yes, our 3 per cent! Our 3 per cent!' deafens the manager's ears even through the window pane of his office. 'Hear them?' asks Bill with a modest but triumphant tone when they are sitting down again at the negotiating table. Since the workers themselves said exactly what the 'workers' voice' had said, the manager cannot dissociate Bill from those he represents and is really confronted with a crowd acting as one single man.

The same is true for the endorphin assay when the dissenter, losing his temper, accuses the Professor of fabricating facts. 'Do it yourself,' the Professor says, irritated but eager to play fair. 'Take the syringe and see for yourself what the assay reaction will be.' The visitor accepts the challenge, carefully checks the labels on the two vials and first injects morphine into the tiny glass chamber. Sure enough, a few seconds later the spikes start decreasing and after a minute or so they return to the base line. With the vial labelled endorphin, the very same result is achieved with the same timing. A unanimous, incontrovertible answer is thus obtained by the dissenter himself. What the Professor said the endorphin assay will answer, if asked directly, is answered by the assay. The Professor cannot be dissociated from his claims. So the visitor has to go back to the 'negotiating table' confronted not with the Professor's own wishes but with a Professor simply transmitting what endorphin really is.

No matter how many resources the scientific paper might mobilise, they carry little weight compared with this rare demonstration of power: the author of the claim steps aside and the doubter sees, hears and touches the inscribed things or the assembled people that reveal to him or to her exactly the same claim as the author.

(3) Trials of Strength

For us who are simply following scientists at work there is no exit from such a setup, no back door through which to escape the incontrovertible evidence. We have already exhausted all sources of dissent; indeed we might have no energy left to maintain the mere idea that controversy might still be open. For us laymen, the file is now closed. Surely, the dissenter we have shadowed will give up. If the things say the same as the scientist, who can deny the claim any longer? How can you go any further?

The dissenter goes on, however, with more tenacity than the laymen. The identical tenor of the representative's words and the answers provided by the represented were the result of a carefully staged situation. The instruments needed to be working and finely tuned, the questions to be asked at the right time and in the right format. What would happen, asks the dissenter, if we stayed longer than the show and went backstage; or were to alter any of the many elements which, everyone agrees, are necessary to make up the whole instrument? The unanimity between represented and constituency is like what an inspector sees of a hospital or of a prison camp when his inspection is announced in advance. What if he steps outside his itinerary and tests the solid ties that link the represented and their spokesmen?

The manager, for instance, heard the roaring applause that Bill received, but he later obtains the foremen's opinion: 'The men are not for the strike at all, they would settle for 2 per cent. It is a union order; they applauded Bill because that's the way to behave on the shopfloor, but distribute a few pay rises and lay off a few ringleaders and they will sing an altogether different song.' In place of the unanimous answer given by the assembled workers, the manager is now faced with an *aggregate* of possible answers. He is now aware that the answer he got earlier through Bill was extracted from a complex setting which was at first invisible. He also realises that there is room for action and that each worker may be made to behave differently if pressures other than Bill's are exerted on them. The next time Bill screams 'You want the 3 per cent don't you?' only a few half-hearted calls of agreement will interrupt a deafening silence.

Let us take another example, this time from the history of science. At the turn of the century, Blondlot, a physicist from Nancy, in France, made a major discovery like that of X-rays.[2] Out of devotion to his city he called them 'N-rays'. For a few years, N-rays had all sorts of theoretical developments and many practical applications, curing diseases and putting Nancy on the map of international science. A dissenter from the United States, Robert W. Wood, did not believe Blondlot's papers even though they were published in reputable journals, and decided to visit the laboratory. For a time Wood was confronted with incontrovertible evidence in the laboratory at Nancy. Blondlot stepped aside and let the N-rays inscribe themselves straight onto a screen in front of Wood. This, however, was not enough to get rid of Wood, who obstinately stayed in the lab asking for more experiments and himself manipulating the N-ray detector. At one point he even surreptitiously removed the aluminium prism which was generating the N-rays. To his surprise, Blondlot on the other side of the dimly lit room kept obtaining the same result on his screen even though what was deemed the most crucial element had been removed. The direct signatures made by the N-rays on the screen were thus made by something else. The unanimous support became a cacophony of dissent. By removing the prism, Wood severed the solid links that attached Blondlot to the N-rays. Wood's interpretation was that Blondlot so much wished to discover rays (at a time when almost every lab in Europe was christening new rays) that he unwittingly made up not only the N-rays, but also the instrument to inscribe them. Like the manager above, Wood realised that the coherent whole he was presented with was an aggregate of many elements that could be induced to go in many different directions. After Wood's action (and that of other dissenters) no one 'saw' N-rays any more but only smudges on photographic plates when Blondlot presented his N-rays. Instead of enquiring about the place of N-rays in physics, people started enquiring about the role of auto-suggestion in experimentation! The new fact had been turned into an artefact.

The way out, for the dissenter, is not only to dissociate and disaggregate the many supporters the technical papers were able to muster. It is also to shake up the complicated set-up that provides graphs and traces in the author's laboratory in order to see how resistant the array is which has been mobilised in order to convince everyone. The work of disbelieving the literature has now been turned into the difficult job of manipulating the hardware. We have now reached another stage in the escalation between the author of a claim and the disbeliever, one that leads them further and further into the details of what makes up the inscriptions used in technical literature.

Let us continue the question-and-answer session staged above between the Professor and the dissenter. The visitor was asked to inject morphine and endorphin himself in order to check that there was no foul play. But the visitor is arguments, we have analysed so far. What was the endorphin tried out by the dissenter? The superimposition of the traces obtained by: a sacrificed guinea pig whose gut was then hooked up to electric wires and regularly stimulated; a hypothalamus soup extracted after many trials from slaughtered sheep and then forced through HPLC columns under a very high pressure.

Endorphin, before being named and for as long as it is a new object, *is* this list *readable* on the instruments *in* the Professor's laboratory. So is a microbe long before being called such. At first it is something that transforms sugar into alcohol in Pasteur's lab. This something is narrowed down by the multiplication of feats it is asked to do. Fermentation still occurs in the absence of air but stops when air is reintroduced. This exploit defines a new hero that is killed by air but breaks down sugar in its absence, a hero that will be called 'Anaerobic' or 'Survivor in the Absence of Air'. Laboratories generate so many new objects because they are able to create extreme conditions and because each of these actions is obsessively inscribed.

This naming after what the new object does is in no way limited to actants like hormones or radioactive substances, that is to the laboratories of what are often called 'experimental sciences'. Mathematics also defines its subjects by what they *do*. When Cantor, the German mathematician, gave a shape to his transfinite numbers, the shape of his new objects was obtained by having them undergo the simplest and most radical trial:[3] is it possible to establish a one-to-one connection between, for instance, the set of points comprising a unit square and the set of real numbers between 0 and 1? It seems absurd at first since it would mean that there are as many numbers on one side of a square as in the whole square. The trial is devised so as to see if two different numbers in the square have different images on the side or not (thus forming a one-to-one correspondence) or if they have only one image (thus forming a two-to-one correspondence). The written answer on the white sheet of paper is incredible: 'I see it but I don't believe it,' wrote Cantor to Dedekind. There are as many numbers on the side as in the square. Cantor creates his transfinites from their performance in these extreme, scarcely conceivable conditions.

The act of defining a new object by the answers it inscribes on the window of an instrument provides scientists and engineers with their final source of strength. It constitutes our **second basic principle**,* as important as the first in order to understand science in the making: scientists and engineers speak in the name of new allies that they have shaped and enrolled; representatives among other representatives, they add these unexpected resources to tip the balance of force in their favour. Guillemin now speaks for endorphin and somatostatin, Pasteur for visible microbes, the Curies for polonium, Payen and Persoz for enzymes, Cantor for transfinites. When they are challenged, they cannot be isolated, but on the contrary their constituency stands behind them arrayed in tiers and ready to say the same thing.

---

*Editor's note: Latour's First Basic Principle states, "the fate of facts and machines is in the later user's hands; their qualities are thus a consequence, not a cause, of a collective action."

(4) Laboratories against Laboratories

Our good friend, the dissenter, has now come a long way. He or she is no longer the shy listener to a technical lecture, the timid onlooker of a scientific experiment, the polite contradictor. He or she is now the head of a powerful laboratory utilising all available instruments, forcing the phenomena supporting the competitors to support him or her instead, and shaping all sorts of unexpected objects by imposing harsher and longer trials. The power of this laboratory is measured by the extreme conditions it is able to create: huge accelerators of millions of electron volts; temperatures approaching absolute zero; arrays of radio-telescopes spanning kilometres; furnaces heating up to thousands of degrees; pressures exerted at thousands of atmospheres; animal quarters with thousands of rats or guinea pigs; gigantic number crunchers able to do thousands of operations per millisecond. Each modification of these conditions allows the dissenter to mobilise one more actant. A change from micro to phentogram, from million to billion electron volts; lenses going from metres to tens of metres; tests going from hundreds to thousands of animals; and the shape of a new actant is thus redefined. All else being equal, the power of the laboratory is thus proportionate to the number of actants it can mobilise on its behalf. At this point, statements are not borrowed, transformed or disputed by empty-handed laypeople, but by scientists with whole laboratories *behind* them.

However, to gain the final edge on the opposing laboratory, the dissenter must carry out a fourth strategy: he or she must be able to transform the new objects into, so to speak, older objects and feed them back into his or her lab.

What makes a laboratory difficult to understand is not what is presently going on in it, but what *has been* going on in it and in other labs. Especially difficult to grasp is the way in which new objects are immediately transformed into something else. As long as somatostatin, polonium, transfinite numbers, or anaerobic microbes are shaped by the list of trials I summarised above, it is easy to relate to them: tell me what you go through and I will tell you what you are. This situation, however, does not last. New objects become **things**: 'somatostatin', 'polonium', 'anaerobic microbes', 'transfinite numbers', 'double helix' or '*Eagle* computers', things isolated from the laboratory conditions that shaped them, things with a name that now seem independent from the trials in which they proved their mettle. This process of transformation is a very common one and occurs constantly both for laypeople and for the scientist. All biologists now take 'protein' for an object; they do not remember the time, in the 1920s, when protein was a whitish stuff that was separated by a new ultracentrifuge in Svedberg's laboratory.[4] At the time protein was nothing but the action of differentiating cell contents by a centrifuge. Routine use however transforms the naming of an actant after what it does into a common name. This process is not mysterious or special to science. It is the same with the can opener we routinely use in our kitchen. We consider the opener and the skill to handle it as one black box which means that it is unproblematic and does not require planning and attention. We forget the many trials we had to go through (blood, scars, spilled beans and ravioli, shouting parent) before we handled it properly, anticipating the weight of the can, the reactions of the opener, the resistance of the tin. It is only when watching our own kids still learning it the hard way that we might remember how it was when the can opener was a 'new object' for us, defined by a list of trials so long that it could delay dinner forever.

This process of routinisation is common enough. What is less common is the way the same people who constantly generate new objects to win in a controversy are also constantly transforming them into relatively older ones in order to win still faster and irreversibly. As soon as somatostatin has taken shape, a new bioassay is devised in which sosmatostatin takes the role of a stable, unproblematic substance in a trial set up for tracking down a new problematic substance, GRF. As soon as Svedberg has defined protein, the ultracentrifuge is made a routine tool of the laboratory bench and is employed to define the constituents of proteins. No sooner has polonium emerged from what it did in the list of ordeals above than it is turned into one of the well-known radioactive elements with which one can design an experiment to isolate a new radioactive substance further down in Mendeleev's table. The list of trials becomes a thing; it is literally *reified*.

This process of reification is visible when going from new objects to older ones, but it is also reversible although less visible when going from younger to older ones. All the new objects we analysed in the section above were framed and defined by stable black boxes which had *earlier* been new objects before being similarly reified. Endorphin was made visible in part because the ileum was known to go on pulsating long after guinea pigs are sacrificed: what was a new object several decades earlier in physiology was one of the black boxes participating in the endorphin assay, as was morphine itself. How could the new unknown substance have been compared if morphine had not been known? Morphine, which had been a new object defined by its trials in Seguin's laboratory sometime in 1804, was used by Guillemin in conjunction with the guinea pig ileum to set up the conditions defining endorphin. This also applies to the physiograph, invented by the French physiologist Marey at the end of the nineteenth century. Without it, the transformation of gut pulsation would not have been made graphically visible. Similarly for the electronic hardware that enhanced the signals and made them strong enough to activate the physiograph stylus. Decades of advanced electronics during which many new phenomena had been devised were mobilised here by Guillemin to make up another part of the assay for endorphin. Any new object is thus shaped by simultaneously importing many older ones in their reified form. Some of the imported objects are from young or old disciplines or pertain to harder or softer ones. The point is that the new object emerges from a complex set-up of sedimented elements each of which has been a new object at some point in time and space. The genealogy and the archaeology of this sedimented past is always possible in theory but becomes more and more difficult as time goes by and the number of elements mustered increases.

It is just as difficult to go back to the time of their emergence *as it is to contest them*. The reader will have certainly noticed that we have gone full circle from the first section of this part (borrowing more black boxes) to this section (blackboxing more objects). It is indeed a circle with a feedback mechanism that creates better and better laboratories by bringing in as many new objects as possible in as reified a form as possible. If the dissenter quickly re-imports somatostatin, endorphin, polonium, transfinite numbers as so many incontrovertible black boxes, his or her opponent will be made all the weaker. His or her ability to dispute will be decreased since he or she will now be faced with piles of black boxes, obliged to untie the links between more and more elements coming from a more and more remote past, from harder disciplines, and presented in a more reified form. Has the shift been noticed? It is now the author who is weaker and the dissenter stronger. The author must now either build a better laboratory in order to dispute the dissenter's claim and tip the balance of power back again, or quit the game—or apply one of the many tactics to escape the problem altogether that we will see in the second part of this book. The endless spiral has traveled one more loop. Laboratories grow because of the number of elements fed back into them, and this growth is irreversible since no dissenter/author is able to enter into the fray later with fewer resources at his or her disposal—everything else being equal. Beginning with a few cheap elements borrowed from common practice, laboratories end up after several cycles of contest with costly and enormously complex set-ups very remote from common practice.

The difficulty of grasping what goes on inside their walls thus comes from the sediment of what has been going on in other laboratories earlier in time and elsewhere in space. The trials currently being undergone by the new object they give shape to are probably easy to explain to the layperson—and we are all laypeople so far as disciplines other than our own are concerned—but the older objects capitalised in the many instruments are not. The layman is awed by the laboratory set-up, and rightly so. There are not many places under the sun where so many and such hard resources are gathered in so great numbers, sedimented in so many layers, capitalised on such a large scale. When confronted earlier by the technical literature we could brush it aside; confronted by laboratories we are simply and literally impressed. We are left without power, that is, without resource to contest, to reopen the black boxes, to generate new objects, to dispute the spokesmen's authority.

Laboratories are now powerful enough to define **reality**. To make sure that our travel through technoscience is not stifled by complicated definitions of reality, we need a simple and sturdy one able to withstand the journey: reality as the latin word *res* indicates, is what *resists*. What does it resist? *Trials of strength.* If, in a given situation, no dissenter is able to modify the shape of a new object, then that's it, it *is* reality, at least for as long as the trials of strength are not modified. In the examples above so many resources have been mobilised by the dissenters to support these claims that, we must admit, resistance will be vain: the claim has to be true. The minute the contest stops, the minute I write the word 'true', a new, formidable ally suddenly appears in the winner's camp, an ally invisible until then, but behaving now as if it had been there all along; Nature.

APPEALING (TO) NATURE

Some readers will think that it is about time I talked of Nature and the real objects *behind* the texts and behind the labs. But it is not I who am late in finally talking about reality. Rather, it is Nature who always arrives late, too late to explain the rhetoric of scientific texts and the building of laboratories. This belated, sometimes faithful and sometimes fickle ally has complicated the study of technoscience until now so much that we need to understand it if we wish to continue our travel through the construction of facts and artefacts.

(1) 'Natur mit uns'

'Belated?' 'Fickle?' I can hear the scientists I have shadowed so far becoming incensed by what I have just written. 'All this is ludicrous because the reading and the writing, the style and the black boxes, the laboratory set-ups—indeed all existing phenomena—are simply *means* to express something, vehicles for conveying this formidable ally. We might accept these ideas of 'inscriptions', your emphasis on controversies, and also perhaps the notions of 'ally', 'new object', 'actant' and 'supporter', but you have omitted the only important one, the only supporter who really counts, Nature herself. Her presence or absence explains it all. Whoever has Nature in their camp wins, no matter what the odds against them are. Remember Galileo's sentence, '1000 Demosthenes and 1000 Aristotles may be routed by any average man who brings Nature in.' All the flowers of rhetoric, all the clever contraptions set up in the laboratories you describe, all will be dismantled once we go from controversies about Nature to what

Nature is. The Goliath of rhetoric with his laboratory set-up and all his attendant Philistines will be put to flight by one David alone using simple truths about Nature in his slingshot! So let us forget all about what you have been writing for a hundred pages—even if you claim to have been simply following us—and let us see Nature face to face!'

Is this not a refreshing objection? It means that Galileo was right after all. The dreadnoughts I studied may be easily defeated in spite of the many associations they knit, weave and knot. Any dissenter has got a chance. When faced with so much scientific literature and such huge laboratories, he or she has just to look at Nature in order to win. It means that there is a *supplement*, something more which is nowhere in the scientific papers and nowhere in the labs which is able to settle all matters of dispute. This objection is all the more refreshing since it is made by the scientists themselves, although it is clear that this rehabilitation of the average woman or man, of Ms or Mr Anybody, is also an indictment of these crowds of allies mustered by the same scientists.

Let us accept this pleasant objection and see how the appeal to Nature helps us to distinguish between, for instance, Schally's claim about GHRH and Guillemin's claim about GRF. They both wrote convincing papers, arraying many resources with talent. One is supported by Nature—so his claim will be made a fact—and the other is not—it ensues that his claim will be turned into an artefact by the others. According to the above objections, readers will find it easy to give the casting vote. They simply have to see who has got Nature on his side.

It is just as easy to separate the future of fuel cells from that of batteries. They both contend for a slice of the market; they both claim to be the best and most efficient. The potential buyer, the investor, the analyst are lost in the midst of a controversy, reading stacks of specialised literature. According to the above objection, their life will now be easier. Just watch to see on whose behalf Nature will talk. It is as simple as in the struggles sung in the Iliad: wait for the goddess to tip the balance in favour of one camp or the other.

A fierce controversy divides the astrophysicists who calculate the number of neutrinos coming out of the sun and Davis, the experimentalist who obtains a much smaller figure. It is easy to distinguish them and put the controversy to rest. Just let us see for ourselves in which camp the sun is really to be found. Somewhere the natural sun with its true number of neutrinos will close the mouths of dissenters and force them to accept the facts no matter how well written these papers were.

Another violent dispute divides those who believe dinosaurs to have been coldblooded (lazy, heavy, stupid and sprawling creatures) and those who think that dinosaurs were warmblooded (swift, light, cunning and running animals).[5] If we support the objection, there would be no need for the 'average man' to read the piles of specialised articles that make up this debate. It is enough to wait for Nature to sort them out. Nature would be like God, who in medieval times judged between two disputants by letting the innocent win.

In these four cases of controversy generating more and more technical papers and bigger and bigger laboratories or collections, Nature's voice is enough to stop the noise. Then the obvious question to ask, if I want to do justice to the objection above, is 'what does Nature say?'

Schally knows the answer pretty well. He told us in his paper, GHRH *is* this amino-acid sequence, not because he imagined it, or made it up, or confused a piece of haemoglobin for this long-sought-after hormone, but because this is what the molecule is in Nature, independently of his wishes. This is also what Guillemin says, not of Schally's sequence, which is a mere artefact, but of his substance, GRF. There is still doubt as to the exact nature of the real hypothalamic GRF compared with that of the pancreas, but on the whole it is certain that GRF is indeed the amino-acid sequence earlier. Now, we have got a problem. Both contenders have Nature in their camp and say what it says. Hold it! The challengers are supposed to be refereed by Nature, and not to start another dispute about what Nature's voice really said.

We are not going to be able to stop this new dispute about the referee, however, since the same confusion arises when fuel cells and batteries are opposed. 'The technical difficulties are not insurmountable,' say the fuel cell's supporters. 'It's just that an infinitesimal amount has been spent on their resolution compared to the internal combustion engine's. Fuel cells are Nature's way of storing energy; give us more money and you'll see.' Wait, wait! We were supposed to judge the technical literature by taking another outsider's point of view, not to be driven back *inside* the literature and *deeper* into laboratories.

Yet it is not possible to wait outside, because in the third example also, more and more papers are pouring in, disputing the model of the sun and modifying the number of neutrinos emitted. The real sun is alternately on the side of the theoreticians when they accuse the experimentalists of being mistaken and on the side of the latter when they accuse the former of having set up a fictional model of the sun's behaviour. This is too unfair. The real sun was asked to tell the two contenders apart, not to become yet another bone of contention.

More bones are to be found in the paleontologists' dispute where the real dinosaur has problems about giving the casting vote. No one knows for sure what it was. The ordeal might end, but is the winner really innocent or simply stronger or luckier? Is the warm-blooded dinosaur more like the real dinosaur, or is it just that its proponents are stronger than those of the cold-blooded one? We expected a final answer by using Nature's voice. What we got was a new fight over the composition, content, expression and meaning of that voice. That is, we get *more* technical literature and *larger* collections in bigger Natural History Museums, not less; *more* debates and not less.

I interrupt the exercise here. It is clear by now that applying the scientists' objection to any controversy is like pouring oil on a fire, it makes it flare anew. Nature is not outside the fighting camps. She is, much like God in not-so-ancient wars, asked to support all the enemies at once. 'Natur mit uns' is embroidered on all the banners and is not sufficient to provide one camp with the winning edge. So what is sufficient?

(2) The Double-Talk of the Two-Faced Janus

I could be accused of having been a bit disingenuous when applying scientists' objections. When they said that something more than association and numbers is needed to settle a debate, something outside all our human conflicts and interpretations, something they call 'Nature' for want of a better term, something that eventually will distinguish the winners and the losers, they did not mean to say that we know what it is. This supplement beyond the literature and laboratory trials is unknown and this is why they look for it, call themselves 'researchers', write so many papers and mobilise so many instruments.

'It is ludicrous,' I hear them arguing, 'to imagine that Nature's voice could stop Guillemin and Schally from fighting, could reveal whether fuel cells are superior to batteries or whether Watson and Crick's model is better than that of Pauling. It is absurd to imagine that Nature, like a goddess, will visibly tip the scale in favour of one camp or that the Sun God will barge into an astrophysics meeting to drive a wedge between theoreticians and experimentalists; and still more ridiculous to imagine real dinosaurs invading a Natural History Museum in order to be compared with their plaster models! What we meant, when contesting your obsession with rhetoric and mobilisation of black boxes, was that *once the controversy is settled, it is*

Nature the final ally that has settled it and not any rhetorical tricks and tools or any laboratory contraptions.'

If we still wish to follow scientists and engineers in their construction of technoscience, we have got a major problem here. On the one hand scientists herald Nature as the only possible adjudicator of a dispute, on the other they recruit countless allies while waiting for Nature to declare herself. Sometimes David is able to defeat all the Philistines with only one slingshot; at other times, it is better to have swords, chariots and many more, better-drilled soldiers than the Philistines!

It is crucial for us, laypeople who want to understand technoscience, to decide which version is right, because in the first version, as Nature is enough to settle all disputes, we have nothing to do since no matter how large the resources of the scientists are, they do not matter in the end—only Nature matters. Our chapters may not be all wrong, but they become useless since they merely look at trifles and addenda and it is certainly no use going on for four other chapters to find still more trivia. In the second version, however, we have a lot of work to do since, by analysing the allies and resources that settle a controversy we understand *everything* that there is to understand in technoscience. If the first version is correct, there is nothing for us to do apart from catching the most superficial aspects of science; if the second version is maintained, there is everything to understand except perhaps the most superfluous and flashy aspects of science. Given the stakes, the reader will realise why this problem should be tackled with caution. The whole book is in jeopardy here. The problem is made all the more tricky since scientists *simultaneously* assert the two contradictory versions, displaying an ambivalence which could paralyse all our efforts to follow them.

We would indeed be paralysed, like most of our predecessors, if we were not used to this double-talk or the two-faced Janus. The two versions are contradictory but they are not uttered by the same face of Janus. There is again a clear-cut distinction between what scientists say about the cold settled part and about the warm unsettled part of the research front. As long as controversies are rife, Nature is never used as the final arbiter since no one knows what she is and says. But *once the controversy is settled*, Nature is the ultimate referee.

This sudden inversion of what counts as referee and what counts as being refereed, although counter-intuitive at first, is as easy to grasp as the rapid passage from the 'name of action' given to a new object to when it is given its name as a thing (see above). As long as there is a debate among endocrinologists about GRF or GHRH, no one can intervene in the debates by saying, 'I know what it is, Nature told me so. It is that amino-acid sequence.' Such a claim would be greeted with derisive shouts, unless the proponent of such a sequence is able to show his figures, cite his references, and quote his sources of support, in brief, write another scientific paper and equip a new laboratory, as in the case we have studied. However, once the collective decision is taken to turn Schally's GHRH into an artefact and Guillemin's GRF into an incontrovertible fact, the reason for this decision is not imputed to Guillemin, but is immediately attributed to the independent existence of GRF in Nature. As long as the controversy lasted, no appeal to Nature could bring any extra strength to one side in the debate (it was at best an invocation, at worst a bluff). As soon as the debate is stopped, the supplement of force offered by Nature is made the explanation as to why the debate did stop (and why the bluffs, the frauds and the mistakes were at last unmasked).

So we are confronted with two almost simultaneous suppositions:

> Nature is the final cause of the settlement of all controversies, *once controversies are settled.*
> As long as they last *Nature will appear simply as the final consequence of the controversies.*

When you wish to attack a colleague's claim, criticise a world-view, modalise a state-ment you cannot *just* say that Nature is with you; 'just' will never be enough. You are bound to use other allies besides Nature. If you succeed, then Nature will be enough and all the other allies and resources will be made redundant. A political analogy may be of some help at this point. Nature, in scientists' hands, is a constitutional monarch, much like Queen Elizabeth the Second. From the throne she reads with the same tone, majesty and conviction a speech written by Conservative or Labour prime ministers depending on the election outcome. Indeed she *adds* something to the dispute, but only after the dispute has ended; as long as the election is going on she does nothing but wait.

This sudden reversal of scientists' relations to Nature and to one another is one of the most puzzling phenomena we encounter when following their trails. I believe that it is the difficulty of grasping this simple reversal that has made technoscience so hard to probe until now.

The two faces of Janus talking together make, we must admit, a startling spectacle. On the left side Nature is cause, on the right side consequence of the end of controversy. On the left side scientists are *realists*, that is they believe that representations are sorted out by what really is outside, by the only independent referee there is, Nature. On the right side, the same scientists are *relativists*, that is, they believe representations to be sorted out among themselves and the actants they represent, without independent and impartial referees lending their weight to any one of them. We know why they talk two languages at once: the left mouth speaks about settled parts of science, whereas the right mouth talks about unsettled parts. On the left side polonium was discovered long ago by the Curies; on the right side there is a long list of actions effected by an unknown actant in Paris at the Ecole de Chimie which the Curies propose to call 'polonium'. On the left side all scientists agree, and we hear only Nature's voice, plain and clear; on the right side scientists disagree and no voice can be heard over theirs.

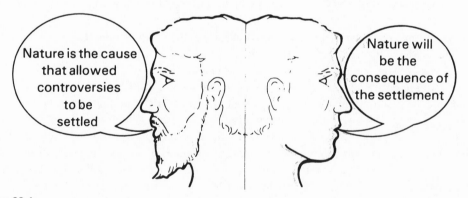

Figure 29.4

(3) The Third Rule of Method

If we wish to continue our journey through the construction of facts, we have to adapt our method to scientists' double-talk. If not, we will always be caught on the wrong foot: unable to withstand either their first (realist) or their second (relativist) objection. We will then need to have two different discourses depending on whether we consider a settled or an unsettled part of technoscience. We too will be relativists in the latter case and realists in the former.

When studying controversy—as we have so far—we cannot be *less* relativist than the very scientists and engineers we accompany; they do not *use* Nature as the external referee, and we have no reason to imagine that we are more clever than they are. For these parts of science our **third rule of method** will read: since the settlement of a controversy is *the cause* of Nature's representation not the consequence, we *can never use the outcome—Nature—to explain how and why a controversy has been settled.*

This principle is easy to apply as long as the dispute lasts, but is difficult to bear in mind once it has ended, since the other face of Janus takes over and does the talking. This is what makes the study of the past of technoscience so difficult and unrewarding. You have to hang onto the words of the right face of Janus—now barely audible—and ignore the clamours of the left side. It turned out for instance that the N-rays were slowly transformed into artefacts much like Schally's GHRH. How are we going to study this innocent expression 'it turned out'?

Using the physics of the present day there is unanimity that Blondlot was badly mistaken. It would be easy enough for historians to say that Blondlot failed because there was 'nothing really behind his N-rays' to support his claims. This way of analysing the past is called Whig history, that is, a history that crowns the winners, calling them the best and the brightest and which says the losers like Blondlot lost simply *because* they were wrong. We recognise here the left side of Janus' way of talking where Nature herself discriminates between the bad guys and the good guys. But, is it possible to use this as the reason why in Paris, in London, in the United States, people slowly turned N-rays into an artefact? Of course not, since at that time today's physics obviously could not be used as the touchstone, or more exactly since today's state is, in part, the *consequence* of settling many controversies such as the N-rays!

Whig historians had an easy life. They came after the battle and needed only one reason to explain Blondlot's demise. He was wrong all along. This reason is precisely what does not make the slightest difference while you are searching for truth in the midst of a polemic. We need, not one, but *many* reasons to explain how a dispute stopped and a black box was closed.[6]

However, when talking about a cold part of technoscience we should shift our method like the scientists themselves who, from hard-core relativists, have turned into dyed-in-the-wool realists. Nature is now taken as the cause of accurate descriptions of herself. We cannot be more relativist than scientists about these parts and keep on denying evidence where no one else does. Why? Because the cost of dispute is too high for an average citizen, even if he or she is a historian and sociologist of science. If there is no controversy among scientists as to the status of facts, then it is useless to go on talking about interpretation, representation, a biased or distorted world-view, weak and fragile pictures of the world, unfaithful spokesmen. Nature talks straight, facts are facts. Full stop. There is nothing to add and nothing to subtract.

This division between relativists and realist interpretation of science has caused analysts of science to be put off balance. Either they went on being relativists even about the settled parts of science—which made them look ludicrous; or they continued being realists even about the warm uncertain parts—and they made fools of themselves. The third rule of method stated above should help us in our study because it offers us a good balance. We do not try to undermine the solidity of the accepted parts of science. We are realists as much as the people we travel with and as much as the left side of Janus. But as soon as a controversy starts we become as relativist as our informants. However we do not follow them passively because our method allows us to document both the construction of fact and of artefact, the cold and the warm, the demodalised and the modalised statements, and, in particular, it allows us to trace with accuracy the sudden shifts from one face of Janus to the other. This method offers us, so to speak, a stereophonic rendering of fact-making instead of its monophonic predecessors!

NOTES

1. I am following here the work of Trevor Pinch (1986).
2. I am following here the work of Mary Jo Nye (1980, 1986).
3. See J. W. Dauben (1974).
4. For the ultracentrifuge see the nice study by Boelic Elzen (forthcoming).
5. I am alluding here to the remarkable work by A. Desmond (1975).
6. This basic question of relativism has been nicely summed up in many articles by Harry Collins. See in particular his latest book (1985).

30

Scientific Visualism

Don Ihde

THE "WEAK PROGRAM": HERMENEUTICS
IMPLICIT WITHIN SCIENCE

Much of the line I have argued with respect to the history and philosophy of science is that Modern to late Modern science is what it is because it has found ways to enhance, magnify, and *modify* its perceptions. Science, as Kuhn and others after him seem to emphasize, is a way of "seeing." Given its explicit late Modern hyper-*visualism*, this is more than mere metaphor. There remains, deep within science, a belief that seeing is believing. The question is one of how one can see. And the answer is: One sees *through, with, and by means of instruments*. It is, first, this perceptualistic hermeneutics that I explore in the weak program.

Scientific Visualism

It has frequently been noted that scientific "seeing" is highly visualistic. This is, in part, because of historical origins, again arising in early Modern times in the Renaissance. Leonardo da Vinci played an important bridge role here, with the invention of what can be called the "engineering paradigm" of vision.[1] His depictions of human anatomy, particularly those of autopsies which display musculature, organs, tendons, and the like—"exploded" to show parts and interrelationships—were identical with the same style when he depicted imagined machines in his technical diaries. In short, his was not only a way of seeing which anticipated modern anatomies (later copied and improved upon by Vesalius) and modern draftmanship, but an approach which thus visualized both exteriors and interiors (the exploded style). Leonardo was a "handcraft imagist."

The move, first to an almost exclusively visualist emphasis, and second to a kind of "analytic" depiction, was faster to occur in some sciences than in others. In astronomy, analytic drawing of telescopic sightings was accurate early on and is being rediscovered as such today.

From Don Ihde, *Expanding Hermeneutics: Visualism in Science*, (Evanston, Ill.: Northwestern University Press, 1999), pp. 158–177. Copyright © 1999 by Northwestern University Press.

The "red spot" on Jupiter was already depicted in the seventeenth century. But here, visual observations and depictions were almost the only sensory dimension which could be utilized. Celestial phenomena were at first open only to visual inspection, at most magnified through optical instrumentation. It would be much later—the middle of the twentieth century—that astronomy would expand beyond the optical and reach beyond the Earth with instruments other than optical ones.

Medicine, by the time of Vesalius, shifted its earlier tactile and even olfactory observations in autopsy to the visualizations à la da Vincian style, but continued to use diagnostics which included palpations, oscultations, and other tactile, kinesthetic, and olfactory observations. In the medical sciences, the shift to the predominantly visual mode for analysis began much later. The invention of both photography and X-rays in the nineteenth century helped these sciences become more like their natural science peers.

Hermeneutically, in the perceptualist style of interpretation emphasized here—the progress of "hermeneutic sensory translation devices" as they might be called—*imaging technologies* have become dominantly visualist. These devices make nonvisual sources into visual ones. This, through new visual probes of interiors, from X-rays, to MRI scans, to ultrasound (in visual form) and PET processes, has allowed medical science to deal with bodies become transparent.[2]

More abstract and semiotic-like visualizations also are part of science's sight. Graphs, oscillographic, spectrographic, and other uses of visual hermeneutic devices give Latour reason to claim that such instrumentation is simply a complex *inscription-making device* for a visualizable result. This vector toward forms of "writing" is related to, but different from, the various isomorphic depictions of imaging. I shall follow this development in more detail later.

While all this instrumentation designed to turn all phenomena into visualizable form for a "reading" illustrates what I take to be one of science's deeply entrenched "hermeneutic practices," it also poses something of a problem and a tension for a stricter phenomenological understanding of perception.

Although I shall outline a more complete notion of perception below, here I want to underline the features of perception which are the source of a possible tension with scientific "seeing" as just described. Full human perception, following Merleau-Ponty, is always *multidimensioned* and *synesthetic*. In short, we *never just see something* but always experience it within the complex of sensory fields. Thus the "reduction" of perception to a monodimension—the visual—is already an abstraction from the lived experience of active perception within a world.

Does this visualizing practice within science thus reopen the way to a division of science from the lifeworld? Does it make of science an essentially reductive practice? I shall argue against this by way of attempting to show that visualization in the scientific sense is a deeply *hermeneutic practice* which plays a special role. Latour's insight that experiments deliver *inscriptions* helps suggest the hermeneutic analogy, which works well here. Writing is language through "technology" in that written language is inscribed by some technologically embodied means. I am suggesting that the sophisticated ways in which science *visualizes* its phenomena is another mode by which understanding or interpretive activity is embodied. Whether the technologies are translation technologies (transforming nonvisual dimensions into visual ones), or more isomorphically visual from the outset, the visualization processes through technologies are science's particular hermeneutic means.

First, what are the epistemological advantages of visualization? The traditional answer, often given within science as well, is that vision is the "clearest" of the senses, that it delivers greater distinctions and clarities, and this seems to fit into the histories of perception tracing all the way back to the Greeks. But this is simply *wrong*. My own earlier researches into auditory phenomena showed that even measurable on physiological bases, hearing delivers within its dimension distinctions and clarities which equal and in some cases exceed those of visual acuity. To reach such levels of acuity, however, skilled practices must be followed. Musicians can detect minute differences in tone, microtones, or quarter tones such as are common in Indian music; those with perfect pitch abilities detect variations in tone as small as any visual distinction between colors. In the early days of auditory instruments, such as stethoscopes, or in the early use of sonar, before it became visually translated, skilled operators could detect and recognize exceedingly faint phenomena, as clearly and as distinctly as through visual operations. Even within olfactory perception, humans—admittedly much poorer than many of their animal cousins—can nevertheless detect smells when only a few molecules among millions in the gas mixture present occur in the atmosphere. In the realms of connoisseurship such as wine tasting, tea tasting, perfume smelling, and the like, specifics such as source, year, and blend—even down to individual ingredients—can be known. It is simply a cultural prejudice to hold that vision is ipso facto the "best" sense.

I argue, rather, that what gives scientific visualization an advantage are its *repeatable Gestalt features* which occur within a technologically produced visible form, and which lead to the rise and importance of *imaging* in both its ordinary visual and specific hermeneutic visual displays. And, here, a phenomenological understanding of perception can actually enhance the hermeneutic process which defines this science practice.

Let us begin with one of the simplest of these Gestalt features, the appearance of a figure against a ground. Presented with a visual display, humans can "pick out" some feature which, once chosen, is seen against the variable constant of a field or ground. It is not the "object" which presents this figure itself—rather, it is in the interaction of visual intentionality that a figure can appear against a ground.

In astronomy, for example, sighting comets is one such activity. Whether sighted with the naked eye, telescopic observation, or tertiary observations of telescopic photographs, the sighting of a comet comes about by noting the movement of a single object against a field which remains relatively more constant. Here is a determined and trained figure/ground perceptual activity. This is also an *interest-determined* figure/ground observation. While, empirically, a comet may be accidentally discovered, to recognize it as a comet is to have sedimented a great deal of previous informed perception.

These phenomenological features of comet discovery stand out by noting that the very structure of figure/ground is not something simply "given" but is *constituted* by its context and field of significations. To vary our set of observables, one could have "fixed" upon any single (or small group) of stars and attended to these instead. Figures "stand out" relative to interest, attention, and even history of perceivability *which includes cultural or macroperceptual features* as well. For example, I have previously referred[3] to a famous case of figure/ground reversibility in the history of aesthetics. In certain styles of Asian painting, it is the background, the openness of space, which is the figure or intended object, whereas the almost abstract tracing of a cherry blossom or a sparrow on a branch in the foreground is now the "background" feature which makes space "stand out."

When one adds to this mix the variability and changeability of instruments or technologies, the process can rapidly change. As Kuhn has pointed out, with increased magnifications in later Modern telescopes, there was an explosion of planet discoveries due to the availability of detectable "disc size," which differentiated planets from stars much more easily.[4]

I have noted Latour, in effect, sees instruments as "hermeneutic devices." They are means by which *inscriptions* are produced, visualizable results. This insight meshes very nicely with a hermeneutic reconstrual of science in several ways.

If laboratories (and other controlled observational practices) are where one prepares inscriptions, they are also the place where objects are made "scientific," or, in this context, *made readable*. Things, the ultimate referential objects of science, are never just naïvely or simply observed or taken, they must be *prepared* or *constituted*. And, in late Modern science, this constitutive process is increasingly pervaded by technologies.

But, I shall also argue that the results are often not so much "textlike," but are more like repeatable, variable *perceptual Gestalts*. These are sometimes called "images" or even pictures, but because of the vestigial remains of modernist epistemology, I shall call them *depictions*. This occurs with increasing sophistication in the realm of *imaging technologies* which often dominate contemporary scientific hermeneutics.

To produce the best results, the now technoconstituted objects need to stand forth with the greatest possible clarity and within a context of variability and repeatability. For this to occur, the conditions of instrumental transparency need to be enhanced as well. This is to say that the instrumentation, in operation, must "withdraw" or itself become transparent so the thing may stand out (with chosen or multiple features). The means by which the depiction becomes "clear" is constituted by the "absence" or invisibility of the instrumentation.

Of course, the instrumentation can never *totally* disappear. Its "echo effect" will always remain within the mediation. The mallet (brass, wood, or rubber) makes a difference in the sound produced. In part, this becomes a reason in late Modern science for the deliberate introduction of *multivariant* instrumentation or measurements. These *instrumental phenomenological variations* as I have called them also function as a kind of multiperspectival equivalent in scientific vision (which drives it, not unlike other cultural practices, toward a more postmodern visual model).

All of this regularly occurs within science practice, and I am arguing that it functions as a kind of perceptual hermeneutics already extant in those practices. I now want to trace out a few concrete examples, focused upon roles within imaging technologies, which illustrate this hermeneutic style.

Galileo's hand-held telescopes undertook "real time" observations, with all the limitations of a small focal field, the wobbliness of manual control, and the other difficulties noted above. And, while early astronomers also developed drawings—often of quite high quality—of such phenomena as planetary satellites, the isomorphism of the observation with its imaged production remained limited.

If, on the other hand, it is the repeatability of the Gestalt phenomenon which particularly makes instrumentally produced results valuable for scientific vision, then the much later invention of *photography* can be seen as a genuine technological breakthrough. Technologies as perception-transforming devices not only magnify (and reduce) referent phenomena, but often radically change parameters either barely noted, or not noted at all.

It would be interesting to trace the development of the camera and photography with respect to the history of science. For example, as Lee Bailey has so well demonstrated, not only was the camera obscura a favorite optical device in early Modern science, but it played a deliberately modeling role in Descartes's notion of both eye and ego.[5] From the camera obscura and its variants to the genuine photograph, there is a three-century history. This history finally focused upon the *fixing* of an image. As early as 1727, a German physician, Johann Schultze, did succeed in getting images onto chalk and silver powders, but the first successfully fixed image was developed by Joseph Niepce in 1826. His successor, Louis Daguerre, is credited usually (in 1839), but Daguerre simply perfected Niepce's earlier process.[6] I shall jump immediately into the early scientific use of photography.

If the dramatic appearance of relative distance (space) was the forefront fascination with Galileo's telescope, one might by contrast note that it is the dramatic appearance of a transformation of *time* which photography brought to scientific attention. The photograph "stops time," and the technological trajectory implicitly suggested within it is the ever more precise micro-instant which can be captured. In early popular attention, the association with time stoppage often took the association between the depiction and a kind of "death" which still photography evoked. Ironically, the stilted and posed earliest photos were necessary artifacts of the state of the technology—a portrait could be obtained only with a minutes-long fixed pose, since it took that long for the light to form the negative on glass covered with the requisite chemical mixture.

Photography, however, was an immediately popular and rapidly developing new medium. And, if portraits and landscapes were early favored, a fascination with motion also occurred almost immediately. The pioneers of stop-motion photography were Eadweard Muybridge and Thomas Eakins at the end of the nineteenth century. Muybridge's studies of horses' gaits served a popular scientific interest. He showed, with both galloping horses and trotters, that all four feet left the ground, thus providing "scientific" evidence for an argument about this issue, considered settled with Muybridge's photos of 1878.[7] Insofar as this is a "new" fact (this is apparently debatable since there are some paintings which purport to show the same phenomenon), it is a discovery which is instrumentally mediated in a way parallel to Galileo's telescopic capture of mountains on their Moon. And, if this time-stop capacity of the technology can capture a horse's gait, the trajectory of even faster time-stop photography follows quickly. By 1888 time-stop photography had improved to the extent that the Mach brothers produced the first evidence of shock waves by photographing a speeding bullet. In this case, the photo showed that the bullet itself penetrated its target, not "compressed air," which was until then believed to advance before the projectile and cause injury.

Here we have illustrations of an early *perceptual hermeneutic* process which yields visually clear, repeatable, convincing Gestalts of the phenomena described. At this level, however, there is a "realism" of visual result which retains, albeit in a time-altered form, a kind of visual *isomorphism* which is a variant upon ordinary perception. It is thus less "textlike" than many other variants which develop later.

The visual isomorphism of early still photography was also limited to surface phenomena, although with a sense of frozen "realism" which shocked the artists and even transformed their own practices.[8] The physiognomy of faces and things was precise and detailed. The stoppage of time produced a *repeatable image of a thing*, which could be analytically observed and returned to time and again.

A second trajectory, however, was opened by the invention of the X-ray process in 1896. Here the "insides" of things could be depicted. Surfaces became transparent or disappeared altogether, and what had been "invisible" or, better, occluded became open to vision. X-ray photos were not so novel as to be the first interior depictions; we have already noted the invention of the "exploded diagram" style practiced by da Vinci and Vesalius. And one could also note that various indigenous art, such as that of Arnhemland Aborigines and Inuit, had an "X-ray" style of drawing which sometimes showed the interiors of animals. But the X-ray photo did to its objects what still photography had done to surfaces—it introduced a time-stop, "realistic" depiction of interior features. In this case, however, the X-ray image not only depicts differently, but produces its images as a "shadow." The X-rays pass through the object, with some stopped by or reduced by resistant material—in early body X-rays, primarily bones.[9]

Moving rapidly, once again a trajectory may be noted, one which followed ever more distinct depiction in the development of the imaging technologies: today's MRI scan, CT tomography, PET scans, and sonograms all are variants upon the depiction of interiorities. Each of these processes not only does its depicting by different means but also produces different visual selectivities which vary what is more or less transparent and what is more or less opaque. (I shall return to these processes in more detail.) This continues to illustrate the inscription or visualization process which constitutes the perceptual hermeneutic style of science.

A third trajectory in visualization is one which continues from the earliest days of optical instrumentation: the movement to the ever more microscopic (and macroscopic) entities. The microscope was much later to find its usefulness within science than the telescope. As Ian Hacking has pointed out, as late as 1800 Xavier Bichet refused to allow a microscope in his lab, arguing that "When people observe in conditions of obscurity each sees in his own way and according as he is affected."[10] In part, this had to do with the features of the things to be observed. Many micro-organisms were translucent or transparent and hard to make stand out even as figures against the often fluid grounds within which they moved. When another device which "prepared" the object for science was invented, *staining processes through aniline dyes*, the microscope could be more scientifically employed.[11]

The trajectory into the microscopic, of course, explodes in the nineteenth and twentieth centuries, with electron microscopes, scanning, tunneling processes, and on to the processes which even produce images of atoms and atom surface structure. Let me include here, too, the famous radio crystallography which brought us DNA structure, as well as today's chromosome and genetic fingerprinting processes.

The counterpart, macro-imaging, occurs with astronomy and the "earth sciences" which develop the measuring processes concerning "whole Earth measurements. "While each trajectory follows a different, exploitable image strategy, the result retains the Gestalt-charactered visualization which is a favored perceptual object within science.

The examples noted above all retain repeatable Gestalt, visualizable, and in various degrees, isomorphic, features. This is a specialized mode of perception and perceptual hermeneutics which plays an important role within science, but which also locates this set of practices within a now complicated lifeworld.

"Textlike" Visualizations

I shall now turn to a related, but different, set of visualizations, visualizations which bear much stronger relations to what can be taken as "textlike" features. Again, Latour is relevant: if the laboratory is science's *scriptorium*, the place where inscriptions are produced, then some of the production is distinctly textlike. A standard text, of course, is perceived. But to understand it one must call upon a specific hermeneutic practice—*reading*, and the skills which go into reading.

Once again, it would be tempting to follow out in more detail some of the history of the writings which have made up our civilizational histories (and which characterize the postmodern penchant for textuality, following Derrida's *On Grammatology*). But as far as written texts are concerned, I want to note in passing only that the histories of writing have tended to converge into an over narrowing set of choices: alphabetical, ideographic, and, for special purposes, simple pictographic forms. Related to this shrinkage of historical forms, I also want simply to note in passing that science follows and exacerbates this trend within its own institutional form, so much so that its dominantly alphabetic actual text preference is even more clearly narrowed to the emerging dominance of English as "the" scientific natural language.

And there is plenty of "text" in this sense within science. The proliferation of journals, electronic publications, books, and the range of texts produced is obvious enough. These texts, however, always remain secondary or tertiary with respect to science, as we have seen from Latour. So this is not the textlike phenomenon I have in mind; instead, I am pointing to those analogues of texts which permeate science: charts, graphs, models, and the whole range of "readable" inscriptions which remain visual, but which are no longer isomorphic with the referent objects or "things themselves."

Were we to arrange the textlike inscriptions along a continuum, from the closest analogue to the farthest and the most abstractly disanalogous, one would find some vague replication of the history of writing. Historians of alphabetic writing, for example, have often traced the letters of alphabetic writing to earlier *pictographic* items in pre-alphabetic inscriptions. Oscar Ogg, for example, shows that our current letter "A" derives from an inverted pictograph of a bull image.[12] Earlier hieroglyphic inscriptions could serve double purposes: as an analog image of the depicted animal or as the representation of a particular phoneme in the alphabetic sense.

The vestigial analog quality noted in the history of writing also occurs in scientific graphics: for example, a typical "translation" technology occurs in oscillography. If a voice is being patterned on an oscilloscope, the sound is "translated" into a moving, squiggly line on the scope. Each sound produces a recognizable squiggle, which highly skilled technicians can often actually "read."[13] The squiggle is no more, nor no less, "like" the sound made than the letter is within a text, but the technical "hermeneutic" can read back to the referent. As the abstraction progresses, often purposefully so that a higher degree of *graphic Gestalt* can be visualized, the reading-perception becomes highly efficient. "Spikes" on a graph, anomalies, upward or downward scatters—all have immediate significance to the "reader" of this scientific "text." Here is a hermeneutic process within normal science. And it remains visualizable and carries now in a more textlike context the repeatable Gestalt qualities noted above as part of the lingua franca of this style of hermeneutic.

Older instrumentation often was straightforward analogous to the phenomenon being measured. For example, columns of mercury within a thermometer embodied the "higher" and "lower" temperatures shown. Or, if a container was enclosed, a glass tube on the outside with piping to the inside could show the amount of liquid therein. Even moves to digital or numeric dials often followed analog representations.

Finally, although I am not attempting comprehensiveness in this location of hermeneutics in my "weak program" within scientific practice, I want to conclude with some *conventions* which also serve to enhance the textlike reading perceptions. Graphs come with conventions: up and down for high and low temperatures or intensities; with the range of the growing uses of "false color" imagery, rainbow spectrum conventions are followed again for intensities, and so on. All of this functions "like" a reading process, a visual hermeneutics which retains its visualizations, but which takes textlike directions.

In the weak program I have been following to this point, I have chosen science activities which clearly display their hermeneutic features. These, I have asserted, include a preference for visualization as the chosen sensory mode for getting to the things. But, rather than serving simply as a reduction of perceptual richness by way of a monosensory abstraction, visualization has been developed in a hermeneutic fashion—akin to "writing" insofar as writing is also a visual display. Thus, if science is separate from the lifeworld, it is so in precisely the same way that writing would not be included as a lifeworld factor.

Second, I have held that the process within science practice which prepares things for visualization includes the instrumentarium, the array of technologies which can produce the display, depiction, graphing, or other visualizable result which brings the scientific object "into view." (I am not arguing, as some have, that *only* instrumentally prepared object may be considered to be scientific objects. But, in the complex late Modern sciences, instrumentation is virtually omnipresent and dominant when compared to the older sciences and their observational practices.)

Third, I am not arguing that these clearly hermeneutic practices within science *exhaust* the notion of science. I have not dealt with the role of mathematization, with forms of intervention which do not always yield visualizable results, or the need to take apart the objects of science, to analyze things. And I do not mean to imply that these factors are also important to science. Rather, I have been making, so far, the weak case that there are important hermeneutic dimensions to science, especially relevant to the final production of *scientific knowledge*. In short, hermeneutics occurs inside, within, science itself.

Moving now from the implicit hermeneutics within science praxis to the more complex practices—increasingly technologically embodied and instrumentally constructed—we are ready to take note of a stronger program.

THE "STRONG PROGRAM": HERMENEUTIC SOPHISTICATION

In the "weak program" I chose to outline what could easily be recognized as hermeneutic features operative within science. As I now turn to a stronger program, I shall continue to examine certain extant features within science practice which relate to hermeneutic activity, but I shall increasingly turn here to forefront modes of investigation which drive the sciences closer to a postmodern variant upon hermeneutics.

Whole Body Perception

It is, however, also time to introduce more fully, albeit sketchily, a phenomenological understanding of perception in action. This approach will be recognizably close to the theory of perception developed by Merleau-Ponty, although taken in directions which include stronger aspects of multistability and polymorphy, which earlier investigations of my own developed.

1. I have already noted some perceptual *Gestalt features*, including the presentation of a perceptual field, within which figure/ground phenomena may be elicited. Following a largely Merleau-Pontean approach, one notes that fields are always complexly structured, open to a wide variety of intentional interests, and bounded by a horizonal limit. Science, I have claimed, in its particular style of knowledge construction, has developed a visualist hermeneutic which in the contemporary sense has fulfilled its interests through *imagery* constituted instrumentally or technologically. The role of repeatable, Gestalt patterns, in both isomorphic and graphic directions, is the epistemological product of this part of the quest for knowledge.

2. In a strong sense, all sensory fields, whether focused upon in reduced "monosensory" fashion, or as ordinarily presented in synthesized and multidimensional fashion, are *perspectival* and concretely *spatial-temporal*. Reflexively, the embodied "here" of the observer not only

may be noted but is a constant in all sensory perspectivalism. This constant may be enhanced only by producing a string of interrelated perspectives, or by shifting into multiperspectival modes of observation. The "ideal observer," a "god's eye view," and nonperspectivalism do not enter a phenomenology of perception.

3. However, while a body perspective relative to the perceptual field or "world" is a constant, both the field and the body are *polymorphic* and *multistable*. In my work in this area, I have shown that multistability is a feature of virtually every perceptual configuration (and the same applies to the extensions and transformations of perception through instrumentation), and that the interrelation of bodily (microperceptual features) and cultural significations (macroperceptual features) makes the polymorphy even more complex. *There is no perception without embodiment; but all embodiment is culturally and praxically situated and saturated.*

4. While I have sometimes emphasized spatial transformations (Galileo's telescope) in contrast to temporal transformations (still photography), all perceptual spatiality is *spatial-temporal*. This space-time configuration may be shown with different effects, as in contrasts between visual repeatability and auditory patterning, but is a constant of all perceptions.

5. All perceptual phenomena are synesthetic and multidimensioned. The "monosensory" is an abstraction—although useful and possible to forefront—and simply does not occur in the experience of the "lived body" (*corps vécu*). The same applies, although not always noted, in our science examples. I will say more below on this feature of perception. The issue of the "monosensory" is particularly acute with respect to the technological embodiments of science, since instruments (not bodies) may be "monosensory." Again, we reach a contemporary impasse which has been overcome only in part. Either we turn ingenious in the ways of "translating" the spectrum of perceptual phenomena into a visual hermeneutic—perhaps the dominant current form of knowledge construction in science—or we find ways of enhancing our instrumental reductions through variant instruments or new modes of perceptual transformation (I am pointing to "virtual reality" developments here).

A "strong program," I am hinting, may entail the need for breakthroughs whereby a fuller sense of human embodiment may be brought into play in scientific investigations. Whereas the current, largely visualist hermeneutic within science may be the most sophisticated such mode of knowledge construction to date, it remains short of its full potential were "whole body" knowledge made equally possible. This would be a second step toward the incorporation of lifeworld structures within science praxis.

Instrumental Phenomenological Variations

In the voice metaphor I used to describe the investigation of things, I noted that the "giving of a voice" entails, actually, the production of a "duet" at the least. But this also means that different soundings may be produced, either in sequence or in array, by the applications of different instruments. This is a material process which incorporates the practice of "phenomenological variations" along with the intervention within which a thing is given a voice. This practice is an increasing part of science practice and is apparent in the emergence of a suite of new disciplines which today produce an ever more rapid set of revolutions in understanding or of more frequent "paradigm shifts."

I use this terminology because it is a theme which regularly occurs in science reporting. A Kuhnian frame is often cast over the virtually weekly breakthroughs which are reported in *Science, Scientific American, Nature*, and other magazines. Challenges to the "standard view" are common. I shall look at a small sample of these while relating the challenges to the instrumental embodiments which bring about the "facts" of the challenges. Here the focus is upon *multiple instrumental arrays* which have different parameters in current science investigation.

1. *Multiple new instruments/more new things.* The development of multivariant instruments has often led to increased peopling of the discipline's objects. And much of this explosion of scientific ontology has occurred since the mid-twentieth century. This is so dramatically the case that one could draw a timeline just after World War II, around 1950 for most instrumentation, and determine new forms for many science disciplines. For example, in astronomy, until this century, the dominant investigative instrumentation was limited to *optical* technologies and thus restricted to the things which produce *light*. With the development of *radio-telescopy*, based upon technologies developed in World War II—as so many fields besides astronomy also experienced—the field expanded to the forms of microradiation which occur along spectra beyond the bounds of visible light. *The New Astronomy* makes this "revolution" obvious. The editors note that

> The range of light is surprisingly limited. It includes only radiation with wavelengths 30 percent shorter to 30 percent longer than the wavelength to which our eyes are most sensitive. The new astronomy covers radiation from extremes which have wavelengths less and one thousand-millionth as long, in the case of the shortest gamma rays, to over a hundred million times longer for the longest radio waves. To make an analogy with sound, traditional astronomy was an effort to understand the symphony of the Universe with ears which could hear only middle C and the two notes immediately adjacent.[14]

Without noting which instruments came first, second, and so on, expanding out from visible light, first to ultraviolet on one side, and infrared on the other, now reaching into the previously invisible-to-eyeball perceptions, but still within the spectrum of optical light waves, the first expansion into invisible light range occurs through types of "translation" technologies as I have called them. The usual tactic here is to "constitute" into a *visible depiction* the invisible light by using some convention of *false color* depiction.

The same tactic, of course, is used once the light spectrum itself is exceeded. While some discoveries in radio astronomy were made by *listening* to the radio "hiss" of background radiation, it was not long before the gamma-to-radio wavelengths beyond optical capacities were also "translated" into visible displays.

With this new instrumentation, the heavens begin to show phenomena previously unknown but which are familiar today: highly active magnetic gas clouds, radio sources still invisible, star births, supernovas, newly discovered superplanets, evidence of black holes, and the like. The new astronomy takes us closer and closer to the "birth" of the universe. More instruments produce more phenomena, more "things" within the universe.

The same trajectory can be found in many other science disciplines, but for brevity's sake I shall leave this particular example as sufficient here.

2. *Many instruments/the same thing.* Another variant, now virtually standard in usage, is to apply a range of instruments which measure different processes by different means to the same object. Medical imaging is a good example here. If some feature of the brain is to be investigated, perhaps to try to determine without surgical intrusion whether a formation is malignant or not, multiple instrumentation is now available to enhance the interpretation of the phenomenon. A recent history of medical imaging, *Naked to the Bone*, traces the imaging technologies from the inception of X-rays (1896) to the present.

As noted above, X-rays allowed the first technologized making of the body into a "transparent" object. It followed the pattern noted of preparing the phenomenon for a scientific "reading" or perception. Early development entailed—to today's retrospective horror—long exposures, sometimes over an hour, to get barely "readable" images sometimes called "shadowgraphs." This is because X-ray imaging relies upon radiation sent through the object to a plate, and thus the degrees of material resistance cast "shadows" which form the "picture." The earliest problems focused upon getting clearer and clearer images.[15]

I have noted that the microscope became useful only when the specimen could be prepared for "reading" through a dye process which enhanced contrasting or differentiated structures (in the micro-organism). This image enhancement began to occur in conjunction with X-rays as early as 1911 with the use of *radioactive tracers* which were ingested or injected into the patient. This was the beginning of nuclear medicine.[16]

Paralleling X-ray technologies, *ultrasound* began to be explored with the first brain images produced in 1937. The quality of this imaging, however, remained poor since bone tended to reduce what could be "seen" through this sounding probe. (As with all technologies, it takes some time before the range of usefulness is discovered appropriate to the medium. In the case of ultrasound, soft tissue is a better and easier-to-define target.)

But, even later than the new astronomy, the new medical imaging does not actually proliferate into its present mode until the 1970s. Then, in 1971 and 1972, several patents and patent attempts are made for magnetic resonance processes (MRIs). These processes produce imagery by measuring molecular resonances within the body itself. At the same moment, the first use of the computer—*as a hermeneutic instrument*—comes into play with the refinement of *computer-assisted tomography* (CT scanning). Here highly focused X-ray beams are sent through the object (brains, at first), and the data are stored and reconstructed through computer calculations and processes. Computer-"constructed" imaging, of course, began in the space program with the need to turn data into depictions. Kevles notes that

> After the Apollo missions sent back computer-reconstructed pictures of the moon, it did not stretch the imagination to propose that computers could reconstruct the images of the interior of the body, which, like pictures from space, could be manipulated in terms of color and displayed on a personal video moniter.

Here mathematics and imagery or constructed perceivable depictions meet. I claim this is important to a strong hermeneutic program in understanding science.

By 1975 the practical use of positron emissions is captured in the PET scan process. These emissions (from positrons within the object) are made visible. This imagery has never attained the detail and clarity of the above technologies but has some advantage in a dynamic situation when compared to the "stills" which are produced by all but ultrasound processes (and which also are limited in clarity). Thus living brain *functions* can be seen through PET instrumentation. Then, in the 1990s, functional MRI and more sophisticated computer tomographic processes place us into the rotatable, three-dimensional depictions which can be "built up" or "deconstructed" at command, and the era of the *whole body image* is attained.

While each of these processes can show different phenomena, the multiple use is such that ever more complete analysis can also be made of single objects, such as tumors, which can be "seen" with differences indicating malignancy or benignity. This, again, illustrates the ever more complex ways in which science instrumentation produces a visible result, a visual hermeneutics which is the "script" of its interpretive activity.

3. *Many instruments/convergent confirmations.* Another variant upon the multiple instrument technique is to use a multiplicity of processes to check—for example, dating—for

greater agreement. In a recent dating of Java *homo erectus* skulls, uranium series dating of teeth, carbon 14, and electron spin resonance techniques were all used to establish dates much more recent (27,000 B.P. +/-53,000 B.P. for different skulls) than previously determined and thus found *homo erectus* to be co-extant with *homo sapiens sapiens*.[17] And, with the recent discovery of 400,000 B.P. javelin-like spears in Germany, one adds thermo-luminescence techniques to establish this new date for human habitation in Europe, at least double or triple the previously suspected earliest date for humans there. (Similar finds, now dated 350,000 B.P. in Siberia, and 300,000 + B.P. finds in Spain, all within 1996–97 discovery parameters, evidence this antiquity.)[18]

4. *Single instrument (or instrumental technique)/widespread multiple results.* Here perhaps one of the most widely used new techniques involves DNA "fingerprinting," which is now used in everything from forensics (rapists and murderers both convicted and found innocent and released), to pushing dates back for human migrations or origins. (The "reading" process which goes with DNA identification entails matching pairs and includes visualizations once again.) *Scientific American* has recently reported that DNA tracing now shows that human migrations to the Americas may go back to 34,000 years (not far from the dates claimed for one South American site, claimed to be 38,000 B.P., which with respect to physical data remain doubtful), with other waves at 15,000 B.P., to more recent waves, included a set of Pacific originated populace around 6,500 B.P.)[19] The now widely cited DNA claims for the origins of *homo sapiens* between 200,000 and 100,000 B.P. is virtually a commonplace.

DNA fingerprinting has also been used in the various biological sciences to establish parentage compared to behavioral mating practices. One result is to have discovered that many previously believed-to-be "monogamous" species are, in fact, not. Similarly, the "Alpha Male" presumed successful at conveying his genes within territorial species has been shown to be less dominantly the case than previously believed.[20]

5. *Multiple instruments/new disciplines.* Beginning with DNA (mitochondrial DNA matching) again, the application of this technique has given rise to what is today called "ancient DNA" studies, with one recent result developed in Germany this year, which purports to show that Neanderthals could not have interbred with modern humans, due to the different genetic makeup of these hominids who coexisted (for a time) with modern humans.[21]

Then, returning to variants upon imaging, the new resources for such disciplines as archeology produce much more thorough "picturing" of ancient sites, activities, and relations to changes in environmental factors. Again, drawing from *Scientific American*—one can draw similar examples from virtually every issue of this and similar science-oriented magazines—the array of instruments now available produces, literally, an "in-depth" depiction of the human past. In part, now drawing from uses originally developed for military purposes, imaging from (1) *Landsat,* which used digital imaging and multispectral scanners from the 1970s, to (2) refinements for *Landsats 4* and *5* in the 1980s, which extended and refined the imaging and expanded to infrared and thermal scanning, (3) to *SPOT,* which added linear-array technologies to further refine imaging, to (4) imaging radar, which actually penetrates below surfaces to reveal details, to (5) *Corona,* which provides spectrographic imagery (recently declassified), the modern archeologist, particularly desert archeologists, can get full-array depictions of lost cities, ancient roads, walls, and the like from remote sensing used now.[22]

Instrumentation on Earth includes (1) electromagnetic sounding equipment, which penetrates up to six meters into the earth, (2) ground-penetrating radar, which goes down to ten meters, (3) magnetometers, which can detect such artifacts as hearths, (4) resistivity instruments, which detect different densities and thus may be used to locate artifacts, and (5) seismic instruments, which can penetrate deeper than any of the above instruments. At a recent meeting in Mexico, an anthropologist reported to me that a magnetometer survey of northern Mexico has shown there to be possibly as many as 86,000 buried pyramids (similar to the largest, Cholula, although the remainder are smaller).[23]

In short, the proliferation of instrumentation, particularly that which yields imagery, is radical and contemporary and can now yield degrees and spans of three-dimensional imagery which includes all three of the image breakthroughs previously noted: early optics magnified the micro- and macro-aspects of barely noted or totally unnoted phenomena through magnification (telescopy and microscopy) but remained bound to the limits of the optically visible, by producing "up close" previously distant phenomena.

Photography increased the detail and isomorphy of imaging in a repeatable produced image, which could then be studied more intensely since "fixed" for observation. It could also be manipulated by "blowups" and other techniques, to show features which needed enhancement. Then, with X-rays, followed by other interior-producing imagery, the possibilities were outlined for the contemporary arrays by which image surfaces are made transparent so that one may see interiors. Then add the instruments which expand thoroughly beyond the previously visible and which now go into previously invisible phenomena through the various spectra which are "translated" into visible images for human observation.

Here we have the decisive difference between ancient science and Modern science: Democritus claimed that phenomena, such as atoms, not only were in fact imperceptible, but were *in principle* imperceptible. A modern, technologized science returns to Democritus to the "in fact" only—that is, the atomic is invisible only until we can come up with the technology which can make it visible. This, I claim, is an instrumental *visual hermeneutics*.

C. Technoconstitution

In the reconstrual of science which I have been following here, I have argued that late Modern science has developed a complex and sophisticated system of *visual hermeneutics*. Within that visualist system, its "proofs" are focused around the things seen. But, also, things are never just or merely seen—the things are *prepared* or made "readable." Scientifically, things are (typically, but not exclusively) *instrumentally mediated*, and the "proof" is often a *depiction* or image.

Interestingly, if in ordinary experience there is a level of *naïve realism* where things are taken simply to be what they are seen to be, similarly, within imaging there is at least the temptation to an imaging naive realism. That naïveté revolves around the intuitive taking of the image to be "like" or to "represent" an original (which would be seen in unmediated and eyeball perceptions). In short, "truth" is taken to be some kind of *isomorphism* between the depiction and the object.

With the issue stated in precisely that way, there are many traps which are set which could lead us back into the issues of modern (that is, Cartesian or seventeenth- and eighteenth-century) epistemologies. But to tackle these would lead us into a detour of some length. It would entail deconstructing "copy theory" from Plato on, deconstructing "representationalism" as the modern version of copy theory, before finally arriving at a more "postmodern" theory which entails both a theory of relativistic intentionality, a notion of perspectivalism, and an understanding of instrumental mediations as they operate within a phenomenological context. (I have addressed these issues in some degree in essays which preceded my formulation here.).[24] I simply want to avoid these traps.

To do so, I shall continue to interpret science in terms of a visual hermeneutics, embodied within an instrumentally realistic—but *critical*—framework in which instruments mediate perceptions. The device I shall now develop will fall within an idealized "history" of imaging, which, while containing actual chronologically recognizable features, emphasizes patterns of *learning to see*.

Isomorphic visions

The first pattern is one which falls into one type of *initial isomorphism* within imaging. As a technical problem, it is the problem of getting to a "clear and distinct" image. Imaging technologies do not just happen, they develop. And in the development there is a dialectic between the instrument and the user in which both a learning-to-see meets an elimination-of-bugs in the technical development. This pattern is one which, in most abstract and general terms, moves from initial "fuzziness" and ambiguity to greater degrees of clarity and distinctness.

Histories of the telescope, the microscope, photography, and X-rays (and, by extension, all the other imaging processes as well) are well documented with respect to this learning-to-see. Galileo, our quasi-mythical founder of early Modern science, was well aware of the need to teach telescopic vision, and of the problems which existed—although he eventually proclaimed the *superiority* of instrumentally mediated vision over ordinary vision. The church fathers, however, did have a point about how to take what was seen through the telescope. Not all of Galileo's observations were clear and easily seen by "any man." The same problem reemerged in the nineteenth century through the observations of Giovanni Schiaparelli, who gave the term the "canals of Mars." Schiaparelli was a well-known astronomer who had made a number of important discoveries, particularly with respect to asteroids and meteor swarms (because, in part, he had a much better telescope than Galileo). But in noting "canali"—which should have been translated into "channels"—which were taken to be "canals"—he helped stimulate the speculations about life on Mars. But neither channels nor canals existed—these, too, were instrumental artifacts.[25]

The dialectic between learning and technical refinement, in the successful cases, eventually leads to the production of clear and distinct images and to quick and easy learning. These twin attainments, however, cover over and often occlude the history and struggle which preceded the final plateaus of relative perfections. Thus, as in the previous illustrations concerning my guests and our Vermont observations of the Moon, once the instrument is focused and set, it literally takes only instants before one can recognize nameable features of its surface. The "aha phenomenon," in short, is virtually immediate today because it is made possible by the advanced technologies. That instantaneity is an accreted result of the hidden history of learning-to-see and its accompanying technical debugging process.

This same pattern occurred with the microscope. Although microorganisms never before seen were detected early, the continued problems of attaining clear and distinct microscopic vision were so difficult that it did not allow the microscope to be accepted into ordinary scientific practice until the nineteenth century. Again, the dialectic of learning how and what to see meets the gradual technical improvement concerning lenses and focusing devices, and finally the application of dying procedures to the things themselves. (This is an overt example of preparing a thing to become a scientific or "readable" object!)

Photography stands in interesting contrast to microscopy—if it took a couple of centuries for microscopy to become accepted for scientifically acceptable depiction, photography was much faster to win the same position. From Niepce's first "fixed" image in 1826, to the more widely accepted date of 1839 for Daguerre's first images, it was less than a half century until, as Bettyann Kevles notes in her history of medical imaging, "By the 1890's photographs had become the standard recorders of objective scientific truth."[26]

The same pattern occurs, but with even greater speed, in the history of early X-rays. In publicizing his new invention, Wilhelm Röntgen made copies of the X-ray of his wife's hand, which showed the bones of her fingers and the large ring which she wore, and sent these to his colleagues across Europe as evidence of his new process. That X-ray (with a long exposure time) was fuzzy, and while easily recognizable as a skeletal hand and ring, contrasts starkly with the radiograph made by Michael Pupin of Prescott Butler's shot-filled (shotgun-injured) hand later the same year. X-rays, duplicated across Europe and America almost immediately after Röntgen's invention, were used "scientifically" from the beginning.

The acceleration of acceptance time (of the learning-technical vision dialectic) similarly applies to the recent histories of imaging, which include, as above, sonograms (1937) and MRIs (1971) in medicine, of remote imaging since the *Tiros* satellite (1965), or of digitally transmitted and reconstructed images from *Mariner 4* (1965) in Earth and space science.

All of the above samples, however, remain within the range of the possible "naive image realism" of visual isomorphisms in which the objects are easily recognizable, even when new to the observer's vision. (Even if Röntgen had never before seen a "transparent hand" as in the case of his wife's ringed fingers, it was "obvious" from the first glimpse what was seen.) The pattern of making clear is an obvious trajectory. Yet we are not quite ready to leave the realm of the isomorphic.

How does one make "clearer" what is initially "fuzzy"? The answer lies in forms of manipulation, what I shall call *image reconstruction*. The techniques are multiple: enlargements (through trajectories of magnification noted before), enhancements (where one focuses in upon particular features and finds ways to make these stand out), contrasts (by heightening or lessening features of or around the objects), and so on. In my examinations I shall try not to be comprehensive, but to remain within the ranges of familiarity (to at least the educated amateur) concerning contemporary imaging. All of these manipulations can and do occur within and associated with simply isomorphic imaging and, for that matter, within its earlier range of black-and-white coloring. Histories of the technical developments which go with each of these techniques are available today and provide fascinating background to the rise of scientific visualism.

The moral of the story is images don't just occur. They are made. But, once made— assuming the requisite clarity and accuracy and certification of origin, etc.—they may then be taken as "proofs" within the visual hermeneutics of a scientific "visual reading." We are, in a sense, still within a Latourean laboratory.

Translation techniques

Much of what can follow in this next step has already been suggested within the realm of the isomorphic. But what I want to point to here is the use in late Modern science of visual techniques which begin ever more radically to vary *away from the isomorphic*.

One of these variables is—if it could be called that—simply the variable use of *color*. Returning to early optics, whatever Galileo or Leeuwenhoek saw, they saw in "true color." And, as we have seen, sometimes that itself was a problem. The transparent and translucent micro-organisms in "true color" were difficult to see. With aniline dyes, we have an early use of "false color." To make the thing into a scientific or "readable" object, we intervene and create a "horse of a different color." "False coloring" becomes a standard technique within scientific visual hermeneutics.

The move away from isomorphism, taken here in gradual steps which do not necessarily match chronologically what happened in the history of science, may also move away from the limits of ordinary perception. As noted above, the "new" late Modern astronomy of midcentury to the present was suddenly infused with a much wider stretch of celestial "reality" once it moved beyond optical and visible limits into, first, the humanly invisible ranges of the still optical or light itself, in the ranges of the infrared and ultraviolet. The instrumentation developed was what I have been calling a *translation technology* in that the patterns which are recordable on the instrumentation can be rendered by "false coloring" into visible images. This same technique was extended later to the full wave spectrum now available from gamma rays (short waves) through the optical to radio waves (long waves), which are rendered in the standard visually gestaltable, but false color, depictions in astronomy. All this is part of the highly technologized, instrumentalized visual hermeneutics which makes the larger range of celestial things into seeable scientific objects.

The "realism" here—and I hold that it is a realism—is a Hacking-style realism: if the things are "paintable"[27] (or "imagable") with respect to what the instruments detect as effects which will not go away, then they are "real." But they have been *made visible* precisely through the technological constructions which mediate them.

Higher level construction

Within the limits of the strong program, I now want to take only two more steps: I am purposely going to limit this attempt to reconstrue science praxis as hermeneutic to contemporary imaging processes which make (natural) things into scientific, and thus "readable," visual objects. I am not going to address the related, but secondary, visual process which entails *modeling*. That process which utilizes the computer as a hermeneutic device is clearly of philosophical interest, but I shall stop short of entering that territory here.

Computers, of course, are integral to many of the imaging processes we have already mentioned. Medical *tomography* (MRI, PET, fMRI, etc.) entails computer capacities to store and construct images. What is a visual Gestalt is built up from linear processes which produce data which have to be "constructed" by the computer. Similarly, the digitally transmitted imagery from distance sensing in satellite, space, and other remote imaging processes also has necessary computer uses. Much of contemporary imaging is computer embodied. And computers open the ways to much more flexible, complex, and manipulable imaging than any previous technology. For the purposes here, however, they will remain simply part of the "black boxes" which produce images which mediate perceptions.

The two higher-level constructive activities I want to point to here entail, first, the refinement of imaging which can be attained through specifically recognizing our technologies as mediating technologies which, in turn, must take into account the "medium" through which they are imaging. I turn again to astronomical imaging: the *Hubble* space telescope has recently captured the most public attention, but it is but one of the instrumental variations which are today exploring the celestial realms.

The advantage *Hubble* has is that it is positioned beyond the effects of the atmosphere with its distortions and interferences—the clarity of *Hubble* vision in this sense is due in part to its extra-Earthly perspective. (Science buffs will recall that at launch it had several defects in operation which were subsequently fixed—thus placing the *Hubble* in the usual pattern of needing technical adjustment to make its images clear!) But, in part by now being able to (phenomenologically) vary *Hubble* with Earth-bound optical telescopes, the move to enhancing Earth-bound telescopy through computer compensations has become possible. Astronomy is moving toward technoconstructions which can account for atmospheric distortions "on the spot" through a combination of laser targeting and computer enhancements. Earth-bound telescopes are today being given new life through these hi-tech upgrades which "read" atmospheric distortions and "erase" these processes which can make clearer new "readable" images. *Science* regularly publishes an "imaging" issue devoted to updating what is taken as the state-of-the-art in imaging (in 1997 it was the 27 June issue). A description of how one "undoes the atmosphere" is included, which entails computer reconstructions, telescopes in tandem, and adaptive optics. This process, *Science* claims, "combat[s] the warping effects of gravity on their giant mirrors . . . reclaims images from the ravages of the atmosphere . . . [and] precisely undoes the atmospheric distortions."[28]

But alongside *Hubble* are the other variants: the infrared space observatory, the *Cosmic Background Explorer*, and other satellite instrumentation which produces imagery from the nonoptical sources. All these technologies are variants upon the same multidimensioned variables which produce readable images, or make things into readable scientific objects.

The final set of instrumental productions I wish to note are the *composites* which produce variants upon "wholes." Earlier I dealt with "whole Earth measurements" which constitute one realm of composite imagery. To determine whether or not sea levels are rising overall, the composite imagery produced combines (1) multiple satellite photo imagery, (2) Earth-bound measurements (such as buoys, laser measurements, and land markers), and (3) computer averaging processes to produce a depiction (false colored) which can, in comparing time slices, show how much the oceans have risen. The composite depiction displays a flat-projection map of the Earth with level plateaus in false color spectra which can be compared between years, decades, and so on.

Similar processes occur in medical imaging. The "whole body imagery" available today on the internet is the result of two full-body "image autopsies," one each of a male and a female, whose bodies through tomographic processes may be seen in whatever "slice" one wishes. The linear processes of tomography show, slice-by-slice, vertically, horizontally, or in larger scans, the full bodies of the corpses used. The dimensions can be rotated, realigned, sectioned, and so on. Tomography also allows one to "peel," layer by layer, the object imaged—from skin, to networked blood vessels, to bones, and so on. (Both the whole Earth and whole body images are probably among the world's most expensive "pictures.") Moreover, all the manipulations which entail enhancements, contrasts, colorings, translations, and the like are utilized in these "virtual" images. Yet, while these virtual "realities" are different from the examination of any actual cadaver, they clearly belong to the visual hermeneutics of science in the strong sense. Things have been prepared to be seen, to be "read" within the complex set of instrumentally delivered visibilities of scientific imaging.

NOTES

1. A more complete discussion of Leonardo's transformation of vision can be found in chapter 1 of my *Postphenomenology* (Evanston, 1993).

2. See Bettyann Holtzmann Kevles, *Naked to the Bone: Medical Imaging in the Twentieth Century* (New Brunswick, N.J., 1997).

3. See my *Experimental Phenomenology* (Albany, 1986), pp. 128–29.

4. Kuhn, *Structure of Scientific Revolutions*, pp. 115–16.

5. See Lee W. Bailey, "Skull's Darkroom: The *Camera Obscura* and Subjectivity," in *Philosophy of Technology* (Dordrecht, 1989).

6. See both Jon Darius. *Beyond Vision* (Oxford, 1984), pp. 34–35, and Peter Poltack, *The Picture History of Photography* (New York, 1977), pp. 65–67.

7. Darius, *Beyond Vision*, pp. 34–35.

8. I have traced some interesting cross-cultural aspects of the imaging of others in pre- compared to post-photographic contexts; see chapter 4 of *Postphenomenology* (Evanston, 1993).

9. See Kevles, *Naked to the Bone*, pp. 3–20.

10. Ian Hacking, *Representing and Intervening* (Cambridge, 1983), p. 193.

11. Ibid.; see the chapter on "Microscopes," pp. 186–209.

12. Oscar Ogg. *The 26 Letters* (New York, 1967), p. 78.

13. These patterns are used by speech pathologists, for example, to show speakers how what they are saying does not, in fact, correspond to the standard form of a native language.

14. Nigel Henbest and Michael Marten, *The New Astronomy*, 2d ed. (Cambridge, 1996), p. 6.

15. Kevles, *Naked to the Bone,* p. 20.

16. Kevles provides a time chart, paralleling the various developments in the multiple imaging instrumentation.

17. *Science* 274 (13 December 1996): 1870–73.

18. *Science* 276 (30 May 1997): 1331–34.

19. *Scientific American* 276 (August 1997): 46–47.

20. *Science* 243 (31 March 1989): 1663.

21. *Science* 277 (11 July 1997): 176–78.

22. *Scientific American* 276 (August 1997): 61–65.

23. The visit to the Cholula pyramid and conversations with anthropologists occurred during the 9th International Conference of the Society for Philosophy and Technology, November 1996.

24. Postmodernism is more thoroughly discussed in Postphenomenology (Evanston 1993).

25. *Micropaedeia, Encyclopaedia Britannica*, 15th ed. (Chicago, 1994), vol. 10, p. 514.

26. Kevles, *Naked to the Bone*, p. 15.

27. I refer to Hacking's "If you can spray them then they are real" in *Representing and Intervening* (Cambridge, 1983), p. 23.

28. *Science* 276 (27 June 1997): 1994. This rhetoric is an example of the more-than-neutral language often employed by science reporting.

31

Should Philosophies of Science Encode Democratic Ideals?

Sandra Harding

EXTERNAL VERSUS INTERNAL DEMOCRACY ISSUES

How are the economic benefits and costs of the production of scientific information distributed within societies and between them? Who receives the social and political benefits and costs? Who gets to make the decisions that produce such distributions? Are the processes responsible for such distributions democratic?

Most people concerned to strengthen the links between modern science and democratic projects have focused on these kinds of questions, raising issues external to sciences' cognitive, technical core. According to this external democracy view, as it will be called, sciences' internal core—its most tested theories, models, methods, descriptions, and explanations of nature's order—is immune to such questions. These people have been concerned, for example, with who has access to mathematical, technological, and science training, who gets to decide which scientific projects should be funded, who gets access to the information and technologies that research makes possible, and who gets to make decisions about the social and environmental risks generated by scientific and technological projects.

Such external issues arise in global as well as national contexts. They have emerged in controversies over the role of sciences (intended or not by scientists) in military matters, environmental destruction, and the negative consequences of development policies and practices on the world's economically and politically least-advantaged groups. For example, U.S. funding for the natural sciences—physics especially—disproportionately has been tied to national security priorities. Yet the military information and technologies produced end up disproportionately used within (or against) third world societies in Latin America, the Middle East, and Africa. In another case, post–World War II development projects were supposed to enable third world societies to reach the higher standards of living available in the first world. Development was to be accomplished through the transfer to the third world of first world sciences, technologies, and their philosophies of rational inquiry and rational organization. Yet the priorities, policies, and practices of these programs have ended up "developing" primarily the already most economically and politically well-positioned groups in the North and in third world countries. They have largely dedeveloped and maldeveloped the great majority of the world's peoples who are already the most economically and politically vulnerable. In most cases development policies and practices redirected the third world's natural resources and human labor to serve the needs of trans-national corporations and "the investing classes" in the first and third worlds, and substituted socially and environmentally destructive lifestyles for less harmful local practices. Modern sciences' agendas often have ended up aligned with antidemocratic projects globally as well as nationally, though this certainly was not the intent of most of the scientists or, in many—perhaps most—cases, of the development administrators. For the most part, preserving and/or advancing desirable cultural, political, and environmental values were simply not on the agenda of development agencies or their funders.

Efforts to promote external democracy, crucially important as they are, do not challenge the idea that social and political neutrality can, does, and should characterize sciences' internal, cognitive, technical cores. They do not challenge the Enlightenment assumption that sciences can be, should be, and in the best existing cases are value-free. Of course modern sciences are conducted within social worlds in that their human and material resources must be provided by the larger social order. Moreover, the amounts, kinds, and sources of such resources have varied from era to era and from one culture to another. But sciences' transcultural, socially neutral theories, models, and methods are believed to enable them to detect the facts about the order of the universe that are everywhere and always the same. According to the externalists, sciences are in society, but society is not *in* sciences, their best theories, models, methods or results of research.

In contrast, cognitive democracy approaches, as they will be called, are concerned with how social and political fears and desires get encoded in that purportedly purely technical, cognitive core of scientific projects. How best to deal with this kind of phenomenon in trying to link sciences more closely to democratic projects is the topic of my analysis here. However, before turning to say more about it in the next section, several possible misunderstandings of this kind of project must be addressed.

Some readers may fear that this cognitive approach adopts a relativist epistemology, or consists in a "flight from reason." There are no sound reasons for such fears. The cognitivists' point—at least as I and the vast majority of such analysts develop it—is not that scientific practices are "nothing but" social, political projects, or that the representations of nature they produce are shaped entirely and only by such projects. Rather the point is that technical, cognitive elements of scientific practices and the information these produce always represent social and political priorities, meanings, and ideals as well as more or less accurate pictures of nature's order. (Central arguments in this approach will be reviewed below.) There are indeed rational theoretical and practical standards for evaluating competing knowledge claims. Modern sciences do in many obvious ways achieve less and less false claims about nature's order.

It is easy to overlook the fact that "less and less false" is not the same as "true." One can never be sure the sciences have arrived at absolutely true claims for two reasons: present claims must be held open to revision in case of the appearance of further empirical evidence, and they must be held open to the need for fruitful conceptual shifts. It is these considerations—these reasons for refusing to claim the production of Truth—that are supposed to distinguish empirical sciences from dogmatic positions. Most of the observations made by medieval astronomers are still facts within astronomy today. Yet the hypothesized relations between these facts and the meanings such facts have in the modern world are vastly different from the relations and meanings attributed to them in the medieval world.

Philosophies of science are one site where social and political fears and desires appear. Such philosophies are produced, used, interrogated, revised, and refined by professional philosophers, but also always by scientists, science policy-makers and managers, and by all of us citizens who consume the products of science in the air we breathe, the food we eat, the technologies of everyday life, and myriad other ways. That is, philosophizing on any topic rarely is monopolized by professionals since other groups usually have their own interests in the kinds of general and prescriptive theories about human activity on which professional philosophers professionally focus.

Concern with philosophies of science may seem to many readers largely irrelevant to the real life projects of encouraging closer links between modern sciences and democratic policies and practices. My point is that it is precisely such encoded ideals that have powerful effects on our daily activities. Such ideals make seem natural, logical, "common sense," and otherwise desirable precisely the kinds of antidemocratic policies and practices of concern to the externalists. There are consistencies between the external antidemocratic policies and practices that shape modern sciences and models or idealizations of such policies and practices that can be found in philosophic aspects of sciences' cognitive cores. Moreover, this argument makes another point: that there are significant scientific costs to such antidemocratic idealizations as well as the more obviously recognizable political costs.

Another caveat. Perhaps such concerns will appear to *introduce* social and political elements into otherwise socially neutral sciences and accounts of them. As will be clear from what follows, however, everyone who reflects on the matter understands that modern sciences already do encode precisely such social and political fears and desires. The question here, instead, is whether they should more effectively encode democratic ideals, how they can do so, and on what grounds could one justify such recommendations. A preoccupation with the futile (and undesirable, the argument will go) project of excluding social and political fears and desires from the sciences' cognitive cores delegitimates and distracts from these kinds of important issues.

There are many respects in which philosophies of science could encode democratic ideals. Here I shall focus on just one. Elsewhere I have argued that the universality ideal in particular is scientifically and politically dysfunctional (Harding, 1998, chapter 10). Pursuit of the universality ideal devalues cognitive diversity, which is now and always has been an important resource for the growth of knowledge. The sections that follow will summarize those arguments, and then propose that something important for democratic sciences can be retained and transformed from the older universality ideal nevertheless. We can retain the recognition that it is valuable to try to universalize one's hypotheses—to see how far from their cognitive and cultural point of origin they have value. However, the desirability of such processes should themselves be universalized. An appropriate philosophy of science should encourage the recognition that many scientific and technological traditions besides those of modern science contain elements that can prove valuable far from their site of origin. Moreover, it is important that no one culture's scientific and technological traditions serve as a gatekeeper on the flow of such processes and the cognitive resources they distribute. Thus, a philosophic model that recommends appreciation of the distinctive strengths of many cultures' knowledge systems for the legacy of human knowledge can link cognitive and political virtues in a more extensive encoding of the democratic ethos within philosophies of science.

ENCODED DEMOCRATIC IDEALS

The practice of trying to identify cultural, social, and political elements in scientific projects brings them to our attention and makes possible critical examination of them. Sciences have an "integrity" with their historic eras, as Thomas Kuhn famously put the point, and as the subsequent post-Kuhnian, feminist, and postcolonial science and technology studies have demonstrated (Harding, 1998). Such practices make these elements objects of our conscious thought, thus preventing them from functioning as unexamined evidence for the apparent reasonableness of one claim over another. We can learn how philosophies of science that guide research already do encode political fears and ideals.

However, not all of such fears and ideals are ones we should want to eliminate. Some are democratic, and these democratic elements are thought to advance the growth of knowledge. For example, everyone regards it as a strength, not a limitation, that philosophies of science insist that the social status of the observer should not provide a standard for evaluating the adequacy of scientific observations or arguments. Modern science is supposed to be a democracy in this sense, not an aristocracy. Galileo argued that anyone could see through his telescope the facts about the heavens. So, too, scientific practices today support the assumption that a graduate student no less than a Nobel Prize winner can turn out to be the one who comes up with an important observation or argument. Of course experienced observers can be expected to come up with more reliable contributions than those less experienced. The constant emergence of valuable observations and arguments from less experienced observers, however, sustains belief in the scientific value of this democratic ideal.

Relatedly, the results of research must be replicable by any individual or group that wishes to undertake the rigorous procedures necessary to produce them. This requires that the results of research must be public; they belong to "humanity," and on scientific grounds may not legitimately be shielded from public view. There is a related external-democracy issue with which this one may be confused. At times it has explicitly been claimed that the results of research should also be available for anyone to use; scientific knowledge should be public in this sense also. Appeal to such democratic principles can still be detected in arguments for funding projects promised to advance "human knowledge," to increase "human welfare," or to contribute to "human progress." And groups who receive too few of the benefits and who bear too many of the costs of scientific research often make this claim. Yet systems of contracts, patents, and licenses now insure that the results of scientific research that have the greatest social consequences are not public in this sense. They are privatized by those groups powerful enough to enforce such monopolies, such as states, corporations, and the research institutions that they sponsor. Thus, in many respects citizens who are not privileged to be party to such contracts, patents, and licenses now have the least access to the results of research that has the greatest consequences for their lives. This is an issue about the social "uses and abuses" of scientific information, however. My point here is that the required publicity of the results of research is an important democratic ideal that is part of scientific method; it is part of the cognitive, technical core of scientific practice.

Again, in early modern science the belief that matter was everywhere composed of the same kinds of materials was perceived to be a radically democratic claim. It challenged the Christian argument that the celestial and terrestrial spheres were composed of fundamentally different kinds of matter that obeyed different kinds of laws. To challenge the hierarchy of matter was to challenge also the political hierarchies that the Christian view of matter modeled. Christian hierarchies began to lose their ability to model the "order of nature," and vice versa, after the advent of early modern science, for the matter composing the earth and the heavens increasingly was seen as identical and thus equal. This case, too, demonstrates that there is nothing new about encoding political ideals in the cognitive core of the sciences—this time in a scientific ontology. This is not an example of political neutrality winning out over a politics of science, but rather of one politics replacing another. The "equality of material bodies" is or should be a political principle of democracies, as feminist and race theorists have been arguing. What is at issue here instead is whether the political ideals currently encoded in the cognitive, technical cores of the sciences are the ones we do and should want.

My point is not that a "politically correct" philosophy of science automatically insures better sciences, a more democratic social order, or a more effective link between them. Rather, the social ideals encoded in scientific thought about nature and in technical and popular thought about the sciences provide justifications for external, political practices, including desirable political practices. The values and interests that appear in sciences' cognitive cores can have positive scientific as well as political consequences, not just the negative ones that the older philosophies of science presumed. The ones with good scientific effects can also make certain democratic ideals attractive; in early modern science they made the equality of observers, the publicity of information, and the equality of apparently different kinds of bodies appear natural, "common sense," and desirable. We shall return below to identify some of the negative scientific as well as political effects created by other political elements in familiar philosophies of science.

First, however, it needs to be noted that what counts as maximally desirable "democratic ideals" is itself a controversial issue.

STANDARDS FOR DEMOCRATIC SOCIAL RELATIONS

There will be scientific standards for adopting democratic social ideals into sciences' cognitive cores: such ideals must promise to advance the growth of scientific knowledge (as the familiar ones identified above do) in addition to whatever political benefits they deliver. But what political standards should one use for selecting such democratic social ideals? One approach would be to identify a (hopefully relatively uncontroversial) general democratic principle, and then try to specify social practices that would conform to it. For such a principle one could take the familiar claim that those who bear the consequences of decisions should have proportionate shares in making them. There will be exceptions made to such a general rule, of course: infants and other very carefully identified groups cannot be expected to be able to make such decisions, or to make them wisely. Specification of the exceptions will itself be an important part of any democratic process. Nevertheless, the general principle is attractive because it has guided so many different kinds of effectively democratic practices. The institutions and procedures through which those proportionate shares would be exercised can be expected to differ in different social contexts: practices appropriate for small, homogeneous, and/or oral cultures would have to be different from those appropriate for large, heterogeneous, and multiply literate cultures.

Political philosophers point to the varying democratic effects of three kinds of more specific principles that have been thought to conform to such a democratic directive. One recommends that the interests of relevant groups should be fairly represented during decision-making processes. In local, national, and trans-national scientific councils, the interests of all the groups who will bear the consequences of scientific and technological decisions should be represented during policy-making processes. While this conception of democratic practices is better than none, critics argue that it does not produce democratic enough effects. Who is to represent such interests? (Should those in power be presumed to be able to represent fairly the interests of those over whom they exercise power?) And how are such councils to be held accountable for identifying who the relevant groups are, how their interests can best be fairly represented, and how democratically to resolve conflicts between competing interests? This principle has been used to justify patently unjust social and political systems; it has often been accused of a paternalism which blocks recognition and appropriate consideration of the interests of the politically and socially least powerful groups.

A stronger proposal is that members of such relevant interest groups should themselves have rights to represent their group's interests in decision-making councils: there should be proportionate numbers of women and men, whites and Blacks, and so on among the groups that design and manage scientific institutions and projects. This practice can go far to correct antidemocratic tendencies in any project. There is ample evidence that not only scientific and technological benefits but also political ones can develop from grassroots organization, "participatory action research," "bottom-up" design, and other such ways of giving "end users" a central voice in the design of scientific and technological projects. Nevertheless, these approaches can be enacted in too conservative ways that reduce both their scientific and political benefits. Widespread experience with "adding women and minorities" to worksites and policy groups where they have heretofore been excluded reveals such limitations. Who decides which groups have a right to be so represented? Will only the least threatening members of minoritized groups be the ones permitted into science policy councils? Must they suppress their "difference"—mute their demands or present only those parts of them that easily fit into prevailing concepts and practices—in order to function effectively within the prevailing standards for organizational behavior in scientific and technology institutions? Can individuals produce powerful enough discourses on behalf of their groups' interests to compete effectively with the prevailing dominant discourses, institutions, and practices of the majorities? After all, it took decades of political and intellectual work by many diverse groups to produce the evidence and arguments demonstrating that so-called development was primarily delivering benefits to already advantaged groups and dedeveloping and maldeveloping the already least advantaged. This understanding was not one that automatically became visible to those initially disadvantaged by development policies.

The strongest proposal for achieving democratic standards is that the latter require real equality among groups in the institutions and societies in which such decisions occur. Until sexism, racism, and class systems no longer are able to distribute social, political and economic resources inequitably, institutions within such societies cannot achieve maximally democratic decision processes. "Real equality," moreover, includes symbolic as well as "material" resources. Thus if a culture's ethnocentric standards—the inequality of groups, their thinking, and traditions—are modeled as ideal in the cognitive cores of sciences, one should not expect real equality among members of such groups in scientific institutions and cultures. Insofar as the politics of the larger society are modeled in a science's cognitive core, they serve as obstacles to efforts to eliminate other ways in which such standards shape scientific practices. Attempts at more democratic overt decision processes in scientific institutions are frustrated by models of ideal scientific practices that broadcast antidemocratic messages. The latter are all the more powerful when they are obscured in a cognitive, technical core of science claimed to be immunized against the possibility of social influence and, thus, of critical social analysis.

"More science and technology" in undemocratic societies cannot be expected to deliver the benefits and costs of research democratically, according to this view. In such conditions, more science and technology is guaranteed to increase social inequality (in spite of the intentions of individual scientists or policy makers). Scientific institutions, their cultures, and practices cannot by themselves counter antidemocratic power distributions within society's other institutions that insure that only those already politically, economically and socially advantaged will be in positions to be able to take advantage of the information scientific and technological work produces. When scientific institutions, cultures, and practices produce conflicting political messages, it is even more unreasonable to expect them to be advancing democratic social relations. There are scientific costs to such antidemocratic conditions also, to which the following sections turn. This situation predicts a rather depressing scenario to most of us who thought that more science and technology always delivered at least some benefits "to humanity," however much other benefits were siphoned off by the already overadvantaged.

The next section turns to review one familiar element of modern philosophies of science which, it has been argued, encodes antidemocratic values and interests—namely, the universality ideal. What is antidemocratic about this ideal, how is it politically and scientifically dysfunctional, and what justification is there for nevertheless retrieving a part of it for prodemocratic projects?

ONE WORLD, ONE TRUTH, ONE SCIENCE?

The "unity of science" thesis is one important configuration of ideas in which the universality ideal has played a central role. This model of the sciences and the world they would describe and explain became popular in the late nineteenth and early twentieth centuries; it became a kind of intellectual movement in the first half of the twentieth century. In modified versions it still holds great appeal for many philosophers, scientists, and the general public. However, today it also is the object of considerable skepticism on the part of philosophers and historians of science (Dupre, 1993; Galison & Stump, 1996).

According to this argument, there is one world, one and only one possible true account of it, and one unique science that can capture that one truth most accurately reflecting nature's own order. Less visible in most articulations of the unity thesis is a fourth assumption: there is just one group of humans, one cultural model of the ideal human, to whom nature's true order could become evident.[1] For early modern scientists and philosophers, the ideal human knower could be found among members of the new educated classes. Such individuals could use distinctive knowledge-seeking procedures, and thereby their theories could come to reflect the true order of nature that God's mind had created, just as the latter had also created human minds "in his own image." As the ideal human mind came to occupy the place in modern philosophy that the soul had occupied in Christian thought, rational man replaced spiritual man as the chosen recipient of the one true vision of the world's order.

There are a number of historical problems with the unity of science thesis and its universal science ideal. For one thing, modern science is plural. There are many distinctive modern sciences with incompatible ontologies, methods, and models of nature and of the research process. If "unity" means singularity rather than simply harmony, the possibility of "reducing" them to one—methodologically, ontologically, theoretically, linguistically—no longer exerts the attraction it did earlier in the century (Hacking, 1996). For example, no longer does it seem reasonable to most philosophers or scientists to try to explain the phenomena of interest to biology referring only to the kinds of natural objects usefully invoked in physics, or by using only the methods, models, or languages useful in physics. Nor are the methods of the sciences reasonably regarded as singular in any interesting sense.

This is not to deny that diverse sciences can usefully share some of their research techniques, models, languages, and other elements. Such borrowings have been a continual source of new insights in every field. Nor is it to deny that they find effective ways to communicate across their differences (Pickering, 1992). Perhaps "unity" should be taken to mean only harmony, as some early defenders of the unity thesis had in mind. However, while many kinds of such harmony certainly do exist among the sciences, such an interpretation of unity undercuts attempts to claim universality for elements of the cognitive cores of sciences. There can be all kinds of "harmonies"—sharings, borrowings, communications—of disparate elements without any elements at all achieving universality.

Moreover, globally, there have been many effective scientific and technological traditions. Elements of these traditions have been borrowed into European sciences, and vice versa (Goonatilake, 1984; Needham, 1954ff; Petitjean, Jami, & Moulin, 1992; Sabra, 1976). Why should anyone, from any culture, value having access to one and only one scientific tradition, as the unity (singularity) of science thesis does, and thereby lose the resources provided by the availability of several or a multiplicity of such traditions? Perhaps there are good reasons for such a value system, but it is worth reflecting on how cognitive diversity in human knowledge-traditions arises, and what is valuable about such diversity.

SOURCES OF COGNITIVE DIVERSITY

It is easy to see why diversity in science and technology traditions will arise, but it is also not difficult to see why such diversity is a valuable resource for the human scientific and technological legacy (Harding, 1998). For one thing, cultures occupy different spatiotemporal locations in nature's heterogeneous order. Some peoples live on deserts, others on fertile plains; some at high altitudes and others at sea level; in warm or in cold climates; in sparse and fragile environments or seemingly bountiful and sturdy ones. Some interact with nature on the sea route from Genoa to the Caribbean; others on the air route from Cape Kennedy to the moon. Each culture that is to survive will have to ask questions relevant to its survival about the environment within which it is located or through which it travels.

Moreover cultures have different interests in the environments with which they interact. Even what is apparently the same environment—a desert, for example—can be the object of different questions by different cultures. Some will want to know how to navigate trade routes across the sands, where camels and their riders can find water, and how both can survive sand storms. Others will want to know how to divert nearby rivers to irrigate a desert for human settlements and even for large-scale agribusiness. Yet other cultures will want information about the geological formations and human populations in particular deserts in order to use these areas as nuclear weapons test ranges. Other cultures will want to know how to mine the minerals and oil that lie beneath a desert's surface. Cultures' different locations in nature's order and their different interests in their environments will lead them to ask different questions and to develop different repositories of knowledge about nature's order. Since a culture's preoccupation with one set of environmental issues can lead it to ignore others, bodies of systematic knowledge are always accompanied by bodies of systematic ignorance: the two are always coproduced.

So far, none of these ways that cultures tend to produce different bodies of knowledge and ignorance would appear to challenge the singularity of science thesis and its universality claim. After all, it is precisely the project of the sciences, the singularity thesis assumes, to try to fit such diverse collections of knowledge into one picture frame that provides the one truth about nature's singular order. The model invoked here is of a jigsaw puzzle in which different sciences, scientific eras, or cultures contribute different pieces of detail to a more and more complete representation. Yet the next two sources of cognitive diversity show how merely analytic was extraction of the first two sources from the social contexts in which they occur. In reality, these first two apparently untroublesome aspects of cultural difference for the singularity thesis defenders turn out to be inextricable from other kinds of incompatible elements of knowledge systems. These incompatible elements create no insurmountable obstacles to harmony between knowledge-systems (sharings, borrowings, communications), but they block the possibility—and the desirability—of singularity.

What parts of nature it is that cultures perceive as desirable to occupy, how they conceptualize their interests, and just what questions they do ask are shaped also by their local discursive resources—the metaphors, models and narratives with which they have come to understand themselves and their environments. For example, consider the different kinds of questions Europeans have been led to ask about their environments over the last five centuries as they conceptualized the world around them. The model of nature as an alive organism ("Mother Earth"), who made available an endless cornucopia of resources, directed scientific research. So too did the Christian conception of nature as a product of God's mind; to practice modern scientific methods provided the opportunity to get to know God's mind in greater detail—an opportunity uniquely available to humans, in whose image God had created them, too. At the same time, scientific work was also directed by the representation of nature as a simple mechanism such as a clock. Passing over the next several centuries, we see today representations of nature's order as a complex mechanism such as a biofeedback computer, and in environmental sciences, of the earth as lifeboat or spaceship.

These models of nature drew scientists' attention to different aspects of nature's order, suggested how fruitfully to enlarge the domain to which their theories could apply, and provided ways to reorder familiar "facts" into more satisfying explanatory patterns. Of course most of medieval observations of the movements of comets, planets, and stars are retained in astronomy today. But the relations between these movements, how they should be explained, and what they mean to people have shifted in light of subsequent scientific theories and the social projects with which they have co-emerged and have been co-constituted. Many of "the facts" are indeed always and everywhere the same, but their significance for scientific and social thought can vastly change. Significant here is that the image of the jigsaw puzzle obscures important aspects of this history. The facts about nature's bountiful order produced within early modern scientific models of nature conflict with those produced by contemporary environmental research. Such conflicting bodies of knowledge cannot be fit neatly together to create a singular scientific picture of nature's order.

Finally, cultures organize differently the production of knowledge about the world around them, and how they do so affects what they can know. This is the fundamental principle of scientific method, of course: different kinds of interactions with nature will produce different bodies of knowledge. In this context, the notion of "method" is usefully expanded to include any distinctive way a group organizes its collection of the information it needs to interact effectively with its environment. Oceanographers and climatologists today produce distinctive bodies of knowledge through their research projects. Careful observation of stars, winds, cloud, and wave patterns enabled Pacific islanders to make voyages in open canoes over thousands of miles of open ocean to New Zealand and Australia, for example, and to return safely home (Watson-Verran & Turnbull, 1995). Arctic inhabitants have developed similar knowledge for navigating snow terrains.

Thus, whichever cultural groups have the power and resources to command searches for the kinds of knowledge they want for their projects will develop distinctive repositories of knowledge. These repositories must continually be revised to adjust to changing environmental conditions, new social interests, and the cognitive resources they gain in exchanges with other cultures and their knowledge systems. Moreover, each such body of knowledge, whether in modern science or other knowledge systems, is accompanied by a matching systematic body of ignorance. In choosing to focus on one set of patterns of nature's regularities, with one set of interests, discourses, and methods of knowledge production, a culture leaves unexamined other patterns and the ways of thinking about them that other interests, discourses, and methods could produce. As postcolonial science and technology theorists often ask, what would modern sciences look like if they had been developed in other parts of the world—China, India, the Middle East or Africa—with the differing interests, discourses, and characteristic ways of organizing human activities found in those local cultures instead of in expansionist Europe? We cannot know the answer to such a question, but contemplating it helps one to appreciate the resources provided by these multiple sources of cognitive diversity.

Against such a background, it is easier to appreciate the political and scientific dysfunctionality of the universality ideal.

COSTS OF THE UNIVERSALITY IDEAL

We are now in a position to count up the political and scientific costs of the universality ideal.

Political Costs

Arguments in the preceding sections have already pointed to some of the most important political costs of the universality ideal. It supports the devaluation of forms of knowledge-seeking that have proved valuable in other cultures; indeed, of ones that today are crucial to the survival of groups effectively delinked from the benefits of international science and technology. Many African and South American cultures, for example, have little or no access to international science and technology, and survive—and in some cases thrive—thanks to the strengths of their local knowledge traditions.

Moreover, to devalue these traditions is to devalue the people and cultures that use them. This legitimates the continuing forcible subjugation of these groups to western projects—

military or commercial. Would the North have so few moral qualms about the sacrifice of third world cultures to purported economic progress if it perceived their knowledge systems as valuable elements for now and for the future of the collectivity of human knowledge? From this perspective, modern sciences have ended up, usually unintentionally, as complicitous with some of the worst genocidal social projects in the name of "human" progress. Would this occur as easily if Northern cultures conceptualized many different centers of human cognitive progress rather than only one—their own? The universality thesis legitimates continuing to move access to nature's resources from those who are already the politically and economically most vulnerable to those who are already the best positioned to take advantage of such access.

Furthermore, the universality ideal supports the construction of models of the rational, the objective, the progressive, the civilized, and the admirably human in terms of distance from the non-European, the economically frugal, as well as the feminine. Moreover, it elevates authoritarianism to a social ideal, for it asserts that it is desirable for everyone to acknowledge the legitimacy of one culture's (the international science culture's) claim to provide the one true account of the world. The authority of the universality ideal is presented as a necessity for the distinctively rational, progressive, civilized, and human. In such respects, the universality ideal is not at all politically neutral.

Yet such political costs are not the only ones exacted by this ideal.

Scientific Costs

To start with, the universality ideal legitimates decreasing cognitive diversity, yet it is just such diversity that has provided continuing resources for the growth of every culture's scientific and technological projects. Without the availability of other systems of knowledge from which to borrow novel understandings of local environments and the resources they can offer, as well as metaphors, models, and narratives of nature and humans' place in nature, new inquiry techniques and ways of organizing the production of knowledge, any knowledge system would be stuck with only what can be generated from within its own "culture." Modern sciences would have been deeply impoverished without the resources gathered into them from the knowledge traditions of the other cultures Europeans encountered. Moreover, we cannot know what knowledge we will need in the future as social and natural environments change, and new needs and desires develop. Different cultures' constantly evolving ways of thinking about nature and social relations will continue to provide valuable resources for each others' projects. It is as foolish to decrease cognitive diversity as it is to decrease biological diversity.

Second, the universality ideal legitimates accepting less-well-supported claims over potentially stronger ones in many cases. If the ontology of a claim (the aspects of the world on which it focuses), the methods used to gather evidence for it, or the models and narratives through which it approaches nature do not fit with the one prevailing one, it can be ranked as less probable than a claim with far weaker empirical evidence that is consistent with prevailing scientific models. My use of terms such as "less-well-supported claims" and "weaker empirical evidence" should not be taken to indicate that I assume universally accepted standards for scientific claims or the infallibility of any empirical reports. My point is that legitimating only one culturally decontextualized set of standards for evaluating evidence can prove to be scientifically costly. Environmental studies that rely exclusively on the analyses of physical sciences cannot recognize as valuable components of "the best explanation" the kinds of analyses that social scientists bring to environmental studies. Of course it is valuable for each culture to "test" the claims of others within the resources of its own knowledge-system. What is problematic is to assume that such a procedure correctly identifies the worth of a claim, rather than only its ability to be confirmed within a favored knowledge-system. There is an important difference between cases where there is overwhelming evidence of a claim's falsity, and where a claim has not yet been fairly tested, or where no adequate explanation is yet available for the effectiveness of a particular kind of intervention in natural or social orders.

Thus, in the third place, the universality ideal legitimates resistance to some of the deepest and most telling criticisms of particular scientific claims. Criticisms that cannot be recognized as coming from within established boundaries of scientific discussion can legitimately be devalued or ignored. Thus feminist analyses are persistently conceptualized as coming from outside science, even when the critics are respected scientists as, for example, with feminist biologists' criticisms of standard interpretations of evolutionary theory and medical representations of women's body processes. Similarly, postcolonial criticisms of western scientific and technical expertise is often rejected as coming from outside science even when the critics are trained in western science. This is so even when the critics' goal is not to reject western scientific expertise wholesale, but rather to better integrate it with insights from local knowledge systems. The ability to detect "rigorous refutations" weakens when rigor is presumed to be the monopoly of the one and only real science.

Next, the universality ideal promotes only narrow conceptions of both nature and science. As long as physics is presumed to be the model for all sciences, whether on historical, ontological (e.g., its focus on primary vs. secondary qualities), methodological, or other grounds, other ways of understanding nature's order will be devalued. To mention just one case, it blocks our ability to bring into focus the social elements—institutions, practices, meanings—in what are often presented as merely natural, scientific, and technological changes.

Another limitation is that the ideal of one true science obscures the fact that any system of knowledge will generate systematic patterns of ignorance as well as of knowledge. Every knowledge system has its limits, since its priorities select which aspects of nature's order to study; which questions to ask; which metaphors, models, narratives, and other discursive resources to use; and which ways to organize the production of knowledge. Knowledge systems are like Thomas Kuhn's paradigms in this respect. They can prove illuminating far from their original sites of production, but they all have their limits, and produce recognizably diminishing returns sooner or later.

Finally, such a model for the natural sciences promotes similar problems in the social sciences that model themselves on the natural sciences, such as physicalist psychologies, rational choice theories in economics, political science, and international relations, and positivistic sociologies. Moreover, the prevalence of such models in the social sciences has bad effects in another respect; social scientists who oppose such models often can see no alternative but to focus entirely on the micro and the local. Relativist epistemological positions start to look far too attractive as long as the universality ideal is the only alternative. As a reaction to naturalistic social sciences' devaluations of the local, these other social research projects get contained by the local. The universalist and relativist positions are really two sides to the same coin. In effect, universalism's conceptual world is advanced in unarticulated forms through the relativist positions.

Thus adherence to the universality ideal brings costly political and scientific consequences.[2]

UNIVERSALIZING UNIVERSALIZING

Yet the universality ideal captures some features of scientific work that deserve to be preserved, whether in international science cultures or in others. My argument is not to abandon the ideal completely. For one thing, it is too deeply a part of our own western Enlightenment legacy to be abandoned so easily. It is an escapist fantasy to imagine we could accomplish such a feat. Setting out to abandon such a central part of a dominant western self-image assures a "bohemian" status for such a traveller. Instead, my goal is to identify what parts of this ideal are still valuable, and how they could be appropriately reconceptualized.

One useful idea here is that while all beliefs and technological practices are generated in the heat of some culturally local project or other, some prove far more useful than others. Beliefs are certainly not automatically more valuable just because they are "local." After all, individuals and cultures lead nasty, short, and brutish lives if they do not figure out how to avoid damage from excessive cold and heat, hurricanes, floods, fires, and ozone holes, exposure to deadly diseases, or poisoning by nicotine or toxic wastes. Some beliefs travel well, persisting and becoming useful at other places and times, and in contexts very different from those of their origination. So one could say that the attempt to *universalize* a belief is simply the attempt to see in what contexts it can gather empirical evidence and prove useful. Restricting this term to its verbal form simply describes an activity common to any knowledge system. It does not commit us to the claim that there is only one truth about the world and one science that can capture that truth.

Nor does it commit us to the fourth assumption of the unity of science claim: that there is one and only one group or culture of humans that can develop that science. As the histories of science reveal, elements of many different cultures' science and technology traditions have found a home in western—now, "international"—science at one point or another, just as elements of western sciences have been integrated into other knowledge systems. All knowledge systems are hybrids; their ability to continue to grow comes from the access they have to continually new cognitive and material—natural and technological—resources. So this universalizing process can and does occur in many different simultaneously existing knowledge systems. We could conceptualize as valuable the universalizing of different cultures' universalizing practices, not just that this process occurs only in modern western sciences. Important elements of many different cultures' inquiry traditions have become valuable within modern biomedicine's different knowledge system, just as elements of Ptolemaic astronomy remained valuable within the vastly different conceptual world of Copernican astronomy.

Such a perspective could lead to prioritizing the development of significantly different knowledge systems rather than of only one perfect system. And here is just one way that philosophies of modern sciences could encode democratic ideals more effectively with benefits both for democratic social movements and for maintaining the cognitive resources every knowledge system needs to flourish.[3]

NOTES

1. Political Philosopher Val Plumwood pointed this out to me.
2. Several of the dysfunctional consequences listed here occur also in John Dupre's (1996) list of problems with what he refers to as the "unity of scientism" thesis.
3. Several themes in this essay have been developed also in Harding, 1998.

Index

Index